D0212890

FLUID BED TECHNOLOGY IN MATERIALS PROCESSING

C.K. Gupta
D. Sathiyamoorthy

CRC Press
Boca Raton London New York Washington, D.C.

Library of Congress Cataloging-in-Publication Data

Catalog information may be obtained from the Library of Congress

International Standard Book Number 0-8493-4832-3
Printed in the United States of America 1 2 3 4 5 6 7 8 9 0
Printed on acid-free paper

Foreword

Fluidization engineering, which owes its origin and credibility to chemical engineering, has now emerged as a separate branch of engineering with a multitude of applications ranging from conventional to advanced engineering. A good deal of progress in terms of research and development has been made in this area over the past two decades, and this elegant unit operation is now being adopted for efficiency enhancement in process and energy industries. The approach of Dr. C.K. Gupta and Dr. D. Sathiyamoorthy of the Bhabha Atomic Research Centre is to painstakingly bring together the wealth of information in this field and to present it concisely in a book on fluid bed technology in materials processing. The aim and scope of this volume as indicated by the title are well achieved.

It is not out of context to note that Dr. Gupta and his colleagues have contributed phenomenally to our materials program. The quality of authorship of Dr. Gupta in particular and of his partners as co-authors on the whole is well reflected in his seven books published to date. Dr. Gupta and his colleagues have been instrumental in the development of several exotic materials and in initiating the materials development program for the Department of Atomic Energy.

I might add here that this kind of research is the first of its kind in India and compares well with that in some of the world-class laboratories with similar objectives. I understand that some excellent experimental work is being carried out, such as work related to distributor design, electrothermal fluid bed chlorination, radioactive waste incineration, and three-phase fluidization. Thus, Dr. Gupta and Dr. Sathiyamoorthy, with all their knowledge and expertise in the theory and practice of fluidization, are very well equipped to competently handle the objectives set forth for the present volume.

The book is divided into six chapters, beginning with the basics of fluidization and then giving an account of its value and use in applied areas in the chapters that follow. The topic of each chapter has been chosen well, and the text is amply supplemented by numerous illustrations and carefully selected references. It is not an exaggeration to comment that this book is replete with valuable information which has hitherto been scattered throughout literature. The book certainly will serve as an important reference source for research scientists in materials processing, practicing process engineers, graduate students, and for fluidization engineers who would like to retrieve good information on this exclusive topic. I am sure this volume will be as useful as the other books which Dr. C.K. Gupta has authored himself and co-authored with his colleagues. I congratulate the authors and wish them success.

Anil Kakodkar
Director, Bhabha Atomic Research Centre
Member, Atomic Energy Commission, India

Preface

The voluminous body of scientific and technical literature published to date on fluidization bears ample testimony to the enormous interest in this field. The chemical engineering discipline has experienced phenomenal gains from the use of this process. This powerful technique has also started making significant inroads in other disciplines, and in this context special mention must be made of the field of materials processing. It has already been established as a peer technique in some processes and has enormous application potential in a number of existing and emerging processes. However, no book on fluidization as applied to materials processing was published until now. It is superfluous to point out the need for a book devoted to this illustriously important field.

For quite some time, we in the Materials Group of the Bhabha Atomic Research Centre have been involved in research and development pertaining to the application of fluidization to materials processing problems. Examples include an assortment of processes such as sulfation roasting, chlorination, fluorination, incineration of radioactive waste, and reduction of metals from their oxidic origins. This involvement, supplemented by our extensive studies on and appreciation of fluidization, led us to undertake the present book.

The presentation is organized into six chapters. Chapter 1 deals with the basics of fluidization. It starts with an introduction to fluid particle systems in general, followed by a description of the relevant fundamental parameters. Various types of fluidization, such as gas–solid, liquid–solid, and three-phase fluidization, are discussed. The general aspects of heat and mass transfer and the end zones of fluidized beds are also dealt with. This opening chapter, in essence, sets the stage for the subsequent chapters.

Chapter 2 is devoted to the applications of fluidization in the extraction and processing of minerals, metals, and materials. The applications of fluidization in drying, roasting, calcination, direct reduction, halogenation, and selective chlorination figure in the presentation.

Chapter 3 describes the importance of fluidization in the nuclear fuel cycle. Areas such as leaching, uranium extraction, and nuclear fuel preparation are covered with reference to selection and application of the fluidization technique. The significant role played by fluidization in the nuclear fuel cycle as a whole and in the processing of nuclear materials in particular is brought out.

Chapter 4 deals with the novel concepts of plasma fluid beds and electrothermal fluidized beds. These fluid bed reactors are described, along with their characteristic behaviors and their applications in high-temperature process metallurgy.

Chapter 5 covers the design aspects of fluidized bed reactors. The prediction and judicious selection of various critical parameters such as operating velocity, aspect ratio, and pressure drop are discussed, and features such as the distributor and its design principles are presented. Modeling aspects of gas fluidized beds and a comparison of the performance of various models are also included in this chapter.

Chapter 6 covers the latest developments in and applications of fluidization in the modern engineering world. Various new techniques of fluidization, such as magnetically stabilized fluidized beds and compartmented fluidized beds, are described. Semifluidized beds are among the other novel topics presented. The essential features of the fluidized electrode cell and its potential uses in the electro-extraction of metals are highlighted in this chapter. The advent of fluidized beds in bioprocessing and the multitude of bed configurations used to carry out bioreactions are covered separately in the closing part of the chapter.

This volume should prove useful to faculties in metallurgy as well as chemical engineering. It can also serve as a reference for professionals who deal with high-temperature materials and nuclear chemical engineering and as a handy source for all interested in this subject. We believe that this book in its own right will invariably find its way into various educational institutions and research centers interested in metallurgy and materials technology.

Lastly, we hope you will enjoy reading this book as much as we enjoyed writing it.

C.K. Gupta
D. Sathiyamoorthy

Dedication

No writer dwells in a vacuum; no author's work is untouched, in one way or another, by associates, friends, and family.

The following warm-hearted people gave so much of themselves — personally and/or professionally — and this book is far better than it could possibly have been without them.

P.L. Vijay, V. Ramani, V.H. Bafna, M.G. Rajadhyaksha, S.M. Shetty, K.P. Kadam, and P.S. Narvekar were especially instrumental, directly or indirectly, in the writing of this book.

The select team of Poonam Khattar, Rajashree Birje, and Yatin Thakur cheerfully and caringly transformed the handwritten material into the typed version and prepared neat and clear drawings and illustrations.

Marsha Baker and Felicia Shapiro, our contacts at the editorial and manuscript-processing levels at CRC Press LLC, helpfully, patiently, and perceptively guided us toward completion of the task we undertook. They were understanding in liberally granting us numerous extensions for submission of the manuscript and in accommodating us in the publication schedule.

P. Mukhopadhyay gave some of his precious time to critically go through the manuscript and offer constructive suggestions.

We owe a great deal to all of them.

Chandrima Gupta, Chiradeep Gupta, the late P.C. Gupta, S. Sasikala, S. Shiva Kumar, S. Srinivas, and the late D. Pappa, our family members, sustained us by their interest and support and by their acceptance, with characteristic cheerfulness, of the sacrifices involved. We are greatly indebted to all of them.

As a token of what we owe to the inspiration provided by this group of people and to express our deepest gratitude and thanks, this work is dedicated to them with due respect, regard, love, affection, and fond reminiscences.

Acknowledgments

The authors gratefully acknowledge the following sources that kindly granted permission to use some of the figures and tables that appear in the book: Elsevier Sequoia, S.A., Lausanne, Switzerland; American Institute of Chemical Engineers, New York; Elsevier Scientific Publishing Company, Amsterdam, The Netherlands; Hemisphere Publishing Corporation, Washington, D.C.; Elsevier Science, Ltd., The Boulevard, Langford Lane, Kidlington, Oxford, U.K.; Canadian Society for Chemical Engineering, The Chemical Institute of Canada, Ottawa; Academic Press, New York; Gordon and Breach Publishers, Langhorne, Pennsylvania; The Metallurgical Society of the AIME, Warrendale, Pennsylvania; John Wiley & Sons, New York; Wiley Eastern Ltd., New Delhi, India; Ann Arbor Science Publishers, Ann Arbor, Michigan; Pergamon Press Ltd., Oxford, U.K.; Pergamon Press Inc., New York; International Union of Pure and Applied Chemistry, Eindhoven, The Netherlands; Butterworths, Australia; American Ceramic Society, Westerville, Ohio; Heywood & Co. Ltd., London; Materials Research Society, Pittsburgh, Pennsylvania; Elsevier Science, Amsterdam, The Netherlands; Gordon and Breach Science Publishers, The Netherlands; and the Institution of Chemical Engineers, England.

The Authors

C.K. Gupta, Ph.D., is Director of the Materials Group at the Bhabha Atomic Research Centre (BARC), Mumbai, India. He received his B.Sc. and Ph.D. degrees in Metallurgical Engineering from Banaras Hindu University, Varanasi, India. He is a research guide for M.Sc. (Tech.) and Ph.D. students at Bombay University, Mumbai.

Dr. Gupta specializes in the field of chemical metallurgy. He is responsible for research, development, and production programs on a wide range of special metals and materials of direct relevance to the Indian nuclear energy program. He is the recipient of a number of awards for the contributions he has made to metallurgical science, engineering, and technology, including setting up production plants.

Dr. Gupta is associated with many professional societies. He is on the editorial board of a number of national and international journals and is a prolific contributor to the metallurgical literature. In addition to seven books, with two more in the pipeline, from publishers such as CRC Press LLC, Elsevier, and Gordon and Breach, he has authored 190 publications, which include research papers, reviews, and popular scientific articles. He has also served as guest editor for a number of special publications.

D. Sathiyamoorthy, Ph.D., is currently Head, Process Engineering Section of the Materials Processing Division of the Materials Group at the Bhabha Atomic Research Centre (BARC), Mumbai, India. He joined the center in 1975.

Dr. Sathiyamoorthy graduated in Chemical Engineering in 1974 from A.C. College, University of Chennai, Chennai, India. He obtained his Ph.D. in Chemical Engineering in 1984 from the Indian Institute of Technology, Mumbai. During 1989–90 he was a research fellow at the University of Queensland, Australia, and during 1990–91 was an Alexander Von Humboldt Research Fellow at Technical University, Clausthal, Germany. He is an invited JSPS fellow (1997–98) in the Department of Chemical Engineering at Tokyo University of Agriculture and Technology under the Japan Society for the Promotion of Sciences.

His professional involvement is with process engineering, operation, and optimization in mineral/extractive metallurgy. His current research is focused on fluidization engineering as applied to process and extraction metallurgy. He has authored and co-authored over 60 technical papers.

Table of Contents

CHAPTER 1

Generalities and Basics of Fluidization

I. INTRODUCTION

Fluidization is a unit operation, and through this technique a bed of particulate solids, supported over a fluid-distributing plate (often called the grid), is made to behave like a liquid by the passage of the fluid (gas, liquid, or gas–liquid) at a flow rate above a certain critical value. In other words, it is the phenomenon of imparting the properties of a fluid to a bed of particulate solids by passing a fluid through the latter at a velocity which brings the fixed or stationary bed to its loosest possible state just before its transformation into a fluidlike bed.

A. Fluidlike Behavior

Let us consider the various situations that could prevail in a bed of particulate solids. When there is no fluid flow in the bed, it remains in a static condition and the variation in pressure across the bed height is not proportional to its height, unlike in a liquid column. When a fluid such as a gas or a liquid is allowed to percolate upward through the voidage of a static bed, the structure of the bed remains unchanged until a velocity known as the minimum fluidization velocity is reached; at this velocity, drag force, along with buoyant force, counteracts the gravitational force. In this situation, the bed just attains fluidlike properties. In other words, a bed that maintains an uneven surface in a static, fixed, or defluidized state now has an even or horizontal surface (Figure 1.1a). A heavy object that would rest on the top of a static bed would now sink; likewise, a light object would now tend to float. The pressure would now vary proportional to the height, like a liquid column, and any hole made on the vessel or column would allow the solid to flow like a liquid. All these features are depicted in Figure 1.1.

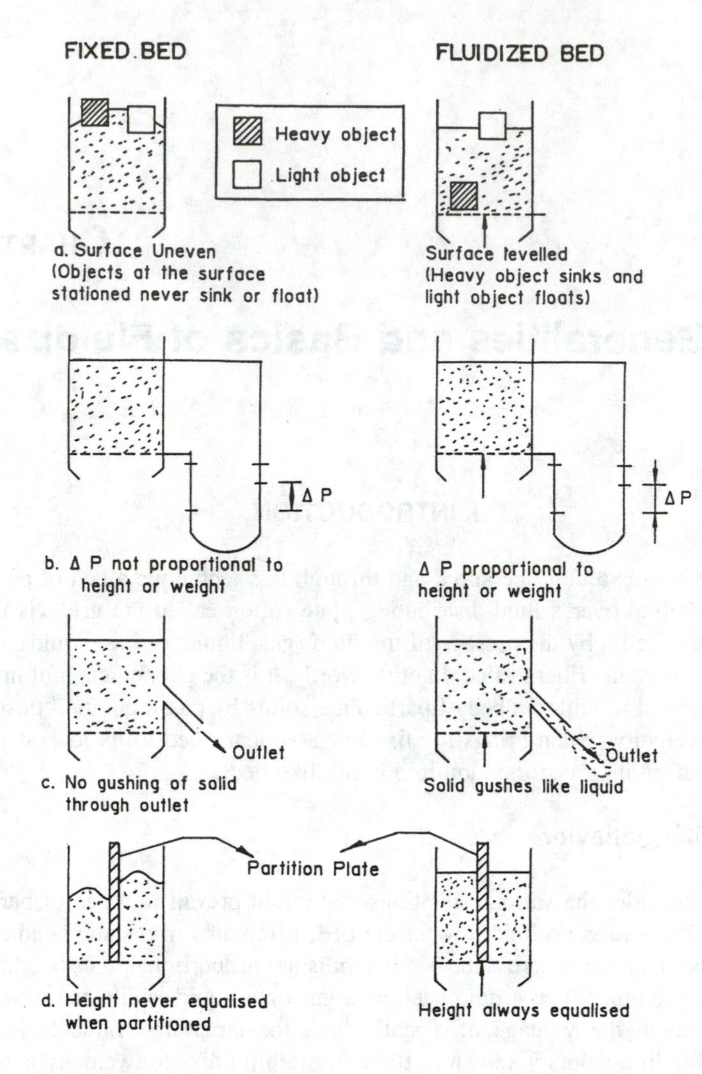

Figure 1.1 Examples of fluidlike behavior of fluidized bed relative to fixed bed.

B. Fluidization State

1. Gas/Liquid Flow

The fluid under consideration which flows upward can be either a gas, a liquid, or both. In general, liquid fluidized beds are said to have a smooth or homogeneous or particulate nature of fluidization. The bed expands depending on the upward liquid flow rate, and due to this expansion the bed can become much higher than its initial or incipient height. In contrast, a gas fluidized bed is heterogeneous or aggregative or bubbling in nature and its expansion is limited, unlike what happens in a liquid fluidized bed. It is seldom possible to observe particulate fluidization in a gas fluidized bed and aggregative fluidization in a liquid fluidized bed. If the fluid flow regimes are such that a bed of particulate solids has a boundary defined by a surface, then it has solid particles densely dispersed in the fluid stream. In other words, a dense-phase fluidized bed is achieved. When the surface is not clearly defined at a particular velocity, the solid particles are likely to be carried away by the fluid. This situation corresponds to a dilute or a lean phase. The situation where solid particles are entrained by the fluid flow corresponds to a state called pneumatic transport.

As the velocity of the liquid in a liquid fluidized bed is increased, homogeneous or particulate fluidization with smooth expansion occurs, followed by hydraulic transport of particles at a velocity equal to or greater than the particle terminal velocity. In the gas fluidized bed, bubbling is predominant. The minimum bubbling velocity is the velocity at which the bubbles are just born at the distributor. The bubbling bed at velocities greater than the minimum bubbling velocity tends to slug, especially in a deep and/or narrow column, and the slugging is due to the coalescence of bubbles. When the bubbles coalesce and grow as large as the diameter of the column, a slug is initiated. Now solids move above the gas slug like a piston, and they rain through the rising slugs. Here the gas–solid contact is poor. The slugging regime, through a transition point, attains a turbulent condition of the bed, and this process is often termed fast fluidization. Pneumatic transport of solid particles by the gas stream occurs at and above the particle terminal velocity. Liquid and gas fluidized beds for various gas flow rates are illustrated in Figure 1.2.

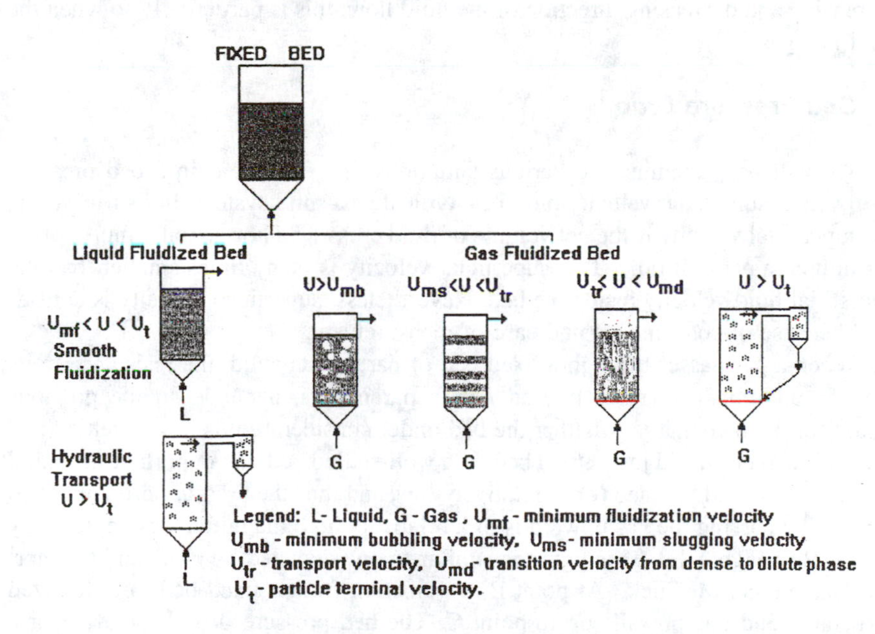

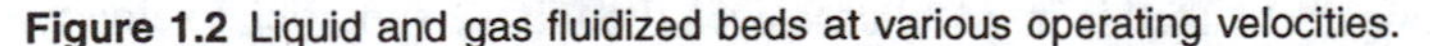

Figure 1.2 Liquid and gas fluidized beds at various operating velocities.

2. Onset of Fluidization

Estimation of the onset of the fluidization velocity is essential because it is the most important fundamental design parameter in fluidization. This velocity determines the transition point between the fixed bed and the fluidized bed. In a fixed bed, solid particles remain in their respective fixed positions while the fluid percolates through the voids in the assemblage of particles. In such a situation, the fluid flow does not affect or alter the voidage or the bed porosity. As the fluid flow is increased, the bed pressure drop increases. At a certain stage, the bed pressure drop reaches a maximum value corresponding to the bed weight per unit area (W/A); at this stage, a channel-free fluidized bed at the ideal condition is obtained.

3. Situation at the Onset of Fluidization

Let us examine the situation at the onset of fluidization; the corresponding fluidization velocity is also referred to as the incipient velocity. When this velocity is just attained, the fixed bed of particles exists in is loosest possible condition without any appreciable increase in bed volume or bed height. In such a condition, the bed weight less the weight equivalent to buoyancy is just balanced by the drag due to the upward flow of the fluid. In other words, the distributor plate or the grid which supports the bed of solids does not experience any load under this condition. The velocity at which the bed is levitated, achieving a state of fluidlike behavior just at the transition of a fixed to a fluidized bed, is called the minimum fluidization velocity. The transition from a fixed to a fluidized bed may not be the same for increasing and decreasing direction of the fluid flow; this is particularly so when the fluid is a gas.

4. Bed Pressure Drop

We will now examine the various situations that can occur in a bed pressure drop versus superficial velocity plot for a typical gas–solid system. In its true sense, the superficial velocity is the net volume of fluid crossing a horizontal (empty) plane per unit area per unit time. This superficial velocity is many times smaller than the interstitial fluid velocity inside the bed. Nevertheless, superficial velocity is considered because of convenience and ease of measurement.

When a gas passes through a fixed bed of particulate solid, the resistance to its flow, in addition to various hydrodynamic parameters, depends on the previous history of the bed, that is, whether the bed under consideration is a well-settled bed or a well-expanded and just settled bed. In a well-settled bed, the important structural parameter, the bed voidage (ϵ), is relatively low, and thus the pressure drop obtained initially by passing the gas upward is of a relatively high magnitude, as depicted by line A–B in Figure 1.3. This figure is similar to one depicted by Zenz and Othmer[1] and Barnea and Mednick.[2] At point B, a transition from a fixed bed to a fluidized bed starts, and this prevails up to point C. The bed pressure drop beyond C for a fluidized bed remains unchanged in an ideal case even though the superficial velocity (U) is increased. The bed pressure drop beyond point D, which corresponds to the particle terminal velocity for a monosized bed of solids, is no longer constant, and it increases in a manner similar to an empty column. This is so because the particles are carried away from the bed or are completely entrained when the superficial velocity equals the particle terminal velocity. In this situation, the bed voidage (ϵ)

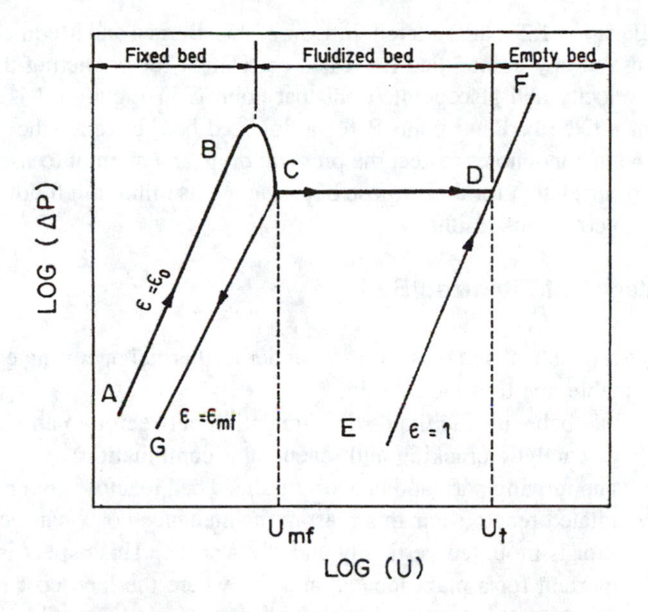

Figure 1.3 Variation in bed pressure drop (ΔP) with superficial velocity (U).

is unity, or the volume fraction of solid particles ($1 - \epsilon$) is zero. Line EDF corresponds to the pressure drop in the empty column. When the upward flow of gas is gradually decreased, the bed pressure drop assumes its original path on a pressure drop versus velocity plot as long as the bed continues to be in a state of fluidization. Retracing the path, as shown in Figure 1.3 by line DCG, indicates that the pressure drop obtained for a fixed bed during its settling is lower than that obtainable for increasing upward flow of gas. Point C on line GCD is the transition point between a fixed and a fluidized bed, and the velocity corresponding to this is the minimum fluidization velocity. It may be observed that during increasing flow through the bed, there is no distinct point marking the transition, except that zone which corresponds to B–C with a peak.

The relatively high value of the pressure drop (ΔP) along line AB in a fixed bed when the flow is in the laminar regime compared to line GC of an expanded settled bed is due to low bed permeability. Let us now look at the fluid dynamic aspects at points A and B. For laminar flow conditions, most correlations for the pressure drop give:

$$\Delta P \propto U(1 - \epsilon)^2/\epsilon^3 \tag{1.1}$$

At point A, the bed porosity (ϵ) corresponds to the static bed value (ϵ_0) (i.e., $\epsilon = \epsilon_0$), and at point B, the bed porosity is equal to the minimum fluidization value (ϵ_{mf}) (i.e., $\epsilon = \epsilon_{mf}$). From Equation 1.1, it follows that:

$$(\Delta P)/[U(1 - \epsilon)^2/\epsilon^3] = \text{Constant} \tag{1.2}$$

and this is true if ϵ is not altered for a fixed bed. However, as indicated in Figure 1.1, points A and B correspond to a fixed bed but to different ϵ values; the proportionality constant in Equation 1.2 is thus altered. In view of this, Barnea and Mednick[2] cautioned against the use of any unmodified fixed bed correlation for predicting the minimum fluidization velocity and also pointed out that point C in Figure 1.3 is the limiting condition for a fixed bed and point B for a fluidized bed. Because the particle concentration and its randomness affect the pressure drop, any attempt to use a fixed-bed pressure drop correlation for the purpose of predicting its minimum fluidization velocity will lead to erroneous results.

C. Advantages of Fluidized Bed

1. A high rate of heat and mass transfer under isothermal operating conditions is attainable due to good mixing.
2. A fluidlike behavior facilitates the circulation between two adjacent reactors (e.g., catalytic cracking and regenerator combination).
3. There is no moving part, and hence a fluidized bed reactor is not a mechanically agitated reactor. For this reason, maintenance costs can be low.
4. The reactor is mounted vertically and saves space. This aspect is particularly important for a plant located at a site where the land cost is high.
5. A continuous process coupled with high throughput is possible.
6. No skilled operator is required to operate the reactor.
7. The fluidized bed is suitable for accomplishing heat-sensitive or exothermic or endothermic reactions.
8. The system offers ease of control even for large-scale operation.
9. Excellent heat transfer within the fluidized bed makes it possible to use low-surface-area heat exchangers inside the bed.
10. Multistage operations are possible, and hence the solids residence time as well as the fluid residence time can be adjusted to desired levels.

D. Disadvantages of Fluidized Bed

1. Fine-sized particles cannot be fluidized without adopting some special techniques, and high conversion of a gaseous reactant in a single-stage reactor is difficult.
2. The hydrodynamic features of a fluidized bed are complex, and hence modeling and scaleup are difficult.
3. Generation of fines due to turbulent mixing, gas or liquid jet interaction at the distributor site, and segregation due to agglomeration result in undesirable products.
4. Elutriation of fines and power consumption due to pumping are inevitable.
5. Sticky materials or reactions involving intermediate products of a sticky nature would defluidize the bed.
6. Limits on the operating velocity regime and on the choice of particle size range are disadvantages of fluidization. Fluidization of friable solids requires careful attention to avoid loss of fines formed due to attrition.
7. Highly skilled professionals in this area are needed for design and scaleup.

8. Erosion of immersed surfaces such as heat-exchanger pipes may be severe.
9. Reactions that require a temperature gradient inside the reactor cannot be accomplished in a fluidized bed reactor.

II. PROPERTIES OF PARTICLES AND THE GRANULAR BED

A particulate solid has several properties which, in addition to the density, size, shape, and distribution, also play a key role in determining stationary bed or fixed bed properties. The roughness and the voidage associated with the particles should also be considered in fluid–particle interactions.

A. Particles

1. Size

Particulate materials or granular solids, whether manufactured or naturally occurring, can never have the same particle size. In other words, particulate solids comprised of uniformly sized particles are very difficult to obtain unless they are sized, graded, or manufactured under extreme control of the operating conditions. Achieving particles close in size in a manufacturing process is not simple; only specific processes like shot powders and liquid-drop injection into a precipitating solution can yield powders of relatively uniform size, but then only in the initial period of large-size particle production. Hence, uniformly sized particles are obtained by several physical techniques such as sieving, sedimentation, microscopy, elutriation, etc.

2. Definition

There are several ways to define particle size. For the purpose of powder characterization, it is not customary to define all the sizes as given by various mathematical functions. A particle size that is the diameter equivalent of a sphere is used along with a shape factor in many hydrodynamic correlations. The shape factor will be discussed and defined later in this section. Several definitions of the mean particle diameter are found in the literature, some of which are presented in the following discussion.

If a sample of a powder of a given mass is constituted of particles of different sizes and if there are n_i particles with a diameter d_{pi} (i = 1 to N), then the diameter can be defined in a variety of ways.

1. Arithmetic mean:

$$\left(d_p\right)_{AM} = \frac{\left(\sum_{i=1}^{N} n_i d_{pi}\right)}{\sum_{i=1}^{N}} \tag{1.3}$$

2. Geometric mean:

$$(d_p)_{GM} = 3(d_{p1} \times d_{p2} \times \ldots \times d_{pn})^{0.5} \tag{1.4}$$

3. Log geometric mean:

$$\log\left(d_p\right)_{GM} = \sum_{i=1}^{N}\left(n_i \log d_{pi}\right) \Big/ \sum n_i \tag{1.5}$$

4. Harmonic mean:

$$\left(d_p\right)_{HM} = \left(\sum_{i=1}^{N} n_i\right) \Big/ \left[\sum_{i=1}^{N}\left(n_i/d_{pi}\right)\right] \tag{1.6}$$

The harmonic mean diameter in principle is related to the particle surface area per unit weight, that is,

$$\frac{\text{Total surface area}}{\text{Weight}} = \frac{\left(\sum_{i=1}^{N} n_i\right) \cdot \pi\left(d_p\right)_{HM}^2}{\left(\sum_{i=1}^{N} n_i\right) \cdot \rho_p \pi\left(d_p\right)_{HM}^3 \big/ 6} = \frac{6}{\rho_p} \cdot \frac{1}{\left(d_p\right)_{HM}} \tag{1.7}$$

where ρ_p is the density of the particle.

5. Mean diameter, d_{pm}:
   50% of particles > d_{pm} 50% of particles < d_{pm}
6. Length mean diameter, $(d_p)_{LM}$:

$$\left(d_p\right)_{LM} = \sum_{i=1}^{N} n_i \cdot d_{pi}^2 \Big/ \sum_{i=1}^{N} d_{pi} \tag{1.8}$$

7. Surface mean diameter, $(d_p)_{SM}$:

$$\left(d_p\right)_{SM} = \left(\sum_{i=1}^{N} n_i\, d_{pi}^2 \Big/ \sum_{i=1}^{N} n_i\right)^{1/2} \tag{1.9}$$

8. Volume mean diameter, $(d_p)_{VM}$:

$$\left(d_p\right)_{VM} = \left(\sum_{i=1}^{N} n_i d_{pi}^3 \Big/ \sum_{i=1}^{N} n_i\right)^{1/3} \tag{1.10}$$

9. Weight mean diameter, $(d_p)_{WM}$:

$$\left(d_p\right)_{WM} = \sum_{i=1}^{N} x_i d_p = \frac{\sum_{i=1}^{N} n_i d_{pi}^4}{\sum_{i=1}^{N} n_i d_{pi}^3} \tag{1.11}$$

where x_i is the weight or volume percent of particles with diameter d_{pi}. For example,

$$x_i = \frac{n_i d_{pi}^3}{\sum_{i=1}^{N} n_i d_{pi}^3} \tag{1.12}$$

10. Volume surface diameter, $(d_p)_{VS}$:

$$\left(d_p\right)_{VS} = \frac{6}{S_w \rho_p} = \frac{\sum_{i=1}^{N} n_i d_{pi}^3}{\sum_{i=1}^{N} n_i d_{pi}^2} \tag{1.13}$$

where S_w is the surface area per unit weight of the particles:

$$S_w = \sum_{i=1}^{N} x_i s_{wi} \quad \text{and} \quad x_i = \frac{n_i d_{pi}^3}{\sum_{i=1}^{N} n_i d_{pi}^3} \tag{1.14}$$

Zenz and Othmer[1] gave an account of industries that are inclined to choose the particle diameter of their own interest and also showed that it is not necessarily meaningful to select any particle diameter as desired. Zabrodsky[3] recommends the use of the harmonic mean diameter in fluidization studies.

Geldart[4] commented that the particle size accepted in packed and fluidized beds is the surface–volume diameter (d_{sv}), defined as the diameter of a sphere that has the same ratio of the external surface area to the volume as the actual particle. For a powder that has a mixture of particle sizes, it is equivalent to the harmonic mean diameter. Another diameter is the volume diameter, which is the diameter of a sphere whose volume is the same as that of the actual particle, that is,

$$d_v = (6M/\rho_p \pi)^{1/3} \tag{1.15}$$

The particle size determined by sieve analysis is the average of the size or opening of two consecutive screens, and it may be referred to as d_p. The surface–volume diameter for most sands[5] is $d_{sv} = 0.87\ d_p$. The ratio d_{sv}/d_p is expected to vary widely depending on the particle shape, and it may not be easy to determine this ratio experimentally for irregularly shaped particles. Based on calculations pertaining to 18 regularly shaped solids, Abrahamsen and Geldart[5] showed that if the volume diameter $(d_v) = 1.127\ d_p$ for a sphericity of 0.773, then $d_{sv} = 0.891\ d_p$.

3. Sphericity

Sphericity (Φ_s) is a parameter that takes into consideration the extent of the deviation of an actual particle from the spherical shape or degree of sphericity. It is defined as the ratio of the surface area of a sphere to that of the actual particle that has the same volume. Let us consider a sphere made up of clay with a surface area S_o. If the sphere is distorted by pressing it, the shape changes; the resultant clay mass has the same volume as it had originally, but now it has a different surface area, say S_v. The ratio of the original surface area (S_o) to the new surface area (S) for the distorted clay sphere is its sphericity. It can be mathematically defined[6] as:

$$\Phi_s = \left(\frac{S_o}{S_v}\right)_{\text{constant volume}} \tag{1.16}$$

If the diameter of a particle (d_p) is defined as the diameter of the sphere that has the same volume as the actual particle, then the shape factor (λ) is given by:

$$\frac{1}{\lambda} = \Phi_s = \frac{1}{5.205}\frac{V_p^{2/3}}{S_p} \tag{1.17}$$

If j is the ratio of the particle volume to the volume of a sphere that has the same surface area as the particle, then

$$\lambda\, j^{2/3} = 1 \tag{1.18}$$

The sphericity of many regularly shaped particles can be estimated by analytical means, whereas it is not easy to estimate the sphericity value analytically for an irregularly shaped particle. Pressure drop correlations for fixed bed reactors incorporate the shape factor. Hence, these correlations are often used to obtain the shape factor for the particle using experimental data on pressure drop and the fluid–solid properties. The shape factor for a regularly shaped particle can be estimated readily. For example, the shape factor for a cube is 1.23. For cylinders, it is a function of the aspect ratio, and for rings it is dependent on the ratio of inside to outside diameter. It should be noted here that the hydraulic resistance or the bed pressure drop correlation used to estimate the shape factor should correspond to the laminar flow ($Re < 10$) regime and there should be no roughness effect.

4. Roughness

The roughness of a particle obviously adds to the friction between particles, and it leads to an increase in bed porosity (loose packing) when the bed settles down. The increase in bed porosity in turn reduces resistance to fluid flow. In other words, the bed pressure drop in a bed of particles with rough surfaces is lower compared to one that has particles with smooth surfaces, which have a tendency to form a less dense or low-porosity bed. Leva[7] estimated that the friction factor values for clay particles are 1.5 times higher than for glass spheres and 2.3 times higher than for fused MgO particles. The particle roughness has to be determined by measuring the friction factor and by comparing with standard reference plots for particles of various known roughness factors.

B. Granular Bed

1. Bed Porosity or Voidage

Bed porosity or voidage is affected by several parameters, such as the size, shape, size distribution, and roughness of the particles; the packing type; and the ratio of the particle diameter to the vessel diameter. Small or fine-sized particles have low settling or terminal velocities and low ratios of mass to surface area. Hence, fine particles, when poured into a vessel, settle slowly and create mass imbalance. As a result of these two factors, the bed has a tendency to form bridges or arches, which causes the bed voidage to increase. A bed with bridges or arches is not suitable for smooth fluidization. A dense bed of fine powders can be obtained by shaking, tapping, or vibrating the vessel.

2. Voidage and Packing

Voidage for spherical particles depends on the type of packing and varies from 25.95% (rhombohedral packing) to 47.65% (square packing). For nonuniform angular particles, voidage can vary widely depending on the type of packing (i.e., whether it is loose, normal, or dense). A bed is in its loosest packed form when the bed material is wet charged (i.e., the material is poured into the vessel containing the liquid or the material and after pouring into an empty container is fluidized by a gas and then settled). The bed will be dense when the container or the column is vibrated, shaken, or tapped for a prolonged period. A representative variation of voidage (ϵ) of packing with uniformly sized particle diameter (d_p) for three packing conditions (viz., loose, normal, and dense) is shown in Figure 1.4.

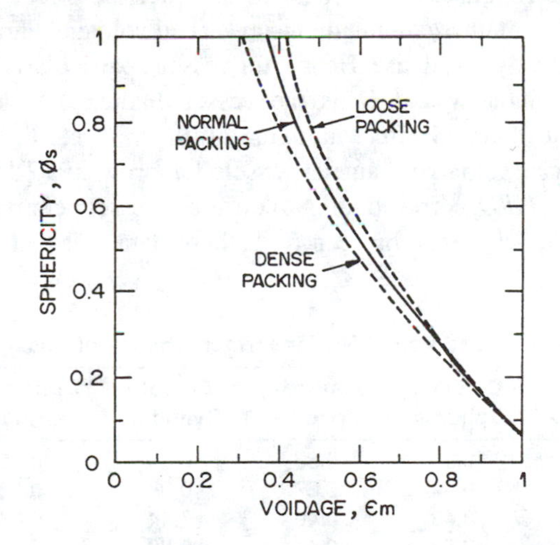

Figure 1.4 Voidage in uniformly sized and randomly packed beds. (From Brown, G.G., Katz, D., Foust, A.S., and Schneidewind, R., *Unit Operation*, John Wiley & Sons, New York, 1950, 77. With permission.)

3. Polydisperse System

Furnas[9] investigated experimentally the voidage of a binary system of varying particle size ratios. His studies showed that if the initial voidages of the individual components of the binary system are not the same, the voidage of the mixture will be less than the volumetric weighted average of the initial voidages. In a binary system of coarse and fine particles that have equal particle density and also equal voidage (ϵ), the volume fraction of the coarse particles is given as $1/(1 + \epsilon)$ at the condition of minimum voidage (i.e., maximum density). This low value of voidage is due to the fine particles that fill the interstices of the coarse material. A third component which is smaller than the second component and also finer relative to the first component may be added to fill the interstices of the second component; theoretically, the process can go on for an infinite number of components. The resultant or the final volume fraction (ϵ_m) of a solid mixture with n components is given by:

$$\left(1-\epsilon_m\right)=\frac{1}{(1+\epsilon)}+\frac{\epsilon}{(1+\epsilon)}+\frac{\epsilon^2}{(1+\epsilon)}+\ldots+\frac{\epsilon^n}{(1+\epsilon)} \tag{1.19}$$

From the above expression, it is possible to obtain the percentage of each component required to produce the minimum voids by multiplying both sides by $100/(1 - \epsilon_m)$.

4. Container Effect

The effect of particle roughness on voidage was discussed in the preceding section on particle roughness. Now we will consider the effect of the ratio of the particle size to the container/column diameter on bed voidage. Particle packing close to the walls is relatively less dense. Hence, for vessels, particularly when d/D_t (where d is the particle diameter and D_t is the vessel diameter) is large, the voidage contribution due to the wall effect is high, and the wall effect for particles away from it is insignificant for large-diameter vessels. Ciborowski's[10] data on bed porosity for various values of d/D_t corresponding to different vessel geometries and materials are shown in Table 1.1. It can be seen from these data that bed porosity increases as d/D_t increases.

Table 1.1 Bed Porosities for Various Shapes of Packings

d/D	Ceramic Spheres	Smooth Spheres	Smooth Cylinders	Raschig Rings
0.1	0.4	0.33	0.34	0.55
0.2	0.44	0.37	0.38	0.58
0.3	0.49	0.40	0.44	0.64
0.4	0.53	0.43	0.50	0.70
0.5	—	0.46	0.56	0.75

5. *Important Properties of Particulate Solids*

a. *Density*

There are in general three types of densities referred to in the literature: true, apparent, and bulk densities. True density is the weight of the material per unit volume, when the volume is considered free of pores, cracks, or fissures. The true density thus obtained gives the highest value of the density of a material. If, on the other hand, the particle volume is estimated by taking into consideration the intraparticle voids, then the density calculated on this basis is the apparent density. If ρ_t is the true or theoretical density and ρ_a is the apparent density for the same mass, then a simple relationship can be deduced between these two parameters. Let ϵ_0 be the intraparticle voids due to pores, cracks, etc.; V_{sv} the volume per unit mass of the particles in the absence of intraparticle voids; and V_a the apparent volume per unit mass; then:

$$\epsilon_I = \frac{V_a - V_{sv}}{V_a} = \frac{\frac{1}{\rho_a} - \frac{1}{\rho_t}}{\frac{1}{\rho_a}} \tag{1.20}$$

Upon rearrangement of Equation 1.20, one obtains:

$$\rho_a / \rho_t + \epsilon_I = 1 \tag{1.21}$$

An equation for estimating ρ_a can also be rewritten as:

$$\rho_a = \frac{1}{V_\epsilon + \frac{1}{\rho_t}} \tag{1.22}$$

where V_ϵ is the pore volume per unit mass. The bulk density (ρ_b) is obtained by considering the weight of granular solids packed per unit volume of a vessel or container. This density is less than the apparent density because the volume under consideration for the same mass of solid is increased in this case due to interparticle voids. If ϵ is the interparticle voidage for a granular bed of porous solids that have an intraparticle voidage ϵ_I, then, as per the preceding procedure:

$$\frac{\rho_b}{\rho_a} + \epsilon = 1 \tag{1.23}$$

Substituting for ρ_a from Equation 1.21 and modifying Equation 1.23, one gets:

$$\frac{\rho_b}{\rho_t} = (1-\epsilon_t)(1-\epsilon) \tag{1.24}$$

The bulk density again depends on certain other parameters, such as the degree of packing. Packing, as discussed earlier, can be of three types and depends on the extent of tapping employed while compacting a granular bed of solids. The bulk density, when measured after compaction by tapping, is called the tap density.

b. Angular Properties

The angular properties that are significant in relation to studies on rheology of powders are

1. Angle of internal friction (α_t)
2. Angle of repose (θ)
3. Angle of wall friction (γ)
4. Angle of rupture (δ)
5. Angle of slide (ω)

III. GROUPING OF GAS FLUIDIZATION

A. Hydrodynamics-Based Groups

The type of fluidization specifically for gas fluidized beds is related to the properties of the gas and the solid. In a gas fluidized bed, the bubbles moving through the dense particulate phase have a strong influence on the quality of fluidization. Hence, it is important to define the types of fluidization with respect to the properties of the gas–solid system.

1. Geldart Groups

In fluidization literature, most of the inferences are drawn from studies on one class of gas–solid system and then extrapolated to another group or class. This could have an adverse effect on scaleup and could result in the failure of the system. Much confusion and many contradictions in the published literature have been pointed out by Geldart,[11] and these have been attributed to extending the data obtained on one powder to another powder. In view of this, Geldart[12] classified powders that have similar properties into four groups and designated them by the letters A, B, C, and D. These groups are characterized by the difference in density between the fluidizing gas (i.e., air and the solid) and the mean size of the particles. A mapping of these groups is shown in Figure 1.5 for air fluidized beds. Of these four groups, the two extreme groups are Group C, which is difficult to fluidize, and Group D, which is spoutable. The intermediate Groups A and B are suitable

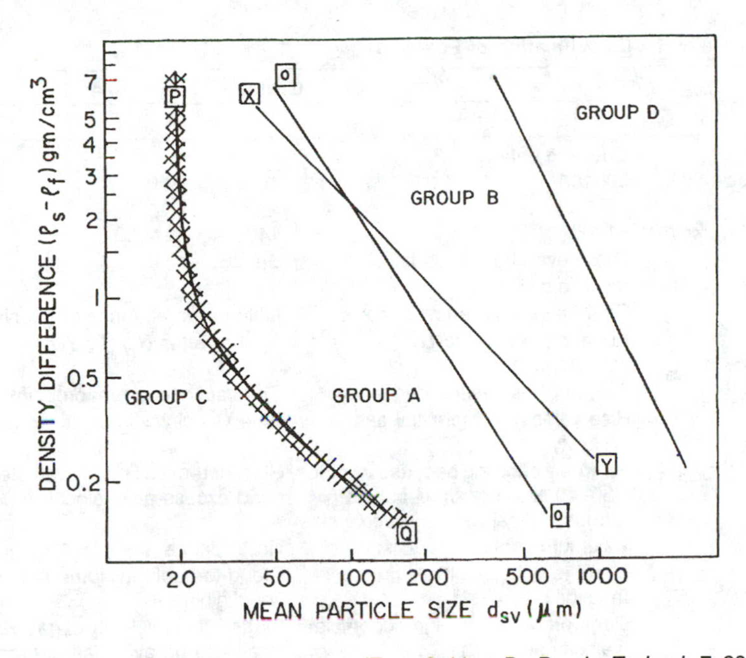

Figure 1.5 Geldart classification of powders. (From Geldart, D., *Powder Technol.*, 7, 285, 1973. With permission.)

for the purpose of fluidization. Of these two groups, Group A powders have dense-phase expansion after minimum fluidization but prior to the commencement of bubbling, whereas Group B powders exhibit bubbling at the minimum fluidization velocity itself. Group A powders are often referred to as aeratable powders and Group B powders as sandlike powders. Detailed characteristics of powders that belong to the four groups are presented in Table 1.2. Geldart[12] developed numerical criteria to differentiate Group A, B, and D powders. The numerical criteria for solid particle size (d_p), density (ρ_s), and fluid density (ρ_f) are

$$1.\quad (\rho_s - \rho_f)\, d_p \le 225 \qquad \text{for Group A} \tag{1.25}$$

Equation 1.25 is the boundary between Group A and Group B powders.

$$2.\quad (\rho_s - \rho_f)\, d_p^2 \ge 10^6 \qquad \text{for Group D} \tag{1.26}$$

Equation 1.26 is the boundary between Group B and Group D powders, where density is expressed in grams per cubic centimeter and the particle diameter (d_p) in micrometers. No numerical criterion or equation for the boundary line between Group A and Group C powders was proposed.

The classification of Geldart groups has been well recognized and is often referred to in the literature, even though several other criteria based on similar conceptual premises were proposed later.

Table 1.2 Geldart[12] Classification of Powders

	Group	
	A	B
Example	Cracking catalyst	Sand
Particle size (d_p), μm	30–100	$40 < d_p < 500$
Density (ρ_s), kg/m³	<1400	$1400 \leq \rho_s \leq 4000$
Expansion	Large even before bubbling	Small
Bed collapse rate	Slow (e.g., 0.3–0.6 cm/s)	Very fast
Mixing	Rapid even with a few bubbles	Little in the absence of bubbles
Bubbles	Appear even before U_{mf} (i.e., $U_{mb}/U_{mf} > 1$)	Appear after U_{mf} ($U_{mb}/U_{mf} \approx 1$)
	Split and recoalesce frequently	Coalescence is predominant
	Rise velocity > interstitial gas velocity	Rise velocity > interstitial gas velocity
	For freely bubbling bed, rise velocity (30–40 cm/s) of small bubble (<4 cm) not dependent on bubble size	Size increases linearly with bed height and excess gas velocity
	Maximum bubble size exists	No evidence
	Cloud-to-bubble-volume ratio is negligible	Cloud-to-bubble-volume ratio not negligible
Slugs	Slugs produced at high superficial velocity and break	Slugs at high velocity of gas, rise along wall and no evidence of breakdown
	Slug size decreases with d_p	

	Group	
	C	D
Example	Finer	Coarse
Particle size (d_p), μm	<60 μm, if $(\rho_s - \rho_g) < 500$ kg/m³, <20 μm, if $(\rho_s - \rho_g) > 1000$ kg/m³	>500
Density (ρ_s), kg/m³	<1400	>1400
Expansion	Powder cohesive in nature; difficult to fluidize	Solid particles are spoutable; hence expansion is similar to spouted bed
Bed collapse rate	Very poor as deaeration is not fast	Fastest of all groups because of dense or large size of particles
Mixing	Particle mixing as well as heat transfer between a surface and bed are poorer than Group A and B	Solid mixing is relatively poor; high particle momentum and little particle contact minimize agglomeration; gas velocity in dense phase is high, and hence backmixing of dense-phase gas is less
Bubbling/ fluidization	As the interparticle forces are greater than the force exerted by fluid, the powder lifts as slug in small-diameter column or channel; hence bubbling is absent or not reported	Bubbles form at 5 cm above the distributor
	Agglomeration due to excessive electrostatic force	Bubbles of similar size to those of Group B are possible at same bed height and excess gas flow rate; largest bubbles rise slower than intertitial gas, and hence gas enters the bubble base and comes out at the top
	Fluidization is generally possible by using agitator or vibrator to break the channels	
	Electrostatic charges removed by using conductive solids or solids with graphite coating or column wall with oxide coating	

2. *Molerus Groups*

In the Geldart classification of powders, one could observe that as the particle density (i.e., the difference in the densities of the solid and the gas [air]) decreases, the boundaries separating the groups shift toward larger particle sizes. The transition between the groups, more specifically from Group C to Group A or from Group A to Group B, is not observed to occur at sharp boundaries. Molerus[13] proposed some criteria to group the powders by taking into account the interparticle force as well as the drag exerted by the gas on the particle. The powder classification diagram of Geldart,[12] superimposed with the criteria of Molerus,[13] results in a mapping of powder groups in the manner depicted in Figure 1.6, wherein sharp boundaries can be seen during the transition from one group to another.

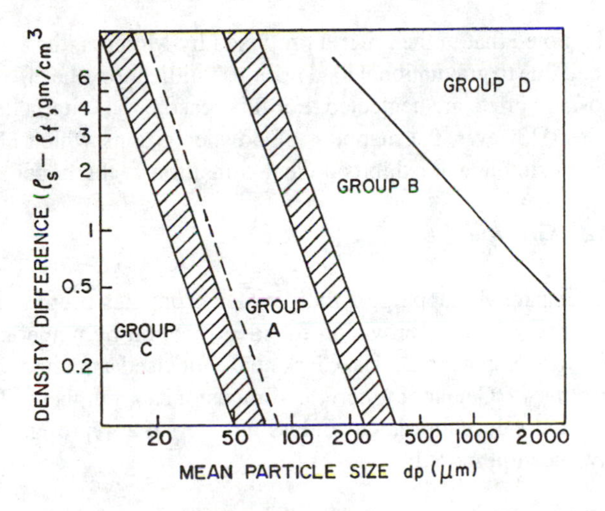

Figure 1.6 Powder classification of Geldart as modified by Molerus. (From Molerus, O., *Powder Technol.*, 33, 81, 1982. With permission.)

1. Criterion for transition from Group A to Group B:

$$\left(\rho_s - \rho_g\right)\frac{\dfrac{\pi d_p^3 g}{6}}{F_e} = K_1 \tag{1.27}$$

The constant K_1 is dependent on the nature of the powder (i.e., soft or hard). Adhesive or cohesive force should be estimated or known for these powders.

2. Criterion for transition from Group C to Group A:

$$\frac{\left(\rho_s - \rho_g\right)d_p^3 g}{F_e} = K_2 \tag{1.28}$$

where K_2 is again dependent on the nature of the powders (soft or hard), and hence two straight lines with a narrow band or strip result in the diagram.

The slope of the straight line obtained for the transition either from Group C to Group A or from Group A to Group B, when plotted on a log scale, is –3, and so the lines are parallel.

3. Criterion for transition from Group B to Group D:

$$(\rho_s - \rho_g)\, d_p\, g = K_3 \tag{1.29}$$

The constant K_3 for a stable spouted bed of sand[14] (d_p = 600 μm and ρ_s = 2600 kg/m^3) is 15.3 N m^{-3}.

It should be noted that in the criteria proposed by Molerus, the effect of consolidation of the bed due to gravitational load prior to fluidization, the effect of humidity, and the electrostatic effect are neglected. For this reason, the criteria developed have some limitations. However, the mapping of powder groups which also satisfy the Geldart criteria is simple and exhibits a clear transition at the boundaries.

3. Clark et al. Groups

The two-dimensional mapping of the powder groups as proposed by Geldart[12] and Molerus[13] suffers from no provision for identification by numerical means that would be useful for computer analysis. Clark et al.[15] devised a method of representing the powder groups of Geldart by certain dimensionless numbers. The following ranges of numerical values for dimensionless numbers that represent different powder groups have been proposed:

Powder number	Powder type	
$@ < 1.5$	C	(1.30)
$1.5 < @ < 2.5$	A	(1.31)
$2.5 < @ < 3.5$	B	(1.32)
$3.5 < @ < 4.5$	D	(1.33)

where the powder number (@) that fits with the two-dimensional mapping of the Geldart and the Molerus groups is complex.

The powder number that fits with the lines in the Geldart mapping[12] is given by:

$$@ = \left\{1.221\left[\log d_p - 2\ \log\left(1 + 0.112(\Delta\rho)^{-1.5}\right)\right] - 0.0885\right\}\frac{1 + \tan h_1}{2}$$

$$+\left[0.901\left(\log(\Delta\rho) + 1.5\ \log\ d_p\right) - 0.526\right] \tag{1.34}$$

$$\left[1 + \tan h(-c_1)\right]\frac{1 + \tan h\, c_2}{4} + \left[0.667\left(\log(\Delta\rho) + 2\ \log\ d_p\right) - 0.5\right]\frac{1 + \tan h(-c_2)}{2}$$

where

$$c_1 = \frac{1000 - (\Delta\rho)d_p^{3/2}}{150} \tag{1.35}$$

$$c_2 = \frac{14,000 - (\Delta\rho)d_p^{3/2}}{3000} \tag{1.36}$$

It can be seen that the equation or the model developed to match the Geldart mapping is rather cumbersome and does not fit very well. The powder number for Molerus powder mapping[13] is relatively simple and is expressed as:

$$@ = 0.454\left[\log(\Delta\rho) + 3\ \log\ d_p\right] - 0.395 - \left(0.9\ \log\ d_p + 2.45\right)\frac{1 + \tan hC_3}{2} \tag{1.37}$$

where

$$C_3 = \left[(\Delta\rho)d_p^3 - 1.73 \times 10^7\right] / 2 \times 10^6 \tag{1.38}$$

The equation(s) for the powder number developed by Clark et al.[15] are dimensionally inconsistent, and the units for particle diameter and particle density are in micrometers and grams per cubic centimeter, respectively. Nevertheless, it has the convenience of determining the powder group without having a recourse to a mapping of powder classification. Clark et al.[15] recommend the development of correlations that relate the powder number to such fundamental parameters as the minimum fluidization velocity and the minimum bubbling velocity, especially for Group A and B powders.

4. Dimensionless Geldart Groups

So far, gas fluidization powder grouping has been discussed on the basis of the two-dimensional powder mapping proposed by Geldart[12] and Molerus[13] and the numerical representation suggested by Clark et al.[15] In view of the difficulties associated with these schemes, Rietema[16] attempted to remap powder classification using dimensionless numbers, namely, the Archimedes number and the cohesion number ($C/\rho_s g d_p$), where C is the cohesion constant. The dimensionless representation of the Geldart classification for powders belonging to Groups A, B, and C is shown in Figure 1.7.

As can be seen from Figure 1.7, the voidage of the packed bed influences the transition from Group A to Group C. The transition or boundary between Group A and Group B powder behavior occurs when the fluidization number, N (= $\rho_s^3\, d_p^4\, g^2/\mu_f^2\, E_M$), is

$$N_{A-B} = \frac{150(1 - \epsilon_0)^2}{\epsilon_0^2(3 - 2\epsilon_0)} \tag{1.39}$$

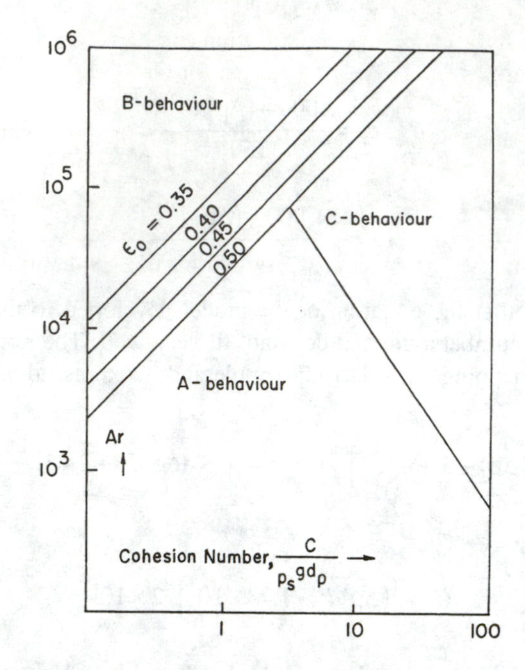

Figure 1.7 Dimensionless representation of Geldart powders. (From Clark, N.N., Van Egmond, J.W., and Turton, R., *Powder Technol.*, 56, 225, 1968. With permission.)

where E_M is the elasticity modulus of the powder bed (N/m^2). For transition from Group A and Group C behavior, one has:

$$N_{A\text{-}C} = \text{Ar}\,(N_{\text{coh}})^{-1} \tag{1.40}$$

where N_{coh} is the cohesion number ($C/\rho_s\, g d_p$). The parameter C is a cohesion constant of powder (N/m^2). The boundary between Group D and any other group powder behavior occurs when:

$$\text{Ar}\,(\rho_g/\rho_s) > 60{,}000 \tag{1.41}$$

One point to note is that the viscosity has a profound effect on the boundary between Group A and Group C powder behavior. Fine cracking catalysts ($d_p \approx 30$ μm), when fluidized by nitrogen or neon gas, behave like Group A powders (i.e., show homogeneous bed expansion), but the behavior changes to that of Group C powders when fluidization is effected by hydrogen. As yet, there is no theory or explanation available for the effect of viscosity on Group C powders.

B. Hydrodynamics- and Thermal-Properties-Based Groups

The classification scheme for powder characterization is generally based on the hydrodynamic characteristics or the fluidization properties of various powders with air at ambient temperature. Saxena and Ganzha[17] pointed out that a particle that is larger according to hydrodynamics classification can be smaller in terms of thermal properties. For example, the Geldart[12] criterion for Group D powders in the case of sand fluidized by air at ambient temperature and pressure leads to a particle size greater than 0.63 mm, and this size obviously falls in the large category according to hydrodynamics classification. On the other hand, they could still be considered to be small particles in terms of thermal characteristics. For example, for large particles, the heat transfer coefficient to the immersed surface increases with particle size. In other words, heat transfer is controlled by the gas-convective component. In the case of fine particles, the heat transfer coefficient decreases with an increase in particle diameter. A size of 1 mm has been proposed[12] to be the demarcation between small and large particle sizes. In view of this, Saxena and Ganzha[17] suggested that a powder should be classified by considering both hydrodynamic and thermal characteristics. The Archimedes number (Ar) along with the Reynolds number at minimum fluidization (Re_{mf}) have been considered for powder grouping due to the fact that Re_{mf} and the Nusselt number at its maximum value are unique functions of Ar. Saxena and Ganzha[17] classified powders into three groups and demonstrated the validity of such a classification by considering heat transfer data for sand and clay at ambient temperature and pressure. This classification takes into consideration the heat transfer correlations and the models for large particles.

The particle groups and the grouping criterion according to Saxena and Ganzha[17] are given in Table 1.3. The quantity Ψ is given by:

$$\Psi = \left[\frac{\mu_g^2}{\rho_g g(\rho_s - \rho_g)} \right]^{1/3}$$

The two subgroups shown in Table 1.3 depict the clear transition between Groups I and II as well as II and III. The classification criterion has a simple dimensionless form and is very useful for setting conditions while computing. However, it cannot identify powder types as originally conceived by Geldart.[12]

Table 1.3 Criteria for Powder Classification Based on Hydrodynamics and Thermal Properties

Group	Criteria or Relation
I	$3.35 \le Ar \le 21{,}700$ or $1.5\ \Psi \le d_p \le 27.9\ \Psi$
IIA	$21{,}700 \le Ar \le 13{,}000$ or $27.9\ \Psi \le d_p \le 50.7\ \Psi$
IIB	$13{,}000 \le Ar \le 1.6 \times 10^6$ or $50.7\ \Psi \le d_p \le 117\ \Psi$
III	$Ar \ge 1.6 \times 10^6$ or $d_p \ge 117\ \Psi$

$Ar = d_p^3\ \rho_f\,(\rho_s - \rho_f)g/d_p^3$. $\Psi = [\mu_f^2/\rho_f\,(\rho_s - \rho_f)g]^{1/3}$.

From Saxena, S.C. and Ganzha, V.L., *Powder Technol.*, 39, 199, 1984. With permission.

C. Variables Affecting Fluidization

In a fluidization bed, the fluidization medium is always comprised of solid particles. The fluidizing medium is any fluid: gas, liquid, or gas and liquid. In normal fluidization, the fluidizing fluid flows upward, thereby counteracting the gravitational force acting on the bed of particulate solid materials. The various parameters that influence the fluidization characteristics can be classified into two major groups comprised of independent variables and dependent variables. The properties of the fluid and the solid, pressure, and temperature are the major independent variables. The dependent variables include Van der Waals, capillary, electrostatic, and adsorption forces. The parameters that influence fluidization behavior can be depicted in a flow diagram (Figure 1.8).

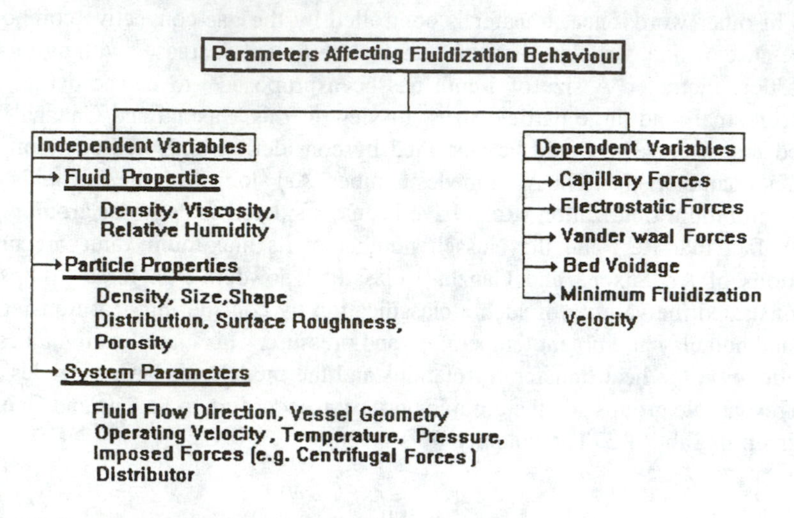

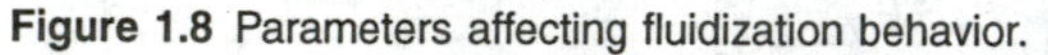

Figure 1.8 Parameters affecting fluidization behavior.

D. Varieties of Fluidization

Depending on the mode of operation and the flow regime, fluidization can be classified in several ways. For example, in normal fluidization, the gas flows upward and the particle density is greater than the fluid density. If a fluidized bed is homogeneous, it is often referred to as particulate. Heterogeneous fluidization is often bubbling or slugging in nature. Particulate fluidization is akin to liquid fluidization. Slugging occurs at high velocities in a gas fluidized bed with a narrow-diameter column and a deep bed.

If the density of the particulate solids is less than that of the fluid, normal fluidization is not possible, and in such a case the fluid flow direction has to be reversed (i.e., the fluid has to flow downward). Such a situation prevails for fluidizing certain polymers. This type of fluidization is termed inverse fluidization.

Fine powders are normally difficult to fluidize due to interparticle forces. In such a case, it is necessary to overcome the cohesive forces by some external forces in addition to the drag force exerted by the fluid flow. Agitation or vibration helps to overcome the interparticle forces. The fluidized bed in such a case is called a vibro fluidized bed.

In all the above types of fluidization, gravitational force plays a key role. There is a minimum fluid flow required to overcome the gravitational force, and the flow rate required just for fluidization may be much more than that demanded by stoichiometry.

This would finally result in wasting the unreacted fluid, which in turn increases the cost of recycling. Furthermore, it is not advisable to use such normal fluidization for a costly gas. One alternative is to use centrifugal fluidization.

A new class of fluidized beds comes into play at very high velocities (much higher than or equal to the particle terminal velocity), where carryover or elutriation of the bed inventory occurs. Here the solids are to be recycled into the bed; such a system is called a circulating fluidized bed.

Particulate solids fluidized by either gas or liquid consist of two phases and belong to the category of two-phase fluidization. A bed of solids fluidized by both gas and liquid is known as three-phase fluidization. Three-phase fluidization is more complex than two-phase fluidization and has additional classifications depending on the flow direction of the gas and the liquid. In general, they can flow either cocurrently or countercurrently. Figure 1.9 shows in a nutshell some common varieties of fluidization.

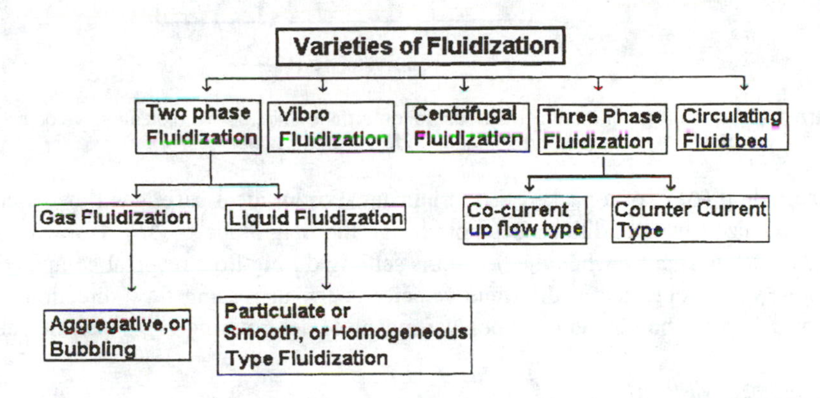

Figure 1.9 Some common varieties of fluidization.

IV. HYDRODYNAMICS OF TWO-PHASE FLUIDIZATION

A. Minimum Fluidization Velocity

1. Experimental Determination

a. Pressure Drop Method

This method of determining the minimum fluidization velocity (U_{mf}) involves the use of data on the variation in bed pressure drop across a bed of particulate solids with fluid velocity. The trend in variation of the bed pressure drop with the superficial gas velocity was shown in Figure 1.3 and discussed in detail. Figure 1.10 depicts the various methods by which the minimum fluidization velocity can be determined. The plot in Figure 1.10A depicts bed pressure drop versus gas velocity. The transition point from the fixed bed to the fluidized bed is marked by the onset of constant pressure. This is also the point at which the increasing trend in the bed

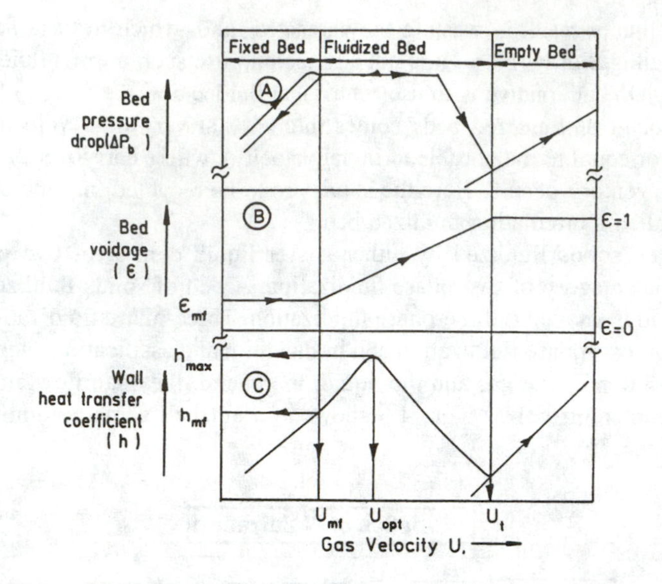

Figure 1.10 Various experimental methods to determine minimum fluidization velocity: (A) pressure drop, (B) bed voidage, and (C) heat transfer.

pressure drop (ΔP_b) of a packed bed terminates. For an ideal case, gas flow reversal in the fluidized bed condition does not change the magnitude of ΔP_b. However, the value of ΔP_b is smaller when the bed starts settling during flow reversal compared to previous values obtained at the same velocity in the increasing flow direction. The pressure drop method is the most popular means of determining U_{mf} experimentally.

b. Voidage Method

Bed expansion in a fixed bed is negligible. Hence, the bed voidage (ϵ) remains constant. When a fixed bed is brought to the fluidized state, the bed voidage increases due to bed expansion. The onset of fluidization corresponds to the point where the voidage just starts increasing with the gas velocity (U). This is shown in Figure 1.10B. The bed voidage becomes constant and is equal to unity when the gas velocity corresponds to the particle terminal velocity. This method of determining U_{mf} is not simpler than the bed pressure drop method, because the bed expansion cannot be accurately determined by any simple (i.e., visual) means.

c. Heat Transfer Method

The variation in the wall heat transfer coefficient (h) with gas velocity (U) forms the basis of one of the interesting methods of determining U_{mf}. The wall heat transfer coefficient increases gradually in a fixed bed as the gas velocity is increased, and it suddenly shoots up at a particular velocity, indicating the onset of fluidization. The velocity that corresponds to the sudden increase in the wall heat transfer coefficient is the minimum fluidization velocity. The trend in the variation of h with U is shown in Figure 1.10C. In this method, another important gas velocity, the optimum gas velocity (U_{opt}), which corresponds to the maximum wall heat transfer coefficient (U_{max}), is also obtained. This method of determining U_{mf} is more expensive than the two aforementioned methods, and it requires good experimental setup to measure the heat transfer data under steady-state conditions. For these reasons, this method is seldom used to determine U_{mf} in fluidization engineering.

2. Theoretical Predictions

The various methods, based on first principles, available in the literature for the theoretical prediction of the minimum fluidization velocity can be broadly classified into four groups. These methods are derived from (1) dimensional analysis, (2) the drag force acting on single/multiparticles, (3) the pressure drop in a fixed bed extendable up to incipient fluidization, and (4) a relative measure with respect to the terminal particle velocity. Although a significant body of literature pertaining to the above methods exists and numerous correlations are listed by various researchers, there has been no classification of these correlations. The aforementioned methods are briefly described and the correlations based on them are listed in the following sections.

a. Dimensional Analysis (Direct Correlation)

This conventional method of developing a correlation takes into consideration the physical properties of the fluid and the solid. In its most general form, the correlation is given as:

$$U_{mf} = K\, d_p^a \left(\rho_s - \rho_f\right)^b \rho_f^c \mu_f^d g^c \left(\phi_s^m \epsilon_{mf}^n\right)\left(1 - \epsilon_{mf}\right)^p \tag{1.42}$$

Correlations of the above type are presented in Table 1.4. It can be seen that the factor ($\phi_s^m \epsilon_{mf}^n$) is variable because of the dependence of ϵ_{mf} on ϕ_s. However, many correlations have been proposed by assuming that the quantity $K\, \phi_s^m \epsilon_{mf}^n\, (1 - \epsilon_{mf})^p$ is a constant. This assumption is valid only within a certain range of experimental parameters. This type of direct correlation has the inherent disadvantage of involving a dimensional constant (K) that changes depending on the system of units used for the variables in Equation 1.42.

b. Drag Force Method

In this method, a force balance is assumed at incipient fluidization. The force balance equation is

$$F_g = \alpha F_D + F_B \tag{1.43}$$

where F_g, F_D, and F_B are, respectively, the force due to gravity, drag, and buoyancy, and α is a correction or multiplication factor for a multiparticle system. Because the

Table 1.4 Correlation for Minimum Fluidization Velocity Based on Dimensional Analysis Method

A. General Form for $U_{mf} = K_o\, d_p^a\, (\rho_s - \rho_f)^b\, \rho_f^c\, \mu_f^d\, g^e \phi_s^m\, e_{mf}^n\, (1 - \epsilon_{mf})^p$
$K = K_o\, \phi^m\, e_{mf}^n\, (1 - e_{mf})^p$

Ref.	a	b	c	d	e	K	m	n	p	Remarks[a]
	Exponential Factor									
Leva[18]	1.82	0.94	–0.06	–0.88	1	7.169×10^{-4}	—	—	—	GF, SI, $Re_{mf} < 5$
Miller and Logwinuk[19]	2	0.9	0.1	–1	1	0.00125	—	—	—	GF, SI units
Rowe and Yacono[20]	1.92	0	0	0	0	0.00085	0	2.92	0.945	U_{mf} = cm/s, d_p = μm
Davidson and Harrison[21]	2	1	0	–1	1	0.0008	—	—	—	All SI units
Yacono[22]	1.77	0	0	0	0	0.000512	0	0	0	U_{mf} = cm/s, d_p = μm
Baerg et al.[23]	1.23	1.23	–1	0	0	0.361	0	0	0	$(\rho_s - \rho_f) = \mu_{sb}$ (bulk density) SI units
Kumar and Sen Gupta[24]	1.34	0.78	–0.22	0.56	0.78	0.005	—	—	—	SI units
Baeyens and Geldart[25]	1.8	0.934	–0.066	–0.87	0.934	9.125×10^{-4}	—	—	—	CGS units
Leva et al.[26]	1.82	0.94	–0.06	—	—	7.39	—	—	—	SI units, $Re_{mf} < 5$

B. General Form for Re_{mf}

$$Re_{mf} = K\,(d_p^3 \rho_f^2 g/\mu_f^2)^x \left(\frac{\rho_s - \rho_f}{\rho_f}\right)^y$$

Ref.	K	x	y	Remarks
Riba[27]	1.54×10^{-2}	0.66	0.7	
Ballesteros[28]	12.56×10^{-2}	1.0523	0.66 + 1.0523	
Doichev and Akhmakov[29]	1.08×10^{-3}	0.947	0.947	
	7.54×10^{-4}	0.98	0.98	$Re_{mf} < 30$
Thonglimp et al.[30]	1.95×10^{-2}	0.66	0.66	$30 < Re_{mf} < 180$

[a] GF = general form.

drag force varies depending on the flow range, there can be several correlations for the prediction of U_{mf} by this method. Taking

$$F_D = \alpha(1/2) C_D \rho_g U_{mf}^2 (\pi/4) d_p^2 \tag{1.44}$$

and using appropriate expressions for F_g and F_B for a particle of diameter d_p fluidized by a fluid of density ρ_g, a general equation can be written as:

$$\frac{3}{4}\alpha \cdot C_D \, \mathrm{Re}_{mf}^2 = \mathrm{Ar} \tag{1.45}$$

The above equation can be solved if an appropriate value of drag coefficient (C_D) is chosen, depending on the flow regime. For Stokes flow ($\mathrm{Re}_{mf} \leq 0.1$), $C_D = 24/\mathrm{Re}_{mf}$. Hence Equation 1.45 becomes

$$\mathrm{Re}_{mf} = \frac{\mathrm{Ar}}{18\alpha} \tag{1.46}$$

For Newton's flow regime (i.e., for turbulent flow condition), $\mathrm{Re}_{mf} > 500$ and $C_D = 0.44$. Hence,

$$\mathrm{Re}_{mf}^2 = \frac{\mathrm{Ar}}{0.33\alpha} \tag{1.47}$$

In the intermediate flow regime, there can be as many as six empirical correlations for C_D, as proposed by Morse.[31] The drag force on a single sphere situated in an infinite expanse of fluid, according to Schiller and Naumann,[32] is

$$F_D = \frac{\pi d_p^2 \rho_g U}{g_c}\left(3/\mathrm{Re} + 0.45 \ \mathrm{Re}^{-0.313}\right) \tag{1.48}$$

for $0.001 < \mathrm{Re} < 1000$. The correlation given by Equation 1.48 fairly covers all three flow regimes and thus can be used in Equation 1.43 to predict Re at U_{mf} conditions for most gas–solid systems (i.e., for Group A to C particles of Geldart's classification).

The factor α in Equation 1.43 has been found to be a function of voidage (ϵ) by Wen and Yu.[33] They found experimentally that for particulate fluidization:

$$\alpha = f(\epsilon) = \epsilon^{-4.7} \tag{1.49}$$

Using Equations 1.48 and 1.49 and the appropriate expressions for F_B and F_g in Equation 1.43, the general form of the correlation for obtaining Re_{mf} in the range 0.001–1000 is

$$2.7 \ \mathrm{Re}_{mf}^{1.687} - 18 \ \mathrm{Re}_{mf} - \epsilon^{4.7}\mathrm{Ar} = 0 \tag{1.50}$$

for $0.818 < \rho_f$ (kg/m^3) < 1135, $6 \times 10^{-4} < d_p$ (μm) $< 25 \times 10^{-2}$, $1060 < \rho_s$ (kg/m^3) $< 11{,}250$, $1 < \mu$ (CP) < 1501, $244 \times 10^{-5} < (d_p/D) < 10^{-1}$.

Assuming ($\epsilon_{mf} = 0.42$, Wen and Yu[33] proposed a simplified form of the above correlation as:

$$159 \ \mathrm{Re}_{mf}^{1{,}687} - 1060 \ \mathrm{Re}_{mf} - \mathrm{Ar} = 0 \tag{1.51}$$

The solution for Re_{mf} using Equation 1.50 or 1.51 is not straightforward. On the other hand, Re_{mf} can be obtained more easily by using appropriate expressions for different flow conditions as given by Equations 1.46 and 1.47.

Using appropriate correlations[33] for F_D and δ, the following relationship can be arrived at:

$$\alpha = \left(\frac{U_t}{U}\right)^{\beta} \tag{1.52}$$

where $\beta = 1$ for laminar flow ($0.001 \leq Re \leq 2$), 1.4 for intermediate flow ($2 < Re < 500$), and 2.0 for turbulent flow ($Re > 500$).

It may be recalled here that Richardson and Zaki[34] experimentally obtained the relationship

$$\frac{U_t}{U} = \epsilon^{-n} \tag{1.53}$$

which indicated that α is a function of voidage only. In light of the above analysis of the development of an expression for Re at U_{mf} conditions, a variety of correlations can be obtained, depending on the range of Re and the corresponding equation for the drag coefficient. Most expressions available for predicting C_D are based on experimental results pertaining to spherical particles, and no single correlation is able to cover the entire (laminar to turbulent) range of flow conditions. The right choice of C_D for a nonspherical particle is often difficult. More recently, Haider and Levenspiel[35] presented explicit equations for predicting C_D for spherical and nonspherical particles. For spherical particles:

$$C_D = \frac{24}{\mathrm{Re}}\left(1 + 0.1806\ \mathrm{Re}^{0.6459}\right) + \frac{0.4251}{\left(1 + 6880.95/\mathrm{Re}\right)} \tag{1.54}$$

for $Re \leq 2.6 \times 10^5$. For nonspherical, isometric particles:

$$\begin{aligned} C_D = {} & \frac{24}{\mathrm{Re}}\left[1 + 8.1716\left(\exp - 4.0655\ \phi_s\right)\right]\mathrm{Re} \\ & + 73.69\ \mathrm{Re}\exp\left(-5.0748\ \phi_s\right)/\left[\mathrm{Re} + 5.378\exp\left(6.2122\ \phi_s\right)\right] \end{aligned} \tag{1.55}$$

for $Re \leq 25{,}000$. A more accurate correlation with rms ≈ 3.1, as given by Turton and Levenspiel,[36] is cumbersome. Equation 1.55 is considered to be a simple form wtih an rms error of 5%.

The drag coefficient in the case of a porous spherical particle falls outside the ambit of the above type of correlations. Masliyah and Polikar[37] reported from their experimental findings that the drag experienced by a porous sphere is less than that experienced by an impermeable sphere of the same diameter and bulk density.

However, at high values of the Reynolds number, the effect of inertia on a porous sphere is found to be greater than that on a comparable impermeable sphere. The drag coefficient proposed by Masliyah and Polikar[37] for a porous sphere is given by:

$$C_D = 24\frac{\Omega}{\text{Re}}\left[1 + K_5\,\text{Re}^{(K_6 - K_7\log_{10}\text{Re})}\right] \tag{1.56}$$

where

$$\Omega = \frac{2\psi^2[1-(\tan h\,\psi)/\psi]}{2\psi^2 + 3[1-(\tan h\,\psi)/\psi]} \quad \text{and} \quad \Psi = d_p\Big/\sqrt{k_{pe}}$$

The values of K_5, K_6, and K_7 are 0.1315, 0.82, and 0.05, respectively (for $0.1 < \text{Re} < 7$), and 0.0853, 1.093, and 0.105, respectively (for $7 < \text{Re} < 120$). A list of correlations based on the drag force principle or with similar forms is given in Table 1.5.

c. Pressure Drop Method

The most general form of expression for the pressure drop through a fixed bed of particulate solids can be given as:

$$(\Delta P_b/H) = f(\rho_g U_g^2/2)\; 6(1-\epsilon)/(\epsilon^3 \Phi_s d_p) \tag{1.57}$$

where the bed friction factor (f) is a function of bed voidage (ϵ) and the particle Reynolds number (Re).

At the incipient fluidization condition, taking $\Delta P_b/H = (\rho_s - \rho_g)(1 - \epsilon_{mf})g$, Equation 1.57 can be transformed to:

$$\text{Ar} = \frac{1}{\phi_s \epsilon_{mf}^3}\text{Re}_{mf}^2\, f_k \tag{1.58}$$

Ergun[42] proposed that the friction factor (f_k) for randomly packed beds can be expressed as:

$$f_k = \frac{150(1-\epsilon)}{\phi_s\,\text{Re}} + 1.75 \tag{1.59}$$

Substituting Equation 1.59 in Equation 1.58 at U_{mf} condition and rearranging:

$$\frac{1.75}{\phi_s \epsilon_{mf}^3}\text{Re}_{mf}^2 + \frac{150}{\phi_s^2 \epsilon_{mf}^3}(1-\epsilon_{mf})\cdot \text{Re}_{mf} = \text{Ar} \tag{1.60}$$

Table 1.5 Correlations for Re_{mf} Based on Drag Force Principle or Similar Form as Derived from Drag Force Principle

Ref.	K_2			Remarks[a]
	A. General Form for Low Re_{mf} **$Re_{mf} = K_2$ Ar**			
Rowe and Henwood[38]	8.1×10^{-3}			GF
Frantz[39]	1.065×10^{-3}			$Re_{mf} < 32$, GF
Wen and Yu[33]	$\epsilon^{-4.7}/18$			$0.001 < Re_{mf} < 2$, GF
Davidson and Harrison[21]	0.00081			GF
Davies and Richardson[40]	7.8×10^{-4}			GF
Pillai and Raja Rao[41]	7.01×10^{-4}			Re < 20, GF
	B. General Form for All Ranges of Re_{mf} **$Re_{mf} = [(4/3)\ (Ar/\alpha)/C_D]^{1/2}$, where $\alpha = \epsilon^{-4.7}$**			
Schiller and Naumann[32]	$C_D = 3/Re_{mf} + 0.45\ Re_{mf}^{-0.313}$			$0.001 < Re_{mf} < 1000$, for spherical particle
Morse[31]	$C_D = C_1/Re + C_2/Re^2 + C_3$			For spherical particle
	C_1	**C_2**	**C_3**	**Re_{mf} Range**
	24	0	0	$Re_{mf} \le 0.1$
	22.73	0.0903	3.69	$0.1 \le Re_{mf} \le 1$
				$1 \le Re_{mf} \le 10$
	29.1667	3.8889	1.222	$10 \le Re_{mf} \le 100$
	46.5	116.67	0.6167	$100 \le Re_{mf} \le 1{,}000$
				$1{,}000 \le Re_{mf} \le 5{,}000$
	98.33	2,778.0	0.3644	$5{,}000 \le Re_{mf} \le 10{,}000$
	148.62	4.75×10^4	0.357	$10^4\ Re_{mf} \le 5 \times 10^4$
				$Re_{mf} \ge 5 \times 10^4$
	490.546	57.87×10^4	0.46	
	1,662.5	5.4167×10^6	0.5191	
	0	0	0.44	
Haider and Levenspiel[35]	See Equation 1.54			$Re_{mf} < 2.6 \times 10^5$ for spherical particle
	See Equation 1.55			$Re_{mf} < 25{,}000$ for nonspherical particle, $\phi \ge 0.67$

[a] GF = general formula.

Wen and Yu[33] were the first to use this type of correlation and to solve it for Re_{mf}. In order to arrive at a suitable solution, Wen and Yu collected the data for ϵ_{mf} and ϕ_s and made the following approximations:

$$\frac{1-\epsilon_{mf}}{\phi_s^2 \epsilon_{mf}^3} \simeq 11 \quad \text{and} \quad \frac{1}{\phi_s \epsilon_{mf}^3} \simeq 14 \tag{1.61}$$

The Wen and Yu[33] correlation expressed using Re_{mf} and Ar is

$$24.5\ \mathrm{Re}_{mf}^2 + 1650\ \mathrm{Re}_{mf} = \mathrm{Ar} \tag{1.62}$$

The above correlation has been found to fit experimental data well in the range 0.001 $< \mathrm{Re}_{mf} <$ 4000, and this form of correlation is cited in the literature frequently. The Wen and Yu[33] correlation has been refined to increase the accuracy of prediction, and several correlations of such type have been introduced by altering the coefficients of Re_{mf}^2 and Re_{mf}.

The friction factor (f_f) for a fluidized or sedimenting suspension can be expressed after Davidson and Harrison[21] as:

$$f_f = \epsilon^3(\rho_s - \rho_f)g/S\rho_f U^2 \tag{1.63}$$

For a packed bed one has:

$$f_p = \left(\frac{\Delta P}{SH}\frac{\epsilon}{1-\epsilon}\right)\Big/\left(\rho_f U^2/\epsilon^2\right) \tag{1.64}$$

Uṣisng Carman's proposed equation[43] for ΔP with a constant of 5, one obtains:

$$f_p = \left[\frac{5(1-\epsilon)^2}{\epsilon^3}\cdot S^2\mu\, H\, U\right]\cdot\frac{1}{SH}\frac{\epsilon}{1-\epsilon}\cdot\frac{\epsilon^2}{\rho_f U^2} \tag{1.65}$$

The above friction factors have been evaluated, and correlations in terms of Re_l based on a linear dimension analogous to the hydraulic mean diameter, i.e., $\epsilon/(1-\epsilon)S$, have been reported as:

$$f_f = 3.36/\mathrm{Re}_L \quad \text{for } \mathrm{Re}_L < 1,\ \epsilon < 0.78 \tag{1.66}$$

$$f_p = 5/\mathrm{Re}_L \quad \text{for } \mathrm{Re}_L < 1,\ \epsilon < 0.78 \tag{1.67}$$

From Equations 1.66 and 1.67, it is obvious that the friction factor for a sedimenting or fluidizing bed is much less compared to that for a fixed bed. The lower value of the friction factor is due to lower resistance in a bed where the particles have freedom of movement relative to one another or can oscillate/rotate. The validity of this inference has been demonstrated by Davidson and Harrison[21] in a plot of the variation of the Richardson–Zaki[34] index n with various values of the Archimedes number wherein the n value predicted using the fixed bed pressure drop equation of Ergun[42] (for the Re_{mf} calculation) is found to be greater than the experimental values. If n is high, then Re_{mf} is high because of ΔP, which is high when calculated for a fixed bed. Hence, it appears that the correlation, which has the form given in Equation 1.60, on solution assumes the form:

$$\mathrm{Re}_{mf} = (A_1 + B_1\ \mathrm{Ar})^{1/2} - A \tag{1.68}$$

The values of A_1 and B_1 depend on the experimental conditions and the range of Re_{mf}. The values of constants A_1 and B_1 in Equation 1.68 that satisfy the various correlations reported in the literature by different researchers are presented in Table 1.6; the ratio $(1 - \epsilon_{mf})/\phi_s$ calculated by using these values is also given in the table. Fletcher et al.[56] reexamined the correlations for the minimum fluidization velocity as applied to Group B powders and explained the variation in U_{mf} with temperature by a correlation. More recent work by Lippens and Mulder[57] on prediction of the minimum fluidization velocity recommends the use of the ratio Ar/Re as the characteristic parameter for the fluidized bed.

Table 1.6 Basic Form of Correlations for Re_{mf} Derived from Pressure Drop Principles

Correlation: $Re_{mf} = (A_1^2 + B_1 Ar)^{1/2} - A_1$, $\alpha = (1 - \epsilon)/\phi_s^2 \epsilon^3$, $\beta = 1/(\phi_s \epsilon^3)$, $m = \alpha/\beta$

Ref.	Constants			Remarks
	A_1	B_1	m	
Ergun[42]	42.85/α	0.57/β	$(1 - \epsilon)/\phi_s$	Modified for fluidized bed
Wen and Yu[33]	33.7	0.0408	0.7857	Water-fluidized, spherical particle
Bourgeois and Grenier[44]	25.46	0.0382	0.594	Spherical particle
Ghosal and Mukherjee[45]	29.2	0.029	0.6814	For spherical and angular ($1 < Re_{mf} < 1000$)
Saxena and Vogel[46]	25.28	0.0571	0.5899	Air fluidized, $6 < Re_{mf} < 102$
Babu et al.[47]	25.25	0.0651	0.5892	High pressure (7000 kPa) for coal gasification
Richardson and Jeromino[48]	25.7	0.0365	0.5997	
Thonglimp et al.[49]	19.9	0.03196	0.4644	Binary system
Chitester et al.[50]	28.7	0.0494	0.6697	High-pressure fluidization (6485 kPa)
Thonglimp et al.[30]	31.6	0.0925	0.7374	Air fluidized
Masaaki et al.[51]	33.95	0.0465	—	Elevated temperature (280–800 K) and pressure (0.1–4.9 MPa)
Agarwal and O'Neil[52]	42.81	0.061	0.999	—
Satyanarayana and Rao[53]	30.10	0.0417	0.7024	Elevated temperature (295–490 K)
Grace[54]	27.2	0.0408	0.635	Correlation from literature
Panigrahi and Murty[55]	32.2	0.0382	0.751	Trial-and-error method adopted to develop general correlation for spherical multiparticle drag coefficient

d. Terminal Velocity Method

In this method, it is necessary to use some experimental technique to precisely evaluate the particle terminal velocity or to rely on standard correlations available in the literature. Most correlations for determining U_t are based on free settling of a single particle. The frequently cited equation[58] for predicting U_t is

$$U_t = \left[\frac{4}{3}\frac{\left(\rho_s - \rho_g\right)}{\rho_g} \cdot \frac{g d_p}{\left(C_D\right)}\right]^{1/2} \tag{1.69}$$

where the drag coefficient (C_D) should be evaluated for use in a multiparticle system. Generally, C_D values are chosen for appropriate flow regimes for a single particle as outlined in the drag force method. Haider and Levenspiel[35] reported that there are over 30 equations in the literature relating C_D to Re of spherical particles falling at their terminal velocities, and they also pointed out the complex nature of the correlations due to the use of several arbitrary constants (as many as 18). Because it is important to evaluate U_t rather than C_D, Haider and Levenspiel[35] recommended correlations to determine U_t for both spherical and nonspherical particles as:

$$U^* = [(18/d^*)^{0.824} + (0.321/d^*)^{0.412}]^{-1.214} \tag{1.70}$$

for Re < 2.6×10^5 (for spherical particles) and

$$U^* = \left[18 d^{*2} + \frac{\left(2.3348 - 1.7439\,\phi_s\right)}{d^{*0.5}}\right]^{-1} \tag{1.71}$$

for Re < 25,000 (for nonspherical particles with $0.5 < \phi_s < 1$). Here, U^* and d^* are dimensionless particle velocity and particle diameter, respectively, and can be expressed as:

$$U^* = U_t\left[\rho_g^2 / g\mu_g\left(\rho_s - \rho_g\right)\right]^{1/3} = \mathrm{Re}_t\,\mathrm{Ar}^{1/3} \tag{1.72}$$

$$d^* = \left(d_p\right)_{\text{spherical}}\left[g\,\rho_g\left(\rho_s \cdot \rho_g\right)/\mu_g^2\right]^{1/3} = \mathrm{Ar}^{1/3} \tag{1.73}$$

It is possible to use Equations 1.72 and 1.73 and to evaluate Re_t; this value can be substituted in Equation 1.50 based on the drag force method or in Equation 1.60 based on the pressure drop method to arrive at the following equations, respectively:

$$2.7\,\mathrm{Re}_t^{1.687} \cdot X^{1.687} - 18\,\mathrm{Re}_t \cdot X - \epsilon_{mf}^{4.7}\mathrm{Ar} = 0 \tag{1.74}$$

$$\frac{1.75}{\phi_s \epsilon_{mf}^3}\mathrm{Re}_t^2 X^2 + \frac{150\left(1-\epsilon_{mf}\right)}{\phi_s^2 \epsilon_{mf}^3}\mathrm{Re}_t \cdot X - \mathrm{Ar} = 0 \tag{1.75}$$

where $X = \mathrm{Re}_{mf}/\mathrm{Re}_t$. The solution for X can be obtained by using the value of Re_t and Re_{mf}.

It should be noted that the solution for X depends on ϵ_{mf} and ϕ_s. The variation of X with the Archimedes number has been calculated and presented graphically by Richardson;[59] it is seen that the ratio X is sensitive to ϵ_{mf} for $Ar \leq 1$. (The magnitude for some typical values of ϵ_{mf} is $X = 64$ for $\epsilon_{mf} = 0.42$, 78 for $\epsilon_{mf} = 0.40$, and 92 for $\epsilon_{mf} = 0.38$.) It drops gradually in the range $1 \leq Ar \leq 10^5$ and attains a constant value above $Ar > 10^5$. Using correlations 1.74 and 1.75, it is possible to generate a plot of X versus Ar, if the values of ϕ_s and ϵ_{mf} are known.

B. Terminal Velocity

1. Definition

The terminal velocity or free-fall velocity of a particle in a fluid bed of uniform size is the maximum allowable velocity for operation, and entrainment or carryover of the particle occurs at this velocity. The terminal velocity of the smallest particle in a polydisperse or mixed-particle assembly limits the operating range of velocity. The terminal velocity of the smallest particle may be just equal to or even less than the minimum fluidization velocity of the largest particle. In such a case, the carryover of fines will take place while the largest particles are fluidized or are kept in a fixed bed condition. Particles that fall between the two size extremes will be in a state of fluidization. The pressure drop calculated for such polydisperse particles by Equation 1.57 will not be correct. The bed height (H) in Equation 1.57 should be used for the fluidized portion, and the bed height for the defluidized portion should be chosen to predict the pressure drop in that static, packed, or fixed bed. It should be noted that expansion in a gas fluidized bed will be maximal for polydisperse particles when the bed is fluidized at the bottom at the minimum fluidization velocity and at the top at the terminal velocity. This could be possible in gas fluidization where the gas expands as the pressure reduces in the upward flow direction. In the case of a liquid fluidized bed, this kind of situation does not arise because the liquid is incompressible (i.e., its density remains unchanged).

2. Mathematical Representation

For a particle of volume V_p and density ρ_p that is moving in a flowing fluid of velocity U_f, the relative velocity (U_r) is

$$U_r = U_p - U_f \tag{1.76}$$

where U_p is the particle velocity. If the motion is one dimensional (i.e., takes into consideration only the vertical upward flow), the rate of change in the momentum of the particle, according to Newton's second law, is

$$V_p \rho_p \frac{dU_p}{dt} = -(F_D + F_B) + F_g \tag{1.77}$$

where the drag force

$$F_D = A_p C_D \left(0.5\, \rho_f U_f^2\right) \tag{1.78}$$

the buoyant force

$$F_B = V_p\, \rho_g\, g \tag{1.79}$$

and the gravitational force

$$F_g = V_p\, \rho_g\, g \tag{1.80}$$

When the fluid velocity (U_f) is constant,

$$\frac{dU_r}{dt} = \frac{dU_p}{dt}$$

and at terminal velocity

$$\frac{dU_p}{dt} = 0$$

Hence, in the case of a spherical particle, Equation 1.77 can be rewritten to solve for U_f (which equals U_p or U_t). U_r is zero when the particle is just entrained or carried by the fluid stream. The solution of Equation 1.77 for U_t is given by:

$$\left(\frac{U_t d_p \rho_g}{\mu_g}\right)^2 = \frac{4}{3}\frac{d_p^3 \rho_g \left(\rho_s - \rho_g\right) g}{C_D \mu_g^2} \tag{1.81}$$

Equation 1.81 can be written using dimensionless numbers Re_t and Ar:

$$C_D\, \mathrm{Re}_t^2 = \frac{4}{3}\mathrm{Ar} \tag{1.82}$$

3. *Drag Coefficient*

It can be seen from the above equations that the product $C_D\mathrm{Re}_t^2$ is a function of the Archimedes number and can be determined if the fluid–particle properties are known. On the other hand, if the variation in $C_D\mathrm{Re}_t^2$ as a function of Re_t is known for a known value of Ar, one can find Re_t. Again, the equation thus derived will be valid only for a spherical particle. For a nonspherical particle, the sphericity factor would come into the picture and $C_D\mathrm{Re}_t^2$ should be known in terms of Re_t and the sphericity (ϕ_s). A family of plots of $C_D Re_t^2$ versus Re_t for different values of sphericity can be generated from experimental data;[8] a typical plot of this type can be found in the literature.[58] In the absence of such data, it is necessary to have reliable correlations to predict the drag coefficient. However, it is not an easy task to generate a single correlation to predict the drag coefficient for the various flow regimes of interest. A list of various correlations available for C_D is presented in Table 1.5B.

a. Evaluation of Drag Coefficient

A standard drag coefficient (C_D) curve[60] obtained under highly idealized conditions is shown in Figure 1.11. The C_D value changes smoothly with Re. When Re approaches 300,000, C_D drops sharply from 0.4 to 0.1, indicating that the flow is in transition to turbulence. The Re value that corresponds to $C_D = 0.3$ is known as a critical Re. In the supercritical regime, the value of C_D increases again, and the extrapolated value of C_D beyond Re $= 9 \times 10^6$ is 0.34. A drag coefficient obtained from a standard drag curve and used to solve for Re_t will lead to erroneous results when the fluid flow is turbulent. Clift and Gauvin[60] predicted the terminal velocity of silica spheres (2.65 g/cc) falling in air at 21°C as a function of relative turbulence and showed that the intensity of turbulence affects the determination of U_t. All these data are reported for spherical particles. For nonspherical particles, the situation is complex.

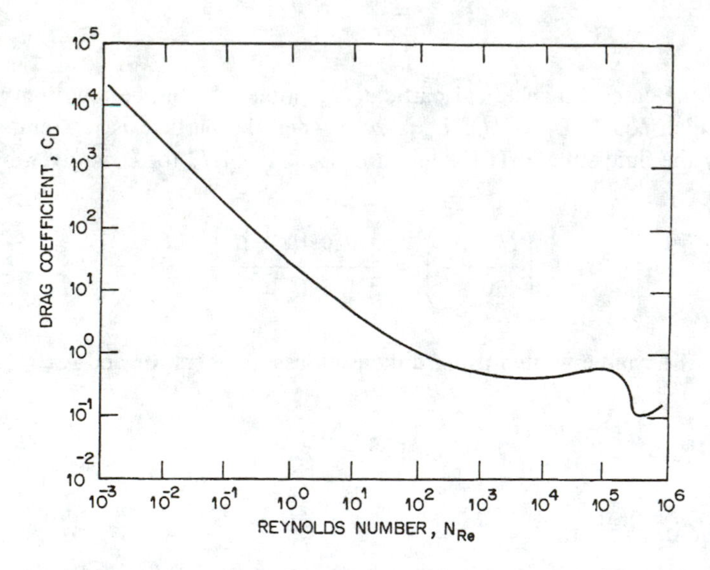

Figure 1.11 Standard drag coefficient curve for a sphere. (From Clift, R. and Gauvin, W.H., *Chemeca 70*, Butterworths, Australia, 1970, 14. With permission.)

b. Correlations for Drag Coefficient

Concha and Almendra[61,62] and Concha and Barrientos[63] carried out a series of investigations to resolve many complex problems encountered in the development of a correlation for C_D. They arrived at a reliable correlation for determination of the settling/terminal velocity in the case of (1) individual spherical particles, (2) suspensions of spherical particles, (3) nonspherical isometric particles, and (4) suspensions of particles of arbitrary shape.

Concha and Barrientos[64] further extended their investigations to nonspherical particles. In the case of particles of arbitrary shape, Concha and Christiansen[65] arrived at complex correlations to predict the drag coefficient and also the particle terminal velocity. Geldart[66] proposed a simple interpolation of the Pettyjohn and Christiansen[67] correlation for calculation of the terminal velocity of nonspherical particles in the intermediate flow regime.

It is beyond the scope of this volume to consider the details of development of various methods for predicting particle terminal velocity. However, some methods for predicting terminal velocity in various situations, from a single spherical particle to suspensions of nonspherical particles, are presented next.

4. Terminal Velocity for Single Spherical Particle

The standard C_D versus Re curve can be segmented, without any loss of accuracy, into three parts, to develop expressions for C_D for three flow regimes:[68]

$$C_D = a/\mathrm{Re}^b \tag{1.83}$$

where a and b assume different values for the three regimes. These values are given in Table 1.7.

Table 1.7 Constants for Evaluating Drag Coefficient for Three Flow Regimes

Range	*a*	*b*	Range $(\mathrm{Ar})^{1/8} = K$
Re < 2, Stokes	24	1	<3.3
2 < Re < 500, intermediate	18.5	0.6	3.3–48.6
500 < Re < 200,000, Newton	0.44	0	43.6–2360

Using Equations 1.82 and 1.83, a general equation for evaluating Re_t can be arrived at:

$$\mathrm{Re}_t = \left[\frac{4}{3}\cdot\frac{\mathrm{Ar}}{a}\right]^{1/(2-b)} \tag{1.84}$$

Correlation 1.83 gives the most general form of the drag coefficient. Research workers in the field of fluid–particle technology have long been looking for a simple correlation to predict the terminal velocity, and attempts have been made in this process to generate correlations for the drag coefficient. The correlations published in the literature to date span a wide range of values of Reynolds number, from laminar flow (i.e., Stokes regime) to turbulent flow (i.e., Newton's law regime).

5. Difficulties in Predicting Particle Terminal Velocity

There are many correlations for the drag coefficient with respect to particle shape. Correlations have been proposed for (1) a spherical particle settling alone in an infinite medium of fluid, (2) a spherical particle settling in an assembly of particles, and (3) particles of irregular shape settling under conditions 1 and 2. Correlations for the drag coefficient which are often referred to in the literature are summarized in Table 1.8. How a correlation is selected to predict the terminal velocity may confuse a beginner in this subject. This kind of difficulty is inevitable because a simple correlation for drag is less accurate even though it is easy to compute. A complex correlation predicts C_D with accuracy, but its use may be cumbersome from the standpoint of computation. We suggest the designer use a correlation that would satisfy the accuracy required for the specific purpose for which the design is being made. Some compromise with regard to accuracy is inevitable for convenience of calculation. At this juncture, it may be helpful to examine a useful graphic for computing the terminal velocity. Figure 1.12 presents a simple form of a graph[79] for dimensionless fluid flux:

$$j^* = j_{fo}\left(\frac{\rho_f^2}{\mu_f g \Delta\rho}\right)^{1/3}$$

versus dimensionless particle size:

$$r^* = r\left(\frac{\rho_f g \Delta\rho}{\mu_f^2}\right)^{1/3}$$

for Reynolds number

$$\mathrm{Re}_s = 2j^* r^* = 2r\, j_{fo}\, \rho_f / \mu_f$$

ranging from 1.5 to 1080. The advantage of this representation is that it takes into account the void fraction and depicts only a simple plot rather than a family of curves. The plot shown in Figure 1.12 is useful for computing the terminal velocity of a spherical particle when the voidage factor is set at unity.

The list of correlations given in Table 1.8 may be used to substitute for the drag coefficient in Equation 1.82, and then the particle terminal velocity can be obtained. However, the choice of correlation for C_D generally poses problems for the designer or for a beginner in the field. For the sake of simplicity or convenience, the preference in most of the literature or textbooks is to use the C_D value given in Equation 1.83. Obviously, however, this may not give the desired accuracy. A direct correlation to predict the settling velocity or terminal velocity is often a difficult task in fluidization or more specifically in fluid–particle technology.

Table 1.8 Drag Coefficient Correlations

Drag Coefficient Correlations	Remarks	Ref.
$C_D = 24/\text{Re}\ (1 + 0.150\ \text{Re}^{0.687})$		
$C_D = 24/\text{Re} + 2$		Rubey[69]
$C_D = 24.4/\text{Re} + 0.4$		Dallavalle[70]
$C_D = (0.63 + 4.8/\text{Re}^{1/2})\xi$		Dallavalle[71]
$C_D = 18.5\ \text{Re}^{-0.6}$		Lapple[72]
$C_D = 24/\text{Re}(1 + 0.197\text{Re}^{0.63} + 0.0026\ \text{Re}^{1.38})$		Torobin and Gauvin[73]
$C_D = 24/\text{Re}(1 + 3/16\ \text{Re})^{1/2}$	$\text{Re} \le 100$	Olson[74]
$C_D = 24/\text{Re}(1 + 0.15\ \text{Re}^{0.687})\epsilon^{-4.7}$	Fluidized bed approach based on single particle	Wen and Yu[33]
$C_D = C_d[1 + (1 - \epsilon)^{1/3}]/\epsilon^3$ $C_d = (0.63 + 4.8/\text{Re}^{1/2})^2$ Re $\text{Re}/\{\epsilon \exp[5(1 - \epsilon)/3\epsilon]\}$	Wide flow range for fluidized bed based on multiparticle approach	Barnea and Mizrahi[75]
$\log[(C_D\text{Re}/24) - 1] = -0.881 + 0.82W - 0.05W^2$	$0.01 < \text{Re} \le 20$	Clift et al.[76]
$\log[(C_D\text{Re}/24) - 1] = -0.7133 + 0.6305W$	$20 < \text{Re} \le 260$	
$\log C_D = 1.6435 - 1.1242W + 0.1558W^2$	$260 < \text{Re} \le 1500$	
$\log C_D = -2.4571 + 2.5558W - 0.9295W^2 + 0.1049W^3$	$1500 < \text{Re} \le 1.2 \times 10^4$	
$\log C_D = -1.9181 + 0.6370W - 0.0636W^2$	$1.2 \times 10^4 < \text{Re} < 4.4 \times 10^4$	
$\log C_D = -4.3390 + 1.5809W - 0.1546W^2$	$4.4 \times 10^4 < \text{Re} \le 3.38 \times 10^5$ $W = \log_{10}\text{Re}$	
$C_D = C_o(\phi_s,\lambda)\ [1 + \delta_o(\phi_s,\lambda)/\text{Re}^{1/2}]^2$ where $C_o = 0.284/0.67\ (5.42 - 4.75\phi_s)\lambda^{-0.0145}$ $\delta_o = 9.06\ [5.42 - 4.75\phi_s/0.795 \cdot \log \phi_s/0.065]^{-1/2}\ \lambda^{0.00725}$	Nonspherical isometric particles ϕ_s = sphericity $\lambda = \rho_p/\rho_f$	Concha and Barrientos[63]
$C_D = C_o f_3(\phi_s,\lambda,\Psi)\ [1 + \delta_o f_1^{1/2}\ (\phi_s,\lambda,\Psi)/\text{Re}_\alpha]^2$ where $f_1(\phi_s,\lambda,\Psi) = [f_B(\phi_s) \cdot f_D(\lambda) f_F(\Psi)]^2$ $f_3(\phi_s,\lambda,\Psi) = [f_A(\phi_s) \cdot f_C(\lambda) f_E(\Psi)]^2$ $f_A(\phi_s) = (a - b\phi_s)/(a - b)$ $f_B(\phi_s) = [0.843\ a - b\phi_s/a - b \log \phi_s/0.065]^{-1/2}$ $f_C(\lambda) = \lambda^m,\ f_D(\lambda) = \lambda^{-m/2}$ $f_E(\Psi) = 1 + C[\Psi/(1 - \Psi)]^d$ $f_F(\Psi) = 1 + e\,\Psi[1 - \Psi]^f$	Ψ = volume fraction of particles $a = 5.42,\ b = 4.75,$ $c = 2.23150,\ d = 0.89957,$ $e = 1.65564,\ f = 1.41911$ Settling velocity of suspensions of particles of arbitrary shape	Concha and Christiansen[65]
$C_D = 24/\text{Re}[1 + \{8.1716 \exp(-4.0655\phi)\} \times \text{Re}^{(0.0964 + 0.5665\phi s)}] + 73.69\ \text{Re} \exp(-5.07)/[\text{Re} + 5.378 \exp(6.2122\phi)]$	Spherical and nonspherical particle	Haider and Levenspiel[35]
$C_D = (24/\text{Re} + 0.44)\epsilon^{-4.8}$	Wide flow range for fluidized beds; model based on single-particle approach	Khan and Richardson[78]
$C_{DE} = (0.63 + 4.8/\text{Re}_\epsilon^{1/2})^2$ $C_{DE} = C_{DE}^{2n},\quad n = 5.026 \log[(A^2 + B^{1/2})^{1/2} - A]$ $A = 3.81\ \text{Re}^{1/2},\ B = 88 + 5664/\text{Re}$ $\text{Re}_\epsilon = \text{Re}\ \epsilon^{-n}$	General multiparticle model for spheres — wide flow ranges	Panigrahi and Murthy[55]
$C_D = 24\Omega/\text{Re}[1 + 0.1315\ \text{Re}^{(0.82 - 0.05W)}] = 24\Omega/\text{Re}[1 + 0.0853\ \text{Re}(1.043 - 0.105W)]$ where $W = \log_{10}\text{Re}$ $\Omega = 2\beta^2[1 - \tanh\beta)/\beta]/2\beta^2 + 3[1 - \tanh\beta)/\beta]$ $\beta = R/\sqrt{K}$	Porous spheres $0.1 < \text{Re} \le 7$ $7 < \text{Re} \le 120$ Ω = correction factor β = normalized spherical radius (m) K = permeability (m^2)	Masliyah and Polikar[37]

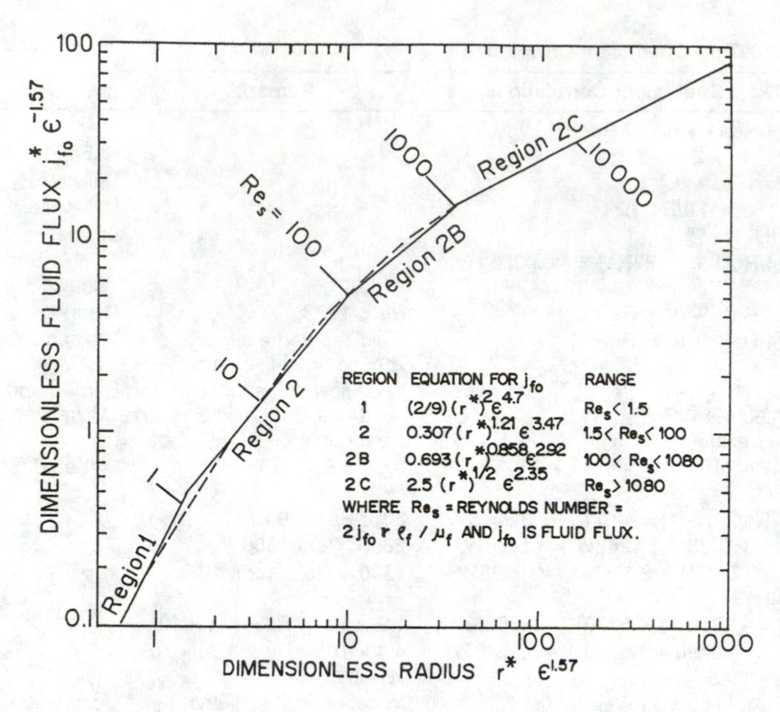

Figure 1.12 Simple form of a graph for dimensionless fluid flux (j_{fo}). (From Wallis, G.B., *Trans. Inst. Chem. Eng.*, 55, 74, 1977. With permission.)

6. Some Advances in Predicting Particle Terminal Velocity

Concha and Almendra[61] proposed a correlation for the settling velocity of a spherical solid particle in a dimensionless form as:

$$U^* = \frac{20.52}{d^*}\left[\left(1+0.0921d^{*3/2}\right)^{1/2} - 1\right]^2 \tag{1.85}$$

where

$$U^* = \frac{u}{Q} = \left(\frac{4}{3}\frac{\text{Re}}{C_D}\right)^{1/3} \tag{1.86}$$

$$d^* = \frac{d}{P} = \left(\frac{3}{4}C_D \,\text{Re}^2\right)^{1/3} \tag{1.87}$$

The notations P and Q are designated as fluid–particle parameters and their definitions are

$$P = \left(\frac{3}{4}\frac{\mu^2}{\Delta\rho\,\rho_f g}\right)^{1/3} \tag{1.88}$$

$$Q = \left(\frac{4}{3} \frac{\Delta \rho \mu}{\rho_f^2} \right)^{1/3} \tag{1.89}$$

Although Equation 1.85 for a single spherical particle is simple, the equation becomes quite complex for a multiparticle system or in a suspension of solids. In such a situation, the use of an appropriate plot would be handy, and such a plot, as proposed by Concha and Almendra,[62] is given in Figure 1.13. Turton and Clark[80]

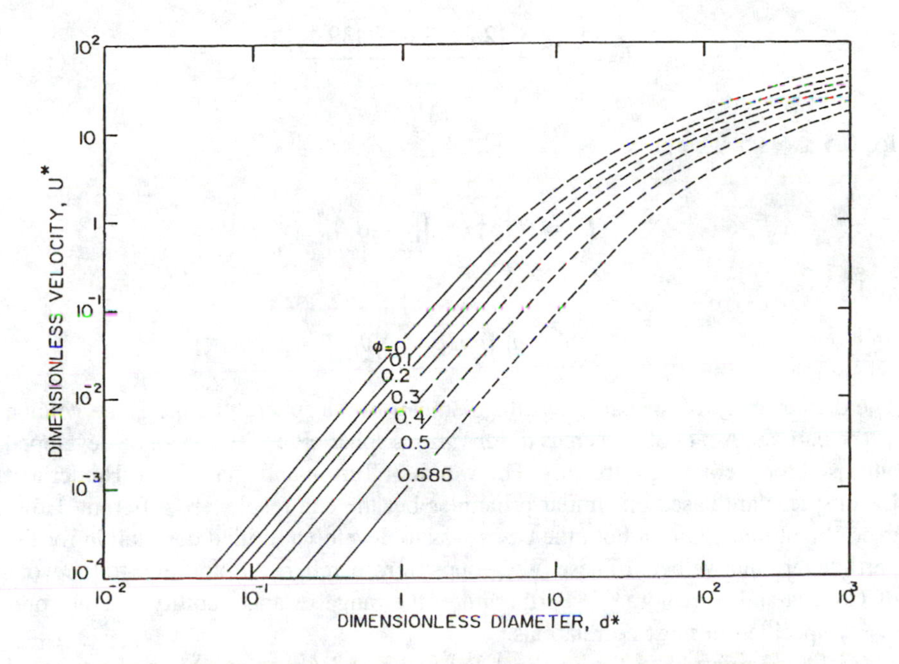

Figure 1.13 U^* versus d^* for different values of sphericity (ϕ). (From Concha, F. and Almendra, E.R., *Int. J. Miner. Process.*, 6, 31, 1979. With permission.)

proposed an explicit relationship to predict the terminal velocity of a spherical particle and did not refer to the work of Concha and Almendra.[61,62] The new form of the correlation is

$$U^* = \left[\frac{1}{\left(\frac{18}{d^{*2}} \right)^{0.824} + \left(\frac{0.321}{d^*} \right)^{0.412}} \right]^{1.214} \tag{1.90}$$

and a correlation of the above type with respect to U_{mf} also has been proposed by Turton and Clark:[80]

$$U^* = U_{mf}\epsilon^{-n}\left(\frac{\rho_f^2}{g\mu\Delta\rho}\right)^{1/2} \tag{1.91}$$

Correlation 1.91 involves the two fundamental parameters ϵ and U_{mf}, which should be known beforehand. Haider and Levenspiel[35] proposed a generalized and simple correlation for predicting the terminal velocities of spherical and nonspherical particles:

$$U^* = \left[\frac{18}{d^{*2}} + \frac{(2.3348 - 1.7439\ \phi_s)}{d^{*0.5}}\right]^{-1} \tag{1.92}$$

for $0.5 \leq \phi \leq 1$ where

$$U^* = U_t\left[\rho_f^2 / g\mu(\rho_s - \rho_f)\right]^{1/3}$$

and

$$d^* = d_v\ [g\rho_f(\rho_s - \rho_f)/\mu^2]$$

The diameter (d_v) of the particle is the diameter of the sphere that has same volume as the particle. A plot of U^* versus d^* for various sphericities (≥0.5) can be developed and used for predicting U_t readily. The works of Turton and Clark[80] and Haider and Levenspiel[35] are based on similar principles but the end results are different. However, the ultimate aim in both the cases was to develop a unified correlation for the particle terminal velocity. These two groups of researchers did not refer to the work of Concha and Almendra.[61,62] Furthermore, the range of applicability was not precisely specified in most correlations.

Geldart[66] referred to the work of Pettyjohn and Christiansen[67] and presented correlations for the free-fall velocity of a nonspherical particle for various flow regimes as:

$$U_t = K_{\mathrm{ST}}\left(\frac{\Delta\rho \cdot g d_v^2}{18\mu}\right) \tag{1.93}$$

for the Stokes regime ($\mathrm{Re}_t < 0.2$)

$$K_{\mathrm{ST}} = 0.843\ \log_{10}(\phi_s/0.065) \tag{1.94}$$

and

$$U_t = \frac{1}{K_N^2}\left(\frac{4}{3}\frac{\Delta\rho}{\rho_f} g\ d_v\right)^{1/2} \tag{1.95}$$

for the Newton regime ($1000 \leq Re_t \leq 3 \times 10^5$)

$$K_N = 5.31 - 4.88\,\phi_s \tag{1.96}$$

It should be noted that $d_v = \phi d_{sv}$ (where d_v is the size of the sphere that has the same volume as the particle, and d_{sv} is the size of the sphere that has the same surface-to-volume ratio as the particle).

For intermediate flow regimes (i.e., between Stokes and Newton flow regimes), Geldart[81] proposed a correlation for the factor K_I:

$$K_I = \left[K_{ST} - \left(\frac{0.43}{K_N}\right)^{1/2}\right]\frac{1000 - Re_t}{1000 - 0.2} + \left(\frac{0.43}{K_N}\right)^{1/2} \tag{1.97}$$

7. Experimental Methods for Determining Particle Terminal Velocity

a. Common Procedures

These methods can be classified into three types and are the same as those mentioned earlier for determining the minimum fluidization velocity except for the selection of the point at which the terminal velocity is located in the relevant plot (see Figure 1.10). The terminal velocity for the particle is determined from a plot of bed pressure drop versus fluid velocity (U) at the point where the bed pressure drop is zero. In other words, it corresponds to a drop across the test column which is just emptied by the terminal velocity. For a monosize particulate bed of solids, the fall of ΔP_b to zero would be abrupt, whereas the fall will be gradual for polydisperse particles. Hence, for precise determination of the particle terminal velocity, a monosize bed of solids is preferable.

The most common method to determine the particle terminal velocity is the expansion of the bed with the fluid velocity and extrapolation of the bed voidage or porosity to unity. This method, in general, is well accepted for liquid fluidized beds.[82] However, with fine particles fluidized by a gas, extrapolation of the value of the porosity to unity in the plot of log ϵ versus log U may result in a value several times larger than the computed terminal velocity.[83]

The method of determining the terminal velocity from the plot of the wall heat transfer coefficient versus the fluid velocity is not as popular as the two preceding methods. However, it seems to be quite an attractive and simple method for estimating the terminal velocity. The terminal velocity in the plot of the wall heat transfer coefficient versus the fluid velocity corresponds to a point where the heat transfer coefficient just starts falling and coincides with the value corresponding to the empty column. Compared to the previous two methods, this method is not simple in the sense that it requires precise measurement of the temperature and the steady-state condition to evaluate the wall heat transfer coefficient. The bed pressure drop versus velocity plot is not widely used for the purpose of determining U_t even though it is the most popular method for determining the minimum fluidization velocity. Measurement of the bed expansion and extrapolation of the voidage to unity in a plot of log U versus log ϵ is a simple technique.

b. Relative Methods

In these methods, well-established correlations for predicting U_{mf} and U_t are used to generate a plot of (U_t/U_{mf}) versus the Archimedes number. Determination of U_t is dependent on the known value of U_{mf}. Extensive discussion of predicting the minimum fluidization velocity was presented in Section IV of this chapter. One of the methods used is a relative technique to calculate U_{mf}, if U_t is readily known. In the context of this method, one can understand the complications involved in determining the terminal velocity. The relative technique for calculating the terminal velocity is useful if a reliable correlation for U_{mf} can be determined or selected. Let us analyze some of the pertinent equations and arrive at a conclusion as to how to choose a correlation for U_{mf} or Re_{mf} which could be used to predict Re_t in terms of Re_{mf}.

The popular correlation used to solve for Re_{mf} is Equation 1.62.[33] The Wen and Yu equation for predicting U_{mf} has been found to be inaccurate for fine powders,[84] for particles of different shapes,[85] and at elevated pressures and temperatures.[86] Our purpose here is not to discuss all the methods and analyze their relative merits and deficiencies. Rather, the objective is to explain how a plot of Re_t/Re_{mf} or (U_t/U_{mf}) versus Ar can be developed and then Re_t determined.

As a first step, a correlation for Re_{mf} has to be correctly chosen with known $A = (1/\phi_s\epsilon_{mf}^3)$ and $B = (1-\epsilon_{mf})/\phi_s^2\epsilon_{mf}^3$. According to Lucas et al.,[85] the values of the constants A and B can be chosen from Table 1.9. The values given in Table 1.9 are valid for a wide range of Reynolds number.

Table 1.9 Constants *A* and *B* from Table 1.6 for Various Shape Factors

		Constants	
Particle Category	**Shape Factor**	**A**	**B**
Round	$0.8 \le \phi \le 1$	16	11
Sharp	$0.5 \le \phi \le 0.8$	10	7.5
Other	$0.1 \le \phi \le 0.5$	8.5	5.0

In the case of fine powders where bubbling and channeling are predominant, selection of an appropriate correlation is difficult. In such cases, more detailed knowledge of the influence of ϕ_s and ϵ_{mf} on U_{mf} is required. For this purpose, an analysis was done by Chen and Pei,[84] and they proposed a voidage function (E_{mf}) given by:

$$E_{mf} = \frac{\phi_s^2\epsilon_{mf}^2}{\left(1-\epsilon_{mf}\right)} = 0.5 \qquad \text{for Ar} < 2 \tag{1.98}$$

$$= 0.5\ \mathrm{Ar}^{-0.11} \quad \text{for } 2 < \mathrm{Ar} < 2 \times 10^4 \tag{1.99}$$

$$= 0.18 \quad \text{for } \mathrm{Ar} > 2 \times 10^4 \tag{1.100}$$

Using Correlations 1.98–1.100 for the viscous-loss-dominated regime of the Ergun equation (neglecting the kinetic loss term), one can obtain:

$$\left(\mathrm{Re}_{mf}/\epsilon_{mf}\right) = \frac{\mathrm{Ar}}{300} \quad \text{for Ar} < 2 \tag{1.101}$$

$$= 0.0037\ \mathrm{Ar}^{0.88} \quad \text{for } 2 < \mathrm{Ar} < 2 \times 10^4 \tag{1.102}$$

$$= \sqrt{85.7^2 + \frac{\mathrm{Ar}}{4.87}} - 85.7 \quad \text{for Ar} > 2 \times 10^4 \tag{1.103}$$

For elevated pressures and temperatures, the method of Yang et al.[86] should be used; it is based on graphical methods using $(\mathrm{Re}^2 \cdot C_D)^{1/3}$ versus $(\mathrm{Re} \cdot C_D)^{1/3}$ plots for various ϵ_{mf} values that are functions of pressure and temperature.

Having selected an appropriate correlation for Re_{mf} and then for Re_t as outlined, it is possible to generate a plot of $\mathrm{Re}_t/\mathrm{Re}_{mf}$ versus Ar. Richardson[59] utilized a similar kind of technique and generated a plot using the Ergun equation for spherical particles and taking $\epsilon_{mf} = 0.4$ to predict Re_{mf}. The following correlations were employed to predict Re_t:

$$\mathrm{Ar} = 18\ \mathrm{Re}_t \quad \text{for Ar} < 3.6 \tag{1.104}$$

$$\mathrm{Ar} = 18\ \mathrm{Re}_t + 2.7\ \mathrm{Re}_t^{1.87} \quad \text{for } 3.6 < \mathrm{Ar} < 10^5 \tag{1.105}$$

$$\mathrm{Ar} = 1/3\ \mathrm{Re}_t^2 \quad \text{for Ar} > 10^5 \tag{1.106}$$

His plots for $\epsilon_{mf} = 0.38$, 0.4, and 0.42 fit well with experimental data and show that:

$$\mathrm{Re}_t/\mathrm{Re}_{mf} = \text{constant} \quad \text{for Ar} < 1 \tag{1.107}$$

where the constant

$$= 64 \text{ for } \epsilon_{mf} = 0.42$$

$$= 78 \text{ for } \epsilon_{mf} = 0.4$$

$$= 92 \text{ for } \epsilon_{mf} = 0.38$$

$$\text{and } \mathrm{Re}_t/\mathrm{Re}_{mf} = \approx 10 \tag{1.108}$$

at different values of ϵ_{mf} (0.38, 0.4, 0.42) for $\mathrm{Ar} > 10^5$.

V. FLOW PHENOMENA

A. Particulate and Aggregative Fluidization

Particulate or smooth fluidization, which is normally exhibited by a liquid–solid system, should be distinguished from aggregative- or heterogeneous-type fluidization that prevails in a gas–solid system. There is not much work available on this subject to enable one to distinguish between particulate and aggregate fluidization in a precise manner. Wilhem and Kwauk[87] proposed a dimensionless group to explain the quality of fluidization as follows:

$$Fr_{mf} < 0.13 \text{ for smooth or particulate fluidization} \tag{1.109}$$

$$> 1.3 \text{ for aggregative or bubbling fluidization} \tag{1.110}$$

where Fr_{mf} is the Froude number ($U_{mf}^2/d_p g$).

It may be recalled here that Equation 1.62 for the determination of U_{mf}, when used for inertial flow or Newton's flow regime, results in the equation

$$Re_{mf}^2 = \frac{Ar}{24.5} \tag{1.111}$$

which can be simplified to:

$$Fr_{mf} = 0.0408\left(\frac{\rho_s - \rho_f}{\rho_f}\right) \tag{1.112}$$

For most gases $\rho_f < 10$ kg/m³, and for liquids $\rho_f > 500$ kg/m³. For glass beads of $\rho_s < 2500$ kg/m³, when fluidized by gas, $\rho_s/\rho_f > 250$, and the values obtained are $Fr_{mf} \geq 10.1$, Hence, the glass bead–gas fluidized bed system is expected to behave like an aggregative fluidization. Applying a similar rule for a water–glass bead system, one gets $\rho_s/\rho_f > 2500/1000$ and $Fr_{mf} \simeq 0.0612$ (which is less than 0.13). Thus the system exhibits smooth fluidization. In the above examples, it is possible to obtain an expression for the Froude number directly by using the densities ρ_s and ρ_f for the flow regime dominated by the inertial force, more precisely for $Re_p > 1000$. For $Re_p < 20$, the viscous force is dominant and the expression for Fr_{mf} is obtained in terms of the density ratio $(\rho_s - \rho_f)/\rho_f$ and Re_{mf}:

$$Fr_{mf} = \frac{1}{1650} \cdot Re_{mf} \cdot \frac{\rho_s - \rho_f}{\rho_f} \tag{1.113}$$

For example, a quantitative determination of Fr_{mf} when $Re_{mf} = 20$ can be carried out using Equation 1.113. For the glass bead–air system ($\rho_s = 2500$ kg/m³ and $\rho_f =$ 1.2 kg/m³) when $Re_{mf} = 20$:

$$\mathrm{Fr}_{mf} = \frac{20}{1650} \cdot \left(\frac{2500}{1.2} - 1 \right) = 0.0121(2082.3) = 25.2$$

Because $\mathrm{Fr}_{mf} > 1.3$, an aggregative type of fluidization occurs for the glass bead–air system.

Similarly, for the glass bead–water system ($\rho_f = 1000$ kg/m^3), when $\mathrm{Re}_{mf} = 20$:

$$\mathrm{Fr}_{mf} = 0.018$$

Because $\mathrm{Fr}_{mf} < 0.13$, smooth fluidization is possible with the glass bead–water system.

The very fact that the quality of fluidization cannot be assessed in terms of a single dimensionless group like the Froude number has prompted researchers to evaluate the quality of fluidization by using a larger number of dimensionless groups. Romero and Johanson[88] proposed four dimensionless groups to assess the quality of fluidization:

$$\mathrm{Fr}_{mf}, \quad \mathrm{Re}_{mf}, \quad \frac{\rho_s - \rho_f}{\rho_f} \text{ and } \frac{H_{mf}}{D_t}$$

The quality of fluidization is related to its stability, which would be affected by bubbling. An increase in each of the four dimensionless groups will lead to poor quality or unstable or bubbling fluidization. From the experimental findings, the criterion for grouping the two modes of fluidization may be assessed by the product of the four dimensionless groups:

$$\left(\mathrm{Fr}_{mf} \cdot \mathrm{Re}_{mf} \cdot \frac{\rho_s - \rho_f}{\rho_f} \cdot \frac{H_{mf}}{D_t} \right) < 100 \tag{1.114}$$

for particulate or smooth fluidization and:

$$\left(\mathrm{Fr}_{mf} \cdot \mathrm{Re}_{mf} \cdot \frac{\rho_s - \rho_f}{\rho_f} \cdot \frac{H_{mf}}{D_t} \right) > 100 \tag{1.115}$$

for aggregative or bubbling fluidization.

Classifying the mode of fluidization in terms of the product of four dimensionless groups seems to be superior to using the single Froude group because almost all variables are considered in the former. However, there is no theoretical proof available to validate the aforementioned criterion; therefore, the criterion for identifying the fluidization mode can be regarded merely as a useful rule of thumb.

B. Regimes of Fluidization

1. Bubbling Bed

In a gas fluidized bed, right from the onset of fluidization, a change in the fluidization regime or the bed behavior would occur with an increase in the gas flow rate. The first fluidization regime prevails when bubbles are formed in the bed from the attainment of a minimum gas velocity designated as the minimum bubbling velocity. Bubbles once formed in the bed start rising, grow in size, coalesce, reach the bed surface, and finally erupt. In a tall column with a small diameter, the bubbles coalesce as they rise up and form slugs whose size or diameter would be the same as the column diameter. Generally, the slugs are followed by a piston of solids; once carried up to the top surface of the bed, the solids rain down along the column wall. When a fluidized bed of a larger diameter is used for a bubbling fluidized bed, the formation of slugs can be avoided or minimized. Further, with a large-diameter fluidization vessel and shallow beds, bubbles have less of a chance to coalesce. The resulting bubbling fluidized bed would then have bubbles of more or less uniform size. The bubbling point varies depending on the powder group (as classified by Geldart[12]). Bubbling in a fine powder system is quite different from one comprised of coarse particles. With fine powders, the bed may expand uniformly without any bubble formation, and particulate fluidization can occur until the first bubble is formed at a gas velocity known as the minimum bubbling velocity (U_{mb}). Experimental measurement of this velocity is not easy and requires careful assessment of the bed behavior. Just at the minimum bubbling point, the expanded particulate bed will show a decrease in its height without a change in the overall voidage of the emulsion phase. However, the structure of the bed is changed. The simplest correlation proposed by Geldart[89] for the minimum bubbling velocity (U_{mb}) is

$$U_{mb} = K_{mb}\, d_{sv} \tag{1.116}$$

where $K_{mb} = 100$ if U_{mb} and d_{sv} are expressed in CGS units and $d_{sv} = 1/\Sigma x_i d_{svi}$ is the particle surface/volume diameter (m^{-1}). Abrahamsen and Geldart[5] determined the ratio of the minimum bubbling velocity (U_{mb}) to the minimum fluidization velocity (U_{mf}) as:

$$\frac{U_{mb}}{U_{mf}} = \frac{2300\rho_g^{0.126}\mu^{0.523}\exp\left(0.716\, F_f\right)}{d_p^{-0.8} g^{0.934}\left(\rho_s - \rho_g\right)^{0.934}} \tag{1.117}$$

It is, in general, observed that the addition of fines (<45 μm) in a bed of coarse particles improves the quality of fluidization. The term F_f in Equation 1.117 is the fines fraction less than 45 μm, and the particle diameter (d_p) is the mean particle size obtained from standard sieve analysis and is equal to $1/\Sigma x d_p$; x is the weight fraction of particle of diameter d_p. Bubbling ends when the bubbles coalesce and form slugs. The onset of slugging can be predicted by the following correlation proposed by Ho et al.:[90]

$$\mathrm{Re}_{ms} = [33.7^2 + 2(0.0408\ \mathrm{Ar})]^{0.5} - 33.7 \tag{1.118}$$

2. Turbulent Bed

Just after the bubbling phenomenon is over, the bed assumes an entirely new structure, known as turbulent fluidization. The transition from the bubbling to the turbulent state of fluidization is marked by the breakdown of larger bubbles into smaller ones.[91] The transition point cannot be determined precisely. At the onset of turbulent flow, the mean amplitude of pressure fluctuation begins to level off. However, the mean amplitude does not fall sharply but shows a continuous drop[92] depending on the size of the equipment and the particle size. Hence, it is difficult to determine the exact point at which the onset of turbulent fluidization occurs.

In recent years, turbulent fluidized beds have been increasingly used in many catalytic and noncatalytic reactions due to the advantage of improved gas–solid contact. The literature on the characterization of turbulent fluidized beds, especially with respect to pressure fluctuations brought about by varying gas velocity, is scant. Lee and Kim[93] statistically analyzed the mean amplitude of pressure fluctuations in a 0.1-m-ID × 3-m-high fluid bed of glass beads (size 0.210–0.417 mm) with a mean diameter of 0.362 mm and a density of 2500 kg/m³. A typical plot of the variation in the mean amplitude of pressure drop fluctuations with changing gas velocity is shown in Figure 1.14. From this plot, the transition from bubbling to turbulent

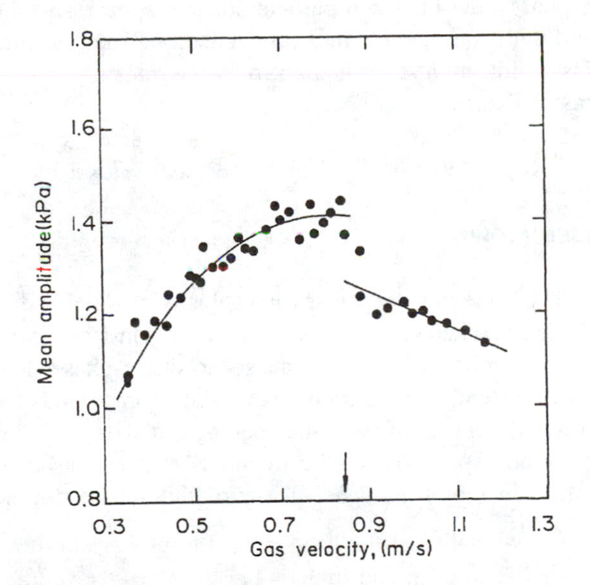

Figure 1.14 Mean amplitude of pressure drop fluctuations with velocity to identify the transition from bubbling to turbulent fluidization. (From Lee, G.S. and Kim, S.D., *J. Chem. Eng. Jpn.*, 21, 515, 1988. With permission.)

fluidization can be assumed to take place just at the velocity where the increase in the mean amplitude of pressure drop fluctuation attains its maximum value. This transition velocity for the glass–air system is 0.85 m/s. It is eight times the minimum fluidization velocity (0.105 m/s) and about 30% of the particle terminal velocity (2.5 m/s). It may be recalled that the operating velocity in the fluidization literature is generally taken to be three times the minimum fluidization velocity, which would, in principle, correspond to a bubbling fluidization condition. There is no valid theoretical reason for the choice of $3U_{mf}$ to fix the operating velocity. In the present example, it can be seen that the bubbling regime would end at $8U_{mf}$ and the mid value is $4U_{mf}$, which is close to $3U_{mf}$, where bubbling fluidization with a large number of small bubbles could be possible.

Lee and Kim[93] proposed a correlation for Re_c based on the transition velocity from bubbling to turbulent fluidization (U_c):

$$Re_c = U_c \rho_g d_p / \mu_g = 0.74\ Ar^{0.485} \quad \text{for } 0.44 \le Ar \le 4.4 \times 10^7 \qquad (1.119)$$

The ratio of the velocities taken at the transition of the bed to slugging and turbulent states for Group A or fine powder is far above unity, and for Group B powders it is close to unity or even smaller. For Group D powders this ratio is much less than unity.

3. *Fast Fluidization*

As the gas velocity is increased in a turbulent fluidized bed, the next stage of fluidization, known as fast fluidization, is reached. Fast fluidization involves a non-slugging dense-phase flow in which particle clusters appear and the particles are conveyed upward with considerable internal circulation. The transition between the turbulent and fast fluidization states is marked by the transport velocity (U_{tr}) which is given in terms of Re_{tr} as:

$$Re_{tr} = 2.916\ Ar^{0.354} \quad \text{for } 1.22 \le Ar \le 5.7 \times 10^4 \qquad (1.120)$$

4. *Dilute-Phase Flow*

The fast fluidization regime terminates at a velocity at which a dilute-phase flow commences at a bed voidage close to 0.95. The transition from fast fluidization to dilute-phase flow is similar to that from dense- to dilute-phase flow in a vertical gas–solid transport system, and the transport velocity corresponds to the condition of choking. For a bed voidage of 0.95, the choking criterion or definition given by Yang[95,96] results in an expression for the Reynolds number based on the velocity (U_{md}) at which the transition from dense-phase to dilute-phase flow will occur. The expression for Re_{md} is

$$Re_{md} = Re_f \left(1 + \frac{7.38}{Fr_t} \right) \qquad (1.121)$$

where Fr_t is the Froude number based on the column diameter (D_t) (i.e., $Fr_t = U_t/\sqrt{g\,D_t}$).

5. Flow Regime Mapping

A quantitative method for flow regime mapping has not been developed yet. Lee and Kim[94] summarized the methods for generating a flow regime map and concluded that they all are only qualitative in nature. They pointed out that there is a need to develop quantitative flow mapping using the Reynolds number and the Archimedes number. A flow regime mapping developed by using the correlations starting with Re_{mf} and ending at Re_{md} is shown in Figure 1.15. It is interesting to note from Figure 1.15 that the diameter of the column plays a key role in determining the dilute-phase flow condition.

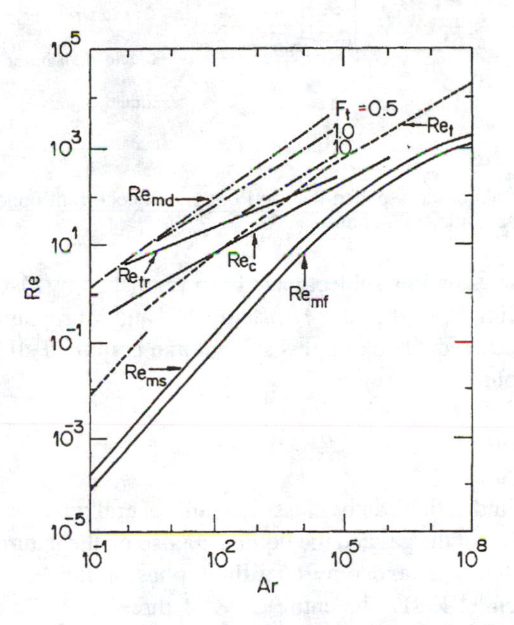

Figure 1.15 Flow regimes map in terms of Reynolds number (Re) with Archimedes number (Ar). (From Lee, G.S. and Kim, S.D., *Powder Technol.*, 62, 207, 1990. With permission.)

VI. THREE-PHASE FLUIDIZATION

A. Introduction

The term three-phase fluidization is generally used for the fluidization of solid particles larger than 200 μm by the upward cocurrent flow of a gas and a liquid with the liquid as the continuous phase. As a result of the past two decades of research on three-phase fluidization, a significant volume of interesting information is now available. This subject derives its importance from its several industrial applications, such as the H–oil process for hydrogenation and hydrodesulfurization of residual oil, H–coal process for desulfurization of residual oil, liquid-phase methanation for catalytic conversion of carbon dioxide to methane, biooxidation process for wastewater treatment, ion exchange, waste gas desulfurization, hydrometallurgy, and antibiotic production. A typical schematic of a three-phase fluidized bed is depicted in Figure 1.16.

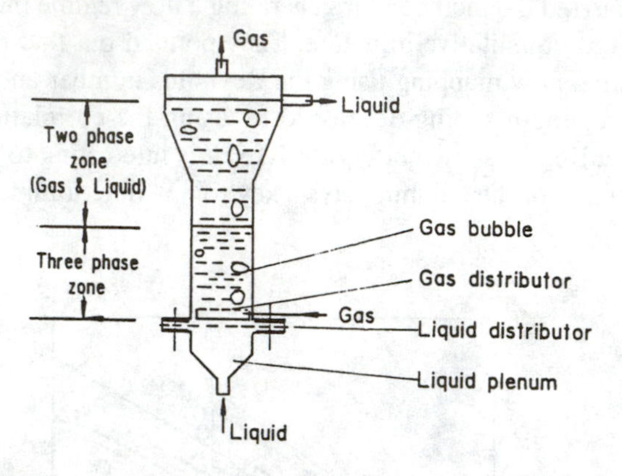

Figure 1.16 Schematic of three-phase fluidized beds with cocurrent upflow of gas–liquid with liquid as continuous phase.

Excellent reviews on this subject have been published by Østergaad,[97-99] Davidson,[100] Kim and Kim,[101] Wild et al.,[102] Epstein,[103,104] and Muroyama and Fan.[105] More recently, Fan[106] authored a book on this subject and compiled all available information in a single source.

B. Classification

Three-phase fluidization can be classified into several types or modes, depending on the flow direction of the gas and the liquid and also on the nature of the continuous phase in the reactor. The taxonomy of a three-phase fluidized bed was originally depicted by Epstein[103] 1981. The entire class of three-phase fluidized beds can be represented as shown in Figure 1.17.

Muroyama and Fan[105] classified the mode of fluidization into four major categories. The first two categories are for cocurrent upflow of gas–liquid for continuous phase of liquid and gas, respectively. The other two categories are similar to the first two except that the gas and the liquid flows are countercurrent.

1. Cocurrent Upflow of Gas–Liquid

In a cocurrent upflow of gas–liquid for fluidizing solid particles, the flow regimes in the case of the liquid as the continuous phase are referred to as being of the first type. The cocurrent upflow is classified into three types: coalesced bubble flow, dispersed bubble flow, and slug flow. It is difficult to identify the exact boundaries of these flow regimes. As the gas flow is increased in the slug flow regime, the

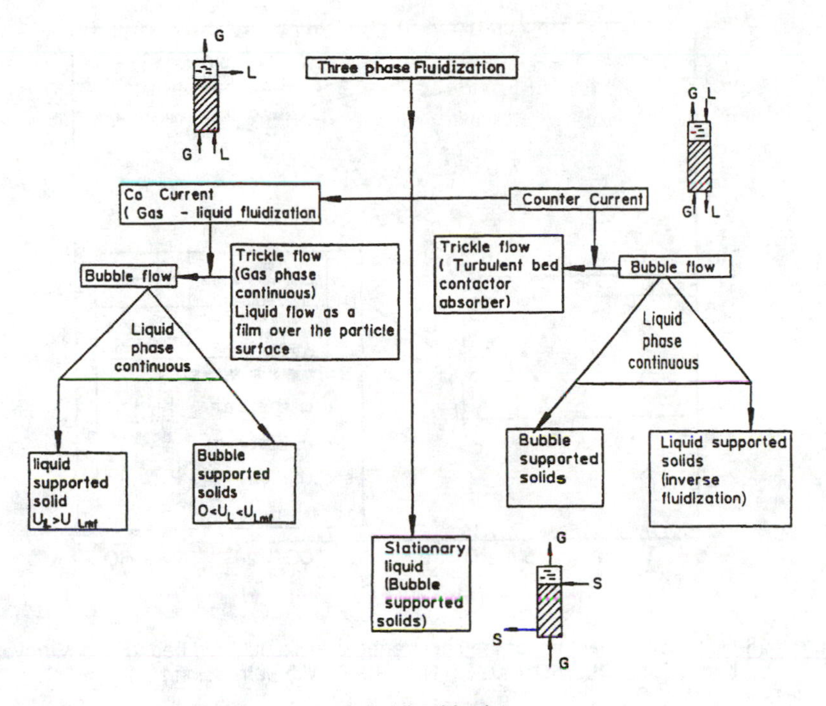

Figure 1.17 Classification of three-phase fluidized beds.

frequency of slugging increases, reducing the liquid holdup of the liquid–solid suspension. At still higher flow rates of the gas, the liquid–solid suspension is dispersed and the gas becomes the continuous phase; this is the second type of three-phase fluidization. No distinct point of transition from the first to the second type of fluidization has been identified yet in the cocurrent upward flow of gas–liquid in three-phase fluidization. A macroscopic pattern of the various flow regimes was, however, presented by Muroyama and Fan,[105] as shown in Figure 1.18.

2. *Countercurrent Flow*

Countercurrent three-phase fluidization with the liquid as the continuous phase is the third type of three-phase fluidization. It is also called inverse three-phase fluidization. Here the liquid fluidizes the floating particles ($\rho_s << \rho_l$). In fluidization of this type, there are four flow regimes: (1) fixed bed with or without bubble dispersion, (2) bubbling fluidized bed, (3) transition regime, and (4) slugging fluidized bed. When the gas flow rate is high in a countercurrent flow type of fluidization and the solids are suspended by the gas, the liquid passes through as a thin film over the solid particles; this is known as the fourth type of three-phase fluidization or as a turbulent contact absorber. Because the hydrodynamic resistance in a turbulent contact absorber is low due to bed expansion, a gas–liquid flow rate higher than that in countercurrent packed beds is achieved. A typical flow regime map[107] for countercurrent three-phase fluidization is presented in Figure 1.19.

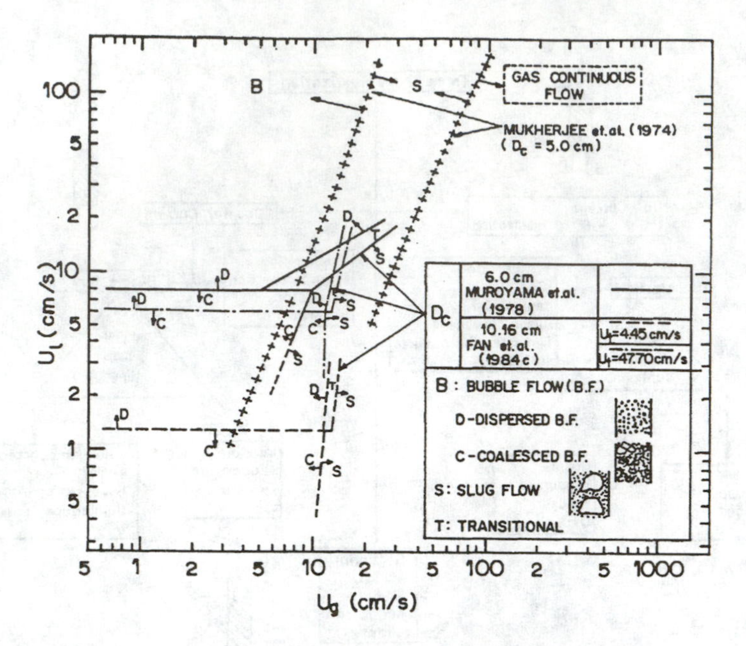

Figure 1.18 Flow regime diagram for the cocurrent gas–solid fluidized bed. (From Muroyama, K. and Fan, L.S., *AIChE J.*, 31(1), 1, 1985. With permission.)

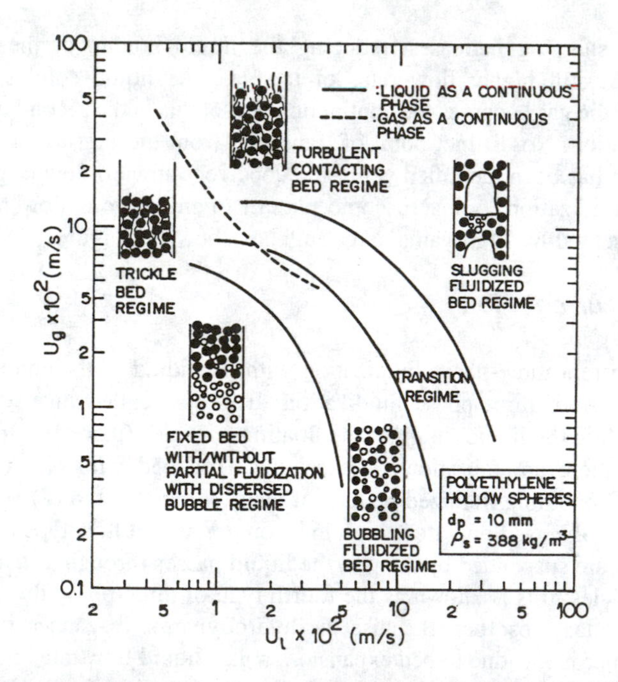

Figure 1.19 Flow regime diagram for the countercurrent gas–liquid–solid fluidized bed. (From Fan, L.S., Muroyama, K., and Chern, S.H., *Chem. Eng. J.*, 24, 143, 1982. With permission.)

C. Hydronamics

1. Parameters

The parameters that affect the hydrodynamics of three-phase fluidization can be classified into three major groups: (1) gas–liquid properties, (2) design parameters, and (3) operating variables. The parameters that belong to each group and the influencing engineering parameters are given in Table 1.10.

Table 1.10 Parameters Affecting Engineering Properties of Three-Phase Fluidization

1. Gas–Liquid Properties	2. Design Parameters	3. Operating Variables	4. Engineering Properties Influenced by 1–3
Density of liquid, gas, and solid; viscosity and surface tension of liquid; size, shape, and distribution of solid particles; wettability	Bed diameter, bed height, bed internals such as baffles or staging, bed shape (e.g., tapered), conditions like semifluidized state and spout fluidized bed	Gas flow rate and liquid flow rate	Heat and mass transfer, flow regime, pressure drop, phase holdup, interphase mixing, and solid entrainment

2. Pressure Drop and Holdup

The pressure drop across a three-phase fluidized bed of height H is given by:

$$\Delta P_b = (\epsilon_g \rho_g + \epsilon_l \rho_l + \epsilon_s \rho_s) Hg \tag{1.122}$$

where ϵ is the holdup per unit volume of the phase under consideration and ρ is its density. The various holdups must satisfy the relationship

$$\epsilon_g + \epsilon_l + \epsilon_s = 1 \tag{1.123}$$

The solid holdup (ϵ_s) can be directly estimated from a known mass of solid M in a three-phase bed of height H_{gsl}:

$$\epsilon_s = M/(\rho_s A H_{gsl}) \tag{1.124}$$

However, calculation of the bed voidage (ϵ) due to the gas and the liquid is not simple and is usually estimated by using empirical correlations. Bhatia and Epstein[108] developed a generalized wake model to predict the liquid holdup and the bed porosity; in developing this model, knowledge of some key parameters is required, such as the ratio (K_w) of the wake voidage to the gas-phase voidage and the ratio (X) of the solid holdup in the wake region to the solid holdup in the liquid fluidized region. These parameters are complex in nature. A more detailed discussion of this topic is outside the scope of this book.

For the purpose of estimating the bed voidage or porosity, the correlations of Dakshinamurthy et al.[109] and Begovich and Watson[110] are recommended. They proposed correlations on the basis of a large volume of data obtained for various particles. The correlation of Dakshinamurthy et al.[109] is

$$\epsilon = (\epsilon_g + \epsilon_l) = a\left(\frac{U_l}{U_t}\right)^b \left(\frac{U_g \mu_l}{\sigma}\right)^{0.08} \tag{1.125}$$

where $a = 2.12$ and $b = 0.41$ for $Re_l < 500$ and $a = 2.65$ and $b = 0.60$ for $Re > 500$. The correlation of Begovich and Watson[110] is

$$\begin{aligned} \epsilon = (\epsilon_g + \epsilon_l) &= (0.371 \pm 0.017) U_l^{(0.271\pm0.01)} U_g^{(0.041\pm0.005)} \\ &(\rho_s - \rho_l)^{(-0.316\pm0.01)} \cdot d_p^{(-0.268\pm0.01)} \mu_l^{(0.055\pm0.008)} \cdot D_t^{(0.033\pm0.013)} \end{aligned} \tag{1.126}$$

where the units used are in the CGS system.

Correlation 1.126 can be extended to liquids with high viscosities ($10 < \mu_l < 124$ [mPa · s]) by replacing the constant term (0.371 ± 0.017) by $2.7 + 13.2\ \mu_l^{0.64}$, as proposed by Nikov et al.[111]

The liquid holdup may be predicted by the correlation proposed by Kato et al.:[112]

$$\epsilon_l = 1 - 9.7\left(350 + Re_t^{1.1}\right)^{-0.5} \left(\rho_l U_g^4 / g\, \sigma\right)^{0.092} \tag{1.127}$$

These correlations (1.125–1.127) are not applicable with a good degree of accuracy to the case of a low liquid velocity. Hence, Saberian-Broudjenni et al.[113] proposed some refined correlations for predicting bed porosity or voidage:

$$\epsilon = \epsilon_0 \left(\frac{U_l}{U_{lmf}^{ls}}\right)^{0.27} \left(1 + 0.07\, Re_l^{0.34}\right) \tag{1.128}$$

where $Re_l = \rho_l\, U_l\, d_p / \mu_l$;

$$\epsilon_g = \epsilon \cdot \frac{U_g - U_{gl}}{U_g + U_l} \tag{1.129}$$

$$\epsilon_l = \epsilon \cdot \frac{U_{gl} - U_l}{U_s + U_l} \tag{1.130}$$

where U_{gl} is the gas–liquid slip velocity;

$$U_{gL}(\text{SI unit}) = 0.017 \left(\rho_l U_g^2\right)^{0.45} \tag{1.131}$$

Correlation 1.131 has been modified by Nikov et al.[111] by replacing the constant 0.017 with 0.034 $\mu_l^{0.1}$. U_{gl} is slightly increased with liquid velocity.

The parameter U_{lmf}^{ls} in Equation 1.128 is determined by the correlations of Grace:[114]

$$U_{lmf}^{ls} = \frac{\mu_l}{d_p \rho_l}\left[(25.25 + 0.0651\ \text{Ar})^{1/2} - 25.25\right] \tag{1.132}$$

3. Holdup Determination by Experiments

The estimation of gas holdup is somewhat difficult, and no unified correlation is available to predict gas holdup as there are several flow regimes. The best way would be to determine it experimentally using a reasonably large-diameter column (not less than 0.015 m for industrial applications). An experimental determination of the individual phase holdup warrants a basic understanding of the system, and no explanation or description of a method is usually given in publications. However, a simple technique has been used by Saberian-Broudjenni et al.[113] We will illustrate here the experimental method to determine ϵ_l and ϵ_g by three figures (Figure 1.20(a–c) which are self-explanatory.

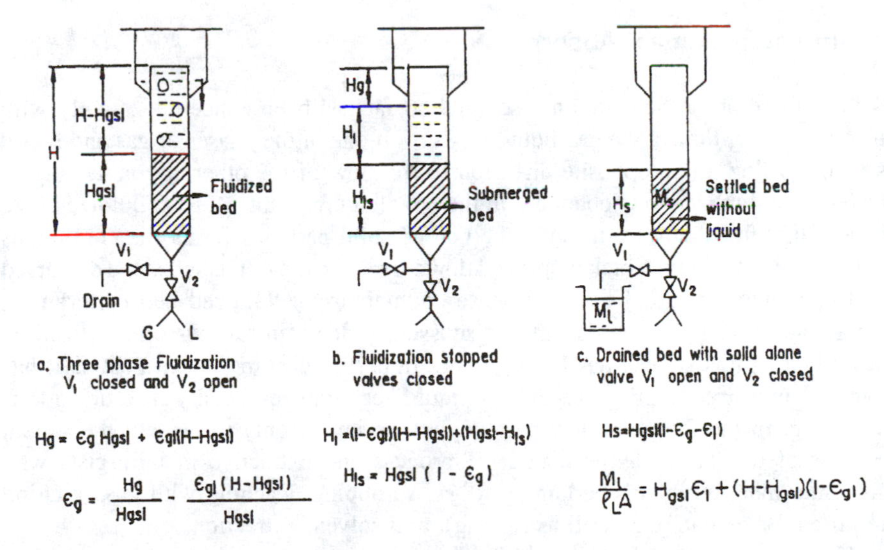

Figure 1.20 Schematic representation of steps to determine phase holdup.

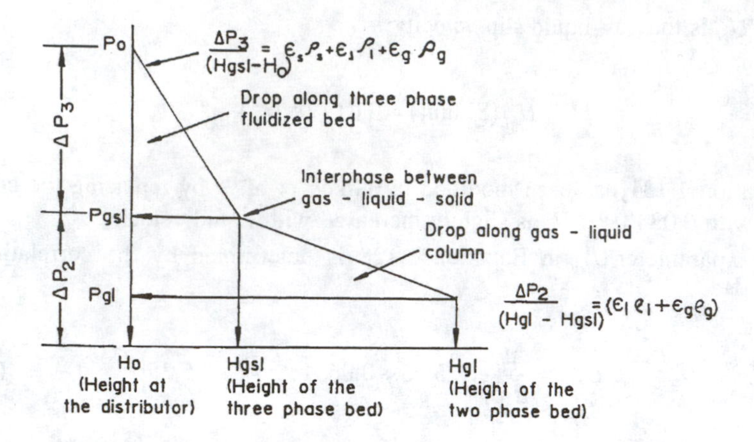

Figure 1.21 Pressure gradient diagram for a two-phase and a three-phase fluidized bed.

Usually the voidage ϵ_{gl} in the gas–liquid zone is determined by measuring the pressure profile and determining its gradient ($\Delta P/\Delta H$) from the plot of P versus H as shown in Figure 1.21. By substituting the value of $\Delta P/\Delta H$ in the pressure balance equation:

$$\left(\frac{\Delta P}{\Delta H}\right)_{gl} = \epsilon_{gl}\rho_g + \left(1 - \epsilon_{gl}\right)\rho_l \tag{1.133}$$

one can obtain ϵ_{gl}.

D. Turbulent Contact Absorber

The three-phase fluidization discussed so far has been concerned mainly with the cocurrent upflow of the gas-liquid mixture. If one of the phases of gas and liquid is made to flow in an opposite direction to the flow of the other, a countercurrent three-phase fluidization is obtained. In the countercurrent three-phase fluidized bed, the gas that flows upward through the bed of solid particles forms the continuous phase, while the liquid which is sparged down from the top of the bed is the dispersed or discontinuous phase. The three-phase countercurrent fluidized bed or turbulent contact absorber (TCA) is not often discussed in depth in reviews on fluidization. Nevertheless, this subject has been gaining importance in the recent past as it has potential industrial applications in gas–liquid reactions, especially in efficient gas scrubbing. In the following few paragraphs, we intend only to present the salient features of the TCA for the benefit of process and extractive metallurgists who encounter many problems pertaining to gas scrubbing, leaching with gas agitation (required for oxidation as well as mixing), and solvent extraction.

Countercurrent gas–liquid–solid fluidization was first applied for gas scrubbing, and the unit used for this purpose was termed a floating bed scrubber. A TCA generally consists of two restraining grids separated by a distance equal to three times the static bed height of spheres of polyethylene, polypropylene, or polystyrene of density ranging from 100 to 400 kg/m³. The bottom grid is normally provided with 70% or more free area in order to allow the liquid to drain through the grid, countercurrent to the flow of the gas which constitutes the continuous phase.

In a TCA, for a given liquid flow, there are three modes of bed operation depending on the rate of the upward flow of the gas. The three modes are (1) static bed, (2) fluidized bed, and (3) flooding bed. Figure 1.22 depicts the dependence of

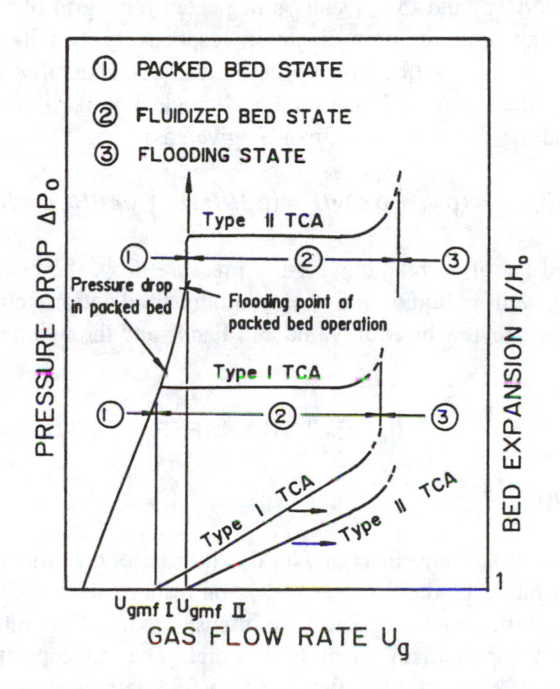

Figure 1.22 Pressure drop and bed height as functions of superficial gas velocity for a given liquid flow rate for a turbulent contact absorber (TCA) (type I flooding $U_g < U_{gmfI}$, type II flooding $U_g = U_{gmfII}$. (From Muroyama, K. and Fan, L.S., *AIChE J.*, 31(1), 1, 1985. With permission.)

the pressure drop and the bed height on the superficial gas velocity for a constant liquid flow rate in a TCA. For a bed of low particle density, fluidization could occur before the flooding state in an equivalent countercurrent bed. On the other hand, fluidization would occur in the flooding state for heavy particles. In both cases, an increase in the gas velocity beyond the flooding velocity would lift the bed, and the upper retaining screen would then arrest the expansion. A packed bed of spheres would be formed below the retaining screen. Ultimately the gas stream would carry the liquid through the exhaust.[115] Due to the intense mixing and the high rate of mass transfer, TCA beds are normally operated in the fluidized-flooding mode unless a very high pressure drop is acceptable.[116]

Fluidization with a shallow bed (that is, when the static bed height is less than the column diameter) is generally uniform. In the case of deep beds, slugging phenomena are commonly observed.[117] It has been noted that the grid open area strongly affects the hydrodynamics of the system.[118,119] Wall effects also play a significant role when the ratio of the column diameter to the packing diameter is small.

The pressure drop[120] across the column of a TCA is given by:

$$(\Delta P)_{TCA} = (\rho_s\epsilon_s + \rho_l\epsilon_l + \rho_g\epsilon_g)gH + \Delta P_{grid} + \Delta P_{\sigma} + \Delta P_{wall} \tag{1.134}$$

The quantities ΔP_{grid} and ΔP_{wall} can be neglected for a grid of free area greater than 8.3% and for a small column diameter, respectively. Details of the quantity ΔP_{σ}, which is concerned with surface tension, are not generally available in the literature. A simplified form of Equation 1.134, neglecting ΔP due to grid, surface tension, wall, and the gas static head, can be given as:

$$(\Delta P)_{TCA} = (\rho_s\epsilon_s + \rho_l\epsilon_l)gH = (\rho_s H/H_o\ \epsilon_s + \rho_l H/H_o\ \epsilon_l)gH_o \tag{1.135}$$

For a detailed list of correlations for the pressure drop, the bed expansion, the holdup of gas as well as liquid, and the minimum fluidization velocity in a TCA column, refer to the review by Muroyama and Fan[105] and the monograph by Fan.[106]

VII. HEAT TRANSFER

A. Introduction

As the specific surface area in a (gas) fluidized bed is of the order of 3000–45,000 m^2/m^3, the temperature gradient in the bed is on a meso-scale.[121] This is due to a high heat transfer surface area coupled with intense mixing. Generally, heat transfer from solid matter to the fluidizing agent is very high. The heat capacity of a fluidized bed is high (about 10^6 J/m^3 °C) and the heat transfer coefficient ranges from 250 to 700 W/m^2 °C.

Heat transfer studies have hitherto by and large focused on problems related to gas fluidized beds because of their commercial importance (mainly in coal combustion/gasification). Research on the heat transfer aspects of gas fluidized beds appears to be more extensive than that carried out in other areas of fluidization engineering. By contrast, not many heat transfer investigations have been conducted on systems like liquid fluidized beds, three-phase fluidized beds, spout fluidized beds, and vibro fluidized beds. More recently, work on heat transfer in turbulent beds, circulating fluidized beds, and three-phase fluidized beds has emerged due to their speciality applications. In this section, we will deal mainly with the basics of heat transfer in gas fluidization because of its relevance to materials processing. The literature on liquid fluidization will also be covered briefly, although not much work seems to have been done in this area.

Heat transfer in the gas fluidized bed, in principle, occurs by all three modes of transference, namely, conduction, convection, and radiation. The relative magnitude of the contribution made by each of these mechanisms depends on factors like the flow condition, the nature of the particles, and the temperature of operation. For example, for Geldart Group A and Group B powders, for which particles are less than 500 μm in size and have a density of 4000 kg/m^3, the gas flow occurs through the interstices of the particles or between the transfer surface and the particle. The flow range in such a case is in the laminar/creep regime. Hence, the gas-convective component becomes insignificant compared to the particle-convective component. In the case of Geldart Group D particulate solids, the situation is reversed: the gas-convective component is more significant than the particle-convective component because the gas flow in the interparticle space is in a transitional turbulent state. When the bed operates at high temperatures (≈900°C), the radiant heat contribution is significant. Heat transfer by all these mechanisms is additive and contributes to the overall heat transfer rate.

B. Groups

Heat transfer in gas fluidized beds can be broadly grouped into three components: gas–particle transfer, bed-to-surface (immersed or wall) transfer, and transfer from bed to immersed banks of tubes. A detailed discussion of these is beyond the scope of this book. Interested readers can refer to pertinent reviews and textbooks.[122-124]

The various mechanisms of heat transfer, as applied to a two-phase gas fluidized bed, are briefly outlined in the following sections.

1. Fluid–Particle Heat Transfer

Ranz and Marshall[125] proposed a correlation to predict the Nusselt number for a single particle suspended in a gas stream; this correlation, modified to be applicable to a multiparticle system, is given by:

$$\mathrm{Nu}_p = 2 + 1.8\,\mathrm{Pr}^{1/3}\,\mathrm{Re}_p^{1/2} \qquad \text{for } \mathrm{Re}_p > 100 \tag{1.136}$$

For $\mathrm{Re}_p < 100$, the contribution of the convective component is negligible and $\mathrm{Nu}_p = 2$ when Re_p tends to zero. The gas velocity relative to a moving particle is less, and this leads to a lower value of Nu_p in a fluidized bed than in a packed bed. Measurements of the temperatures of individual particles and of the gas are not possible by using a bare thermocouple, which can measure only the mean temperature of the emulsion packet. Therefore, experimental determination of Nu_p values is difficult.

Two types of experimental techniques were proposed by Kunii and Levenspiel[58] to predict the fluid–particle temperature, and these techniques are briefly discussed below.

a. Steady State

In this method, the heat lost by the gas stream is assumed to be equal to the heat gained by the solids. For example, the heat balance for a section of a bed of height dH is

$$\underset{\text{(Heat lost by gas)}}{-C_{pg}U_g\rho_g dT_g} = \underset{\text{(Heat gained by solids)}}{h_{gp}(T_g - T_p)dH} \tag{1.137}$$

The value of the heat transfer coefficient (h_{gp}) as determined by the above equation is only an apparent value, if axial conduction is not considered. In order to arrive at the true value, the right-hand side of the equation, that is, the heat gained by the solids, should be reduced by an amount equivalent to the axial conduction loss.

b. Unsteady State

In this method, the heat lost by the gas stream over a differential section of the fluid–solid mixture is considered to be the same as the heat accumulated by the solids. The final equation which could be used to estimate h_{gp} by this method is

$$\ln\left(\Delta T/\Delta T_i\right) = -\frac{C_{pg} U_g \rho_g A}{W C_{ps}}\left[1 - \exp\frac{h_{gp} H}{C_{pg} U_g \rho_g}\right] t \tag{1.138}$$

Experimental investigations by Chen and Pei[126] on fluid-to-particle heat transfer in fluidized beds of mixed-size particles showed that the addition of coarse particles to a fine-particle system improves the fluidization and hence increases the fluid–particle heat transfer coefficient. The most effective concentration of the coarse particles is about 25% and the optimum particle diameter ratio is ≈7. The correlation proposed by Chen and Pei[126] to predict fluid–particle heat transfer is

$$j_H (1 - \epsilon) = 0.000256\, \mathrm{Ar}^{0.4}\, \mathrm{Pr}^{2/3} \tag{1.139}$$

for $10 < \mathrm{Ar} < 10^5$ and $0.1 < \mathrm{Re}_p < 50$ where j_H = Colburn's factor = $\mathrm{Nu}/\mathrm{Re}\,\mathrm{Pr}^{1/3}$.

2. Bed–Wall Heat Transfer

The overall bed–wall heat transfer coefficient (h_w) is more useful for design purposes than the local heat transfer coefficient. Heat transfer measured at various locations or at the same location at different times usually fluctuates. It is thus necessary to be able to estimate the local heat transfer coefficient in order to develop a suitable model that could help in predicting the overall mean value. In a gas fluidized bed, h_w in principle increases with increasing gas velocity until it reaches a maximum value; thereafter h_w falls with further increase in gas velocity. The trend in the plot of h_w versus U is the same as that shown in Figure 1.10C. Thus, in a gas fluidized bed, at gas velocities beyond U_{mf}, the plot of h_w versus U has two sections: (1) a rising branch up to h_{max}, corresponding to a velocity designated as U_{opt}, and (2) a falling branch beyond $U = U_{opt}$.

C. Models

Various models have been proposed to explain the variation in h_w with gas velocity, and this topic was reviewed in detail by Gutfinger and Abuaf,[127] Bartel,[128] and Saxena and Gabor.[129] Among these models, three distinct groups may be identified, as indicated in the following sections.

1. *Film Model*

A thin fluid film adjacent to the heat transfer wall is considered to offer the main resistance to heat transfer and this resistance is reduced by the moving particles which scour away the fluid film. Steady-state heat transfer is assumed. This model/hypothesis was supported by Van Heerden et al.[130] and Ziegler and Brazelton.[131] However, the results based on this model were subsequently found to be in poor agreement with many other experimental data. The drawback of this model lies in its failure to consider the properties of solid particles, which play a vital role in the transport of heat in a fluidized bed.

2. *Modified Film Model*

Transient conduction of heat between the surface and the particles (single/multiparticles)[131–134] is assumed in a particulate type of fluidization. The solid particle properties are considered, and the heat transfer rate is assumed to change depending on the mobility of the solid particles and their concentration on the heat transfer surface. As this model cannot be applied to a bubbling bed, a third type of model, known as the packet model, has emerged.

3. *Emulsion Packet Model*

A packet of emulsion (a mixture of solid and gas corresponding to the incipient state of fluidization) contacts the heat transfer surface periodically, undergoes transient conduction during its stay, moves away (making way for a new packet), and transports the sensible heat to the bulk of the bed. Unsteady-state approach and surface renewable rate of the packets are the main features of this packet. Mickley and Fairbanks[135] were the first to propose a model of this kind.

The emulsion packet model does not consider the individual properties of the gas and the solid. Instead, it considers the properties of the homogenous gas–solid mixture at the incipient state. Botterill and Williams,[136] in their single-particle model, considered the individual properties of the gas and the solid. The single-particle model has the limitation of a short contact time during which heat cannot penetrate into a second particle; for this reason, this model has been extended to a two-particle layer by Botterill and Butt[137] and further to a chain of particles of unlimited length by Gabor.[138]

D. Predictions of Heat Transfer Coefficient

There are numerous correlations available in the literature for predicting the wall heat transfer coefficient, and these correlations were developed under different conditions. Zabrodsky,[139] therefore, recommended that care be taken to use these correlations only under the conditions for which they were proposed.

The applicability of any mechanism warrants detailed knowledge of the transient contact characteristics between the emulsion and the heat transfer surface on a local basis. The literature on this subject is scant. Selzer and Thomson[140] pointed out that a model should clearly explain the heat penetration depth and the contact time criteria. They also stressed the need to test the applicability of these models for different types of distributors.

A detailed listing of all the available correlations to predict the wall heat transfer coefficient can be found in the literature. In general, the heat transfer coefficient is a function of (1) the properties of the gas and the solid (ρ_g, K_g, C_{pg}, μ_g, ρ_s, C_{ps}, K_s), (2) the relevant geometric factors (d_p, D_t), (3) the fluidization parameters (θ_D, f_L, ϵ_p, ϵ_b), (4) the system variables (U, U_{mf}), (5) the location of the heat transfer surface (immersed) inside the bed, and (6) the type and design of the grid/distributor plates.

1. Additive Components

The wall heat transfer coefficient, also called the convective heat transfer coefficient, is generally constituted of two additive components: the particle-convective component (h_{pc}) and the gas-convective component (h_{gc}). The particle-convective component arises due to the motion or mobility of the particulate solids over the heat transfer surface and is more pronounced for fine particles than for coarse particles. The gas-convective component is due to the gas percolating through the bed and to gas bubbles. The gas-convective component is negligible in fine particle fluidization but is predominant in coarse or heavy particle fluidization; the reason is the requirement for high flow rates to achieve fluidization in the latter case. The overall convective heat transfer is given by:

$$h_c = h_{pc} + h_{gc} \tag{1.140}$$

Prediction of h_{pc} is difficult due to the complex hydrodynamics of fluidization and the interaction of the lean and the dense phases over the heat transfer surface.

a. Particle-Convective Component

The particle-convective component (h_{pc}), in the absence of any film resistance, is h_p and should be evaluated from the knowledge of the instantaneous heat transfer coefficient (h_i) and the age distribution of the packet of emulsion over the transfer surface. Mickely and Fairbanks[135] proposed a model to evaluate h_p as:

$$h_p = \frac{1}{\tau}\int_0^t h_i dt \tag{1.141}$$

where the instantaneous heat transfer coefficient (h_i) is given by:

$$h_i = (K_e \rho_e C_{ps} / \pi t) \tag{1.142}$$

and the packet residence time (τ) is given by:

$$\tau = 0.44\left[\frac{d_p g}{U_{mf}^2\left(\frac{U}{U_{mf}}-1\right)^2}\right]^{0.14}\frac{d_p}{D_t} \tag{1.143}$$

If the conditions correspond to slugging, the packet residence time (τ) over a short vertical surface of length L is

$$\tau = L/(U - U_{mf}) \tag{1.144}$$

Under all practical conditions, the heat transfer is effective only when the packet contacts the surface. For the fraction of surface which is shrouded by gas bubbles (f_b), h_p is negligible. Hence, h_{pc}, in the absence of any film resistance, can be expressed as:

$$h_{pc} = h_p(1 - f_b) \tag{1.145}$$

where

$$f_b = 0.33\left[\frac{U_{mf}^2\left(U/U_{mf} - A\right)}{d_p g}\right]^{0.14} \tag{1.146}$$

The accuracy of the determination of the average value of h_{pc} can be improved to better than 5% by taking into consideration the film resistance ($1/h_f$) which is in series with the packet resistance ($1/h_p$). When this is done, the particle-convective component of heat transfer (h_{pc}) is given by:

$$h_{pc} = \frac{1}{\left(1/h_p\right)+\left(1/h_f\right)}\left(1-f_b\right) \tag{1.147}$$

The film heat transfer coefficient (h_f) is given by:

$$h_f = m_c K_g / d_p \tag{1.148}$$

where m_c has varying values; the range of these values under different experimental conditions was enumerated by Xavier and Davidson.[141] They also suggested that m_c = 6 may be taken for design purposes.

In order to evaluate h_{pc}, it is necessary to know the effective thermal conductivity of the particulate bed (K_e), which is the sum of the effective thermal conductivity of the bed when there is no fluid flow (K_e^o) and the conductivity due turbulent diffusion (K_e^t). Thus,

$$K_e = K_e^o + K_e^t \tag{1.149}$$

K_e^o can be evaluated[142,143] by using the expression

$$\frac{K_e^o}{K_g} = \left(\frac{K_s}{K_g}\right)^{0.28-0.757} \log_{10} \epsilon^{-0.057} \log_{10} \frac{K_s}{K_g} \tag{1.150}$$

The contribution of turbulent heat diffusion (K_e^t) in determining the effective thermal conductivity is given by:

$$K_e^t / K_g = (\alpha_1, \beta_1) \mathrm{Re}_p \, \mathrm{Pr} \tag{1.151}$$

Here, α_1 is the ratio of the mass velocity of the fluid in the direction of heat flow to the superficial mass velocity in the direction of fluid flow. The parameter β_1 is the ratio of the distance between the two adjacent particles to the mean particle diameter. The product $\alpha_1\beta_1$ for small values of d/D is equal to 0.1 for normal packing of spheres. Hence, the effective conductivity of the particulate phase or the emulsion packet is

$$K_e = K_e^o + 0.1 \, \rho_g C_{pg} d_p U_{mf} \tag{1.152}$$

where the value of U_{mf} can be predicted as per the correlations presented in Tables 1.4–1.6. The particle-convective heat transfer coefficient can be calculated from Equation 1.147 by using the appropriate expression for h_p (Equations 1.141–1.146 and Equation 1.148).

b. Gas-Convective Component

Because the interphase gas-convective component (h_{gc}) cannot be determined directly, Baskakov et al.[145] proposed from an analogy to mass transfer that:

$$\mathrm{Nu}_{gc} = \frac{h_{gc} d_p}{K_g} = 0.009 \, \mathrm{Ar}^{1/2} \, \mathrm{Pr}^{1/3} \tag{1.153}$$

For most gases, Pr is constant and for the inertial flow regime $\mathrm{Re}_{mf} \alpha \, \mathrm{Ar}^{1/2}$. Hence, $\mathrm{Nu}_{gc} \alpha \, \mathrm{Re}_{mf}$. In view of this, it has been assessed that the value of h_{mf} obtained at U_{mf} is approximately the same as that of h_{gc}. However, for a bubbling fluidized bed at a high fluid flow rate, the equation

$$h_{gc} = h_{mf}(1 - \epsilon_B) + h_B \epsilon_B \tag{1.154}$$

can be approximated to $h_{gc} \approx h_{mf}$.

Botterill and Denloye[146] proposed the following correlation for predicting h_{gc} by taking into account the heater length (L):

$$h_{gc}\sqrt{dL}/K_g = 0.3\,\text{Ar}^{0.39} \quad \text{for } 10^3 < \text{Ar} < 2\times 10^6 \tag{1.155}$$

c. *Radiative Component*

The radiative heat transfer coefficient (h_r) between a fluidized bed at a temperature T_b and a heater surface like an immersed tube at a temperature T_s is

$$h_r = q_r/(T_b - T_s) = \sigma_s\, E_{bs}(T_b^2 + T_s^2)(T_b + T_s) \tag{1.156}$$

where q_r is the radiant heat flux, σ_s is the Stefan–Boltzmann constant (5.67×10^{-8} W/m^2 K^4), and E_{bs} is the generalized emissivity factor given by:

$$E_{bs} = (1/E_b + 1/E_s - 1) \tag{1.157}$$

The generalized emissivity factor depends on the shape, disposition, and emissivity of the radiating and receiving bodies.

Radiation plays a major role in heat transfer when the temperature of the radiating body is above 800–900°C. The convective heat transfer coefficient increases due to the component h_r, which is significantly high when the temperature of the radiating body is above 1000°C. In the case of many a fluidized bed combustion, the bed temperature (T_b) itself is around 1000°C. At high temperatures or at high values of the difference between T_b and T_s, it is important to evaluate the intermediate mean temperature at which the physical properties of the fluid and the solid must be determined. The intermediate temperature[147] (T_e) is given by:

$$\frac{T_b - T_e}{0.5\,R_r} = \frac{T_b - T_w}{0.5\,R_r + R_w} \tag{1.158}$$

R_r is the mean thermal resistance of the packet, that is,

$$R_r = \sqrt{\frac{\pi t}{K_e C_{ps} \rho_B}}$$

where $C_s = C_p\,(1 - \epsilon_{mf})$, ρ_B is the bulk density of the packet = ρ_{bmf}, and R_w is the contact resistance of the zone near the wall and equals δ_w/K_{ew}.

2. *Overall Heat Transfer Coefficient*

The overall heat transfer coefficient is the sum of the individual convective components that are due to bubbles (h_b), particles (h_{pc}), interphase gas (h_{gc}), and radiant heat transfer (h_r). Thus, the overall or total heat transfer coefficient (h) is given by:

$$h = h_b f_b + (h_{pc} + h_{gc})(1 - f_b) + h_r \tag{1.159}$$

In the case of a dense fluidized bed where there are no bubbles and the fraction of the immersed surface covered by bubbles (f_b) is zero, the total heat transfer coefficient (h) is given by:

$$h = h_{pc} + h_{gc} + h_r \tag{1.160}$$

For large particles, due to the requirement of a high fluidization velocity, it can be assumed that $h_b \approx h_{gc}$, so that:

$$h = (1 - f_b)\, h_{pc} + h_{gc} + h_r \tag{1.161}$$

In the case of solid particles whose emissivity lies in the range of 0.3–0.6, Baskakov[148] proposed the following expression for the overall heat transfer coefficient under maximum condition:

$$h_{max} = K_g \left(0.85\ \mathrm{Ar}^{0.19} + 0.006\ \mathrm{Ar}^{0.5}\ \mathrm{Pr}^{0.33}\right)/d_p + 7.3\, E_p E_s T_s^3 \tag{1.162}$$

E. Heat Transfer to Immersed Surfaces

The immersed surfaces inside a fluidized bed can be of any object meant for transferring the heat either from the bed or to the bed, depending on whether the bed is to be cooled or heated. Generally, tubes or tube banks are immersed in large-scale fluidized beds for exchanging the heat. Leva[18] reported that, except at high flow rates of the fluidizing fluid/gas (that is, for $U_{mf} >> 1$), the heat transfer rate obtainable with immersed surfaces is fourfold higher than that between the bed and the external wall. There have been many investigations[1,7,149–156] into this subject. The heat transfer coefficients relevant to heat transfer to immersed objects are broadly of two types: local heat transfer coefficients and total or overall heat transfer coefficients. Because the latter coefficients are useful for design purposes, this topic is given special emphasis in the following discussion.

1. Vertical Surfaces

Studies[135] on vertically immersed electrical heaters have shown that the heat transfer coefficients are smaller than those corresponding to horizontal surfaces under similar experimental conditions. This has been attributed to the small size of the heater over which large bubbles pass during their upward flow. The correlations often cited in the literature for predicting the wall-to-bed heat transfer coefficient for vertically immersed surfaces are those of Vreedenberg[157,158] and Wender and Cooper.[159]

2. *Horizontal Surfaces*

Horizontally immersed tubes in a fluidized bed are exposed to more cross-flow of solids than are vertical tubes, and thus relatively higher heat transfer rates could be possible with these. However, a single horizontal tube poses some problems, mainly due to the hydrodynamic environment around a single horizontal tube whose top portion is piled with stagnant solid particles that reduce the local heat transfer rate at the top. On the other hand, the bottom portion of the tube becomes shrouded with rising bubbles, creating a solid free zone. This condition is opposite that prevailing at the top surface of the tube. Lateral parts of the tubes are frequently contacted by solid particles, and hence the maximum local heat transfer occurs around this lateral region.

Many reports are available in the literature[160–162] on the measurement of local heat transfer coefficients. In a gas fluidized bed, the minimum occurs at the top or the upward face of a horizontally immersed tube (i.e., in the downstream side of the flow). The maximum[162,162] has been found to occur on the lateral side, which tends to shift toward the downward direction in a large-diameter tube. In a liquid fluidized bed, unlike a gas fluidized bed, a uniform temperature around a horizontally immersed tube was observed by Blenke and Neukirchem.[163] In order to improve the heat transfer or to arrive at a uniform temperature around the tube, the use of tube bundles is recommended.

Heat transfer from a horizontal tube is lowered by the presence of unheated tubes placed on one side or both sides of a heated tube. For a horizontally immersed loop type of heater, heat transfer is poor if it is a single tube.[164] This has been attributed to the stable void of gas between the tubes of a loop-type heater, which would disappear in the case of tube bundles.[165] Generally, it has been observed[164,165] that with a well-immersed tube bundle, the heat transfer obtained is independent of the location of the bundle inside the fluidized bed. There are numerous correlations[157,167–172] available in the literature for predicting the total heat transfer coefficient for an immersed horizontal tube. The correlation proposed by Vreedenberg[157,158] is widely accepted. Subsequently, the need to incorporate the volumetric heat capacity of the solid and an appropriate correction for extending the application of the correlation to tube bundles were suggested by Saxena and Grewal.[172]

F. Effects of Operating Variables

1. *Effect of Velocity*

a. *Heat Transfer Coefficient versus Velocity*

The effect of velocity on heat transfer was discussed in Section IV. It was also mentioned that there exists an optimum velocity at which the heat transfer coefficient is maximum (h_{max}).

A typical plot depicting the heat transfer coefficient as a function of the fluidizing velocity is shown in Figure 1.23 at a bed temperature of 600°C. A list of the variables

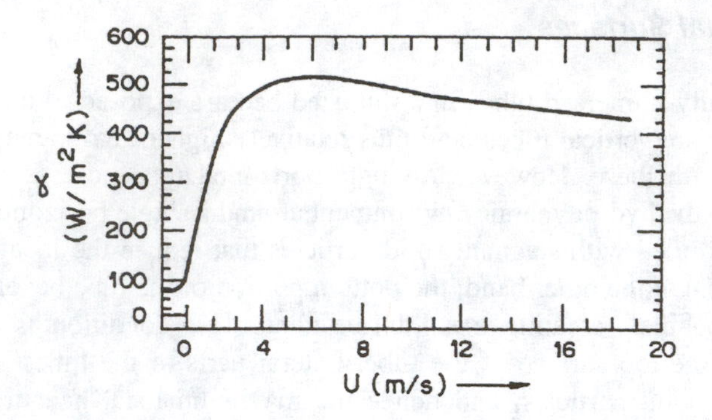

Figure 1.23 Heat transfer coefficient as a function of the fluidizing velocity. (d_p = 0.55 mm, H_o = 57–60 cm. For $U < U_{mf}$, T_p = 43–52°C, T_b = 600°C; for $U > U_{mf}$, T_p = 98–134°C.) (From Draijer, W., in *Fluidized Bed Combustion*, Radovanovic, M., Ed., Hemisphere Publishing, Washington, D.C., 1986, 211. With permission.)

that influence the heat transfer in a gas fluidized bed is given in Table 1.11. Among the various variables, bed temperature and pressure are important in studies on advanced fluidized bed reactors for coal combustion or fluidized bed boilers. These effects have not been examined in depth for gas–solid reactions in mineral or materials processing.

b. Flow Regime Effect

Molerus[185] recently gave a detailed account of the effects of gas velocity on the heat transfer coefficient for three different flow regimes: laminar, intermediate, and turbulent. A typical plot presented by Molerus is shown in Figure 1.24 for the powders classified by Geldart.[12] Curve 1 in Figure 1.24 corresponds to the laminar flow regime (i.e., laminar flow Ar number, Al = $[(d_p^3\, g)^{0.5}(\rho_s - \rho_g)/\mu_g = 159]$); this curve also corresponds to a typical Geldart Group A powder for which bubbles flow at low velocity. Hence, h_{max} is attained only at a high velocity, about three times greater than that obtainable with Group B powders (especially in the intermediate flow regime).

Curve 2 is for the intermediate flow regime (e.g., $10^3 \leq Ar \leq 10^5$) and is typical of Group B powders for which the advantage of unlimited bubble causes h to rise steeply just after U_{mf} and brings about the attainment of h_{max} at $2U_{mf}$. In the intermediate flow regime, both particle- and gas-convective components are important in determining the overall heat transfer coefficient. Curve 3 in Figure 1.24 corresponds to the turbulent flow regime ($Ar \geq 10^5$), where gas-convective heat transfer is predominant. Hence, the transition from a fixed bed to a fluidized bed does not have a significant effect on the h values in this case.

Borodulya et al.[186] reported that the maximum heat transfer coefficient increases exponentially at elevated temperatures and varies linearly at high pressures.

Table 1.11 Influence of Various Parameters on Heat Transfer in Gas Fluidized Bed

Sl. no.	Variable	Influences	Ref.
1	Fluid velocity (U)	Heat transfer increases above U_{mf} up to an optimum velocity (U_{opt}) and then decreases	173
2	Particle diameter	Heat transfer coefficient (h) increases with fine-sized particle and decreases with coarse size; for large size of particle, h increases mainly due to increase in convective component in heat transfer at high fluid velocity	173
3	Thermal conductivity of solid (K_p)	No influence on h	174, 175
4	Specific heat of solid (C_p)	h is proportional to C_p^n where $0.25 < n < 0.8$	176, 177
5	Specific heat of fluid (C_f)	Data are contradictory at moderate pressure and velocity; at high pressures, h is increased by C_f or the volumetric specific heat $C_f \rho_f$	178, 179
6	Thermal conductivity of fluid (K_f)	$h \alpha K_f^n$, n = 0.5–0.66; as bed temperature increases, h increases, attributed mainly to increase in K_f	180
7	Fluid bed height (H)	For a well-developed fluidized bed and a well-immersed heat transfer surface, h is not dependent on H	159, 176, 181
8	Fluidized grid zone	Grid zone affects h depending on grid type and its free area; for low free area, h is relatively higher than high free area grid	182
9	Fluidized bed diameter	No qualitative or quantitative information available	
10	Length of heat transfer surface (L)	h is independent of L	152
11	Heat transfer tube diameter (d_t)	h increases with decrease in d_t	179, 183
12	Vertical versus horizontal heat transfer tubes	h for vertical tubes is 5–15% higher than for horizontal tubes; construction and technological condition decide the tube orientation	152, 180
13	Tube bundles		
a	Vertical	Ratio of tube spacing to tube diameter affects h; when the ratio is reduced, h_{max} drops; for closely packed bundles, h reduces even by 35–50%	152
b	Horizontal	Almost similar effect as vertical bundle; for same ratio of tube spacing to diameter, horizontal bundles occupy greater portion of bed cross-section	152
14	Bed temperature	Gas-convective component increases for small (d_p < 0.5 mm) particle and decreases for coarse (d_p > 0.5) particle; particle-convective component decreases; for 1-mm particle, net effect is a reduction	184
15	Bed pressure	Particle-convective component not affected; gas-convective component is enhanced and is proportional to square root of gas density	141

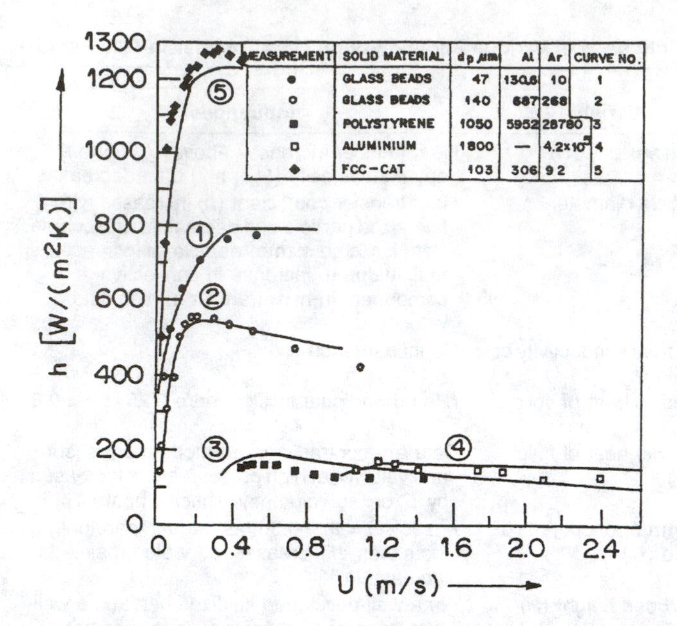

Figure 1.24 Comparison of measurements and predictions for heat transfer versus superficial gas velocity (fluidizing agent helium). Curve 1: A powder, curve 2: B-powder laminar flow region, curve 3: intermediate Archimedes number, curve 4: turbulent flow region, curve 5: B-powder laminar flow region. (Al is laminar flow Archimedes number. Al = $(d_p^3 g)^{1/2}(\rho_s - \rho_g)/\mu_g$.) (From Molerus, O., *Powder Technol.*, 15, 70, 1992. With permission.)

2. Optimum Velocity

The occurrence of a maximum in the heat coefficient could be anticipated because of the opposing effects of the particle velocity and the bed voidage, both of which increase with the fluidizing velocity. The attainment of a maximum heat transfer coefficient and a corresponding optimum velocity has been confirmed by numerous research workers, and their results show that the maximum heat transfer depends mainly on fluid–solid properties. A list of the correlations that can be applied to gas–solid fluidized beds to predict the maximum heat transfer coefficient and the corresponding optimum fluidizing gas velocity is given in Table 1.12. Although there are many correlations for Nu_{max} or h_{max}, they are not, in principle, functions of the optimum velocity because U_{opt} is also a function of gas–solid properties. Figure 1.25 shows the plot of the variation in Re_{opt} with the Galileo number for a wide range of values of the latter (Archimedes number and Galileo number are the same). The correlations for predicting Re_{opt} are also given in Table 1.12. The best fit of experimental data was made by Sathiyamoorthy et al.[182] for predicting the optimum Reynolds number (Re_{opt}) with respect to the maximum heat transfer coefficient; the correlations proposed are

$$\mathrm{Re}_{opt} = 0.00812\ \mathrm{Ar}^{0.868} \quad \text{for } 1 \le \mathrm{Ar} < 1000 \tag{1.163}$$

Table 1.12 Correlations for Predicting Maximum Heat Transfer Coefficient and Optimum Reynolds Number in Gas Fluidized Bed

Correlations	Remarks	Ref.
$h_{max} = 33.7\ \rho_s^{0.2} K_f^{0.6} d_p^{-0.36}$	—	Zabrodsky[123]
$Nu_{max} = 0.86\ Ar^{0.2}$	$Re_{opt} = 0.118\ Ar^{0.5}$	Varygin and Martyushin[187]
$h_{max} = 239.5\ [\log(7.05 \times 10^{-3}\ \rho_{SB})]/d_p$	ρ_{SB} = nonfluidized bed density	Baerg et al.[188]
$Nu_{max} = 0.64\ Ar^{0.22}\ Z/d_t$	Z = distance between axes of tubes, d_t = heater diameter	Gelprin et al.[189]
$Nu_{max} = 0.0087\ Ar^{0.42}\ Pr^{0.33}(C_{ps}/C_{pf})^{0.45}$	$Re_{opt} = 0.12\ Ar^{0.5}$ (laminar region)	Sarkits[173]
$Nu_{max} = 0.019\ Ar^{0.5}\ Pr^{0.33}(C_{ps}/C_{pf})^{0.1}$	$Re_{opt} = 0.66\ Ar^{0.5}$	Sarkits[173]
$Nu_{max} = 0.021\ Ar^{0.4}\ Pr^{0.33}(D/d_p)^{0.13}(H_o/d_p)^{0.16}$	$Re_{opt} = 0.55\ Ar^{0.5}$	Traber et al.[178]
$h_{w,max} = h_o(1-\epsilon)\ [1 - \exp(-p\ k_f)]$	h_o = constant, p = constant	Jakob and Osberg[190]
$Nu_{max} = 0.0017\ Re_{opt}^{0.8}\ \ Pr^{0.4} dp^{-0.69}$	$Re_{opt} = 0.20\ Ar^{0.52}$	Chechetkin[191]
$Nu_{max} = Re_{opt}^{0.423}\ Ar^{0.14}\ Pr^{1/8}$	$Re_{opt} = 0.09\ Ar^{0.58}$	Ruckenstein[192]
$Nu_{max} = 0.3\ Ar^{0.2}\ Pr^{0.4}$	$1 < Ar \leq 220$	Richardson and Shakiri[193]
$Nu_{max} = 0.0843\ Ar^{0.15}$	$10^3 \leq Ar \leq 10^6$	Botterill and Denloye[146]
$Nu_{max} = 1.304\ Ar^{0.2}\ Pr^{0.3}$	$Re_{opt} = 0.113\ Ar^{0.53}$	Kim et al.[194]
$Nu_{max} = 0.064\ Ar^{0.4}$	$5 \times 10^4 \leq Ar \leq 5 \times 10^8$	Brodulya et al.[186]
$Nu_{max} = 2.0\ (1-\epsilon)\ (Ar\ d_{12.7}/d_t)0.21\ (C_{ps}/C_{pg})^{45.5}\ Ar^{-0.7}$	$75 < Ar < 20{,}000$	Grewal and Saxena[195]
$Nu_{max} = 0.85\ Ar^{0.19} + 0.006\ Ar^{0.5}\ Pr^{0.33} + d_p/k_g\ (7.3\ \sigma\ E_p E_s T_s^3)$	$100 < d_p < 5000$ μm for horizontal tubes and plates	Baskakov and Panov[196]
$Nu_{max} = \alpha\ Ar^{\beta}\ Pr^{1/3}\ (C_{vs}/C_{vg})^{1/8}$	For $20 < Ar < 20{,}000$, $\alpha = 0.074$, $\beta = 020$; for $2 \times 10^4 < Ar < 10^7$, $\alpha = 0.013$, $\beta = 0.37$	Chen and Pei[197]
$Nu_{max} = 0.14\ Ar^{0.3}(d_p/d_t)^{0.2}(\rho_p/\rho_o)^{-0.07}$	Heat transfer between particle (p) and freely circulating object (o), ϕ = shape factor	Palchenok and Tamarin[198]
$Nu_{max} = 0.99\ Re_{opt}^{0.25}\ Pr^{0.33}\ (C_{vs}/V_{vg})^{0.09}$	$Re_{opt} = 0.00812\ Ar^{0.868}$ $1 \leq Ar \leq 3000$	Sathiyamoorthy et al.[182]
Nu_{max} = const. $Fr^{0.42}(\alpha \cdot \beta \cdot Re_{mf})^{0.5}(Re_{opt}/Re_{mf})Pr^{0.3}$	$\alpha = (1-\epsilon_{mf})/\epsilon_{mf}$, $\beta = C_{vs}/C_{vg}$, const. = 0.0755 when $Fr = U_{mf}^2/(g \cdot d_t)$, const. = 0.0144 when $Fr = U_{mf}^2/(g \cdot d_t)$	Sathiyamoorthy and Raja Rao[199]
1. Overall a. $Nu_{max} = 0.13\ Al^{0.6}\ \alpha^{-1}$ $Al \leq 300$ b. $Nu_{max} = 0.54\ Al^{0.34}\ \alpha^{-1}$ $Al \geq 300$	Al = laminar flow, $Ar = \sqrt{d_p^3 g}\ \Delta\rho/\mu$, $\alpha = [1 + a + k_g/(2C_p\mu)]$, a = constant to take care of additional resistance due to thick and porous oxide layer over metal powder; 0 for ceramic, 0.5 for Cu	Molerus[185, 200]
2. Particle convective a. $Nu_{pc} = 0.69\ (Al\ \Delta\rho/\rho_g)^{0.1}\ \alpha^{-1}$ b. $Nu_{pc} = 9\ \alpha^{-1}$	$10^3 \leq Ar \leq 10^5$ $Ar \geq 10^5$	Molerus[185, 200]
3. Gas convective a. $Nu_{gc} = 0.4527\ Ar^{0.2323}\ Pr^{0.33}$ b. $Nu_{gc} = 0.024\ Ar^{0.4304}\ Pr^{0.33}$	$10^3 \leq Ar \leq 10^5$ $Ar \geq 10^5$	Molerus[185, 200]

$$Re_{opt} = 0.13\ Ar^{0.52} \qquad \text{for } 3000 \leq Ar < 10^7 \tag{1.164}$$

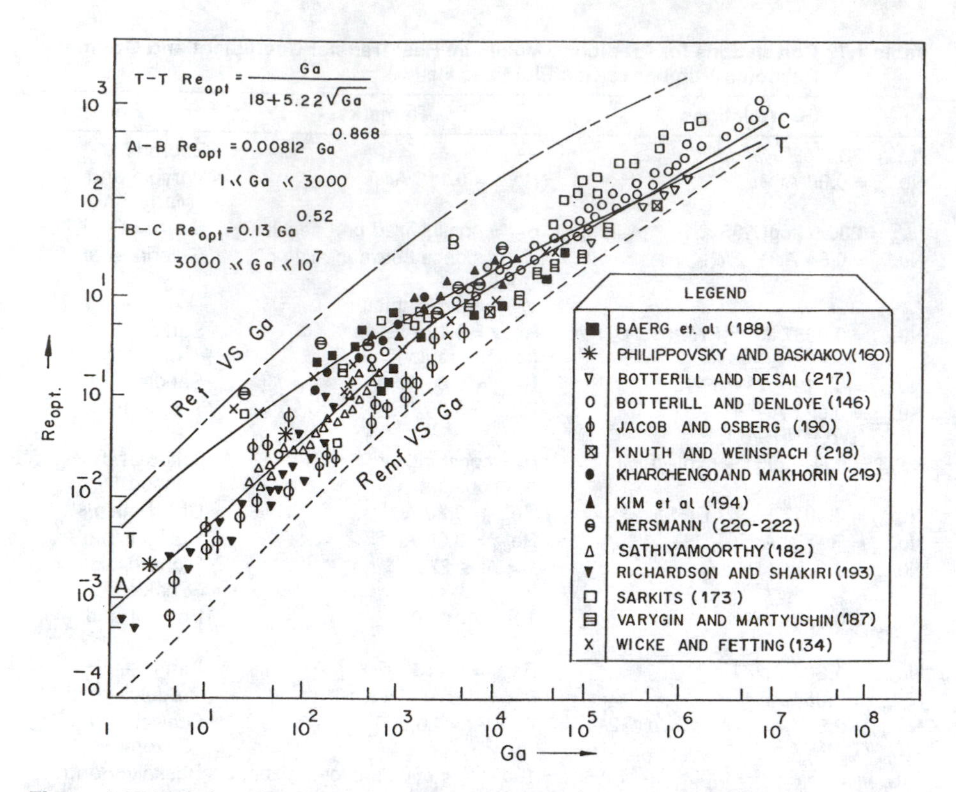

Figure 1.25 Variation of optimum Reynolds number (Re_{opt}) with Galileo number (Ga) (Ga is same as the Archimedes number). (From Sathiyamoorthy, D., Sridhar Rao, C.H., and Raja Rao, M., *Chem. Eng. J.*, 37, 149, 1988. With permission.)

The experimental results of Sarkits[173] show a strong inverse dependence for h_{max} on the diameter of fine particles (i.e., for $d_p < 2.5$ mm) and a weak direct dependence on the diameter of coarse particles ($2.5 \leq d_p \leq 4.5$ mm).

3. *Distributor Effects*

The gas–solid mixing in a fluidized bed is influenced by the distributor type and its design. Many research workers do not mention the types of distributors used in their studies. However, Baerg et al.[188] used perforated plate-type and Sarkits[173] used screen-type distributors. Saxena and Grewal[201] found that the free open area of a distributor affects the maximum heat transfer coefficient and also the corresponding optimum gas velocity. Sathiyamoorthy et al.[202] confirmed that the free area in a multiorifice distributor affects the value of h_{max}, which increases as the distributor free-flow area is reduced. Many correlations for predicting Nu_{max} or h_{max} can be found in the literature, and a list of these is given in Table 1.12. A model based on first principles and incorporating all the possible parameters for the prediction of h_{max} has yet to evolve. A model for predicting the maximum heat transfer coefficient which incorporates the distributor parameters and defines the stable state of fluidization was developed by Sathiyamoorthy and Raja Rao,[199] and the relevant correlations are given in

Table 1.12. Although many correlations are available for predicting h_{max}, the choice of the most appropriate correlation depends on the specific requirements and the designer's skill.

G. Heat Transfer in Liquid Fluidized Beds

1. *Differences with Gas–Solid Systems*

Few studies have been carried out on heat transfer in liquid fluidized beds, unlike gas fluidized beds. In contrast to the latter, heat transfer in liquid fluidized beds increases with increasing particle size. Liquid fluidization is generally particulate in nature. The mixing process and the heat transfer characteristics of a liquid fluidized bed are not exactly similar to those associated with a gas fluidized bed. For a given temperature difference, the quantity of heat transferred through a liquid (film) to a solid particle could be thousands-fold higher than what could be transferred through a gas film.[203]

2. *Heat Transfer*

In a liquid fluidized bed, the temperature gradient extends more into the bed than it does in a gas fluidized bed. According to Wasmund and Smith,[204] the ratio of bed resistance to total resistance increases with increasing solid concentration and decreases with increasing particle size for a constant bed porosity. They suggest that the particle-convective component and the effect of the thermal conductivity of the solid particles are negligible. Generally, it has been established that heat transfer in a liquid fluidized bed is influenced by bed voidage. At constant voidage, an increase in particle size increases heat transfer. On the basis of experimental data, Romani and Richardson[205] proposed a correlation that has limited application. The mechanism of heat transfer in a liquid fluidized bed was analyzed by Krishnamurthy and Sathiyamoorthy[206] with the aim of investigating the reason why the heat transfer rate should increase with increasing particle size in liquid fluidized beds.

Krishnamurthy and Sathiyamoorthy[206] tested a model and arrived at the conclusion that for fine-sized particles in a liquid fluidized bed, the film thickness on the heat transfer surface increases. In a typical case, it was found that the film thickness varies from 1.6 to 36% for just a twofold increase in particle size but when liquid velocity is less than 10 cm/s. Particle size seems to have a relatively greater influence compared to liquid velocity in altering the magnitude of the heat transfer rate. Although the applicability of the few available correlations for practical uses is limited, the correlation of Mishra and Farid[207] may be used for purposes of prediction:

$$Nu = 0.24\,(1 - \epsilon)\,\epsilon^{-2/3} Re^{0.8}\, Pr^{0.33} \tag{1.165}$$

Heat transfer in liquid fluidized beds has been shown to have enhanced rates[208] (as much as fourfold) on pulsating the fluidization.

H. Heat Transfer in Three-Phase Fluidized Bed

1. Heat Transfer Coefficient

Heat transfer data on three-phase fluidized beds have not been published extensively and only limited information is available. Studies on the heat transfer between a fluidized bed and a heat transfer surface were reported by Østergaard,[209] Viswanathan et al.,[210] Armstrong et al.,[211] Baker et al.,[212] and Kato et al.[213] In general, all these studies reveal that the extent of heat transfer increases with the gas flow rate and is higher than in corresponding gas–liquid (bubble column) or liquid–solid (two-phase fluidization) systems.

Heat transfer coefficients obtained in an air–water–glass-bed system have been found to increase with the gas flow rate. Figure 1.26 shows the results on the dependence of the heat transfer coefficient on the gas flow rate, and it can be seen that the trend is similar to that associated with a bubble column or a bubbling fluidized bed. Heat transfer coefficients are very high in three-phase fluidized beds (or the order of 4000 W/m² K) and are reported to be two to three times greater than those attainable even with efficient processes such as boiling and condensation.

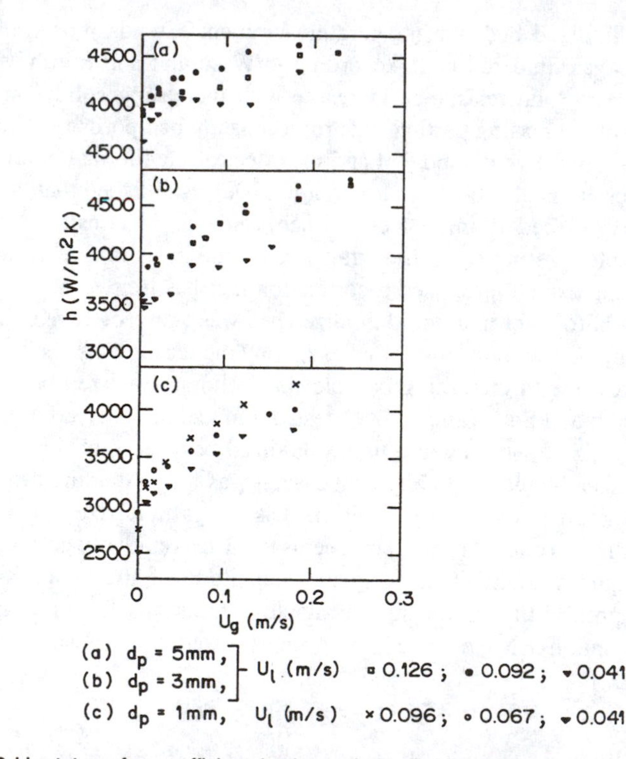

Figure 1.26 Heat transfer coefficient in three-phase fluidized beds. Air–water–glass beads. Data of Armstrong et al.[211] (From Armstrong, E.A., Baker, C.G.J., and Bergognou, M.A., in *Fluidization Technology,* Keairns, D.L., Ed., Hemisphere Publishing, New York, 1976, 453. With permission.)

2. Particle Size Effect

Heat transfer increases for particle diameters greater than 1 mm (Armstrong et al.[211]). With smaller particles in a bed of 0.12 m ID, Kato et al.[213,214] observed a local maximum followed by a minimum during the transition from a fixed bed to a fully fluidized bed. This type of behavior is attributed to the heterogeneity of fluidization with fine particles when working at low liquid velocities.[215] No such behavior was reported by Kato et al.[214] for beds of coarse particles.

3. Correlation

The wall heat transfer coefficient can be estimated by using the correlation of Kato et al.,[214] which is given as:

$$h\frac{d_p}{K_l}\cdot\frac{\epsilon_1}{1-\epsilon_1}=0.44\left[\frac{d_pU_1\rho_1}{\mu_1}\cdot\frac{C_{pl}\mu_l}{k_1}\frac{\epsilon_1}{1-\epsilon_1}\right]^{0.78}+2\left[\frac{U_g}{gd_p}\right]^{0.17} \tag{1.166}$$

This correlation is recommended for a wide range of liquid–phase Prandtl numbers. The literature on three-phase fluidized beds with bed internals such as cooling or heating coils/pipes and studies on high velocities is scant.

VIII. MASS TRANSFER

A. Introduction

Fluidized beds in general are considered to be in transition between the states of packed bed and pneumatic transport. As a result, the correlations for the transport processes in all three units (viz., packed bed, fluidized bed, pneumatic transport reactor) are almost similar in type. Mass transfer data or the driving forces in a fluidized process must be evaluated quantitatively to arrive at a sound design of the reactor. Processes that are carried out in a fluidized bed can be either surface area based (like coating and agglomeration) or volume based (like crystallization and freezing). The transport processes of the surface-area-dependent type are complex in the sense that data on the probability of particle-to-particle or particle-to-object interaction are difficult to obtain. However, in volume-based processes, adequate information on the transport steps is available. In a fluidized bed, the particles act as turbulence promoters to enhance mass transport and reduce the hydrodynamic boundary layer. In a heat transfer process, however, the particles perform an extra function by carrying the heat. The conventional approach of drawing an analogy between heat transfer and mass transfer is valid for a fluidized bed also due to the fact that the particle-to-particle or particle-to-object contact time is low and is not much different for the heat and mass transport steps. However, there is some complexity in fluidized beds in which segregation (i.e., aggregative or bubbling fluidization) occurs.

B. Mass Transfer Steps

The mass transport steps in a fluidized bed can be broadly regarded to be (1) transfer between fluidized bed and object or wall, (2) transfer between particle and fluid, and (3) transfer between segregated phases (e.g., lean to dense phase).

1. Mass Transfer Between Fluidized Bed and Object or Wall

a. Correlations

The j_M factor in mass transfer studies is defined as:

$$j_M = \text{St Sc}^{2/3} = (K_{gw}/u)\ \text{Sc}^{2/3} = \text{constant Re}^{-p} \quad (1.167)$$

where K_{gw} is mass transfer between a fluidized bed and an object or wall. The above equation can be used for a particulate type of fluidized bed by considering the bed as having several channels formed by the irregular contour of the wall of the particles. In the specific case of mass transfer between a fluidized bed and the wall or an object, the transfer can be viewed to occur between the two walls of a channel, one made up of the particle layer itself and the other formed by the object. The hydraulic diameter of the channel (d_H) is

$$d_H = \frac{6\epsilon}{S} = \frac{\epsilon d_p}{(1-\epsilon)} \quad (1.168)$$

The velocity of the fluid through the channel is assumed to be proportional to the interstitial velocity (u) (i.e., U/ϵ). If the Reynolds number in Correlation 1.167 is expressed in terms of U and d_H, then,

$$\frac{K_{gw}}{U} \epsilon\, Sc^{2/3} = \text{constant}\left[\frac{U d_p \rho}{\mu(1-\epsilon)}\right]^{-p} \quad (1.169)$$

The constant and the exponent p in Equation 1.169 vary depending on the state and the type of the bed. Typical values are tabulated in Table 1.13.

Table 1.13 Magnitude of Constant, Bed Voidage (ϵ), and the Exponent (m) of the Correlation

$(K_{gw}/U)\ \epsilon \text{Sc}^{2/3} = \text{Const}\ [Udp/\mu\ (1-\epsilon)]^{-p}$

Bed Type	Ref.	ϵ	Constant in Equation 1.169	m	Re	Sc
Packed	223	0.35–0.45	0.7	0.4	50–500	0.9–218 450–3,700
Liquid fluidized bed	224	0.5–0.9	0.43	0.38	200–24,000	1,300
Gas fluidized bed	225	0.5–0.95	0.7	0	300–12,000	2.57

b. Influencing Parameters

Most studies on mass transfer do not provide useful data on voidage (ϵ), and hence comparison of results is often difficult. The degree of particle turbulence in a gas or liquid fluidized bed is not the same even if an average dynamic similarity is achieved. In addition to particle mean velocity and porosity, data on turbulence are necessary for an accurate prediction of mass as well as heat transfer coefficients. Measuring techniques play a key role. For example, mass transfer data can be obtained by simple experimental techniques more easily in a liquid than in a gas fluidized bed. To obtain reproducible results, gas fluidized beds operated at incipient fluidization are preferred.

c. Role of Voidage

It is useful for the data on voidage to correspond to an optimum situation where mass and/or heat transfer rates are high. The effect of the operating velocity on attainment of the maximum heat transfer coefficient was explained in the preceding section, and various correlations for predicting U_{opt} were presented. By analogy, one can choose the value of U_{opt} for heat transfer for mass transfer also. However, data on ϵ_{opt} are also required, as can be seen from Equation 1.169. The voidage at U_{opt} does not have a unique value. In fact, the occurrence of the maximum does not correspond to any particular velocity, as pointed out in the section on heat transfer. In other words, the maximum can occur over a range of U. The voidage[226] at the optimum condition lies in the range 0.6–0.75. In many industrial practices, attainment of maximum conversion and minimization of effluent heat content and convective heat loss are more important than U_{opt} values.

The various useful ranges of values for ϵ_{opt}, U_{opt}/U_{mf}, and U_{opt}/U_t and the range of values for the drag coefficient (C_D) for a single spherical particle were discussed by Beek.[227] These data are presented in Table 1.14. It can be seen from the table that the useful operating velocity for attaining the maximum mass (heat) transfer rate falls in the range of three to five times the minimum fluidization velocity, and these values are used in many designs as a rule of thumb.

Table 1.14 Optimum Parameter for Mass Transfer for Various Drag Coefficients (C_D) of a Single Spherical Particle

ϵ_{opt}	U_{opt}/U_{mf}	U_{opt}/U_t	C_D
$(0.65 \pm 0.08)\ C_D^{1/6}$	$(5 + 0.5)\ C_D^{3/4}$	$(0.3 + 0.1)\ C_D^{1/4}$	<10
(0.8 ± 0.04)	36	0.6	>10

2. *Mass Transfer Between Particle and Fluid*

a. Comparison of Mass Transfer from Single Particle and Fixed Bed to Fluid

A comparison of correlations for mass transfer from a solid particle to a flowing fluid (K_{pl}) can be made from Table 1.15 for three cases: (1) a single sphere moving through a fluid, (2) a fixed bed, and (3) a liquid fluidized bed.

Table 1.15 Correlation Comparison for Particle-to-Fluid Mass Transfer Coefficient (K_p)

Condition	Correlations for Sherwood no. (Sh)	Remarks	Ref.
Single sphere to fluid	$2 + 0.65Sc^{1/3}\,Re^{1/2}$ $Re = d_p\,U_o\,\rho_f/\mu$ $Sc = \mu/\rho_f$	Particle moving through the fluid at relative velocity (U_o)	228
Fixed bed	$2 + 1.85\,C^{1/3}\,Re_p^{1/2}$	For coarse solids (i.e., Re_p > 80), U_o is superficial velocity of fluid	229
Liquid fluidized bed	$2 + 1.5\,Sc^{1/3}$ $[(1 - \epsilon)Re_p]^{1/2}$	For laboratory-size bed exhibiting particulate fluidization, $5 < Re_p < 120$ $\epsilon \leq 0.84$	230

For large particles (Reynolds number, $Re_p > 80$), it can be generalized from the data given in Table 1.15 that:

$$K_{p,\text{ fixed bed}} > K_{p,\text{ fluidized bed}} > K_{p,\text{ single sphere}} \tag{1.170}$$

The constant 2 in all the correlations (Table 1.15) is due to molecular diffusion and represents the theoretical minimum when the fluid is stagnant (i.e., $Re_p = 0$). For a fixed bed of fine solids, measurement of mass transfer parameters is difficult due to rapid attainment of equilibrium brought about by the large surface area of the particles. For industrial-scale liquid fluidized beds, particulate fluidization need not necessarily prevail.

b. Complexity in Measurement

In a gas fluidized bed, measurement of the mass transfer coefficient poses practical problems due to features like high surface area, bubble formation, backmixing, and gas bypassing. Attainment of equilibrium for a gas inside the bed is so rapid that the gas has to pass through a bed of a height equal to only a few particle diameters to reach equilibrium. The height required to reach equilibrium is often referred to as the height of a transfer unit. In order to measure the mass transfer coefficient in a gas fluidized bed, a very shallow bed is necessary because there is no bubble generation and gas backmixing or bypassing in shallow beds. It is not possible to attain truly ideal conditions in a fluidized bed. The mass transfer data of Rebnick and White[231] and Kettenring et al.[232] were analyzed by Kunii and Levenspiel;[233] they showed that a sharp decrease in the Sherwood number occurs when the Reynolds number is lowered. The decrease in the Sherwood number is lower than that obtained under corresponding conditions in a fixed bed or a single sphere. It was found that the Sherwood number in a fluidized bed can be less than 2, a value which is the minimum for a single particle as well a fixed bed (see Table 1.15). This could result from the hydrodynamic conditions which may not be conducive for mass transfer from particle to fluid.

c. Correlations

Richardson and Szekely[234] proposed correlations for the Sherwood number in the case of a shallow fluidized bed of a height equivalent to five times the particle diameter. The correlations are

$$\text{Sh} = 0.374\ \text{Re}^{1.18} \tag{1.171}$$

for $0.1 < \text{Re}_p < 15$ and

$$\text{Sh} = 2.01\ \text{Re}_p^{0.5} \tag{1.172}$$

for $15 < \text{Re}_p < 250$. It may be observed from these correlations (1.171 and 1.172) that the Sherwood number attains a value far below 2 for Re_p values smaller than 15. According to Richardson and Szekely,[234] the low values of Sh (<2) are due to the decrease in the value of the dispersion coefficient, which drops 200-fold when Re is lowered 100-fold. However, this trend does not appear to be consistent, as pointed out by Kunii and Levenspiel.[233] A bubbling bed model could explain this situation better and is treated in Chapter 5.

Because there are practical problems in measuring the particle-to-fluid mass transfer rate, it has not been possible to propose a single correlation to predict this parameter. However, attempts have been made to group the data available in the literature, and correlations were proposed by Beek.[227] For liquid fluidized beds,

$$\text{St}\ \text{Sc}^{2/3} = K_d\ \epsilon\ \text{Sc}^{2/3}/U\ (0.81 \pm 0.05)\ \text{Re}_p^{-0.5} \tag{1.173}$$

for $5 < \text{Re}_p < 500$, $10^2 < \text{Sc} < 10^3$, $0.43 < \epsilon < 0.63$. For gas and liquid fluidized beds,

$$\text{St}\ \text{Sc}^{2/3} = (0.6 \pm 0.1)\ \text{Re}_p^{-0.43} \tag{1.174}$$

for $50 < \text{Re}_p < 2000$, $0.6 < \text{Sc} < 2000$, $0.43 < \epsilon < 0.75$.

For all practical purposes, the values of St $\text{Sc}^{2/3}$ (i.e., $K_d\ \epsilon\ \text{Sc}^{2/3}/U$) lie between 0.02 and 0.4. This range of values is useful in computing the number of particles passing through a height of one transfer unit, which is defined as the height at which equilibrium between the particle and the fluid is reached.

d. Height of Transfer Unit

The height of a transfer unit (HTU) is defined as:

$$\text{HTU} = \frac{U}{S\ K_d} = U\ d_p/6(1-\epsilon) \tag{1.175}$$

For the range of St $Sc^{2/3}$ between 0.02 and 0.4, Equation 1.175 can be transformed into:

$$0.4\,\epsilon\,Sc^{2/3} < (1-\epsilon)\,HTU/d_p < 8\,\epsilon\,Sc^{2/3} \qquad (1.176)$$

The number of particles that the fluid comes across in the transfer unit is $(1 - \epsilon)\,HTU/d_p$, and the range is given by Equation 1.176. For example, $Sc \approx 1$ for a gas, and for ϵ lying between 0.43 and 0.75, the number of particles, as predicted by this equation, is very small. Therefore, measurement of the particle-to-gas mass transfer coefficient in either a fluidized bed or a packed bed is very difficult. For liquid fluidized beds ($Sc \approx 1000$), when Re_p ranges from 5 to 2000, the number of particles that are passed by the liquid in one HTU lies between 20 and 500. Therefore, for low Reynolds numbers (<5), when particulate fluidization is prevalent, the mass transfer coefficient has to be evaluated using a bed whose height is of the order of 10–20 particle diameters. In case the liquid flow rate is high, the assumption of particulate fluidization and plug flow is no longer valid.

Wakao[235] reevaluated the data pertaining to the particle–fluid mass transfer rate and developed a correlation for a wide range of Re_p values:

$$Sh = 2 + 1.1\,Sc^{1/3}\,Re_p^{0.6} \quad \text{for } 3 < Re_p < 3000 \qquad (1.177)$$

It has been pointed out that J_M factor type of correlations (i.e., J_M versus Re_p) are normally misleading as to their validity at low values of Re_p. Wakao[235] suggested the use of Correlation 1.177 for the design and analysis of packed bed reactors if axial and radial dispersion are taken into consideration.

In general, the K_d value obtained for a particulate fluidized bed is lower than that for a packed bed with the same material and fluid. This was already shown qualitatively. However, the comparison can be presented better by using the HTU in both cases. Thus, the ratio of the HTU for a fluidized bed to that for a packed bed is $(\epsilon/1-\epsilon)/(\epsilon_{mf}/1-\epsilon_{mf})$. It shows that the HTU for a fluidized bed has to be higher because $\epsilon > \epsilon_{mf}$. In other words, to achieve the same degree of equilibrium, a fluidized bed of a larger height should be used, and this height is increased further if the operating velocity is increased (because the expansion is high).

e. *Velocity Effect*

The mass transfer coefficient (K_d), as can be seen from the correlations for St $Sc^{2/3}$, is proportional to ϵ^{-1} and $U^{0.57}$. Again, according to Richardson and Zaki,[34] the bed porosity is proportional to $U^{0.42\text{-}0.33}$. Hence, K_d is proportional to $U^{0.15\text{-}0.24}$. This result is useful for predicting the variation in K_d with U. It is evident that the mass transfer coefficient increases only slightly with U, but decreases sharply if U is decreased. Beek[227] pointed out that any attempt to operate a liquid fluidized bed at a high Reynolds number and assuming the bed to be in a state of particulate fluidization for the purpose of evaluating the mass transfer coefficient may lead to erroneous results.

f. Experimental Technique

Methods of predicting the mass transfer coefficient were discussed in the preceding section. Experimental determination of this parameter has not been described in any textbook. One of the simplest experiments for doing so for a liquid fluidized bed is the dissolution of benzoic acid in water. Such an experiment can be carried out for the determination of the mass transfer coefficient in particulate fluidization. If the amount (W) of benzoic acid dissolved is known over a specific period, then the solid-to-fluid mass transfer coefficient is given by the equation

$$K_d = \frac{W}{A \, \Delta C_{lm}} \tag{1.178}$$

where

$$\Delta C_{lm} = \frac{\left(C^* - C_1\right) - \left(C^* - C_2\right)}{\ln \dfrac{\left(C^* - C_1\right)}{\left(C^* - C_2\right)}}$$

The concentrations at the inlet and the outlet are, respectively, C_1 and C_2. The inlet concentration is usually zero from the weight loss of benzoic acid. The equilibrium concentration (C^*) is taken from literature. Results of experiments carried out on the benzoic acid–water fluidized bed system were reported by Damronglerd et al.[236] They developed a correlation for predicting the Sherwood number as a function of Reynolds number, Schmidt number, bed voidage, Galileo number, and the density ratio. They do not favor a correlation that incorporates the voidage parameter (ϵ). Their contention is that the bed voidage (ϵ) is not a fundamental parameter and can be predicted if the basic variables are fixed. They predicted a critical voidage of $\epsilon = 0.815$ at which the trend in the variation of Sh $Sc^{1/3}$ with ϵ exhibits a sharp change. There seems to be no explanation for this kind of change and the attainment of a critical voidage. Another important observation is that a perforated plate type of distributor performs well and improves the mass transfer coefficient slightly and is better in this regard than porous distributors. An explanation for this observation was not provided by these authors.[236]

The various experimental methods for the measurement of the particle–fluid mass transfer coefficient were briefly reviewed by Wakao[235] and Wakao and Funazkri.[237]

3. Mass Transfer Between Segregated Phases

a. Gas-Exchange Hypothesis

In a gas–solid fluidized bed, the presence of gas bubbles and the surrounding emulsion of particulate solids enables the whole system to be assumed to be constituted of two phases. The transfer or exchange of the gas between the particle-lean bubbles and the particle-rich emulsion phase has been the subject of extensive research over several years. Davidson and Harrison[21] analyzed the behavior of the bubbles in a gas fluidized bed and proposed a mechanism for gas exchange between the two phases. According to their mechanism, gas from a bubble is exchanged by percolation through the bubble voids as well as by diffusion across the frontal surface of the bubble; this is similar to what happens in bubbles present in a gas–liquid system. Subsequent to the development of the above model, Kunii and Levenspiel[238] proposed another theory according to which the transfer of gas from a bubble to the emulsion occurs through a cloud whose dimensions can be obtained by the Davidson and Harrison theory. They attempted to evaluate the rate constants for the exchange of gas from the bubble to the cloud and from the cloud to the emulsion. Several other models have also been proposed. They are all based on concepts similar to those pertaining to the above models but differ as to the assumptions made with regard to the fluid-mechanical conditions around the bubble. For example, Chiba and Kobayashi[239] assumed that transfer of gas from bubbles occurs mainly by convective diffusion, and their analysis is similar to Murray's stream function and bubble geometry.[240] In developing their model, Partridge and Rowe[241] made use of the analogy of mass transfer from a solid sphere placed in a flowing fluid and proposed the occurrence of a boundary layer type of convective diffusion.

b. Gas Exchange Between Lean and Dense Phases (Two-Phase Model)

When a fluidized bed is operated at superficial gas velocity (U) with the gas entering at an inlet concentration of C_o, the whole gas in the bed may be assumed to be present in two phases. One phase is made up of bubbles only (without any particles) and the other is made up of the solid particles (without any bubbles in them). The emulsion phase (also called the dense phase due to the presence of most of the solid particles in it) is assumed to behave like a fluidized bed at the incipient state. Hence, its voidage is at ϵ_{mf} and the flow of gas is at the minimum fluidizing velocity. If the whole phase is assumed to be in a perfectly mixed condition, then the gas is present throughout the emulsion phase at a concentration C_p. This phase can be assumed to be in plug flow, in which case C_p will vary along the bed height. The gas in the lean or the bubble phase is always assumed to be in the plug flow condition. Hence, the concentration of gas in this phase (C_{bz}) will also vary along the bed height. Detailed coverage of two-phase flow is provided in Chapter 5 as part of the discussion on the modeling aspects of a fluidized bed. Mass transfer aspects are mainly discussed in the following paragraphs.

Gas exchange from the bubble phase is important because most of the fluidizing gas is present in this phase. This gas must effectively pass into the particulate solid phase for the gas–solid reaction to occur.

Davidson and Harrison[21] assumed that the volumetric exchange rate (Q) of a gas between a single bubble of volume V_b and the dense phase is made up of two additive components. The first component is due to the convective flow of gas (q, m^3/s) and the second is due to diffusion ($K_G S$). Thus, Q is mathematically expressed as:

$$Q = q + K_G \cdot S \tag{1.179}$$

where S is the frontal surface area of a spherical bubble of equivalent diameter (D_e) and K_G is the average mass transfer coefficient (m/s) that can be estimated by analogy with liquid-film-controlled diffusion from a rising bubble. The correlation for K_G is

$$K_G(\text{in CGS units}) = 0.975\, D_G^{1/2} D_e^{-1/4} g^{1/4} \tag{1.180}$$

The diffusion coefficient (D_G) is a basic parameter and can be obtained from the literature on the diffusion of the species in the gas bubble.

In designing a fluidized bed reactor, it is necessary to know the mass exchange rate. It is not often easy to estimate Q, due to the term q associated with the bulk flow. Hence, experimental determination may be required. Davidson and Harrison[21] theoretically estimated q as:

$$q = 0.75\, \pi\, U_{mf} D_e^2 \tag{1.181}$$

It is normally found[242] that for matertial with low U_{mf}, q is lower than $K_G S$.

c. *Measurement of Gas-Exchange Rate*

Experimentally, Q is determined by injecting a tracer gas that appears as bubbles in an incipiently fluidized bed. Stephens et al.[242] reviewed the pertinent experimental techniques and developed an experimental method using mercury vapor as the tracer gas. They also developed a theoretical model which is briefly outlined below.

For an element of bed height dz, mass balance for a bubble stream with N bubbles gives:

$$\left(U - U_{mf}\right)\frac{dC_b}{dz} + N_b Q\left(C_{bz} - C_p\right) = 0 \tag{1.182}$$

and for the dense phase one has:

$$U_{mf}\frac{dC_p}{dz} + N_b Q\left(C_p - C_{bz}\right) = 0 \tag{1.183}$$

Eliminating C_{bz} from Equations 1.182 and 1.183, a differential equation can be formed as:

$$\frac{\alpha\, U_{mf}}{N_b Q}\frac{d^2 C_p}{dz^2} + \frac{d\, C_p}{dz} = 0 \tag{1.184}$$

where $\alpha = 1 - U_{mf}/U$.

Integration of Equation 1.184 with the boundary conditions at $z = 0$, $C_{bz} = C_{bo}$, $C_p = 0$ and at $z = \infty$, $C_{bz} = C_p = \alpha C_{bo}$ gives:

$$C_p = \alpha\, C_{bo}\left[1 - \exp\left(-\frac{N_b Q z}{U_{mf}\alpha}\right)\right] \tag{1.185}$$

Equation 1.185 can be rewritten in logarithmic form as:

$$\log\left(\alpha\, C_{bo} - C_p\right) = \log \alpha\, C_{bo} - \frac{N_b Q z}{2.303\, U_{mf}\alpha} \tag{1.186}$$

If $\log(\alpha C_{bo} - C_p)$ is plotted against z, the gradient of the line is

$$\frac{-N_b Q U}{2.303\left(U - U_{mf}\right) U_{mf}}$$

Assuming continuity of the bubble phase, one must have:

$$(U - U_{mf}) = N_b\, V_b\, U_A \tag{1.187}$$

Thus, the gradient can be evaluated by eliminating N_b by use of Equation 1.187. The term Q can be finally evaluated if all the other parameters such as U_{mf}, U, U_A, and V_b are known. Because the bubble volume is a function of the diameter of the bubble, it is necessary to know this diameter.

d. Bubble Diameter

There are several correlations in the literature for prediction of the diameter of the bubble inside the bed and also its initial value near the distributor. The correlations for predicting the bubble diameter in the bed and the initial bubble diameter (d_{bo}) above the distributor are presented, respectively, in Tables 1.16 and 1.17. It may be noted that determination of the bubble diameter in a gas fluidized bed has been the subject of extensive research for several years because of its importance in assessing the bubble volume and also the mean average bubble velocity (U_{AV}). The expression for U_{AV} as proposed by Davidson and Harrison[21] is

$$U_{AV}\ (\text{CGS units}) = (U - U_{mf}) + 0.71 g^{1/2} V_b^{1/6} \tag{1.188}$$

It has been pointed out[242] that injection of a tracer gas bubble to measure the gas-exchange parameters is not a good technique, and an alternative technique to complement this method is warranted.

e. Mass Transfer Derived from Bubbling Bed Model

In the bubbling bed model, the gas from a bubble is assumed to be transferred through the cloud of solid particles surrounding the bubble into an emulsion phase.

Table 1.16 Correlations for Bubble Diameter in Gas Fluidized Beds

Correlations (CGS Units)	Remarks	Ref.
$d_b = 2.05\ \rho_p d_p (U/U_{mf}^{-1})^{0.63} h$	Porous distributor, D_t = 10, 15 cm, U_{mf} = 2.85 Glass beads, FCC d_p = 20–580 µm, $0.8 < (U - U_{mf}) < 3.2$ (D_t = column diameter)	244
$d_b = 0.34\ (U/U_{mf})^{0.33} h^{0.54}$	Square cross-sectional bed 0.1, 0.37, 1.5, 5.9 m^2 silicious sand, d_p = 70–300 µm	245
$d_b = 1.4\ \rho_p d_p (U/U_{mf})^h$	$2 < (U - U_{mf}) < 9.7$, porous distributors 14 materials $41 < d_p < 450$ µm	247
$d_b = 1.4\ \rho_p d_p (U/U_{mf}) h + d_{bo}$ $d_{bo} = (6G/\Pi)^{0.4}/g^{0.2}$ $G = (U - U_{mf})/N$	Modified correlation of Kobayashi et al.[247]	248
$d_b = 33.3\ d_p^{1.5}(U - U_{mf})^{0.77} h$	D_t = 10 cm, coke d_p = 70–300 µm	246
$d_b = 0.027\ (U - U_{mf})^{0.94} h + d_{bo}$ $d_{bo} = 1.43\ G^{0.4} g^{0.2}$	D_t = 30.8, sand d_p = 40–350 µm $2.6 < U - U_{mf} < 7.7$	11
$d_b = d_{bo}[(2^{7/6} - 1)(h - d_{bo})/d_{bo} + 1]^{2/7}$ $d_{bo} = (6G/\Pi K_b g^{1/2})^{2/5}$	D_t= 10–29 cm silica gel, mild steel, micros beads cat/d_p = 67–443 µm	249
$d_b = d_{bm} - (d_{bm} - d_{bo})\exp(-0.3\ h/D_t)$ $d_{bm} = 0.652\ [A_T(U - U_{mf})]^{2/5}$ $d_{bo} = 0.347\ G^{2/5}$ for perforated plate $d_{bo} = 0.00376\ (U - U_{mf})^2$ for porous plate	$D_t < 130$ cm $60 < d_p < 450$ µm $0.5 < U_{mf} < 20$ cm/s $(U - U_{mf}) < 48$ cm/s	250
$d_b = 1.1(U - U_{mf})^{0.6} h^{0.6} D^{0.1}/K_B^{0.67} g^{0.3}$	D_t = column diameter $K_B = U_b/\sqrt{g\ d_b}$	251
$d_b = (U - U_{mf})^{1/2}(h + h_o)^{3/4}/g^{1/4}$ h_o = 0 for porous plate	$2.5 < U_{mf} < 8$ cm/s Alumina d_p = 210 µm $1.3 < U - U_{mf} < 15.4$	252
$d_b = 0.54(U - U_{mf})^{0.4}(h + 4\sqrt{A_o})^{0.8}/g^{0.2}$	—	253
$d_b = 0.853\ [1 + 0.272\ (U - U_{mf})]^{1/3}\ [1 + 0.684\ h]^{1.21}$	For porous distributor	254
$d_b = 0.853\ [1 + 0.272\ (U - U_{mf})]^{1/3}\ [1 + 0.684\ (h - h_o - h_j)]^{1.21}$	h_o = height where bubble size is same as commercial reactor, h_j = jet length	255

Table 1.17 Correlations for Initial Bubble Diameter (d_{bo}) Above the Distributor Plate

Correlation for d_{bo} (in CGS units)	Remarks	Ref.
$d_{bo} = 1.381\ G^{0.4}/g^{0.2}$	Derived on the basis of analogy for gas bubble formation in liquid	256
$d_{bo} = 1.295\ G^{0.4}/g^{0.2}$	G = volumetric flow rate of gas per orifice	21
$d_{bo} = 1.295\ (U - U_{mf}/N)^{0.4}/g^{0.2}$	N = number of orifices per unit area	248
$d_{bo} = 1.43\ (U - U_{mf}/N)^{0.4}/g^{0.2}$	For perforated plate For porous plate N = 0.1 hole/cm^2	11
$d_{bo} = 1.295\ (G/K_b)^{0.4}/g^{0.2}$	K_b is dependent on bed material (0.6–0.95); bubble formed at the terminal point of spout	257
$d_{bo} = 0.347\ (U - U_{mf}/N)^{0.4}$ $= 0.00376\ (U - U_{mf})^2$	For perforated plate For porous plate	258
$d_{bo} = 1.08\ (U - U_{mf})^{0.33}$	For bubble-up distributor (61 caps on 2.78 spacing, each cap 1.8 cm high and 0.9 cm in diameter)	259
$d_{bo} = 0$	—	252
$d_{bo} = 1.63\ [(U - U_{mf})\ A_o/g^{1/2}]^{2/5}$	A_o = area of plate per hole A_o = 0.56 cm^2 for porous plate	253

Solids are assumed to be present in the bubble, the cloud, and the emulsion. If the gas is to be absorbed by the solids, part of it is first absorbed by the fraction of solids present in the bubble and the remainder is transferred to the cloud, where a portion is absorbed and the unabsorbed component, after transfer to the emulsion phase, is finally absorbed by the solids present there. Kunii and Levenspiel[243] presented the details of the effect of the particles in the bubbles on the fluidized bed mass and heat transfer kinetics.

In the bubbling bed model, the overall absorption of a gas by all three phases (bubble, cloud, and emulsion) can be mathematically expressed. Details of the bubbling bed model are presented in Chapter 5. The numerous correlations available in the literature to predict the bubble-to-emulsion mass transfer coefficient (K_{be}) are listed in Table 1.18.

Table 1.18 Correlations for Mass Transfer Coefficient from Bubbles to Emulsion Phase

Correlations (CGS Units)	Remarks	Ref.
$K_{be} = 0.75\ U_{mf} + 0.975\ (DG^{1/2} g/d_b^{1/4})^{1/4}$	Two-phase resistance in bubble	21
$K_{be} d_c/D_G = 2 + 0.69 N_{sc}^{1/3} (d_c U_b \rho_g/\mu_g)^{1/2}$	—	241
$K_{be} = 0.303\ D_G^{1/2} g^{1/4} d_b^{1/4}$	$3 < U < 5$ cm/s	260
	$65 < d_p < 142$ μm	
$K_b = 0.75\ U_{mf} + 0.975(D_G^{1/2} g^{1/4} d_b^{-1/4})$	—	58
$K_e = 1.128\ (\epsilon_{mf}^2\ D_G U_{bo}/d_b)^{1/2}$		
$1/K_{be} = 1/K_b + 1/K_e;\ U_{bo} = U - U_{mf} + 0.711 \sqrt{g d_b}$		
$K_c = 1/[1 + 2\ \epsilon_{mf}/(\alpha - 1)] \cdot A\ \epsilon_{mf} A/\Pi\ d_b)$	Two dimension bed (40 × 2)	261
$K_D = 1.02\ \epsilon_{mf}/1 + 2\epsilon_{mf}/(\alpha - 1)\ (U_b D_G/d_b \cdot \alpha - 1/\alpha$ $\sqrt{\alpha} + 1/\alpha - 2)^{1/2}$	Height 1.3 m, $149 < d_p < 10$ μm	
$K_{be} = K_c + D_{Dt}\ \alpha = U_b\ \epsilon_{mf}/U_{mf}$		
$K_{be} = 1.833$	$D_t = 84$ cm	262
	$177 < d_p < 250$ μm	
	$3 < U < 18$ cm/s	
	$U_{mf} = 2.1$ cm/s	
$K_b = 1.128\ (D_G U_b/d_b)^{1/2}$	—	263
$K_e = 1.128\ (m D_{eff} U_b/d_b)^{1/2}$		
$1/K_{be} = 1/K_b + 1/(\beta_r\ K_e)$		
$K_{be} = 1.128\ (\epsilon_{mf}^2\ D_G U_b/d_b)^{1/2}\ (\alpha - 1/\alpha)^{2/3}$	$D_t = 10$ cm	264
$\alpha = (U_b \epsilon_{mf})/U_{mf}$	$140 < d_p < 210$ μm	
$d_b' = 2/3\ d_b + 0.5\ (D_t/d_b)^2\ (L - L_{mf})\ U_b/(f_b\ L)$	$U = U_{mf}$	
f_b = frequency of bubble (1/s)	$3.1 < U_{mf} < 5$ cm/s	
$K_{be} = 3\Pi\ d_b^2\ U_{mf}/4\ S_b + 1.128\ (D_G U_b/L_b)^{1/2}$	$D_t = 15.4$ cm	265
	$80 < d_p < 105$ μm	
	$U_{mf} = 0.5$ cm/s	
$K_{be} = 9\ \delta\ \xi\ U_{mf}/4\ d_b$	For bubble growth zone	266
$[\sqrt{\Pi}/2 \sqrt{(8 D_t/D_G \epsilon_b U_b)} + 4/3\ U_{mf}]^{-1}$	For growthless bubble zone	
$\delta = 1.4\ d_p \rho_p U/U_{mf}$	$\xi = 1$ cm	
$K_{be} = 1.05$	—	1
$K_{be} = 1.19\ U_{mf} + 0.91\ (D_G^{1/2} g^{1/4}/D_b^{1/4})\ (\epsilon_{mf}/1 + \epsilon_{mf})$	—	267

C. Mass Transfer in Three-Phase Fluidized Beds

1. Driving Forces

In a three-phase fluidized bed (i.e., a gas–liquid–solid fluidized bed), the transfer of material from a gas bubble to the liquid phase is usually controlled by the liquid-side mass transfer coefficient. The gas-side mass transfer coefficient is relatively small. The flux of the material transferred from the gas to the liquid phase is

$$N_f = K_l \, a_I \, \Delta c \tag{1.189}$$

where the driving force is Δc, K_l is the liquid-side mass transfer coefficient, and a_I is the interfacial surface area across which the transfer of material takes place. The presence of solid particles in a three-phase fluidized bed can enhance the mass transfer rate either by altering the turbulence to reduce the liquid film thickness or by penetration through the liquid film by particles of very fine size (<5 μm). Inert particles[268] usually do not affect K_l, unlike chemically reactive particles, which accelerate the mass transfer rate.

2. Volumetric Mass Transfer Coefficient

a. Description

Generally, it is difficult to individually measure the liquid-side mass transfer coefficient (K_l) for gas–liquid mass transfer and gas–liquid interfacial area per unit volume of reactor a. Hence, the product K_la, known as the volumetric mass transfer coefficient, is usually measured and relevant correlations are proposed. K_la is affected by particle size. Experimental investigations[269] on the physical absorption of carbon dioxide in a three-phase fluidized bed of inert particles showed that the K_la value obtained with a 1-mm particle size is 20% of that associated with a solid free bubble column. The value of K_la with 6-mm particles is twice that achievable in a bubble column. These observations were rationalized on the basis of bubble coalescence with fine (i.e., 1-mm) particles and bubble disintegration with coarse (i.e., 6-mm) particles. The zone near the distributor, often called the grid zone, influences[269,270] K_la to a great extent. For example, K_la falls sharply near the grid zone for fine particles, whereas with coarse particles (i.e., 6 mm) K_la increases, reaches a maximum, and then decreases. Hence, a shallow bed of coarse particles is expected to have a high mass transfer rate. A three-phase fluidized bed with a grid zone that has a high mass transfer coefficient and a bulk zone (constituting 80% of the bed) that has a poor mass transfer rate was reported by Alvarez-Cuenca and Nerenberg.[271]

b. Measurement Techniques

The overall mass transfer for gas absorption is related to the liquid-phase concentration profile by the mass balance equation:[272,273]

$$E_l \frac{d^2 C_l}{d z^2} - U_l \frac{d C_l}{d z} = K_l a \left(C_g - C_l \right) \tag{1.190}$$

where E_l is the liquid-phase dispersion coefficient. In the gas phase, where plug flow is usually assumed, dispersion is very small. Hence,

$$U_g \frac{d\,C_g}{d\,h} = -K_g\left(C_g - C_l\right) \tag{1.191}$$

The overall mass transfer coefficient ($K_c a$) is given by:

$$\frac{1}{K_c a} = \frac{m}{K_g a} + \frac{1}{K_l a} \tag{1.192}$$

where the Henry constant (m) is small when the concentration gradient within the gas is negligible, K_c is the mass transfer coefficient for gas–liquid mass transfer, a is the gas–liquid interfacial area per unit volume of reactor, and the subscripts g and l refer to the gas and liquid sides, respectively. For small values of m,

$$K_c a \approx K_l a \tag{1.193}$$

The use of Equations 1.190 and 1.191 to predict $K_c a$ requires the concentration profile data in the liquid and the gas phases, respectively. In order to obviate this requirement, the use of a liquid-phase chemical reaction that would take place at a rate which would not influence gas absorption but would be adequate to react with all the absorbed gas in the bulk is advised.[274] In such a situation, Equation 1.191 can be simplified to:

$$U_g \frac{d\,C_g}{dh} = -K_l\, a\, m\, C_{gi} \tag{1.194}$$

where C_{gi} is the concentration of the gas at the interphase. Oxidation of aqueous sodium sulfite with oxygen gas is a typical example. Equation 1.194, after integration between the limits of the measured values of the inlet mole fraction (X_o) and the exit concentration (X_e) over the height (Δh), gives:

$$K_l a = \frac{\ln\left(X_e / X_o\right)}{m\,\Delta\,h} \cdot U_g \tag{1.195}$$

Equation 1.195 is useful for predicting $K_l a$.

3. Influencing Factors

a. Effect of Flow Regimes

The volumetric mass transfer coefficient ($K_l a$) is influenced by the gas velocity (U_g), liquid velocity (U_l), and particle size (d_p). Depending on the magnitude of the velocity of the gas or the liquid, the flow regimes vary and each flow regime has its own K_la. Therefore, a particular expression of K_la may not be valid under all conditions. Gay et al.[275] proposed three regimes based on the variation of K_la with gas velocity and the liquid velocity ratio (U_l/U_{mf}). The schematic diagram in Figure 1.27 shows the trend in the variation of K_la with U_g and U_l/U_{mf}.

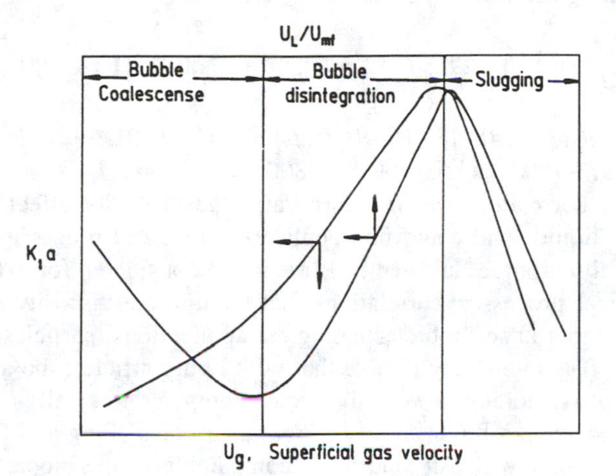

Figure 1.27 Trend in variation of volumetric mass transfer coefficient (K_{la}) with superficial gas velocity (U_g) and ratio of liquid to incipient fluidization velocity (U_l/U_{mf}).

In the bubble coalescence regime, K_la drops as the interfacial area for the transfer is small. In this regime, the gas and/or liquid velocities are low and the effect of particle inertia is small; hence, bubble coalescence is predominant. Large particle sizes would help to increase K_la in this regime. In the case of low liquid velocities, K_la is not affected due to the insignificant effect of low U_l on the hydrodynamic conditions. At moderate liquid flow rates, the gas is transported, reducing its holdup and also K_la. However, when U_l is 2.5–3 times U_{mf}, significant agitation occurs; this causes an increase in the surface area and also in K_la. At high U_l values, buoyancy forces imparted to the particles reduce the inertia, thereby resulting in slugging and lowering of K_la. In the case of increased U_g, bubble coalescence ceases due to the inertial forces and the bubbles disintegrate, enhancing the interfacial surface area and increasing K_la. At high U_g values, the gas holdup increases to form slugs of gas; this reduces the interfacial area and also K_la.

b. *Effect of Properties of Gas and Liquid*

The size and density of particles have a pronounced effect on K_la, which can decrease or increase with these parameters depending on the flow regime. It may be noted that various forces including surface tension play a role in controlling the gas–liquid interface area. In a dilute aqueous solution, where viscous forces are negligible considering solely the surface tension (σ_{gl}), a dimensionless group known as the Weber number ($We_p = \rho_s U_A^2 d_p/\sigma_{gl}$) can be formed and used to identify the flow regimes. Thus, $We_p = 3$ corresponds to a critical particle diameter of 2.5 mm and demarcates the transition from bubble coalescence to bubble disintegration.

The effect of viscosity on $K_l a$ still is not clearly understood. A comprehensive study on this aspect was carried out by Patwari et al.[276] and it was found that $K_l a$ decreases with increasing viscosity. An extended investigation on gas–liquid mass transfer in a three-phase fluidized bed was carried out, using viscous pseudoplastic liquids, by Schumpe et al.[277] and a correlation was proposed. The correlation for $K_l a$ incorporating the viscosity is

$$K_l a\ \left(s^{-1}\right)/D_l^{0.5}\left(\mathrm{m}^2/\mathrm{s}\right) = 2988\ U_g^{0.44}(\mathrm{m/s})U_l^{0.42}(\mathrm{m/s})\mu_{eff}^{-0.34}(\mathrm{Pa\cdot s})U_t^{0.71}(\mathrm{m/s}) \quad (1.196)$$

for $0.017 \le U_g$ (m/s) ≤ 0.118, $0.03\ U_l$ (m/s) ≤ 0.16, $0.001 \le \mu_{\mathrm{eff}}$ (Pa · s) ≤ 0.119, $0.08 \le U_t$ (m/s) ≤ 0.60, and $5.5 < \mathrm{Re}_t \le 4800$.

The above correlation which incorporates the viscosity effect is useful for pseudoplastic liquids and could find application in many processes that involve biomedia. As three-phase fluidized beds are being considered for extensive use in biotechnological processes, correlations that incorporate viscosity effects are of considerable importance. In biotechnological applications, particles usually have low densities. It is interesting to note that with light particles, the axial variation of individual phase holdups is very significant compared to axially constant-phase holdups that occur in a fluidized bed of heavy particles. Tang and Fan[278] investigated the effect of low-density particles, comparable to the bioparticles used in biotechnological processes, and found that an increase in liquid velocity causes $K_l a$ to increase significantly without much increase in gas holdup, thus indicating the significant effect of U_l on K_l. It has also been observed that $K_l a$ decreases with increasing solid concentration and with increasing particle terminal velocity.

c. *Effect of Distributor Plate*

The distributor plate[279,280] was reported to influence the mass transfer rate. The grid region was shown to have an excellent mass transfer rate relative to 80% of the bulk region in the upper part of the bed. Several correlations for predicting $K_l a$ can be found in the literature.[281] These correlations lack general applicability due to the complex dependence of the volumetric mass transfer coefficient on the bubble flow pattern. It is wise to determine K_l and a independently. Experimental investigations of Dhanuka and Stepanek[282] on the determination of $K_l a$, K_l, and a, using the method of simultaneous chemical absorption of carbon dioxide and desorption of oxygen, showed that the particle size has a pronounced effect on a, an insignificant effect on K_l, and a considerable effect on $K_l a$. Liquid velocity has little effect on $K_l a$ and a in beds of coarse-sized particles (4.08 and 5.86 mm), whereas both $K_l a$ and a increase with increasing U_l in a bed of fine-sized particles (1.98 mm).

d. *Effect of Bubble Population*

Increasing the bubble population inside a two- or three-phase fluidized bed by means of bed internals such as baffles or a screen is general practice, particularly in large beds, to enhance the fluid–solid contact surface area and hence overall process efficiency. However, bed internals cause the attrition of particles and increase power consumption as well as equipment costs. A new concept is to use floating bubble breakers. Floating bubble breakers have been used in three-phase fluidized beds, and their effect on individual phase holdups,[283] bubble properties,[284] and heat transfer coefficients[285] has been investigated. Kim and Kim[286] studied the effect of the ratio of the volume of the floating bubble breakers to that of the solid particles on K_la and obtained an increase of up to 30% in K_la, with the maximum occurring at a floating bubble breaker to solid particle volume ratio of 0.15. K_la values were evaluated from the surface renewal frequency of liquid microeddies. The correlation proposed for K_la is

$$K_la\left(\mathrm{s}^{-1}\right) = 0.73\ U_g^{0.87}(\mathrm{m/s})U_l^{0.45}(\mathrm{m/s})d_p^{0.71}(\mathrm{m}) \left[1+0.036\left(V_f/V_s\right)^{1.11} - 1.348\times10^{-3}\left(V_f/V_s\right)^{2.09}\right] \tag{1.197}$$

for $0.02 \le U_g \le 0.20$ m/s, $0.02 \le U_l \le 0.1$ m/s, $1.0 \le d_p \le 6.0$ mm, $0 \le V_f/V_s < 2.0$.

4. *Liquid–Solid Mass Transfer*

Estimation of the liquid–solid mass transfer coefficient ($K_l a_s$) is important because it is a useful quantity to be compared with the gas–liquid volumetric mass transfer coefficient (K_la). If K_la_s (where $a_s = 6\epsilon_s/d_p$) is greater than K_la, then gas–liquid mass transfer is the limiting factor. The liquid–solid mass transfer coefficient in a three-phase fluidized bed (gas–liquid–solid system) is greater[287] than that obtained in a liquid–solid fluidized bed. Similar to a liquid–solid fluidized bed, a maximum[227] in K_la_s exists corresponding to a given liquid velocity and porosity. Batchelor[288] developed a theory for the rate of mass transfer from a particle in a turbulent fluid and proposed a correlation for K_s when the Reynolds number and the Sherwood number are low. The correlation is

$$\nu\frac{K_s d_p}{D_t} = 0.693\left(\frac{\nu_l}{D_t}\right)^{1/3}\left(\frac{d_p^4 e}{\nu_l^3}\right)^{1/6} \tag{1.198}$$

where e is the energy input per unit mass and is equal to $U_g g$ for an aerated suspension. Liquid–solid mass transfer was explored by Arters and Fan[289] using cylindrical particles of benzoic acid (d_p = length = 1.5–4 mm) fluidized by air and water. Their investigations showed that the Sherwood number increases with gas velocity. K_la_s in a three-phase fluidized bed is higher than that obtained in a liquid fluidized bed at the same liquid velocity and is independent of liquid velocity, as in the case of liquid (two-phase) fluidized beds.

IX. END ZONES

A. Grid Zone

1. Introduction

A fluidized bed can, in general, be sectioned or viewed as consisting of zones of different hydrodynamics. The dynamics of a gas fluidized bed just above the distributor or the grid is not the same as in the rest of the fluidized bed. This zone is strongly influenced by the distributor, whose design, selection, and type affect the dynamics of the fluidized bed very significantly. Thus, the grid zone dictates the performance characteristics. Among the three zones[290–292] long known in a gas fluidized bed, the grid zone has been given less than its due importance. In fact, its effect on model development has been studied only to a limited extent. The two other zones above the grid zone which occur in series are the constant bed density zone and the bubble erupting zone at the surface of the fluidized bed.

In a gas fluidized bed, depending on the fluid–solid properties, the operating gas velocity, and the distributor, the fluidizing gas may emerge as tiny bubbles or as jets with enough kinetic energy to even pierce the bed. Selection and design of the distributor are very important in achieving smooth or good quality fluidization. A comprehensive account of distributor design and its effects on the quality of fluidization is provided in Chapter 5. The discussion in this section is restricted to the zone near the grid where bubbles and/or jets are prevalent.

2. Gas Jets

a. Description

Gas jets, which emerge from gas-issuing orifices or nozzles, terminate at a point above the grid plate. The jet length is the distance between the plate and the jet-terminating point. The grid zone extends up to the length of the jets. However, the jets may be permanent or nonpermanent. Permanent jets are formed in a bed of nonfluidized solids. Thus, permanent jets are possible with multiorifice distributors because the defluidized solids between the orifices form stable walls for the jets. Rowe et al.[293] suggested that the proximity of a jet to a flat surface could stabilize it. Yang and Keairns[294] showed that a permanent jet cannot be formed in a semicircular unit when heavy particles are fluidized. Permanent jets[295] can form when the gas density approaches the fluidized bed density. It has been pointed out that whether the jet is pulsating (or nonpermanent) or permanent, a region of intense mixing with good solid contact exists in the grid zone. For modeling or simulation of this zone, it is necessary to develop correlations to predict the jet length. Jet length data are essential in most cases when an exothermic reaction is carried out, more so in a commercial-type reactor. In such a reaction, the grid region, active with intense mixing, generates much heat. Proper positioning of the heat-exchanger tubes without damaging them by jet erosion and elimination of hot spots to save the catalytic activity are essential.

b. Correlations

It is necessary to have reliable correlations to predict the jet penetration depth. A list of correlations available in the literature for predicting the jet penetration depth is presented in Table 1.19.

Table 1.19 Correlations to Predict Jet Penetration Depth in Gas Fluidized Bed

Correlations	Remarks	Ref.
$L/D_o = \{[\log(0.817\ \rho\ \sqrt{U_o})] - 1.3\}/(0.0144)$	Visual observation for horizontal and vertical jets; cracking catalyst (50 μm), sand (50–200 μm), mill scale (170 μm)	296
$L/D_o = 13\ \rho\ U_o/(\rho_s \sqrt{g\ d_p})^{1/2} - 1/2 \cot \theta$	d_p = 3200 μm, θ = half jet angle	297
$L/D_o = [0.919\ d_p/(0.0007 + 0.566\ d_p)]\ u_o^{0.35}/d_o^{0.35}$	u_o = gas velocity at orifice, cracking catalyst $65 < d_p < 540$ μm, radioisotope technique used, correlation for vertical jet penetrating an incipient fluidized bed; D_t = 0.5 mm	298
$L/D_o = 5.2\ (\rho\ d_o/\rho_s d_p)^{0.3}[1.3\ (u_o^2/g\ d_o)^{0.2} - 1]$ $L/D_o = 1/2 \cot \theta\ [1.3\ (U_o^2/g\ d_o)^{0.2} - 1]$	Two-dimensional column (0.3 × 0.012 m), single jet of water into lead shots, θ = half jet angle	299
$L/D_o = 6.5\ \mathrm{Fr}^{0.5}$	$\mathrm{Fr} = \rho\ u_o^2(\rho_s - \rho_f)\ d_o g$; semicircular column; hollow epoxy spheres fluidized by air	294
$L/D_o = 15.0\ \mathrm{Fr}^{0.187}$	Based on data collected from literature and own data; modified the previous correlation; $0.2 < \mathrm{Fr} < 200$; satisfactory only for a single jet	294
$L/D_o = 7.5[\rho(U_o - U_{mf})^2/(\rho_s - \rho)g\ d_o]^{0.187}$	Modified Yang and Keairns[294] correlations for multijets	300
$L/D_o = B\ \mathrm{Fr}^{*0.472}$	For single-jet semicylindrical column $\mathrm{Fr}^* = U_{mf}/(U_{mf})_{atm} - \rho\ U_o^2/(\rho_s - \rho)g\ d_o$; Froude number modified to correct pressure effects on the viscous and inertial forces; B = constant (1.81–4.21) dependent on solid	301
$L/D_o = 87.0(u_o^2/g\ d_o)0.29\ (\rho/\rho_p)^{0.48}\mathrm{Re}_p^{-0.17}$	Data collected from 11 sources	302
$L/D_o = 6.5[\epsilon\rho_f u_o^2 + (1 - \epsilon)\rho_s U_o^2/(\rho_s - \rho_f)g\ d_o]^{1/2}$	Effect of loading of solid particles in jets considered useful when solids are injected along with gas; u_o and U_o relative velocity of gas and solid, respectively	303
$L/d_o = 1.3(u_o^2/g\ d_p)^{0.38}$ $(\rho\ d_p u_o/\mu)^{0.13}(\rho/\rho_p)^{0.56}(d_o/d_p)^{0.25}$ (two-dimensional systems) $L/d_o = 1.15 \times 10^4[u_o - u_{omf})^2/\mu]^{-0.42}(d_p/d_o)^{0.60}$ (three-dimensional system)	Reassessed data from literature to predict particle size effect	304

It can be seen from Table 1.19 that there exist a number of correlations for the jet length. The manner in which the jet penetration depth is defined and measured seems to be responsible for the introduction of several correlations. A unique definition for the jet penetration length has not yet been agreed upon. The methods for measuring the jet length are usually based on a variety of techniques, such as visual or X-ray, wall erosion, radioisotope, capacitance probe, etc. Because reliable data for both bidimensional and three-dimensional beds are scarce, many correlations are proposed in the literature, and the accuracy of prediction differs widely from one to another.

c. *Density Near the Grid*

Jet interactions near the grid zone were studied by Kozin and Baskakov,[305] who measured the density of the bed at various heights above the grid. The bed density near the grid was found to be highest for the bubble-cap-type distributor, and this is attributed to the presence of defluidized solids. The density decreases gradually above the cap and remains constant in the core of the bed. The zone between the caps exhibits a low density. Gas issuing from the bubble caps coalesces at the boundary of the bubble caps and jets emanate, causing intense solid circulation. In view of this severe solid circulation, static bed heights 30–35 times the exit opening of the cap are recommended; such heights are 1.5–1.8 times greater than those used with perforated-plate-type distributors at a gas velocity of 1.3 times U_{mf}.

d. *Jet versus Spout*

Gas is generally entrained[306] into a jet that emerges from an orifice and disentrained beyond a certain height above the jet. For an orifice type of distributor, detachment of a jet has been observed to occur at a distance equal to ten times the orifice diameter for gas velocities ranging from 30 to 90 m/s and an orifice diameter of 1 cm. The subject of jets/spouts has now become a special topic in fluidization and is known as the spout fluid bed. This aspect is discussed in Chapter 6. The shape of a jet from an orifice is similar to that of a diverging channel. In a shallow fluidized bed, a rectilinear expansion of jets with jet angles of 15–30° was approached.[307] It is not easy to determine the shape of a spout. Theoretical estimates show that the spout diameter can fall in the range of one to three times the size of the orifice.[308,309] Bell-shaped profiles of axial gas velocity in jets[310] and along the radius in the case of spouts[311] have been reported. The gas velocity along the axis of jets[312] and spouts[313] decreases with increasing distance above the orifice, and this has been attributed to the increase in cross-section of jets and to gas disentrainment from spouts. Voidage along the axis decreases up to 0.7 for a jet; for a spout, it can attain a value much beyond that obtainable at the maximum spoutable bed height.[314]

e. *Bubbles*

An investigation[315] on the rate of bubble formation over the grid using a mini-capacitance probe revealed that bubbles detach at the end of the jet and the frequency of bubble formation was found to be 20 Hz, which matches the results of Hsiung and Grace.[316] Studies on a large-diameter (i.e., 1 m) fluidized bed were carried out by Werther,[317] who pointed out that the influence of the distributor on the bubble size distribution is mostly confined to the grid zone. His extensive investigations on bubble flow showed that visible bubble flow in the case of a perforated distributor accounts for only 67% of the predictions of the two-phase flow model and that this fraction could increase by another 13% with a porous type of distributor.

3. *Dead Zone*

An important feature that must be considered is the dead zone above the grid plate. Leakage of solids through grids is an important parameter, as pointed out by Brien et al.[318] The dead zone above the distributor in a two-dimensional fluidized bed[319] consists of three zones: (1) a top zone where an intermittent exchange of solid particles takes place due to bubble drift and disturbances caused by oscillating gas jets, (2) an intermediate or quasi-dead zone where little exchange occurs except for the solid particles sliding down the dead zone of solids, and (3) a fixed and completely dead zone of solids. Figure 1.28 depicts the details of these three zones.[320] The dead zone can be eliminated or minimized by adjusting the orifice spacing, the orifice diameter, and the gas flow rate through the orifice. Reduction in the dead zone height was achieved by increasing the gas flow rate through the orifice, reducing the orifice spacing, and increasing the orifice diameter. Elimination of a dead zone formed by fine particles is difficult due to their cohesive nature, if the spacing between the orifices is large.[321]

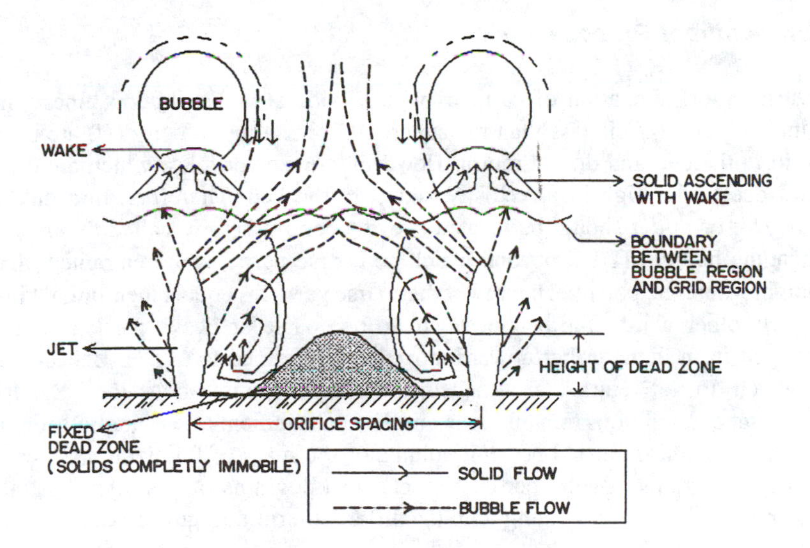

Figure 1.28 Schematic of dead zone pattern above the grid plate. (From Wen, C.Y., Krishnan, R., and Kalyanaraman, R., in *Fluidization*, Grace, J.F. and Matsen, J.M., Eds., Plenum Press, New York, 1980, 405. With permission.)

Particles of fine size normally form clusters and cause the formation of fixed dead zones. From a design standpoint, it is necessary to be able to predict the critical excess flow of gas, $(Q - Q_{mf})_c$, which could completely eliminate the dead zone of solid particles. Wen et al.[322] showed that $(Q - Q_{mf})_c$ for a cap-type distributor is less than that required for an orifice plate. It has also been established by experiments that better mixing, with the same free area and gas flow rate, is possible with a cap-type distributor than with an orifice plate. In a three-dimensional fluidized bed, elimination of the dead zone by $(Q - Q_{mf})_c$ depends on its variation along the radial position, in addition to other influencing parameters such as orifice spacing and orifice diameter.

B. Elutriation

1. *Definition*

For many fluidized bed operations, the post-reactor equipment, such as cyclones, electrostatic precipitators, filter bags, etc., is designed and selected based on knowledge of the amount of carryover of particles from the fluidized bed. The term elutriation mostly refers to the selective removal of fines from coarse particles of the same density or of light particles close in size with different densities. In fluidization, it refers in general to the entrainment or carryover of fines from the bed during the fluidization of polydisperse particles. The space above the surface of a fluidized bed up to the outlet or exit point is called the freeboard space, and the corresponding height is called the freeboard height.

2. *Entrainment Process*

During the fluidization of particles with a wide size distribution, fines whose terminal velocity (U_t) is less than the superficial operating velocity (U) are carried away from the bed and do not return. However, coarse particles, which are carried upward due to the high local velocity, return to the bed. Therefore, in a fluidized bed the exit or outlet should be located above a certain height, called the transport disengaging height (TDH), beyond which no coarse particles are entrained; this is the height where the potential energy of the coarse particles equals their initial kinetic energy. In other words, the height at which the coarse or heavy particles are just disengaged from their further upward movement is termed the TDH. Because only fines which do not return are entrained, the carryover rate above the TDH for a specified set of conditions remains constant. The entrainment rate is normally defined as the mass of solids carried per unit volume or the mass of fluidizing gas per unit time (i.e., kilograms of solid per cubic meter or kilograms of gas). An illustration of the entrainment process, along with the different terminologies used, is shown in Figure 1.29.

3. *Entrainment Rate*

a. *Prediction by Bubble Dynamics Method*

A design engineer estimates the entrainment rate on the basis of experience or from knowledge of the data generated through research. There are several mechanisms by which elutriation or entrainment occurs in a gas fluidized bed. This topic was reviewed by Kunii and Levenspiel,[58] Leva and Wen,[323] Geldart,[324] and Wen et

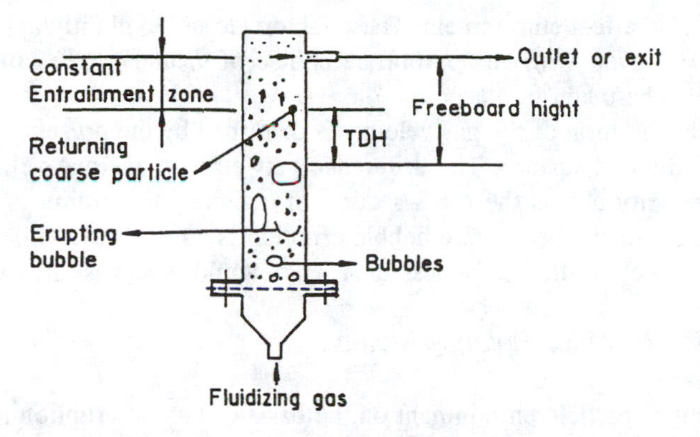

Figure 1.29 Entrainment process and related terminology. Above TDH only fines are carried out, below TDH fines and coarse particles are present, and at TDH coarse particles are disengaged and returned back to the bed.

al.[304] Some salient features are briefly presented here for the benefit of designers. The solids carried by clouds and wakes of the bubbles are exchanged[325] between the emulsion and dispersed phases in a gas fluidized bed. When the gas bubbles reach the fluidized bed surface, they erupt, throwing the solids present in the wake into the freeboard. From knowledge of the number of such bubbles erupting and the fraction of wake occupying the bubble, the entrainment rate has been computed in the following manner.

If N_b is the number of bubbles and f_w is the fraction of wake volume, the solids entrained per unit area of the fluidizing column of cross-sectional area A is given by:

$$W_s = \left(\frac{N_b}{A}\right)\frac{\pi d_b^3}{6} f_w\left(1-\epsilon_{mf}\right)\rho_p \tag{1.199}$$

where d_b can be computed from the data presented in Table 1.16. The term N_b/A can be evaluated on the basis of the assumption that the total bubble volume is equal to $A(U - U_{mf})$. Thus,

$$N_b = \frac{A\left(U-U_{mf}\right)}{\left(\pi d_b^3/6\right)} \tag{1.200}$$

Using Equations 1.200 and 1.199, the entrainment rate can be expressed as:

$$W_s = f_w(1 - \epsilon_{mf})\, \rho_p\, (U - U_{mf}) \tag{1.201}$$

Use of Equation 1.201 requires knowledge of f_w, which can be determined along the lines suggested by Rowe and Partridge.[325] According to them, f_w is 30%. It can be seen that W_s is influenced by U, U_{mf}, and f_w. In addition to these parameters, the bed diameter, the distributor, and the bed internals, which affect the hydrodynamics of fluidization, also affect entrainment. These factors do not explicitly appear in Equation 1.201. However, with an appropriate choice of d_b and f_w, the effects of these parameters can be taken into account.

The radial variation of the gas velocity is disturbed by the presence of elutriated solids above the bed surface. The difference between the average velocity and the maximum velocity across the cross-section of a fluidizing column is rather small near the surface of the bed where bubble eruption is continuous. This would cause the reduction in elutriation to be less than what would otherwise be expected.

b. Prediction by Mass Balance Method

Prediction of particle entrainment on the basis of bubble eruption and the associated particles in the bubble wake requires thorough understanding of bubble dynamics and reliable and pertinent data. Another method for predicting elutriation is by applying mass balance considerations to a continuous or batch system. Let us now evaluate algebraically the elutriation rate (E_i) for the *i*th component in a multicomponent system. In this connection, reference may be made to Figure 1.30.

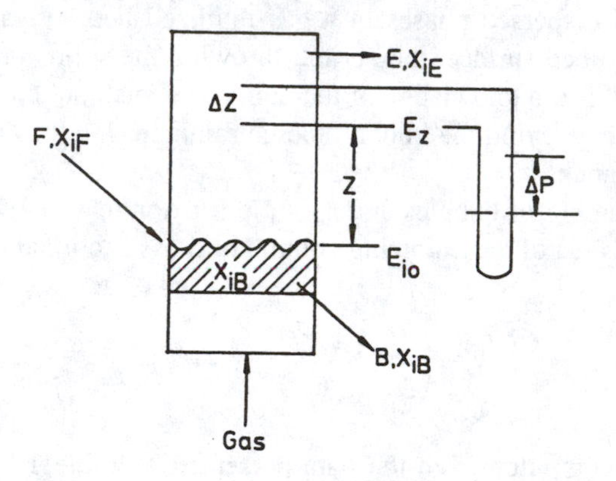

Figure 1.30 Illustration of parameters used in mass balance method.

The mass balance for a continuous system with the feed rate (F), the discharge rate (B) at the bottom, and the elutriation rate (E) is

$$F = E + B \quad \text{(kg/s)} \tag{1.202}$$

For a component i with its mass fraction X_{iF}, X_{iE}, and X_{iB}, respectively, in F, B, and E,

$$FX_{iF} = EX_{iE} + BX_{iB} \quad \text{(kg/s)} \tag{1.203}$$

Equations 1.202 and 1.203 can be used to calculate E, if all the other parameters (i.e., the mass fraction of component i and the feed or discharge rate) are known.

Details of the algebraic method of evaluating the elutriation rate can be found elsewhere in the literature. Only the gist will be presented here so as to understand the basics of the method.

The elutriation rate of the *i*th component from mass M is given by:

$$E_i = \frac{d}{dt}\left(MX_{iB}\right) = -K_E A X_{iB} \tag{1.204}$$

where K is the elutriation rate constant and A is the area of cross-section. From Equation 1.204, $X_{iB,t}$ at any time t in the bed is

$$X_{iB,t} = X_{iB,o} \exp\left(-\frac{K_E A t}{M}\right) \tag{1.205}$$

In other words, the mass of the *i*th component at time t is

$$M_{iB,t} = M_{iB,o}\left[1 - \exp\left(\frac{-K_E A}{M} t\right)\right] \tag{1.206}$$

where $M_{iB,o}$ is equal to $MX_{iB,o}$. The total entrainment (E) at the exit is

$$E = \sum E_i = \sum\left(-K_E\, A\, X_{iB}\right) \tag{1.207}$$

and the solids concentration ($C_{s,E}$) at the exit is

$$C_{s,E} = \frac{E}{AU} \quad \left(\text{kg solids/m}^3\text{gas}\right) \tag{1.208}$$

Thus, knowledge of E is useful in predicting the solids concentration at the exit and also for the design of post-reactor equipment such as cyclones or settling chambers. The other important parameters required to predict the conversion in the freeboard are particle velocity and its residence time. These parameters can be estimated from knowledge of the pressure drop across a section (Δz) in the freeboard. Thus, for a section Δz if ΔP is known, the solid holdup density (ρ_{sz}) is given by:

$$\rho_{sz} = \rho_s(1 - \epsilon) = \frac{\Delta P}{g\, \Delta z} \tag{1.209}$$

The particle residence time (t_z) can be expressed as:

$$t_z = \frac{A\, \Delta Z\, \rho_s}{E_z} \quad \frac{\text{kg}}{\text{kg/s}} \tag{1.210}$$

and the particle velocity as:

$$U_{pz} = \frac{\Delta z}{t_z} \tag{1.211}$$

Knowing that $E_z = C_{sz}AU$ and substituting for t_z in Equation 1.211,

$$U_{PZ} = C_{sZ}\, U/\rho_s \tag{1.212}$$

Equation 1.212 is useful in predicting the particle slip velocity (U_{sl}), which is given by:

$$U_{sl} = \frac{U}{\epsilon} - U_{pz} \tag{1.213}$$

c. Kunii and Levenspiel Method

The mechanism of entrainment discussed above is based on bubble dynamics. An entirely different method of evaluating entrainment, as proposed by Kunii and Levenspiel,[326] can also be used. This model was developed based on the mass balance of three distinct phases: (1) gas stream with dispersed solids (phase 1) moving upward and receiving solids from descending as well as ascending agglomerates, (2) ascending agglomerates (phase 2) losing solids by transfer to an upward-moving gas stream (phase 1) and also to downward-moving agglomerates, and (3) descending agglomerates (phase 3) gaining solids from phase 2 and losing solids to phase 1. According to this model, under normal entrainment, solids are elutriated from the bed and also returned to it. The entrainment rate is

$$E = E_o e^{-aH} \tag{1.214}$$

where E_o is the entrainment at $H = 0$ (i.e., just above the bed surface) and the constant a is a function of the rate coefficients for the transfer of solids among the three phases and the associated phase velocities. Both E_o and H can be computed for a specific system from a plot of ln E versus H.

Kunii and Levenspiel[327] later analyzed their complex model and evaluated the decay constant a by using data from the literature. They also attempted to explain the fast fluidization phenomenon by using their predictive model and presented several sample calculations of their model for practical applications.

d. Complexity of Parameter Determination

In order to evaluate the elutriation rate from Equation 1.204, the elutriation rate constant (K_E) should be known. Geldart[324] presented a list of correlations in this regard. A simple correlation in a dimensionless form, as proposed by Geldart,[324] is

$$\frac{K_E}{\rho_g U} = 23.7 \exp\left[-5.4 \frac{U_t}{U}\right] \tag{1.215}$$

Correlation 1.215 is picked up from the literature only to demonstrate the simplicity of its use. However, the limited range of applicability of this correlation makes designers aware of their restricted choice. The reason why numerous correlations have been proposed to date lies in the complex nature of the parameters that influence entrainment or elutriation. The elutriation rate is affected by the gas flow rate, the size distribution of the solids, the column diameter, and the freeboard height. The effects of several other parameters, such as particle shape, fluid viscosity, and particle surface morphology, have not yet been systematically studied. The effect of the bed diameter also has yet to be clearly established, although it is known that the quality of fluidization is affected by the bed diameter. A small bed diameter can give rise to slugs and can thus increase the elutriation rate. As far as gas dispersion is concerned, elutriation with a dispersed gas is expected to be high. With good quality fluidization, where the gas distribution is appropriate and there are several small gas bubbles, elutriation rates were found to be low.[328] The use of mechanical stirrers and inserts inside the fluidized bed was found to reduce the elutriation rate.[329]

4. *Transport Disengaging Height*

The transport disengaging height (TDH) can be predicted theoretically if the particle velocity above the bed surface is known. The velocity (V) of a spherical particle, when traveling upward in the bed without collision in a gas whose average velocity[323] is U, is given by:

$$V = V_o e^{-\alpha t} - (U_t - U)t \tag{1.216}$$

where $\alpha = 18\mu/\rho_p d_p^2$ and V_o is the velocity at the surface. The height above the bed surface at which the particle would rise in time t is

$$z = \int_0^t V\,dt = \left(1 - e^{-\alpha t}\right)\left(V_o + U_t - U\right) - \left(U_t - U\right)t \tag{1.217}$$

Because the TDH is the height at which coarse particles (where $U_t > U$) return to the bed, z = TDH when $U = 0$ and $t = \tau$. From Equation 1.216, at $V = 0$, τ can be obtained. Using $t = \tau$, in Equation 1.217, z = TDH, i.e.,

$$\mathrm{TDH} = \frac{1}{\alpha}\left[V_o - \left(U_t - U\right)\,\ln\left(1 + \frac{V_o}{U_t - U}\right)\right] \tag{1.218}$$

In order to use Equation 1.218,[323] V_o should be known, and one must depend on data available in the literature for this. Empirical correlations are usually the final choice.

TDH can be thought of in two alternative ways, and this leads to two types of TDH. One is TDH(C) at which coarse particles where $U_t > U$ fall back to the bed, and the other is the height TDH(F) above which entrainment remains constant.

Although both types of TDH apparently seem to have the same value, this is not so in a true sense. Hence, the method of evaluating TDH should always be mentioned if a correlation is proposed for it.

For Group D particles, Soroko et al.[330] proposed the following correlation:

$$\mathrm{TDH(C)} = 1200\, H_o\, \mathrm{Re}_p^{1.55} \mathrm{Ar}^{-1.1} \tag{1.219}$$

for $15 < \mathrm{Re}_p < 300$, $19.5 < \mathrm{Ar} < 6.5 \times 10^5$, $H_o < 0.5$ m.

For Group A particles, the correlation proposed by Horio et al.[331] is

$$\mathrm{TDH(F)} = 4.47\, De_s^{1/2} \tag{1.220}$$

where De_s is the equivalent bubble diameter at the surface. Zenz and Weil[296] provided a graph to compute TDH at various vessel diameters for different operating velocities; their predictions show that TDH(F) increases with increasing bed vessel diameter, and no correlation appears to account for this effect.

5. Some Useful Remarks

We have seen that several mechanisms and methods have been made use of in connection with the evaluation of entrainment. The ultimate goal is to estimate the total entrainment outside the reactor so that post-reactor equipment can be properly designed for efficient gas–solid separation. One of the important applications of data on entrainment is prediction of the conversion in the freeboard region. The solids dispersed in the gas phase in the freeboard can exhibit significant conversion if the bed contains a major fraction of fines. When the reaction is highly exothermic, it is necessary to provide adequate cooling or heat control in the freeboard, and knowledge of solids entrainment is required for this purpose. From the preceding discussion on entrainment, a design engineer can appreciate that estimating the entrainment rate is essential and cannot be neglected in the final design of an efficient reactor. However, it is clear that much work remains to be carried out in this area. The subject of entrainment has been viewed mainly in the context of gas–solid systems. For liquid–solid systems, the mechanisms described earlier cannot automatically be assumed to be valid. In many mineral operation systems, liquid and gas fluidization are used for physical separation. The field of entrainment or elutriation has considerable relevance for such applications. In the context of three-phase fluidization, the subject of elutriation is in the process of evolving and is gaining importance in light of the biotechnological applications of this type of fluidization.

NOMENCLATURE

A	area of cross-section (m^2)
a	constant in Equation 1.214
a_I	interfacial surface area (m^2/m^3)

A_1, B_1	constants in Equation 1.68
A_p	projected surface area of the particle (m^2)
Ar	Archimedes number, $d\rho_s^3\, \rho_f\, (\rho_s - \rho_f) g/\mu_g^2$ (–)
B	discharge rate (kg/m^2)
c_1, c_2	constants in Equation 1.34
C	cohesion constant (N/m^2)
C_1	inlet concentration (mol/l)
C_2	outlet concentration (mol/l)
C_3	constant in Equation 1.37
C_7	equilibrium concentration (mol/l)
C_{bo}	reactant concentration at the distance $z = 0$ (mol/m^3)
C_{bt}	concentration in bubble phase at elevation t (mol/m^3)
C_D	drag coefficient (–)
C_g	reactant concentration in the liquid (mol/l)
C_{gi}	concentration of the gas at the interphase (mol/l)
C_l	reactant concentration in the liquid (mol/l)
C_p	gas concentration in the emulsion phase (mol/m^3)
C_{pg}	specific heat of gas at constant pressure (J/kg · K)
C_{pl}	specific heat of liquid (J/kg · K)
C_{ps}	specific heat of solid particle (J/kg · K)
$C_{S,E}$	solid concentration at the exit of the column (kg)
D	molecular diffusion coefficient (m^2/s)
d	diameter of a spherical particle (m)
d^*	dimensionless particle diameter in Equation 1.73
d_b	bubble diameter (m)
d_{bo}	initial bubble diameter (m)
D_e	equivalent diameter (m)
De_s	equivalent bubble diameter at the surface (m)
D_G	diffusion coefficient (m^2/s)
d_H	hydraulic diameter of the channel (m)
D_L	diffusion coefficient in liquid (m^2/s)
d_p	particle diameter (m)
$(d_p)_{AM}$	arithmetic mean particle diameter (m)
$(d_p)_{GM}$	geometric mean particle diameter (m)
$(d_p)_{HM}$	harmonic mean particle diameter (m)
$(d_p)_{LM}$	length mean particle diameter (m)
$(d_p)_{SM}$	surface mean particle diameter (m)
$(d_p)_{VS}$	volume surface diameter (m)
$(d_p)_{WM}$	weight mean particle diameter (m)
d_{pi}	diameter of particles in ith section (m)
d_{pm}	mean diameter (m)
d_{sv}	surface volume diameter (m)
D_t	diameter of tube or column or vessel (m)
d_v	volume diameter (m)
E	elutriation rate (kg/m^2 · s)

e	energy per unit mass (J/kg)
E_b	emissivity factor of radiating/receiving bodies (–)
E_{bs}	generalized emissivity factor (–)
E_l	liquid-phase dispersion coefficient (m^2/s)
E_M	elasticity modulus of powder bed (N/m^2)
E_o	entrainment just above the bed surface ($kg/m^2 \cdot s$)
E_p	emissivity of the particle (–)
E_s	emissivity of the surface (–)
E_s	emissivity factor due to shape (–)
F	feed rate (kg/m^2)
f	bed friction factor (–)
F_B	force due to buoyancy (N)
f_b	fraction of surface shrouded by gas bubbles (–)
F_D	drag force (N)
F_e	adhesion force transmitted in particle contact (N)
f_f	friction for fluidized bed (–)
F_f	parameter dependent on fines fraction, Equation 1.117 (–)
F_g	force due to gravity (N)
f_k	friction factor for packed bed (–)
f_L	fractional bubble-phase contact time (–)
Fr_{mf}	($= U_{mf}^2/d_p g$), Froude number at minimum fluidization (–)
Fr_t	Froude number based on U_t and D_t, $U_t/\sqrt{(gD_t)}$ (–)
f_w	fraction of wake volume (–)
g	gravitational constant (m^2/s)
g_c	conversion factor, $g_c = 1\ kg \cdot m/N \cdot s^2 = 32.2\ lb \cdot ft/lbf \cdot s^2 = 9.8\ kg \cdot m/kg \cdot wt \cdot s^2$ (–)
H	height up to the bed surface (m)
h	overall heat transfer coefficient ($W/m^2 \cdot K$)
h	bed height below H (m)
h_B	heat transfer coefficient in the bubble-covered surface ($W/m^2 \cdot K$)
h_c	convective heat transfer coefficient ($W/m^2 \cdot K$)
h_f	film heat transfer coefficient ($W/m^2 \cdot K$)
h_{gc}	gas-convective component of heat transfer ($W/m^2 \cdot K$)
h_{gp}	gas-to-particle heat transfer coefficient ($W/m^2 \cdot K$)
H_{gsl}	height occupied by gas–solid–liquid phases (m)
h_i	instantaneous heat transfer coefficient ($W/m^2 \cdot K$)
h_{max}	maximum heat transfer coefficient ($W/m^2 \cdot K$)
h_{max}	maximum heat transfer coefficient corresponding to U_{opt} ($W/m^2 \cdot K$)
h_{mf}	heat transfer coefficient at U_{mf} ($W/m^2 \cdot K$)
H_{mf}	bed height at U_{mf} (m)
H_o	static bed height (m)
h_p	particle-convective component of heat transfer coefficient ($W/m^2 \cdot K$)
h_{pc}	convective heat transfer coefficient due to particle (–)
h_r	radiative component of heat transfer coefficient ($W/m^2 \cdot K$)
h_w	bed-to-wall heat transfer coefficient ($W/m^2 \cdot K$)
i	ith component (–)

J^*	dimensionless fluid flux (–)
j_{fo}	fluid flux (superficial velocity) relative to particles (m/s)
j_H	Colburn's factor (= Nu/Re $Pr^{1/3}$)
j_M	Colburn j factor for mass transfer (–)
K	constant in Equation 1.42 (–)
K_1	constant in Equation 1.27 (–)
K_2	constant in Equation 1.28 (–)
K_3	constant in Equation 1.29 (–)
k	thermal conductivity (W/m · K)
k_5, k_6, k_7	constants in Equation 1.56 (–)
K_c	overall mass transfer coefficient (m/s)
k_c	effective thermal conductivity (W/m · K)
K_d	mass transfer coefficient (m/s)
K_E	elutriation rate constant (–)
k_e^f	conductivity due to turbulent diffusion (–)
k_e^o	effective thermal conductivity of the bed with no fluid (W/m · K)
k_{ew}	effective thermal conductivity near wall zone (W/m · K)
K_G	average mass transfer coefficient (m/s)
k_g	thermal conductivity of gas (W/m · K)
K_g	gas-side mass transfer coefficient in a gas–liquid mass (–)
K_{gw}	mass transfer coefficient between fluidized bed (m/s)
K_I	constant as defined by Equation 1.97 (–)
K_l	liquid-side mass transfer coefficient in a gas–fluid mass transfer (m/s)
k_l	thermal conductivity of liquid (W/m · K)
K_{la}	volumetric mass transfer coefficient (s^{-1})
K_{mb}	constant in Equation 1.116 (–)
K_N	Newton's constant, Equation 1.96 (–)
k_p	mass transfer from particle to particulate bed or spherical particle, Equation 1.170 (s)
k_{pe}	permeability (m^2)
K_{pl}	mass transfer between particle and fluid (s)
k_s	thermal conductivity of solid (W/m · K)
k_s	solid–solid mass transfer coefficient (–)
K_{ST}	Stokes' constant, Equation 1.93 (–)
K_w	ratio of wake voidage to gas-phase voidage (–)
L	length of vertical surface (m)
M	mass of solid (kg)
m	Henry constant (atm/mol fraction) (m/s)
m_e	constant in Equation 1.148 (–)
N	fluidization number (= $\rho_s^3 d_p^4 g^2/\mu^2 E$), Equation 1.39 (–)
N_b	number of bubbles (–)
N_{coh}	cohesion number ($C/\rho_s g d_p$) (–)
N_f	flux of the material transferred from the gas to the liquid phase (mol/s)
n_i	number of particles of diameter d_{pi} (m)

Nu	Nusselt number based on h (–)
Nu_{gc}	Nusselt number for gas convection, Equation 1.153 (–)
Nu_{max}	Nusselt number at h_{max} (–)
Nu_p	Nusselt number based on particle diameter, d_p, hd_p/k (–)
p	constant in Equations 1.167 and 1.169 (–)
Pr	Prandtl number (C_p μ/kg) (–)
Q	volumetric exchange rate of gas between a single bubble and dense phase (m^3/s)
q	volumetric exchange rate due to convective flow of gas (m^3/s)
q_r	radiant heat flux (J/m^2)
$(Q - Q_{mf})_c$	critical excess flow of gas (m^3/s)
r^*	dimensionless particle size (–)
Re	particle Reynolds number, $d_p U_f \rho_f / \mu_f$ (–)
Re_c	Reynolds number based on U_c (–)
Re_l	Reynolds number based on liquid velocity (U_l) (–)
Re_L	Reynolds number based on linear dimension (–)
Re_{md}	Reynold number based on U_{md} (–)
Re_{mf}	Reynolds number at minimum fluidization velocity (U_{mf}) (–)
Re_{ms}	Reynolds number at the onset of slugging (–)
Re_{opt}	Reynolds number based on U_{opt} (–)
Re_p	Reynolds number based on particle diameter (d_p) (–)
Re_s	Reynolds number defined as $2j^*r^*$ (–)
Re_t	Reynolds number based on U_t (–)
Re_{tr}	Reynolds number based on U_{tr} (–)
R_r	mean thermal resistance of the packet ($m^2 \cdot K/W$)
R_w	contact resistance near the wall zone ($m^2 \cdot K/W$)
S	surface area per unit volume (m^2/m^3)
S_b	frontal surface area of a spherical bubble (m^2)
Sc	Schmidt number, $\mu\rho_g/D$ (–)
Sh	Sherwood number, $k_{pl}d_p/D$ (–)
S_i	surface area per unit weight of *i*th component (m^2/kg)
S_o	surface area of a spherical particle (m^2)
S_p	surface area of particle of diameter d_p (m^2)
S_v	surface area of a nonspherical particle of constant volume (m^2)
S_w	surface area per unit weight of the particles (m^2/kg)
St	Stanton number, $k_{gw}\,\epsilon/u$ (–)
t	time (s)
T_b	fluidized bed temperature (K)
T_e	intermediate temperature (K)
T_g	temperature of gas (K)
T_p	temperature of particle (K)
T_s	temperature of the surface (K)
T_w	wall temperature (K)
t_z	particle residence time (s)
U	superficial gas velocity (m/s)

u	fluid velocity (m/s)
U^*	dimensionless particle velocity (Equation 1.72) (–)
U_A	absolute bubble velocity (m/s)
U_{AV}	mean average bubble rise velocity (m/s)
U_c	transition velocity from bubbling to turbulent (m/s)
U_f	fluid velocity (m/s)
U_g	gas velocity (m/s)
U_{gl}	gas–liquid slip velocity (m/s)
U_l	liquid velocity (m/s)
u_l	particle fluid velocity in a suspension (m/s)
U_{lmf}^{ls}	minimum fluidization velocity for liquid in a liquid–solid (two-phase) system (m/s)
U_{mb}	minimum bubbling velocity (m/s)
U_{md}	velocity of transition from dense phase to dilute phase (m/s)
U_{mf}	minimum fluidization velocity (m/s)
U_{op}	optimum gas velocity (m/s)
U_{opt}	optimum operating velocity corresponding to h_{max} (m/s)
U_p	particle velocity (m/s)
U_r	relative velocity (m/s)
U_{sl}	particle slip velocity (m/s)
U_t	single-particle terminal velocity (m/s)
U_{tr}	transport velocity (m/s)
V	particle velocity (m/s)
V_ϵ	pore volume per unit mass (m^3/kg)
V_b	volume of bubble (m^3)
V_f	volume of the floating bubble breakers (m^3)
V_o	particle velocity at the bed surface (m/s)
V_p	volume of particle (m^3)
V_s	volume of solid particle (m^3)
V_{sv}	volume per unit mass of the particle (m^3/kg)
W	weight (kg)
W_s	solids entrained per unit area (kg/m^2)
W/A	bed weight per unit area (kg/m^2)
We_p	Weber number ($= \rho_s U_s^2 d_p / \sigma_{gl}$) (–)
x	solids weight fraction (–)
x	ratio of the solid holdup in the wake region to that in liquid fluidized bed region (–)
X_e	exit mole fraction (–)
X_i	volume fraction (m)
X_{iB}	mass fraction of ith component in discharged solid (–)
X_{iE}	mass fraction of ith component in elutriated solid (–)
X_{iF}	mass fraction of ith component in feed (–)
X_o	inlet mole fraction (–)
z	distance above distributor (m)
@	powder number (Equations 1.34 and 1.37)

Greek Symbols

α	correction factor in Equation 1.43
α_f	angle of internal fraction (°)
α_1	ratio of the mass velocity of the fluid in the direction of heat flow to the superficial mass velocity in the direction of fluid flow (–)
β	constant in Equation 1.52 (–)
β_1	ratio of distance between two adjacent particles to the mean particle diameter (–)
γ	angle of wall friction (°)
$\Delta\rho$	difference in density of solid and gas (kg/m³)
$\Delta\rho_\sigma$	pressure drop due to surface tension (Pa · s)
Δc	concentration driving force (mol/m³)
ΔC_{lm}	log mean concentration difference (mol/l)
ΔP_b	bed pressure drop (Pa · s)
ΔT	mean temperature difference (K)
ΔT_i	mean temperature difference across the *i*th section (K)
δ	angle of rupture (°)
δ_w	film thickness on the wall (m)
ϵ	bed voidage (–)
ϵ_B	voidage due to bubble phase (–)
ϵ_b	bubble volume fraction (–)
ϵ_g	holdup due to gas per unit volume of three-phase fluid bed (–)
ϵ_{gl}	holdup of gas in gas–liquid column per unit volume of two-phase fluidized bed (–)
ϵ_I	intraparticle voids due to pores, cracks, etc. (kg/m³)
ϵ_l	liquid holdup (–)
ϵ_{mf}	bed voidage at minimum fluidization velocity (–)
ϵ_o	static bed porosity (–)
ϵ_{opt}	voidage at optimum velocity (U_{opt}) (–)
ϵ_p	solid volume fraction (–)
ϵ_s	solids volume fraction (–)
θ	angle of repose (°)
θ_D	dense-phase root square average residence time (s)
λ	shape factor (–)
μ	viscosity (Pa · s)
μ_{eff}	effective viscosity (Pa · s)
μ_f	viscosity of fluid (Pa · s)
μ_g	viscosity of gas (Pa · s)
μ_l	viscosity of liquid (Pa · s)
ρ_a	apparent density (kg/m³)
ρ_B	bulk density of the bed material (kg/m³)
ρ_b	bed density (kg/m³)
ρ_{bmf}	bulk density of the bed at U_{mf} (kg/m²)
ρ_e	density of emulsion phase (kg/m³)
ρ_f	density of fluid (kg/m³)

ρ_g	density of gas (kg/m^3)
ρ_l	density of liquid (kg/m^3)
ρ_p	particle density (kg/m^3)
ρ_s	density of solid particle (kg/m^3)
ρ_{sz}	solid holdup density at an elevation of z (kg/m^3)
ρ_t	theoretical density (kg/m^3)
σ	surface tension (N/m)
ϕ	sphericity of a particle (–)
ω	angle of slide (°)

REFERENCES

1. Zenz, F.A. and Othmer, D.F., *Fluidization and Fluid Particle Systems,* Van Nostrand-Reinhold, Princeton, NJ, 1960, 113.
2. Barnea, E. and Mednick, L., Correlation for minimum fluidization velocity, *Trans. Inst. Chem. Eng.*, 53, 278, 1975.
3. Zabrodsky, S.S., *Hydrodynamics and Heat Transfer in Fluidized Beds*, MIT Press, Cambridge, MA, 1966, chap. 1.
4. Geldart, D., Estimation of basic particle properties for use in fluid particle process calculations, *Powder Technol.*, 60, 1, 1990.
5. Abrahamsen, A.R. and Geldart, D., Behavior of gas fluidized beds of fine powders. I. Homogeneous expansion, *Powder Technol.*, 26, 35, 1980.
6. Wadell, H., Volume shape and roundness of rock particles, *J. Geol.*, 15, 443, 1982.
7. Leva, M., *Fluidization*, McGraw-Hill, New York, 1959, chap. 34.
8. Brown, G.G., Katz, D., Foust, A.S., and Schneidewind, R., *Unit Operation*, John Wiley & Sons, New York, 1950, 77.
9. Furnas, C.C., Grading aggregates. I. Mathematical relations for beds of broken solids of maximum density, *Ind. Eng. Chem.*, 23, 1052, 1931.
10. Ciborowski, J., *Fluidyzacja*, PWT, Warsaw, 1957.
11. Geldart, D., The effect of particle size and size distribution on the behaviour of gas fluidized beds, *Powder Technol.*, 6, 201, 1972.
12. Geldart, D., Types of fluidization, *Powder Technol.*, 7, 285, 1973.
13. Molerus, O., Interpretation of Geldart's Type A, B, C, and D powders by taking into account interparticle cohesion forces, *Powder Technol.*, 33, 81, 1982.
14. Mathur, K.B., Spouted bed, in *Fluidization*, Davidson, J.F. and Harrison, D.H., Eds., Academic Press, London, 1971, chap. 17.
15. Clark, N.N., Van Egmond, J.W., and Turton, R., A numerical representation of Geldart's classification, *Powder Technol.,* 56, 225, 1968.
16. Rietema, K., Powders, what are they? *Powder Technol.,* 37, 20, 1984.
17. Saxena, S.C. and Ganzha, V.L., Heat transfer to immersed surfaces in gas fluidized beds of large particles and powder characterization, *Powder Technol.,* 39, 199, 1984.
18. Leva, M., *Fluidization*, McGraw-Hill, New York, 1959, 64.
19. Miller, Co. and Logwinuk, A.K., Fluidization studies of solid particles, *Ind. Eng. Chem.*, 41, 1135, 1949.
20. Rowe, P.N. and Yacono, C., The distribution of bubble size in gas fluidized beds, *Trans. Inst. Chem. Eng.*, 53, 59, 1975.
21. Davidson, J.F. and Harrison, D., *Fluidized Particles*, Cambridge Press, London, 1963.
22. Yacono, C., Ph.D. thesis, University of London, 1975.

23. Baerg, A., Klassen, J., and Gishler, P.E., Heat transfer in a fluidized solids beds, *Can. J. Res.*, F28, 287, 1950.
24. Kumar, A. and Sen Gupta, P., Prediction of minimum fluidization velocity for multicomponent mixtures, *Indian J. Technol.*, 12(5), 225, 1974.
25. Baeyens, J. and Geldart, D., Predictive calculations of flow parameters in gas fluidized beds and fluidization behaviour of various powders, in Proc. Int. Symp. on Fluidization and Its Application, Toulouse, France, 1973, 263.
26. Leva, M., Shirai, T., and Wen, C.Y., La prevision due debut de la fluidization daus les lits solides granulaires, *Genie Chim.*, 75(2), 33, 1956.
27. Riba, J.P., Expansion de couches fluidisees par des liqudes, *Can. J. Chem. Eng.*, 55, 118, 1977.
28. Ballesteros, R.L., These de Docteur Ingenieur, Universite de Toulouse, France, 1980.
29. Doichev, K. and Akhmakov, N.S., Fluidization of polydisperse systems, *Chem. Eng. Sci.*, 34, 1357, 1979.
30. Thonglimp, V., Higuily, N., and Leguerie, C., Vitesse minimale decouches fluidize par un gas, *Powder Technol.*, 38, 233, 1984.
31. Morse, R.D., Fluidization of granular solids–fluid mechanics and quality, *Ind. Eng. Chem.*, 41, 117, 1949.
32. Schiller, L. and Naumann, A.N., *Z. Ver. Dtsch. Ing.*, 77, 318, 1935.
33. Wen, C.Y. and Yu, Y.H., Mechanics of fluidization, *Chem. Eng. Prog. Symp. Ser.*, 62(2), 100, 1966.
34. Richardson, J.F. and Zaki, W.N., Sedimentation and fluidization. I, *Trans. Inst. Chem. Eng.*, 32, 35, 1954.
35. Haider, A. and Levenspiel, O., Drag coefficient and terminal velocity of spherical and non-spherical particles, *Powder Technol.*, 58, 63, 1989.
36. Turton, R. and Levenspiel, O., A short note on the drag correlation for sphere, *Powder Technol.*, 47, 83, 1986.
37. Masliyah, J.B. and Polikar, M., Terminal velocity of porous sphere, *Can. J. Chem. Eng.*, 58, 299, 1980.
38. Rowe, P.N. and Henwood, G.A., Drag forces in a hydraulic model of a fluidized bed. I, *Trans. Inst. Chem. Eng.*, 39, 43, 1961.
39. Frantz, J.F., Minimum fluidization velocities and pressure drop in fluidized beds, *Chem. Eng. Prog. Symp. Ser.*, 62, 21, 1966.
40. Davies, L. and Richardson, J.W., Gas interchange between bubbles and the continuous phase in a fluidized bed, *Trans. Inst. Chem. Eng.*, 44(8), 293, 1966.
41. Pillai, B.C. and Raja Rao, M., Pressure drop and minimum fluidization velocities in air fluidized beds, *Indian J. Technol.*, 9, 225, 1971.
42. Ergun, S., Fluid flow through packed columns, *Chem. Eng. Prog.*, 48, 89, 1952.
43. Carman, P.C., Fluid flow through granular beds, *Trans. Inst. Chem. Eng.*, 15, 150, 1937.
44. Bourgeois, P. and Grenier, P., The ratio of terminal velocity to minimum fluidization velocity for spherical particles, *Can. J. Chem. Eng.*, 46, 325, 1968.
45. Ghosal, S.K. and Mukherjee, R.N., Momentum transfer in solid–liquid fluidized beds. 1. Prediction of minimum fluidization velocity, *Br. Chem. Eng.*, 17, 248, 1972.
46. Saxena, S.C. and Vogel, G.J., The measurement of incipient fluidization velocities in a bed of coarse dolomite at high temperature and pressure, *Trans. Inst. Chem. Eng.*, 55, 184, 1977.
47. Babu, S.P., Shah, B., and Talwalkar, A., Fluidization correlations for coal gasification materials–minimum fluidization velocity and fluidized bed expansion, *AIChE Symp. Ser.*, 74, 176, 1978.

48. Richardson, J.F. and Jeromino, M.A.S., Velocity voidage relations for sedimentation and fluidization, *Chem. Eng. Sci.*, 39, 1419, 1979.
49. Thonglimp, V., These Dr. Ingenier, Institut National Polytechnique de Toulouse, France, 1981.
50. Chitester, D.C., Kornosky, R.M., Fan, L.S., and Danko, J.P., Characteristics of fluidization at high pressure, *Chem. Eng. Sci.*, 39, 253, 1984.
51. Masaaki, N., Hamada, Y., Toyama, S., Fouda, A.E., and Capes, C.E., An experimental investigation of minimum fluidization velocity at elevated temperatures and pressures, *Can. J. Chem. Eng.*, 63, 8, 1985.
52. Agarwal, P.K. and O'Neil, B.K., Transport phenomena in multi-particle system. I. Pressure drop and friction factors unifying the hydraulic radius and submerged object approaches, *Chem. Eng. Sci.*, 43, 2487, 1988.
53. Satyanarayana, K. and Rao, P.G., Minimum fluidization velocity at elevated temperatures, *Ind. Chem. Eng.*, 31, 79, 1989.
54. Grace, J.R., Fluidized bed hydrodynamics, in *Handbook of Multiphase Systems*, Hestroni, G., Ed., Hemisphere Press, New York, 1982, chap. 8.1.
55. Panigrahi, M.R. and Murty, J.S., A generalized spherical multiparticle model for particulate systems: fixed and fluidized beds, *Chem. Eng. Sci.*, 46(7), 1863, 1991.
56. Fletcher, J.V., Deo, M.D., and Hanson, F.V., Reexamination of minimum fluidization velocity correlations applied to group B sands and coked sands, *Powder Technol.*, 69, 147, 1992.
57. Lippens, B.C. and Mulder, J., Prediction of the minimum fluidization velocity, *Powder Technol.*, 75, 67, 1993.
58. Kunii, D. and Levenspiel, O., *Fluidization Engineering*, John Wiley & Sons, New York, 1979, chap. 3.
59. Richardson, J.F., Incipient fluidization and particulate system, in *Fluidization*, Davidson, J.F. and Harrison, D., Eds., Academic Press, New York, 1971, chap. 2.
60. Clift, R. and Gauvin, W.H., The motion of particle in turbulent gas streams, *Chemeca 70*, Butterworths, Australia, 1970, 14.
61. Concha, F. and Almendra, E.R., Settling velocities of particulate systems. 1. Settling velocities of individual spherical particles, *Int. J. Miner. Process.*, 5, 349, 1979.
62. Concha, F. and Almendra, E.R., Settling velocities of particulate systems. 2. Settling velocities of suspensions of spherical particles, *Int. J. Miner. Process.*, 6, 31, 1979.
63. Concha, F. and Barrientos, A., Settling velocities of particulate systems. 3. Power series expansion for the drag coefficient of sphere and prediction of the settling velocity, *Int. J. Miner. Process.*, 9, 167, 1982.
64. Concha, F. and Barrientos, A., Settling velocities of particulate systems. 4. Settling of nonspherical isometric particles, *Int. J. Miner. Process.*, 18, 297, 1986.
65. Concha, F. and Christiansen, A., Settling velocities of particulate systems. 5. Settling velocities of suspensions of particles of arbitrary shape, *Int. J. Miner. Process.*, 18, 309, 1986.
66. Geldart, D., Estimation of particle properties for use in fluid particle process calculations, *Powder Technol.*, 60, 1, 1990.
67. Pettyjohn, E.S. and Christiansen, E.B., Effect of particle shape on free settling rates of isometric particles, *Chem. Eng. Prog.*, 44, 157, 1948.
68. McCabe, W.L. and Smith, J.C., *Unit Operations of Chemical Engineering*, McGraw-Hill, New York, 1967, chap. 7.
69. Rubey, W.W., Settling velocities of gravel, sand and silt particles, *Am. J. Sci.*, 225, 325, 1933.

70. Dallavalle, J.M., *Micromeritics*, Pitman Publishing, New York, 1943.
71. Dallavalle, J.M., *Micromeritics, The Technology of Fine Particles*, 2nd ed., Pitman Publishing, New York, 1948, chap. 2.
72. Lapple, C.E., *Fluid and Particle Mechanics*, University of Delaware Publ., Newark, 1951, 284.
73. Torobin, L.B. and Gauvin, W.H., Fundamental aspects of solid–gas flow. Parts 1 and 2, *Can. J. Chem. Eng.*, 37, 129, 1959.
74. Olson, R., *Essentials of Engineering Fluid Mechanics*, International Text Book, Scranton, PA, 1961.
75. Barnea, E. and Mizrahi, J., Generalized approach to the fluid dynamics of particulate systems. I. General correlation for fluidization and sedimentation in solid multiparticle systems, *Chem. Eng. J.*, 5, 171, 1973.
76. Clift, R., Grace, J.R., and Weber, M.E., *Bubble, Drops and Particles*, Academic Press, New York, 1978.
77. Flemmer, R.L.C. and Banks, C.L., On the drag coefficient of a sphere, *Powder Technol.*, 48, 217, 1986.
78. Khan, A.R. and Richardson, J.F., Pressure gradient and friction factor for sedimentation and fluidization of uniform spheres in liquids, *Chem. Eng. Sci.*, 45, 255, 1990.
79. Wallis, G.B., A simple correlation for fluidization and sedimentation, *Trans. Inst. Chem. Eng.*, 55, 74, 1977.
80. Turton, R. and Clark, N.N., An explicit relationship to predict spherical particle terminal velocity, *Powder Technol.*, 53, 127, 1987.
81. Geldart, D., *Gas Fluidization Technology*, Wiley, Chichester, 1986, chap. 6.
82. Lewis, E.W. and Bowerman, E.W., Fluidization of solid particles in liquid, *Chem. Eng. Prog.*, 48, 603, 1952.
83. Lewis, W.K., Gilliland, E.R., and Bauer, W.C., Characteristics of fluidized particles, *Ind. Eng. Chem.*, 41, 1104, 1949.
84. Chen, P. and Pei, D.C.T., Fluidization characteristics of fine particles, *Can. J. Chem. Eng.*, 62, 464, 1984.
85. Lucas, A., Arnaldos, J., Casal, J., and Pulgjaner, L., Improved equation for the calculations of minimum fluidization, *Ind. Eng. Chem. Process Des. Dev.*, 25, 426, 1986.
86. Yang, W.C., Chitester, D.C., Kornos, R.M., and Keairn, D.L., A generalized methodology for estimating minimum fluidization velocity at elevated pressure and temperature, *AIChE J.*, 31, 1086, 1985.
87. Wilhelm, R.H. and Kwauk, M., Fluidization of solid nonvesicular particles, *Chem. Eng. Prog.*, 44, 201, 1948.
88. Romero, J.B. and Johanson, L.N., Factors affecting fluidized bed quality, *Chem. Eng. Prog. Symp. Ser.*, 58(38), 28, 1962.
89. Geldart, D., The fluidized bed as a chemical reactor: a critical review of the first 25 years, *Chem. Ind.*, p. 1474, September 2, 1967.
90. Ho, T.C., Fan, L.T., and Walawender, W.P., The onset of slugging in gas–solid fluidized beds with large particles, *Powder Technol.*, 35, 249, 1983.
91. Lanneau, K.P., Gas solids contacting in fluidized beds, *Trans. Inst. Chem. Eng.*, 38, 125, 1960.
92. Rhodes, M. and Geldart, D., Measurement of axial solid flux variation in the riser of a circulating fluidized bed, in *Circulating Fluidized Bed Technology II*, Basu, P. and Large, P., Eds., Pergamon Press, Oxford, 1988, 155.
93. Lee, G.S. and Kim, S.D., Pressure fluctuations in turbulent fluidized beds, *J. Chem. Eng. Jpn.*, 21, 515, 1988.

94. Lee, G.S. and Kim, S.D., Bed expansion characteristics and transition velocity in turbulent fluidized beds, *Powder Technol.*, 62, 207, 1990.
95. Yang, W.C., Criteria for choking in vertical pneumatic conveying lines, *Powder Technol.*, 35, 143, 1983.
96. Yang, W.C., A mathematical definition of choking phenomenon and a mathematical model for predicting choking velocity and choking voidage, *AIChE J.*, 21, 1013, 1975.
97. Østergaad, K., *Studies of Gas–Liquid Fluidization*, Danish Technical Press, Copenhagen, 1969.
98. Østergaad, K., Three phase fluidization, in *Fluidization*, Davidson, J.F. and Harrison, D. Eds., Academic Press, New York, 1971, chap. 18.
99. Østergaad, K., Three phase fluidization, in *Recent Advances in the Engineering Analysis of Chemically Reacting Systems*, Doraiswamy, L.K., Ed., Wiley Eastern Ltd., New Delhi, 1984, chap. 22.
100. Davidson, J.F., Three phase systems, in *Chemical Reactor Theory — A Review*, Lapidus, L. and Amundson, N.R., Eds., Prentice-Hall, Englewood Cliffs, NJ, chap. 10.
101. Kim, S.D. and Kim, C.H., Three phase fluidized beds: a review, *Hwahak Konghak*, 18(5), 313, 1980.
102. Wild, G., Saberian, M., Schwartz, J.L., and Charpentier, J.C., Gas–liquid–solid fluidized bed reactors, state of the art and industrial perspectives, *Entropie*, 106, 3, 1982; English translation in *Int. Chem. Eng.*, 24(4), 639, 1984.
103. Epstein, N., Three phase fluidization: some knowledge gaps, *Can. J. Chem. Eng.*, 59, 649, 1981.
104. Epstein, N., Hydrodynamics of three phase fluidization, in *Handbook of Fluids in Motion*, Cheremisinoff, N.P. and Gupta, R., Eds., Ann Arbor Science, Ann Arbor, MI, 1983, 1165.
105. Muroyama, K. and Fan, L.S., Fundamentals of gas–liquid–solid fluidization, *AIChE J.*, 31(1), 1, 1985.
106. Fan, L.S., *Gas–Liquid–Solid Fluidization Engineering*, Butterworths, Boston, 1989.
107. Fan, L.S., Muroyama, K., and Chern, S.H., Hydrodynamics of inverse fluidization in liquid–solid and gas–liquid–solid systems, *Chem. Eng. J.*, 24, 143, 1982.
108. Bhatia, V.K. and Epstein, N., Three phase fluidization; a generalized wake model, in *Fluidization and Its Applications*, Angelino, H., Couderc, J.P., Gibert, H., and Laguerie, Co., Cepadues Editions, Toulouse, France, 1974, 380.
109. Dakshinamurthy, P., Subrahmanyam, V., and Rao, J.N., Bed porosities in gas–liquid fluidization, *Ind. Eng. Chem. Process Des. Dev.*, 10, 322, 1971.
110. Begovich, J.M. and Watson, J.S., Hydrodynamic characteristics of three phase fluidized beds, in *Fluidization*, Davidson, J.F. and Keairns, D.L., Eds., Cambridge University Press, Cambridge, 1978, 190.
111. Nikov, I., Grandjean, B.P.A., Carreau, P.J., and Paris, J., Viscosity effect in cocurrent three phase fluidization, *AIChE J.*, 36(10), 1613, 1990.
112. Kato, Y., Uchida, K., and Morooka, S., Liquid hold-up and heat transfer coefficient between bed and wall in liquid–solid and gas–liquid–solid fluidized beds, *Powder Technol.*, 28, 173, 1981.
113. Saberian-Broudjenni, M., Wild, G., Charpentier, J.C., Fortin, Y., Euzen, J.P., and Patoux, R., Contribution to the hydro-dynamics study of gas–liquid–solid fluidized bed reactors, *Int. Chem. Eng.*, 27(3), 423, 1987.
114. Grace, J.R., Fluidization, in *Handbook of Multiphase Systems*, Hestroni, G., Ed., McGraw-Hill–Hemisphere, Washington, D.C., 9, 1982, 8.1.17.
115. Douglas, H.R., Snider, I.W.A., and Tomlinson, G.H., II, The turbulent contact absorber, *Chem. Eng. Prog.*, 59(12), 85, 1963.

116. Visvanathan, C. and Leung, L.S., On the design of a fluidized bed scrubber, *Ind. Eng. Chem. Process Des. Dev.*, 24, 677, 1985.
117. Barile, R.G., Dengler, J.L., and Hertwig, T.A., Performance and design of a turbulent bed cooling tower, *AIChE Symp. Ser.*, 70, 154, 1975.
118. Gelperin, N.I., Savchenko, V.I., and Grishko, V.Z., Some hydrodynamics laws of absorption apparatus packed with fluidized spheres, *Theor. Found. Chem. Eng.*, 2, 65, 1968.
119. Kito, M., Tabei, K., and Murata, K., Gas and liquid holdups in mobile bed under the countercurrent flow of air and liquid, *Ind. Eng. Chem. Process Des. Dev.*, 17, 568, 1978.
120. Wozniak, M., Pressure drop and effective interfacial area in a column with a mobile bed, *Int. Chem. Eng.*, 17, 553, 1977.
121. Draijer, W., Heat transfer in fluidized bed-boilers, in *Fluidzed Bed Combustion,* Radovanovic, M., Ed., Hemisphere Publishing, Washington, D.C., 1986, 211.
122. Saxena, S.C., Grewal, N.S., Gabor, J.D., Zabrodsky, S.S., and Galershtein, D.M., Heat transfer between a gas fluidized bed and immersed tubes, in *Advances in Heat Transfer*, Vol. 14, Irvine, T.F. and Hartnett, J.P., Eds., Academic Press, New York, 1978, 149.
123. Zabrodsky, S.S., *Hydrodynamics and Heat Transfer in Fluidized Beds*, MIT Press, Cambridge, MA, 1966, chap. 7–10.
124. Botterill, J.S.M., *Fluid Bed Heat Transfer*, Academic Press, London, 1975.
125. Ranz, W.E. and Marshall, W.R., Evaporation from drops. I, *Chem. Eng. Prog.*, 48, 141, 1952.
126. Chen, P. and Pei, D.C.T., Fluid particle heat transfer in fluidized beds of mixed-sized particles, *Can. J. Chem. Eng.*, 62, 469, 1984.
127. Gutfinger, C. and Abuaf, N., in *Advances in Heat Transfer*, Vol. 10, Hartnett, J.P. and Irvine, T.F., Jr., Eds., Academic Press, New York, 1974, 167.
128. Bartel, W.J., Heat transfer from a horizontal bundle of tubes in an air fluidized bed, Ph.D. thesis, Montana State University, Bozeman, 1971.
129. Saxena, S.C. and Gabor, J.D., Mechanism of heat transfer between a surface and a gas fluidized bed for combustion application, *Prog. Energy Combust. Sci.*, 7(2), 73, 1981.
130. Van Heerden, C., Nobel, A.P., and van Krevelen, D., Studies on fluidization. I. The critical mass velocity, *Chem. Eng. Sci.*, 1, 37, 1951.
131. Ziegler, E.N. and Brazelton, W.T., Mechanism of heat transfer to a fixed surface in a fluidized bed, *Ind. Eng. Chem. Fund.*, 3(2), 94, 1964.
132. Botterill, J.S.M., The mechanism of heat transfer to fluidized beds, in *Proc. Int. Symp. on Fluidization*, Drinkenburg, A.A.H., Ed., Netherlands University Press, Amsterdam, 1967, 183.
133. Yamazaki, R. and Jimbo, G., Heat transfer between fluidized beds and heated surfaces, *J. Chem. Eng. Jpn.*, 3(1), 44, 1970.
134. Wicke, E. and Fetting, F., Warmeubertragung in GaswirbelSchichten, *Chem. Ing. Tech.*, 26, 301, 1954.
135. Mickley, H.S. and Fairbanks, D.F., Mechanism of heat transfer to fluidized beds, *AIChE J.*, 1, 374, 1965.
136. Botterill, J.S.M. and Williams, J.R., The mechanism of heat transfer to fluidized beds, *Trans. Inst. Chem. Eng.*, 41, 217, 1963.
137. Botterill, J.S.M. and Butt, M.H.D., Achieving high heat transfer rates in fluidized beds, *Br. Chem. Eng.*, 13(7), 1000, 1968.

138. Gabor, J.D., Wall to bed heat transfer in fluidized beds and packed beds, *Chem. Eng. Prog. Symp. Ser.*, 66(105), 76, 1970.
139. Zabrodsky, S.S., *Hydrodynamics and Heat Transfer*, MIT Press, Cambridge, MA, 1966, chap. 8.
140. Selzer, V.W. and Thomson, W.J., Fluidized bed heat transfer — the packet theory revisited. *AIChE Symp. Ser.*, 73(161), 29, 1977.
141. Xavier, A.M. and Davidson, J.F., Convective heat transfer in fluidized bed, in *Fluidization*, 2nd ed., Davidson, J.F., Clift, R., and Harrison, D., Eds., Academic Press, London, 1985, chap. 13.
142. Krupiczka, R., Analysis of thermal conductivity in granular materials, *Int. Chem. Eng.*, 7, 122, 1967.
143. Ranz, W.E., Friction and transfer coefficients for single particles and packed beds, *Chem. Eng. Prog.*, 48(3), 247, 1952.
144. Baron, T., Generalized graphical method for the design of fixed bed catalytic reactors, *Chem. Eng. Prog.*, 48(3), 118, 1952.
145. Baskakov, A.P., Vitt, O.K., Kirakosyan, V.A., Maskaev, V.K., and Filippovsky, N.F., Investigation of heat transfer coefficient pulsations and of the mechanism of heat transfer from a surface immersed into a fluidized bed, in *Proc. Int. Symp. on Fluidization and Its Applications*, Cepadues Editions, Toulouse, France, 1974, 293.
146. Botterill, J.S.M. and Denloye, A.O.O., Bed to surface heat transfer in a fluidized bed of large particles, *Powder Technol.*, 19, 197, 1978.
147. Gelprin, N.I. and Einstein, V.G., Heat transfer in fluidized beds, in *Fluidization*, Davidson, J.F. and Harrison, D., Eds., Academic Press, New York, 1971, chap. 10.
148. Baskakov, A.P., *Process of Heat and Mass Transfer in Fluidized Beds*, Metallurgia, Moscow, 1978, 184.
149. Kunii, D. and Levenspiel, O., Eds., *Fluidized Engineering*, John Wiley & Sons, New York, 1968, chap. 9.
150. Botterill, J.S.M., Ed., *Fluid Bed Heat Transfer*, Academic Press, London, 1975, 81.
151. Zabrodsky, S.S., Ed., *Hydrodynamics and Heat Transfer in Fluidized Beds*, MIT Press, Cambridge, MA, 1966, chap. 10.
152. Gelprin, N.I. and Einstein, V.G., Heat transfer in fluidized beds, in *Fluidization*, Davidson, J.F. and Harrison, D., Eds., Academic Press, New York, 1971, chap. 10.
153. Grewal, N.S., Experimental and Theoretical Investigations of Heat Transfer Between a Gas Solid Fluidized Bed and Immersed Tubes, Ph.D. thesis, University of Illinois at Chicago, 1979.
154. Gutfinger, C. and Abuaf, N., Heat transfer in fluidized beds, in *Advances in Heat Transfer*, Vol. 10, Irvine, T.F., Jr. and Hartnett, J.P., Eds., Academic Press, New York, 1964, 167.
155. Zabrodsky, S.S., Antonishin, V.N., and Paranas, A.L., On fluidized bed surface heat transfer, *Can. J. Chem. Eng.*, 54, 52, 1976.
156. Grewal, N.S. and Saxena, S.C., Effect of surface roughness on heat transfer from horizontal immersed tubes in a fluidized bed, *Heat Transfer*, 101(3), 397, 1979.
157. Vreedenberg, H., Heat transfer between a fluidized bed and a horizontal tube, *Chem. Eng. Sci.*, 1(1), 52, 1958.
158. Vreedenberg, H., Heat transfer between a fluidized bed and verticle tube, *Chem. Eng. Sci.*, 11, 274, 1960.
159. Wender, L. and Cooper, G.T., Heat transfer between fluidized bed and boundary surfaces: correlation data, *AIChE J.*, 1(1), 15, 1958.

160. Baskakov, A.P., Berg, B.V., Vitt, O.K., Fillippovsky, N.F., Kiroskosyan, V.A., Goldobin, J.M., and Maskaev, V.K., Heat transfer to objects immersed in fluidized beds, *Powder Technol.*, 8, 273, 1973.
161. Dahloff, B. and Von Brachel, H., Warmeubergang, and horizontal cenem fluidat bett ange ordneten rohrbundelu, *Chem. Ing. Tech.*, 40, 373, 1968.
162. Noak, R., Lokaler Warmeubergang, an horizontalen Fohren in Wirbelschichten, *Chem. Ing. Tech.*, 42(6), 371, 1970.
163. Blenke, H. and Neukirchem. B., Gastaltung horizontaler rohrbundel in gas-wirbelschichtreaktorn kach warmetchnischen, *Chem. Ing. Tech.*, 45(5), 307, 1973.
164. McClaren, J. and Williams, D.F., Combustion efficiency, sulphur retention and heat transfer in pilot fluidized bed combustion, *J. Inst. Fuel*, 42, 303, 1969.
165. Lese, J.K. and Kermode, R.I., Heat transfer from a horizontal tube to a fluidized bed in the presence of unheated tubes, *Can. J. Chem. Eng.*, 50, 44, 1972.
166. Bartel, W.J. and Genetti, W.F., Heat transfer from a horizontal bundle of bare and finned tubes in an air fluidized bed, *Chem. Eng. Prog. Symp.*, 69(128), 85, 1973.
167. Andeen, B.R. and Glicksman, L.R., Heat Transfer to Horizontal Tubes in Shallow Fluidized Beds, Heat Transfer Conf. Paper No. 76 H-67, ASME-AIChE., St. Louis, August 9–11, 1976.
168. Petrie, J.C., Freeby, W.A., and Buckham, J.A., In bed exchangers, *Chem. Eng. Prog.*, 64(67), 45, 1968.
169. Zabrodsky, S.S., *Hydrodynamics and Heat Transfer in Fluidized Beds*, MIT Press, Cambridge, MA, 1966, 270.
170. Genetti, W.E., Sehmall, R.A., and Grimmett, E.S., The effect of tube orientation on heat transfer with bare and finned tubes in a fluidized bed, *Chem. Eng. Prog. Symp. Ser.*, 67(116), 90, 1971.
171. Ternovskaya, A.N. and Korenbrg, Y.G., *Pyrite Kilning in a Fluidized Bed*, Izd. Khimiya, Moscow, 1971.
172. Saxena, S.C. and Grewal, N.S., Heat transfer between a horizontal tube and a gas solid fluidized bed, *Int. J. Heat Mass Transfer*, 23, 1505, 1980.
173. Sarkits, V.B., *Heat Transfer from Suspended Beds of Granular Materials to Heat Transfer Surfaces* (Russian), Leningrad Technol. Inst., Lensovieta, 1959.
174. Miller, C.O. and Logwink, A.K., Fluidization studies of solid particles, *Ind. Eng. Chem.*, 43, 1220, 1951.
175. Wicke, E. and Hedden, F., Stromungsformen and warmeubertragung in von luft aufgewirbelten schuttgutschichten, *Chem. Ing. Tech.*, 24(2), 82, 1952.
176. Dow, W.M. and Jakob, M., Heat transfer between a vertical tube and a fluidized bed air mixture, *Chem. Eng. Prog.*, 47, 637, 1951.
177. Gaffney, B.J. and Drew, T.B., Mass transfer from packing to organic solvents in single phase flow through a column, *Ind. Eng. Chem.*, 42, 1120, 1950.
178. Traber, P.G., Pomaventsev, V.M., Mukhenov, I.P., and Sarkits, V.B., Heat transfer from a suspended layer of catalyst to the heat transfer surface, *Zh. Prikl. Khim.*, 35, 2386, 1962.
179. Baskakov, A.P., High speed non oxidative heating and heat treatment in a fluidized bed, *Metallurgia* (*Moscow*), 1968.
180. Vreedenberg, H.A., Heat transfer between fluidized beds and vertically inserted tubes, *J. Appl. Chem.*, 2(Suppl. issue), 26, 1952.
181. Brotz, W., Utersuchungen uber transportvorgange in durchstrometem gekontem Gut, *Chem. Ing. Techn.*, 28, 165, 1956.

182. Sathiyamoorthy, D., Sridhar Rao, Ch., and Raja Rao, M., Effect of distrubutor on heat transfer from immersed surfaces in gas fluidized beds, *Chem. Eng. J.*, 37, 149, 1988.
183. Baskakov, A.P. and Berg, B.V., *Inzh. Fiz Zh. Akad. Nauk Belorussk SSR6*, 8, 3, 1966.
184. Botterill, J.S.M., Teoman, Y., and Yüregir, K.R., Temperature effect on the heat transfer behaviour of gas fluidized beds, *AIChE Symp. Ser.*, 77(208), 330, 1984.
185. Molerus, O., Heat transfer in gas fluidized beds. 2. Dependence of heat transfer on gas velocity, *Powder Technol.*, 15, 70, 1992.
186. Borodulya, V.A., Ganzh, V.A., and Podberlzsky, A.I., Heat transfer in a fluidized bed at high pressure, in *Fluidization*, Grace, J.R. and Matsen, J.M., Eds., Plenum Press, New York, 1980, 201.
187. Varygin, N.N. and Martyushin, I.G., A calculation of heat transfer surface area in fluidized bed equipment, *Khim Mashinostr. (Moscow)*, 5, 6, 1959.
188. Baerg, A., Klassen, J., and Grishler, P.E., Heat transfer in fluidized solid beds, *Can. J. Res. Sect. F*, 28, 324, 1964.
189. Gelprin, N.I., Einstein, V.G., and Romanova, N.A., *Khim. Prom. (Moscow)*, 5, 6, 1959.
190. Jakob, A. and Osberg, G.L., Effect of gas thermal conductivity on local heat transfer in a fluidized bed, *Can. J. Chem. Eng.*, 35, 5, 1957.
191. Chechetkin, A.V., *Vysokotemperature (High Temperature Carriers)*, Gosenergoizdat, Moscow, 1962.
192. Ruckenstein, E., *Zh. Prikl. Khim.*, 35, 71, 1962; as cited by Gutfinger, C. and Abuaf, N., Heat transfer in fluidized beds, in *Advances in Heat Transfer*, Irvine T.F. and Hartnett, J.P., Eds., Academic Press, New York, 1964.
193. Richardson, J.F. and Shakiri, K.J., Heat transfer between a gas–solid fluidized bed and a small immersed surface, *Chem. Eng. Sci.*, 34, 1019, 1979.
194. Kim, K.J., Kim, D.J., and Chun, K.S., Heat and mass transfer in fixed and fluidized bed reactors, *Int. Chem. Eng.*, 8(5), 176, 1971.
195. Grewal, N.S. and Saxena, S.C., Maximum heat transfer coefficient between a horizontal tube and a gas–solid fluidized bed, *Ind. Eng. Chem. Process Res. Dev.*, 20, 108, 1981.
196. Baskakov, A.P. and Panov, O.P., Comparison between measured maximum heat transfer coefficients to surfaces in fluidized beds, *Inzh. Fiz. Zh.*, 45, 896, 1983.
197. Chen, P. and Pei, D.C.T., A model heat transfer between fluidized beds and immersed surfaces, *Int. J. Heat Mass Transfer*, 28(3), 675, 1985.
198. Palchenok, G.I. and Tamarin, A.I., Study of heat transfer between model particles and fluidized beds, *Inzh. Fiz. Zh.*, 45, 427, 1983.
199. Sathiyamoorthy, D. and Raja Rao, M., Prediction of maximum heat transfer coefficient in gas fluidized bed, *Int. J. Heat Mass Transfer*, 25(5), 1027, 1992.
200. Molerus, O., Heat transfer in gas fluidized beds. I, *Powder Technol.*, 70, 1, 1992.
201. Saxena, S.C. and Grewal, N.S., Effect of distributor design on heat transfer from an immersed horizontal tube in a fluidized bed, *Chem. Eng. J.*, 18, 197, 1979.
202. Sathiyamoorthy, D., Sridhar Rao, Ch., and Raja Rao, M., Heat transfer in gas fluidized beds using multi-orifice distributors, *Indian J. Technol.*, 25, 219, 1987.
203. Frantz, J.F., Fluid to particle heat transfer in fluidized beds, *Chem. Eng. Prog.*, 57, 65, 1961.
204. Wasmund, B. and Smith, J.W., Wall to fluid heat transfer in liquid fluidized bed, *Can. J. Chem. Eng.*, 45, 156, 1967.
205. Romani, M.N. and Richardson, J.F., Heat transfer from immersed surfaces to liquid fluidized beds, *Lett. Heat Mass Transfer*, 1, 55, 1974.

206. Krishnamurthy, N. and Sathiyamoorthy, D., Mechanism of heat transfer in liquid fluidized beds, in Proc. National Heat and Mass Transfer Conf., IISc Bangalore, India, Ind. Soc. Heat Mass Trans., Madras, 1987, 81.
207. Mishra, P. and Farid, M.M., Wall to bed heat transfer in liquid fluidized beds, *Ind. Chem. Eng.*, 28(3), 30, 1986.
208. Hog, G.W. and Grimmett, E.S., Effects of pulsation on heat transfer in a liquid fluidized bed, *Chem. Eng. Prog. Symp. Ser.*, 62(67), 51, 1966.
209. Østergaard, K., Discussion in the paper of R. Turner, Fluidization, Soc. for the Chem. Ind., London, 1964, 58.
210. Viswanathan, S.A., Kakar, A.S., and Murti, P.S., Effect of dispersing bubbles into liquid fluidized beds on heat transfer holdup at constant bed expansion, *Chem. Eng. Sci.*, 20, 903, 1964.
211. Armstrong, E.A., Baker, C.G.J., and Bergougnou, M.A., Heat transfer and hydrodynamics studies on three phase fluidized beds, in *Fluidization Technology*, Keairns, D.L., Ed., Hemisphere Publishing, New York, 1976, 453.
212. Baker, C.G.J., Armstrong, F.R., and Bergongnou, M.A., Heat transfer in three phase fluidized beds, *Powder Technol.*, 21, 195, 1978.
213. Kato, Y., Uchida, K., Kago, T., and Morooka, S., Liquid holdup and heat transfer coefficient between bed and wall in liquid–solid and gas–liquid–solid fluidized beds, *Powder Technol.*, 28, 173, 1981.
214. Kato, Y., Kago, T., and Morooka, S., Heat transfer from wall to packed and fluidized bed for gas–liquid co-current up-flow, *Kagaku Kogaku Ronbunshu*, 4, 328, 1978.
215. Ermakova, A., Ziganskin, G.K., and Slinko, M.G., Hydrodynamics of a gas–liquid reactor with a fluidized bed of solid matter, *Theor. Found. Chem. Eng.*, 4, 84, 1970.
216. Phillippovski, N.F. and Baskakov, A.P., Investigation of the temperature field in a fluidized bed close to a heated plate and heat transfer between them, *Int. Chem. Eng.*, 13, 5, 1953.
217. Botterill, J.S.M. and Desai, M., Limiting factors in gas fluidized bed, *Powder Technol.*, 6(2), 231, 1972.
218. Knuth, M. and Weinspach, P.M., Experimentelle unter-Schchung, des warme and Stoffubergangs and die Partikeln einer wirbelschicht beider desubbimation, *Chem. Ing. Tech.*, 48, 893, 1976.
219. Kharchenko, N.V. and Makhorin, K.E., The rate of heat transfer between a fluidized bed and an immersed body at high temperatures, *Int. Chem. Eng.*, 4, 650, 1964.
220. Mersmann, A., Determination of fluidizing velocity in fluidized beds by heat transfer measurements, *Chem. Ing. Tech.*, 38(10), 1095, 1966.
221. Mersmann, A., Heat transfer in fluidized beds, *Chem. Ing. Tech.*, 39(56), 349, 1967.
222. Mersmann, A., Zum Warmeuvergang Vuischen dispersen zueiphasensysten unl senkrechten heizflachen in erdschwereseld, *Verfahrenstechnik*, 10(10), 641, 1976.
223. Thones, D. and Kramers, H., Mass transfer from spheres in various regular packings to a flowing fluid ionic mass transfer in the presence of fluidized solids, *Chem. Eng. Sci.*, 8, 271, 1958.
224. Jaganadharaju, G.J.V. and Rao Venkata, C., Ionic mass transfer in the presence of fluidized solids, *Indian J. Technol.*, 3, 201, 1965.
225. Zeigler, E.N. and Holmes, J.T., Mass transfer from fixed surfaces to gas fluidized beds, *Chem. Eng. Sci.*, 21, 117, 1966.
226. Einstein, V.G. and Gelpin, N.I., Heat transfer between a fluidized bed and a surface, *Int. Chem. Eng.*, 6, 67, 1966.
227. Beek, W.J., Mass transfer in fluidized beds, in *Fluidization*, Davidson, J.F. and Harrison, D., Eds., Academic Press, New York, 1971, chap. 9.

228. Froessling, N., Gerland Beitr, *Geophys. J.*, 52, 170, 1938.
229. Ranz, W.E., Friction and transfer coefficients for single particles and packed beds, *Chem. Eng. Prog.*, 48, 247, 1952.
230. Fan, L.T., Yang, C.Y., and Wen, C.Y., Mass transfer in semi fluidized beds for solid liquid system, *AIChE J.*, 6, 482, 1960.
231. Rebnick, W.E. and White, R.R., Mass transfer in systems of gas and fluidized solids, *Chem. Eng. Prog.*, 45, 377, 1949.
232. Kettenring, K.N., Manderfield, E.L., and Smith, J.M., Heat and mass transfer in fluidized systems, *Chem. Eng. Prog.*, 46, 139, 1950.
233. Kunii, D. and Levenspiel, O., *Fluidization Engineering,* John Wiley & Sons, New York, 1969, 200.
234. Richardson, J.F. and Szekely, J., Mass transfer in a fluidized bed, *Trans. Inst. Chem. Eng.*, 39, 212, 1961.
235. Wakao, N., Particle to fluid heat/mass transfer coefficients in packed bed catalytic reactors, in *Recent Advances in the Engineering Analyses of Chemically Reacting Systems*, Doraiswamy, L.K., Ed., Wiley Eastern Ltd., New Delhi, 1984, chap. 2.
236. Damronglerd, S., Couderc, J.P., and Angelino, H., Mass transfer in particulate fluidization, *Trans. Inst. Chem. Eng.*, 53, 175, 1975.
237. Wakao, N. and Funazkri, T., Effect of fluid dispersion coefficients on particle to fluid mass transfer coefficients in packed beds — correlation of Sherwood number, *Chem. Eng. Sci.*, 33, 1375, 1978.
238. Kunii, D. and Levenspiel, O., Bubbling bed model — model for flow of gas through a fluidized bed, *Ind. Eng. Chem. Fundam.*, 7, 446, 1968.
239. Chiba, T. and Kobayashi, H., Gas exchange between the bubble and emulsion phases in gas fluidized beds, *Chem. Eng. Sci.*, 25, 1375, 1970.
240. Murray, J.D., On the mathematics of fluidization. 1. Fundamental equations and wave propagation, *J. Fluid Mech.*, 21, 465, 1965; 2. Steady motion of fully developed bubbles, *J. Fluid Mech.*, 22, 57, 1965.
241. Partridge, B.A. and Rowe, P.N., Chemical reaction in a bubbling gas fluidized bed, *Trans. Inst. Chem. Eng.*, 44, T335, 1966.
242. Stephens, G.K., Sinclair, R.J., and Potter, O.E., Gas exchange between bubbles and dense phase in a fluidized bed, *Powder Technol.*, 1, 157, 1967.
243. Kunii, D. and Levenspiel, O., Effect of particles in bubbles on fluidized bed mass and heat transfer kintics, *J. Chem. Eng. Jpn.*, 24(2), 183, 1991.
244. Yasui, G. and Johanson, L.N., Characteristics of gas packets in fluidized beds, *AIChE J.*, 4, 445, 1958.
245. Whitehead, A.B. and Young, A.D., Fluidization performance in large scale equipment. I, in *Proc. Int. Symp. on Fluidization*, Drikenburg, A.A.H., Ed., Netherlands University Press, 1967, 284.
246. Park, W.H., Kang, W.K., Capes, C.E., and Osberg, G.L., The properties of bubbles in fluidized beds of conducting particles as measured by an electro resistivity probe, *Chem. Eng. Sci.*, 24, 851, 1969.
247. Kobayashi, H., Arai, F., and Chiba, T., Behaviour bubbles in gas fluidized beds, *Chem. Eng. Sci.*, 21, 239, 1967.
248. Kato, K. and Wen, C.Y., Bubble assemblage model for fluidized bed catalytic reactors, *Chem. Eng. Sci.*, 24, 1351, 1969.
249. Chiba, T., Terashima, K., and Kobayashi, H.J., Lateral distribution of bubble sizes in two dimensional gas fluidized bed, *Chem. Eng. Jpn.*, 8, 167, 1975.
250. Mori, S. and Wen, C.Y., Estimation of bubble diameter in gaseous fluidized bed, *AIChE J.*, 21, 109, 1975.

251. Hirma, T., Ishida, M., and Shirai, T., *Kagaku Kogaku Ronbunshu,* 1, 272, 1975.
252. Rowe, P.N., Prediction of bubble size in a gas fluidized bed, *Chem. Eng. Sci.*, 31, 285, 1976.
253. Darton, R.C., La Nauze, R.D., Davidson, J.F., and Harrison, D., Bubble growth due to coalescence in fluidized beds, *Trans. Inst. Chem. Eng.*, 55, 274, 1977.
254. Werther, J., Mathematical modelling of fluidized bed reactors, *Chem. Ing. Tech.*, 50, 850, 1978.
255. Werther, J., Bubbles in gas fluidized beds. I and II, *Trans. Inst. Chem. Eng.*, 52, 149, 1974.
256. Davidson, J.F. and Schuler, B.O.G., Bubble formation at an orifice in inviscid liquid, *Trans. Inst. Chem. Eng.*, 38, 335, 1960.
257. Chiba, T., Terashima, K., and Kobayashi, H., Behaviour of bubbles in gas–solid fluidized beds; initial formation of bubbles, *Chem. Eng. Sci.*, 27, 965, 1972.
258. Miwa, K., Mori, S., Kato, T., and Muchi, I., Behaviour of bubble in a gaseous fluidized bed, *Int. Chem. Eng.*, 12, 187, 1972.
259. Fryer, C. and Potter, O.E., Experimental investigation of models for fluidized bed catalytic reactors, *AIChE J.*, 22, 38, 1976.
260. Davis, L. and Richardson, J.F., Gas interchange between bubbles and continuous phase in a fluidized bed, *Br. Chem. Eng.*, 12, 1223, 1967.
261. Toei, R., Matsuno, R., Nishitani, K., Miyagawa, H., and Komagawa, Y., Gas transfer between a bubble and the continuous phase in a gas fluidized bed, *Int. Chem. Eng.*, 9, 358, 1969; *Kogaku Kogaku*, 32, 565, 1968.
262. Kobayashi, H., Arai, F., and Chiba, T., *Kagaku Kogaku*, 31, 239, 1967.
263. Miyauchi, T. and Morooka, S., Mass transfer rate between bubble and emulsion phase in fluid bed, *J. Chem. Eng. Jpn.*, 33, 880, 1969.
264. Chiba, T. and Kobayashi, H., Gas exchange between the bubble and emulsion phases in gas fluidized beds, *Chem. Eng. Sci.*, 25, 1375, 1970.
265. Calderbank, P.H., Pereira, J., and Burgss, M.M., The physical and mass transfer properties of bubbles in fluidized beds of electrically conducting particles, in Int. Conf., Pacific Grove, CA, 1976, 261.
266. Mori, S. and Muchi, I., Theoretical analysis of catalytic reaction in a fluidized bed, *Kogaku Kogaku*, 5, 251, 1972.
267. Davidson, J.F., Harrison, D., Darton, R.C., and LaNauze, R.D., in *Chemical Reactor Theory — A Review*, Lapidus, L. and Amundson, N.R., Eds., Prentice-Hall, Englewood Cliffs, NJ, 1977.
268. Alper, E., Wichtendahl, B., and Deckwer, W.D., Gas absorption mechanism in catalytic slurry reactors, *Chem. Eng. Sci.*, 35, 217, 1980.
269. Østergaard, K. and Suchozebrski, W., Gas–liquid mass transfer in gas liquid fluidized beds, in *Proc. Eur. Symp. React. Eng.*, Pergamon Press, Oxford, 1969, 21.
270. Østergaard, K. and Fosbol, P., Transfer of oxygen across the gas–liquid interface in gas–liquid fluidized beds, *Chem. Eng. J.*, 3, 105, 1972.
271. Alvarez-Cuenca, M. and Nerenberg, M.A., The plug flow model for mass transfer in three phase fluidized beds and bubble columns, *Can. J. Chem. Eng.*, 59, 739, 1981.
272. Dankwerts, P.W., *Gas–Liquid Reactions*, McGraw-Hill, New York, 1970, 15.
273. Geankopolis, C.J., *Mass Transport Phenomena*, Holt, Reinhart and Winston, New York, 1972, 316.
274. Sharma, M.M. and Danckwerts, P.V., Chemical method of measuring interfacial area and mass transfer coefficients in two-fluid systems, *Br. Chem. Eng.*, 524, 208, 1970.
275. Gay, F.V., Holtman, C.B., and Sheppard, N.F., Mass Transfer in Three Phase Fluidized Beds, ORNL Report No. ORNL/MIT-283, Oak Ridge, TN, December 21, 1978.

276. Patwari, A.N., Nguyen-tien, K., Schumpe, A., and Deckwer, W.D., Three phase fluidized beds with viscous liquid: hydrodynamics and mass transfer, *Chem. Eng. Commun.*, 40, 49, 1989.
277. Schumpe, A., Deckwer, W.D., and Nigam, K.D.P., Gas fluid mass transfer in three phase fluidized beds with viscous pseudoplastic liquids, *Can. J. Chem. Eng.*, 67, 873, 1989.
278. Tang, W.T. and Fan, L.S., Gas–liquid mass transfer in a three phase fluidized bed containing low density particles, *Ind. Eng. Chem. Res.*, 29, 128, 1990.
279. Alvarez-Cuenca, M., Oxygen Mass Transfer in Bubble Columns and Three Phase Fluidized Beds, Ph.D. thesis, University of Western Ontario, Canada, 1979.
280. Alvarez-Cuenca, M., Oxygen mass transfer in three phase fluidized beds working at large flow rates, *Can. J. Chem. Eng.*, 61, 5, 1983.
281. Muroyama, K. and Fan, L.S., Fundamentals of gas–liquid–solid fluidization, *AIChE J.*, 31(1), 1, 1985.
282. Dhanuka, V.R. and Stepanek, J.B., Gas–liquid mass transfer in a three phase fluidized bed, in *Fluidization*, Grace, J.R. and Matson, G.M., Eds., Plenum Press, New York, 1980, 261.
283. Kim, S.D. and Chang, H.S., Hydrodynamics and bubble breakage in three phase fluidized beds, *Hwahak Konghak*, 17, 407, 1979.
284. Kim, J.O. and Kim, S.D., Bubble characteristics in three phase fluidized beds of floating bubble breakers, *Particulate Sci. Technol.*, 5, 309, 1987.
285. Kim, S.D., Lee, Y.J., and Kim, J.O., Heat transfer and hydrodynamic studies on two and three phase fluidized beds of floating bubble breakers, *Exp. Thermal Fluid Sci.*, 1, 237, 1988.
286. Kim, J.O. and Kim, S.D., Gas–liquid mass transfer in a three phase fluidized bed with floating bubble breakers, *Can. J. Chem. Eng.*, 68, 368, 1990.
287. Morooka, S., Kusakabe, K., and Kato, Y., Mass transfer coefficient at the wall of a rectangular fluidized bed for liquid–solid and gas–liquid–solid systems, *Kogaku Kogaku Ronbunshu*, 5, 162, 1979; *Int. Chem. Eng.*, 20, 433, 1980.
288. Batchelor, G.K., Mass transfer from small particles suspended in turbulent fluid, *J. Fluid Mech.*, 98, 609, 1980.
289. Arters, D. and Fan, L.S., Liquid–solid mass transfer in a gas–liquid–solid fluidized bed, presented at AIChE Meeting, San Francisco, November 25–30, 1984.
290. Groshe, E.W., Analysis of gas fluidized solid systems by X-ray absorption, *AIChE J.*, 1(3), 358, 1955.
291. Bakker, P.I. and Heertjes, P.M., Porosity measurement in fluidized beds, *Br. Chem. Eng.*, 3, 240, 1958.
292. Fan, L.T., Chan, J.L., and Baile, R.C., Axial solid distribution in gas–solid fluidized beds, *AIChE J.*, 8(2), 239, 1962.
293. Rowe, P.N., MacGillivray, H.J., and Chessman, D.J., Gas discharge from an orifice into a gas fluidized bed, *Trans. Inst. Chem. Eng.*, 57, 194, 1979.
294. Yang, W.C. and Keairns, D.L., Design and operating parameters for a fluidized bed agglomerating combustor gasifier, in *Fluidization*, Davidson, J.F. and Keairns, D.L., Eds., Cambridge University Press, Cambridge, 1978, 208.
295. Yang, W.C. and Keairns, D.L., Estimating the jet penetration depth of multiple grid jets, *Ind. Eng. Chem. Fundam.*, 18, 317, 1979.
296. Zenz, F.A. and Weil, N.A., A theoretical-empirical approach to the mechanism of particle entrainment from fluidized beds, *AIChE J.*, 4, 472, 1958.
297. Shakova, N.A., Outflow of turbulent jets into a fluidized bed, *Inzh. Fiz. Zh.*, 14, 6, 1968.

298. Basov, V.A., Markhevka, V.I., Melik-Akhnazarov, T.Kh., and Orochko, D.I., Investigation of the structure of a non-uniform fluidized bed, *Int. Chem. Eng.*, 9, 263, 1969.
299. Merry, J.M.D., Penetration of vertical jets into fluidized beds, *AIChE J.*, 21, 507, 1975.
300. Deole, N.R., Study of Jets in Three Dimensional Gas Fluidized Beds, M.S. thesis, West Virginia University, Morgantown, 1980.
301. Yang, W., Jet penetration in a pressurised fluidized bed, *Ind. Eng. Chem. Fundam.*, 2, 297, 1981.
302. Blake, T.R., Wen, C.Y., and Ku, C.A., The correlation of jet penetration measurements in fluidized beds using non-dimensional hydrodynamic parameters, *AIChE Symp. Ser.*, 80(234), 42, 1984.
303. Kececioglu, I., Yang, W., and Keairns, D.L., Fate of solids fed pneumatically through a jet into a fluidized bed, *AIChE J.*, 30, 99, 1984.
304. Wen, C.Y., Deole, N.R., and Chen, L.H., A study of jets in a three dimensional gas fluidized bed, *Powder Technol.*, 31, 1975, 1982.
305. Kozin, V.E. and Baskakov, A.P., An investigation of the grid zone of a fluidized bed above cap-type distributors, *Int. Chem. Eng.*, 8(2), 257, 1968.
306. Filla, M., Massimilla, L., and Vaccaro, S., Gas jets in fluidized beds and spouts: a comparison of experimental behaviour and models, *Can. J. Chem. Eng.*, 61, 370, 1983.
307. Donadono, S., Maresca, A., and Massimilla, L., Gas injection in shallow beds of fluidized coarse solids, *Ing. Chim. Ital.*, 16, 1, 1980.
308. Lofroy, G.A. and Davidson, J.F., The mechanics of spouted beds, *Trans. Inst. Chem. Eng.*, 47, T120, 1969.
309. Littman, H., Morgan, M.H., III, Vukovic, D.V., Zdanski, F.K., and Graveie, Z.B., A method for predicting the relationship between the spout and inlet tube radii in a spouted bed at its maximum spoutable height, in *Fluidization*, Davidson, J.F. and Keairns, D.L., Eds., Cambridge University Press, Cambridge, 1978, 381.
310. Yang, W.C. and Keairns, D.L., Momentum dissipation and gas entrainment in a gas–solid two-phase jet in a fluidized bed, in *Fluidization*, Grace, J.R. and Matsen, J.M., Eds, Plenum Press, New York, 1980, 305.
311. Mamuro, T. and Hattori, H., Flow pattern of fluid spouted beds, *J. Chem. Jpn.*, 1, 5, 1968.
312. Donsi, G., Massimilla, L., and Colantuonio, L., The dispersion of axisymmetric gas jets in fluidized beds, in *Fluidization*, Grace, J.R. and Matsen, J.M., Eds., Plenum Press, New York, 1980, 297.
313. Becker, H.A., An investigation of laws governing the spouting of coarse particles, *Chem. Eng. Sci.*, 13, 245, 1961.
314. Lim, C.J. and Mathur, K.B., Modelling of particle movement in spouted beds, in *Fluidization*, Davidson, J.F. and Keairns, D.L., Eds., Cambridge University Press, Cambridge, 1978, 104.
315. Ching, Ho. T., Yutani, N., Fan, L.T., and Walawender, W.P., Stochastic modelling of bubble formation on the grid in a gas–solid fluidized bed, *Can. J. Chem. Eng.*, 61, 654, 1983.
316. Hsiung, T.P. and Grace, J.R., Formation of bubbles at an orifice in fluidized beds, in *Fluidization*, Davidson, J.F. and Keairns, D.L., Eds., Cambridge University Press, Cambridge, 1978, chap. 21.
317. Werther, J., Effect of gas distributor on the hydrodynamics of gas fluidized beds, *Ger. Chem. Eng.*, 1, 166, 1978.
318. Brien, C.L., Bergougnou, M.A., and Baker, C.G.J., Grid leakage (weeping, dumping, particle back flow) in gas fluidized beds, in *Fluidization*, Grace, J.R. and Matsen, J.M., Eds., Plenum Press, New York, 1980, 413.

319. Wen, C.Y., Krishnan, R., Dutta, S., and Khosravi, R., Dead zone height near the grid region of fluidized bed, in Proc. 2nd Eng. Foundation Conf., Cambridge, England, 1978, 32.
320. Wen, C.Y. and Dutta, S., Research needs for the analysis, design and scale up of fluidized beds, *AIChE Symp. Ser.*, 73(161), 2, 1977.
321. Wen, C.Y., Krishnan, R., Khosravi, R., and Dutta, S., Dead zone height near the grid of fluidized beds, in *Fluidization*, Davidson, J.F. and Keairns, D.L., Eds., Cambridge University Press, Cambridge, 1978, chap. 2–3.
322. Wen, C.Y., Krishnan, R., and Kalyanaraman, R., Particle mixing near the grid region of fluidized beds, in *Fluidization*, Grace, J.R. and Matsen, J.M., Eds., Plenum Press, New York, 1980, 405.
323. Leva, M. and Wen, C.Y., Elutriation, in *Fluidization*, Davidson, J.F. and Harrison, D., Eds., Academic Press, London, 1971, chap. 14.
324. Geldart, D., Elutriation, in *Fluidization*, Davidson, J.F., Clift, R., and Harrison, D., Eds., Academic Press, London, 1985, chap. 11.
325. Rowe, P.N. and Partridge, B.A., X-ray study of bubbles in fluidized beds, *Trans. Inst. Chem. Eng.*, 43, T157, 1965.
326. Kunii, D. and Levenspiel, O., *Fluidization Engineering*, John Wiley & Sons, New York, 1979, chap. 10.
327. Kunii, D. and Levenspiel, O., Entrainment of solids from fluidized beds. 1. Hold up in the free board, 2. Operation of fast fluidized beds, *Powder Technol.*, 61, 193, 1990.
328. Kraft, W.W., Ulrich, W., and O'Connor, W., in *Fluidization*, Othmer, D.F., Ed., Reinhold, New York, 1956, 194.
329. Lewis, W.K., Gilliland, E.R., and Lang, P.M., Entrainment from fluidized beds, *Chem. Eng. Prog. Symp. Ser.*, 58(68), 65, 1962.
330. Soroko, V., Mikhalev, M., and Mukhlenov, I., Calculation of the minimum height of the space above the in fluidized bed contact equipment, *Int. Chem. Eng.* 9, 280, 1969.
331. Horio, M., Taki, A., Hsieh, Y.S., and Muchi, I., Elutriation and particle transport through the freeboard of gas–solid fluidized bed, in *Fluidization*, Grace, J.R. and Matsen, J.M., Eds., Plenum Press, New York, 1980, 509.

CHAPTER 2

Applications in Mineral, Metal, and Materials Extraction and Processing

I. DRYING

A. Types of Fluid Bed Dryers

1. Classification

Fluidized bed dryers can be successfully and efficiently employed for drying of wet particulate materials as long as the bed of such materials can be kept in a fluidized condition. Fluidized bed dryers are easy to construct and operate at high thermal efficiencies. The dryer size can range from small units such as those used for fine chemicals and pharmaceuticals to larger units such as those for drying coal and minerals. A single fluidized bed drying unit can replace many processes such as evaporation, crystallization, filtration, drying, and pulverization. Fluid bed dryers can be classified into various types depending on flow characteristics, feed and discharge operations, heating mode, geometry of the fluidizing vessel, and mode of operation utilizing vibration or spout action in conjunction with fluidization. The subject of drying constitutes an advanced field in modern process engineering. Many advances are continually emerging with the advent of new unit operation equipment. Fluid bed dryers are now closely competing with many conventional dryers in commercial-scale applications. The need for fluidized bed dryers and their selection can best be exploited if the basics and the various types of fluid bed dryers are understood by the user. In this context, the types of fluidized bed dryers can be classified in the manner shown in Figure 2.1. Various types of fluid bed dryers and the commercial companies that use such dryers were well reviewed by Romankov.[1] Kunii and Levenspiel[2] briefly described various fluidized dryers and their applications. The drying fundamentals and the pertinent models were presented by Reay and Baker.[3] The objective of this section is to give a brief account of fluidized bed

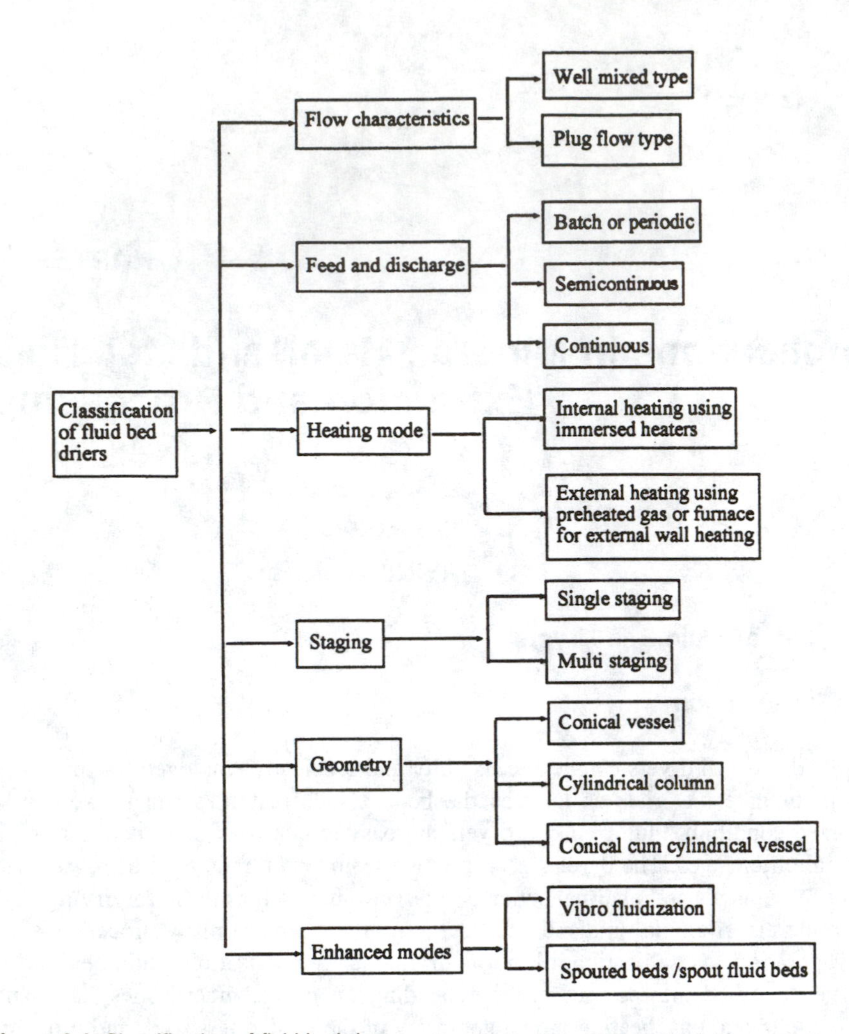

Figure 2.1 Classification of fluid bed dryers.

dryers and the relevant principles that govern the process of drying. Schematic diagrams of various types of dryers and fluid–solid contacting techniques are given in Figure 2.2.

2. Description of Dryers

A single-stage batch/continuous fluidized bed reactor (as shown in Figure 2.2a) approximates the characteristics of a well-mixed reactor in which the solid residence time distribution is not very close to the ideal condition. As a result, after drying, the wet particles will have varying moisture contents. Hence, solid particles which are to be just surface dried may be used in a single-stage dryer. Wet solids can be fed directly into the fluidized bed of drying solid particles. The bed temperature of such a reactor is normally around the vaporization temperature of moisture, and

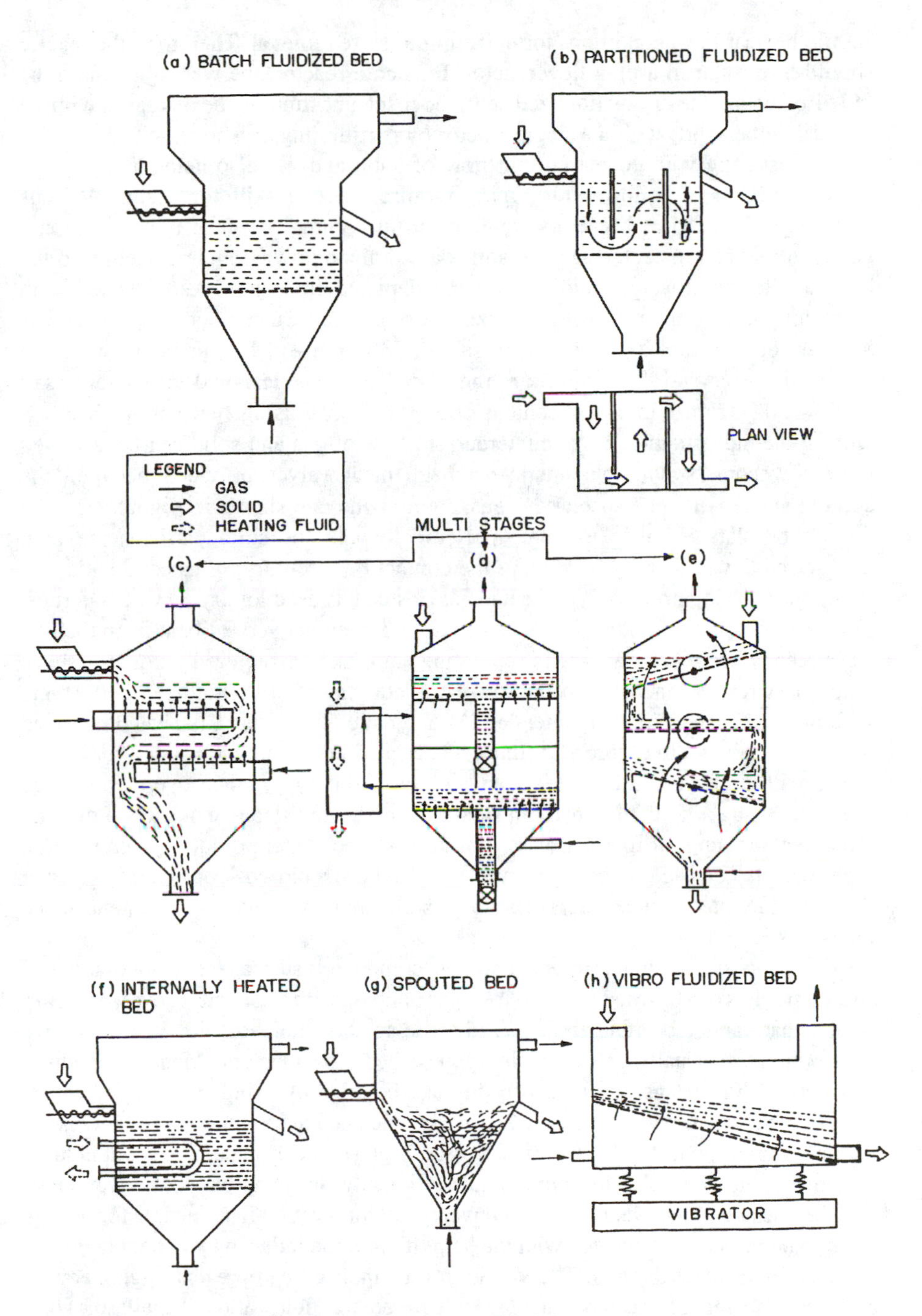

Figure 2.2 Schematics of various types of fluidized bed dryers.

waste hot flue gas can be used efficiently for this purpose. Particles that are to be dried completely require a long residence time, and the residence time distribution should be near ideal conditions for uniform moisture removal. Therefore, the reactor should be similar to a plug flow reactor. In such a reactor, the wet solid cannot be fed directly into the drying fluidized bed. The residence time can be increased without bypassing the solids within a single reactor by partitioning, as shown in Figure 2.2b.

Multistaging with countercurrent flow of solid and gas also helps in achieving uniform drying of solids. Multistaging requires special skill for design. Various multistage fluidized bed reactors are depicted in Figure 2.2c–e. In Figure 2.2c, hot gas is introduced in each stage and solids are drained down countercurrently to the hot gas. The multistage fluidized reactor depicted in Figure 2.2d is used with preheating arrangements for the fluidizing (drying) gas. Such reactors were used as long as four decades ago[4] for drying salt. A 2400-mm- (8-ft-) diameter two-stage fluidized salt crystal dryer with a rating of 5 ton/h was designed to dry table salt (–30 +100#) from a moisture content of 3% to 0.03% using fuel gas for heating. During the multistaging, easy countercurrent flow of gas and solid can be achieved with downcomers with pneumatic or mechanical valves provided between the adjaunt stages. In some special designs, such as the one shown in Figure 2.2e, the distributor plate is made to rotate periodically and the solid particles are then transferred down, thereby improving the contact between drying gas and solid.

In most drying processes, preheated gas or steam is used for drying wet materials. This involves pumping huge amounts of gas and then recovery of heat from the off-gas streams. Postreactor gas–solid separating equipment is required. In order to have a compact reactor and also to minimize the equipment cost, fluidized bed dryers with internal heaters are recommended. A schematic of an internally heated fluidized bed dryer is shown in Figure 2.2f. Internally heated dryers are well suited for drying solids with high moisture content, and dryers of this type, when operated at high pressures using a fluidizing medium such as superheated steam, can show a thermal efficiency far superior to other classes of dryers. The steam produced from the first dryer can also be used in the next dryer, so as to have a closed-loop-operating dryer. Thus, an internally heated dryer has its own advantages if the water content of the feedstock is high.[5]

Often the solids contain vapors of organic compounds such as methanol or toluene, and such solids can be dried using an inert gas coupled with a solvent recovery system.[6] Solids that cannot be fluidized so easily due to their unfavorable shape and size distribution are usually processed in spouted-bed-type reactors. Many agricultural products such as beans, wheat, and paddy can be dried by using spouted bed dryers similar to the one shown in Figure 2.2g. In a spouted bed dryer, the gas used for drying is injected as a spout and the dryer is necessarily of a conical shape. The spout induces violent mixing and helps in drying solids efficiently. In many cases, the wet solids may be sticky or cohesive, rendering conventional fluidization inapplicable for drying. Such materials can be fluidized with the help of vibration induced by either pneumatic or electromagnetic vibrators. The vibration thus induced can prevent agglomeration and can also reduce the amount of gas required for fluidization. Details of vibro fluidized bed dryers and the basics of this technique are given later in this section. The schematic of a simple vibro fluidized bed dryer is shown in Figure 2.2h. Here the solid is fed at one end, fluidized by hot air, conveyed along the dryer by the fluidizing gas, and the dried solid is finally discharged at the other end. Thus, a vibro fluidized bed can simultaneously do the job of drying as well as conveying solids. Vibro fluidized beds have found extensive application in the chemical, foodstuffs, mineral, and plastics industries. Vibro fluidization has the dual merits of a moving bed (i.e., plug flow) and a conventional fluidized bed (i.e., perfect mixing). Hence, it is very useful in continuous drying of solids without short-circuiting the feed to the product. Particles of fine size that are to be only surface dried can be dried in a dilute-phase fluidized bed reactor. The residence time for the solid particles in such a reactor is very low. This type of reactor is also called a flash reactor.

B. Multistaging of Dryers

For fluidized bed dryers with several stages, solid flow down the reactor countercurrent to the gas flow must be controlled; therefore, the multistaging has to be constructed wisely. There are several ways to construct a multistage fluidized bed, and the method varies mainly with regard to the manner by which the solid is transferred from one stage to another. In a single-column multistage reactor, solids can be transferred from one stage to another by using a fluidized dipleg in combination with a nonmechanical valve. Figure 2.3a depicts such a method of solid transfer in a multistage fluidized bed. Multistaging can be done using two columns, either made up of two columns of the same size joined at the wall to transfer the solids alternately between each stage of a column (see Figure 2.3b) or with two coaxial columns as illustrated in Figure 2.3c. Figure 2.3d shows two separate multistaging columns interconnected for transfer of solids between them. Multistaging of a fluidized bed reactor using pneumatically controlled downcomer was described by Liu and Kwauk[7] and Liu et al.[8] Kwauk[9] consolidated the various configurations of such multistage fluidized beds and illustrated them. The theory behind the nonmechanical valve operation and the pneumatic control of solid flow is complex and beyond the scope of this book. Further details on this topic can be found in the literature.[10]

C. Drying Basics

1. Drying Rate

The drying of a moist granular material depends on many factors, such as the characteristics of the material, the height of the charge, the fluidization velocity, the particle size, the humidity, and the temperature of the fluidizing gas. Let us examine the process of drying when the drying gas is flowing through a wet granular solid. Initially, the moisture present at the surface of the particle is driven away by evaporation. During this condition, the particle surface is at the wet bulb temperature (T_{wb}) and the partial pressure of the vapor at this condition is p_{wb}. If at a constant rate period N is the moles of vapor evaporating per unit area from the surface of the particle at T_{wb} to the gas flowing at the bulk temperature (T_b) then:

$$N\left(\frac{\text{mol}}{\text{m}^2\text{s}}\right) = K\left(p_{wb} - p_b\right) = \frac{h_p}{\lambda}\left(T_b - T_{wb}\right) \tag{2.1}$$

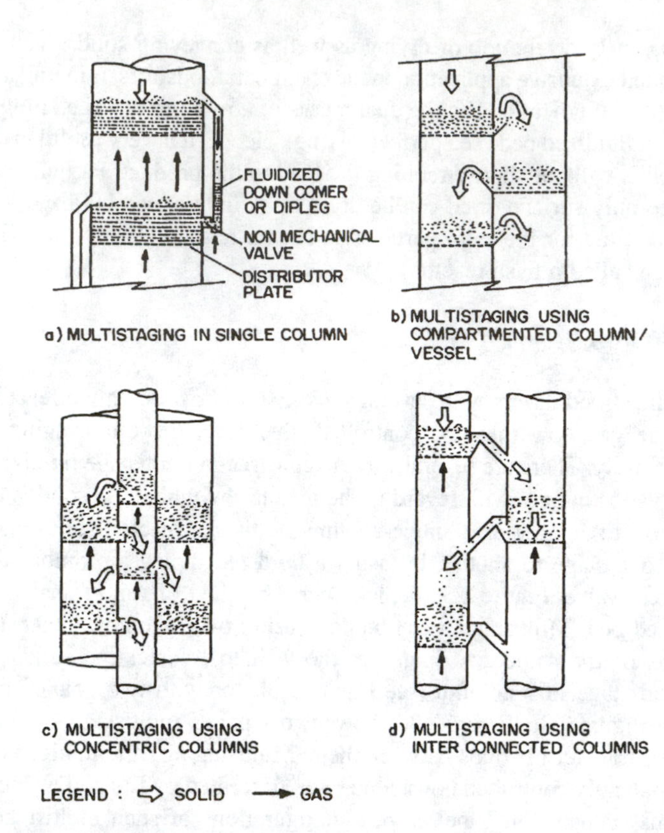

Figure 2.3 Various multistaging techniques.

where K is the mass transfer coefficient based on partial pressure, h_p is the particle-to-gas heat transfer coefficient, and λ is the latent heat of evaporation. The value of N remains constant until the surface of the particle is saturated with moisture. When the moisture at the surface is depleted, drying continues with transport of moisture from the interior of the particle. Now the flux (N) starts falling. The point at which N starts falling corresponds to the critical drying flux (N_{cr}). In other words, the surface moisture ceases to exist from this point onward. If the moisture content at any time is designated as X (i.e., weight of evaporating liquid per unit weight of dry solid), X will approach a minimum (X_e) at the equilibrium corresponding to the relative humidity and the temperature of the inlet gas. It can be inferred that the moisture content present in a wet solid at any time is $X - X_e$. The plot of the variation of X with time and the plot of the derivative dX/dt versus $X - X_e$ are the characteristics of drying for a given material and the drying conditions. This can be depicted as shown in Figure 2.4.

2. *Moisture Transport*

While drying at a constant rate can be determined from the vaporized moisture transport across the boundary layer enveloping the particle, the same cannot be

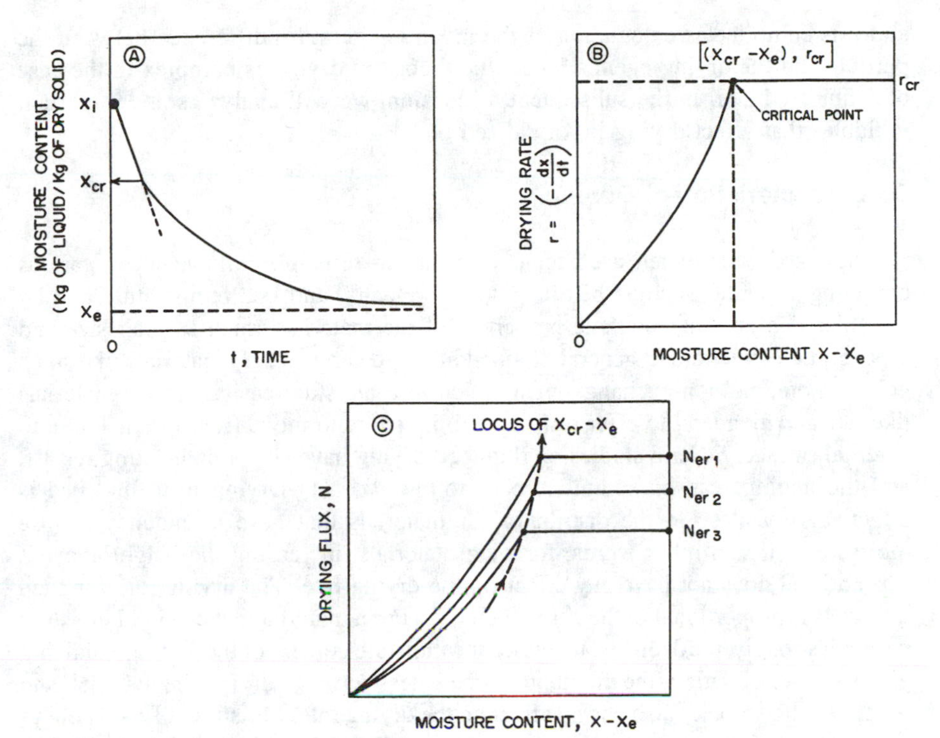

Figure 2.4 Drying curves: (A and B) batch drying and (C) varying external drying conditions.

predicted during the falling rate period. This is mainly due to the unsaturated condition that prevails at the surface of the particle. During the falling rate of drying, moisture is transported from the core of the particle to the surface. Several factors influence the transport process: the nature of the particles (hygroscopic or cellular), the porous structure, capillary action, and vapor diffusion. The internal structure of the material can offer resistance to moisture transport. Capillary action plays a role initially when the pores are full of moisture. Vapor diffusion takes place toward the end of the process and encounters resistance thereby complicating the mechanism of moisture removal from the particle. As a result, prediction of the drying rate from knowledge of vapor transport across the boundary layer is theoretically impossible. Hence, it is essential to obtain the drying curves for the falling rate only through experiments. The drying flux during the falling rate is reduced to a certain fraction (f) of the critical drying flux (N_c), and f is dependent on the characteristics of the material. Drying is a complex process which involves simultaneous heat and mass transport. The value of N_c as determined by Equation 2.1 is based on mass transfer and depends on the parameter K, which is influenced by gas–solid contacting and the driving force (p_{wb}–p_b). The heat transfer coefficient (h_p) is also influenced by the disturbance of the hydrodynamic boundary layer by the vapor flow. The mean temperature of the particle is difficult to evaluate because of the varied moisture and temperature profile within the particle. During drying, the temperature of both the gas and the particle increases within the fluidized bed, and this complicates calculation of the mean temperature difference between the particle and the drying agent. Hence, the theory of drying is complex in the case of a fluidized bed. In the subsequent discussion, we will analyze some important variables that affect drying in a fluidized bed.

D. Characteristics of Dryers

Fluidized beds, when used for drying, are in principle influenced by various operating variables such as bed height, gas velocity, and bed temperature and by variables that are characteristic properties of the materials, such as their size and type. Particle types are in general grouped into two varieties: (1) materials like silica gel, iron ore, and ion-exchange resin which lose moisture easily and (2) materials like wheat which tend to offer resistance to moisture to move from internal core to external surface. Several studies on fluid bed drying have shown interesting results, and the findings can be summed up as follows:[11-13] (1) Drying in a fluid bed is effective very close to the distributor for materials that have a tendency to lose moisture easily. In other words, for such materials, increasing the height above a certain level does not have any effect on the drying rate. The drying rate for such materials is proportional to the gas velocity. (2) For a material that can hold moisture within it strongly, the deep bed increases the moisture content of the exiting fluidizing gas. In such a situation, the distributor zone is less active in drying the material, and the fluidizing velocity also has no effect on the drying rate. Moisture diffusion obeys Fick's law, and the moisture diffusivity is generally of the order 10^{-10} m^2/s. (3) The bed temperature has a strong effect in drying all materials. (4) For particles ranging in size from 106 to 2247 μm, the drying rate is proportional to $d_p^{0.1}$, and this shows a weak dependency on particle size of the drying rate. In a gas fluidized bed, the gas at the inlet is assumed to divide into two portions; one portion passes through a dense phase and the other through a bubble phase. The drying gas passing through the bubble phase is not saturated with moisture to the extent of saturation of the gas that passes through the dense phase. The degree of saturation is dependent on the nature of the material present in the bed.

1. Well-Mixed Type

In a well-mixed fluid bed reactor that carries out drying continuously at constant temperature, tracer particles of the same initial moisture content and size distribution as that of the feed material are fed into the reactor; at the exit, the weight fraction of such colored particles and their moisture content are analyzed at regular intervals. From the data on the moisture content, $X(t)$, at any time and the residence time distribution function, $f(t)$, for perfect mixing, the mean moisture content (X_m) of the dried solid can be evaluated from the equation

$$X_m = \int_0^\infty X(t) f(t) dt \tag{2.2}$$

where $f(t)$ is given by:

$$f(t) = \frac{1}{t_m}\exp(-t/t_m) \tag{2.3}$$

2. Plug Flow Type

A fluidized bed reactor, in principle, is not an ideal reactor. Hence, it is not easy to achieve ideal conditions corresponding to a plug flow reactor. If plug flow behavior is achieved, the moisture content of the particle can be easily predicted from knowledge of the residence time. However, plug flow is influenced by several parameters, such as particle properties, operating conditions, and bed geometry. Plug flow is characterized by a dimensionless parameter known as the axial dispersion coefficient (α_H), which is the ratio of the product of particle diffusivity (D) and its residence time to the square of the bed height (H) (i.e., $\alpha_H = Dt_m/H^2$). This parameter is used in the equation that relates the change in concentration rate ($\partial C/\partial t$) of the particle to the axial concentration gradient ($\partial C/\partial x$):

$$\frac{\partial C}{\partial t} = D\frac{\partial C}{\partial x^2} \tag{2.4}$$

Taking $\theta = t_m$, $\varphi = x/H$, the dimensionless form of Equation 2.4 is

$$\frac{\partial C}{\partial \theta} = \alpha_H \frac{\partial C}{\partial \varphi^2} \tag{2.5}$$

Equation 2.4 has to be modified by subtracting from the right-hand side the term $\partial C/\partial x$ if the particles have a net flow velocity u. For the case of plug flow, $\alpha_H = 0$, and for perfect mixing $\alpha_H \to \infty$. The value of α_H is close to 0.1 for most plug flow fluid bed dryers. The mean standard deviation[3] of the residence time distribution (i.e., $\sigma_\theta = \sigma/t_m = \sqrt{2\alpha_H}$) is $\sqrt{0.2}$. A deviation from plug flow is overcome by several perfectly mixed reactors connected in series. In deeper beds, where particle circulation cells may be formed, α_H values will tend to increase, thereby indicating a departure from plug flow characteristics. In other words, shallow beds can approach the plug-flow-type reactor. It has been well documented in the literature that a fluidized bed of particles with magnetic properties can be operated without bubbles over a wide range of gas flow rates by the use of a magnetic field. The fluidlike behavior of bubbleless fluidization under the influence of an external magnetic field is often termed a magnetically stabilized fluidized bed (MSFB).[14] These beds have less turbulence even at relatively high velocities, thereby allowing operation of the bed without particle attrition and elutriation. Furthermore, the MSFB can behave like a plug flow reactor, as axial and radial dispersion can be suppressed. Geuzens and Thoenes[15] reported that radial mixing in an MSFB is comparable to a packed bed and axial mixing is comparable to something intermediate between a packed and fluidized bed. Thus, an MSFB reactor has the advantages of a packed as well as a fluidized bed reactor. Literature on drying in this class of reactors is scant.

3. Models

a. Definition of Models

There are several models to assess the performance of a fluidized bed for a chemical reaction. The models and their assumptions as applied to a fluidized bed reactor are presented in Chapter 5. The models of a fluidized bed as applied to drying are different. In fact, the process of drying has usually been dealt with in terms of thermal balance or mass transport of moisture from a wet solid to a dry gas. For a well-fluidized bed, it is widely accepted that the fluidizing gas finds it way through the reactor by dividing itself in the form of bubbles and as interstitial gas through the solid particle-rich dense or emulsion phase. The bubble which rises along with the bed is assumed to have either a boundary layer due to a gas film or a cloud due to a mixture of dense gas–solid phase. A mathematical model for fluidized bed drying was proposed by Alebregtse.[16] The model is based on hydrodynamics and mass transfer for the powders which can be fluidized well. The model assumes three phases: (1) the solid phase which contains the moisture, (2) the dense phase which is a mixture of gas and solid (also known as the emulsion phase), and (3) the bubble phase (also known as the lean phase). Figure 2.5 illustrates the model definition usually considered in the mass transport of moisture in a three-phase model.

b. Mass (Moisture) Transport

The moisture from the wet solid particles diffuses to the surface, and the moisture thus emerged is found in the emulsion-phase gas of a fluidized bed dryer. During the diffusion of moisture from the solid, the internal diffusion resistance may be neglected in some situations; in others, it must be considered. During constant rate drying, the diffusional resistance for moisture transport can be neglected. Palanez[17] proposed a three-phase model that neglects the heat and mass transport inside the particle as the limiting factor. However, he emphasized the need for a refined approach if any transport resistance for moisture within the particle is encountered. Hoebink and Rietema[18] proposed a three-phase model for drying of a particle that has internal diffusion as the limiting factor. Hence, particle drying is assumed to be a slow process. Once diffused to the surface, the moisture is assumed to have negligible resistance for transport into the emulsion gas. This is because of the large surface area of the solid particle available for transport of moisture to the emulsion gas, thereby bringing an equilibrium quickly. The moisture content of the emulsion gas then can be assumed to be transferred to the bubble-phase gas. During such a transfer, the resistance can be assumed to be offered either by the boundary layer of gas or by a cloud of gas–solid mixture that exists around a rising bubble. In a model developed for continuous drying of solids, Verkooijen[19] proposed a simplified approach for drying smaller particles (i.e., $d_p < 120$ μm). The transport of moisture is assumed to take place through a boundary layer. The boundary theory of Chiba

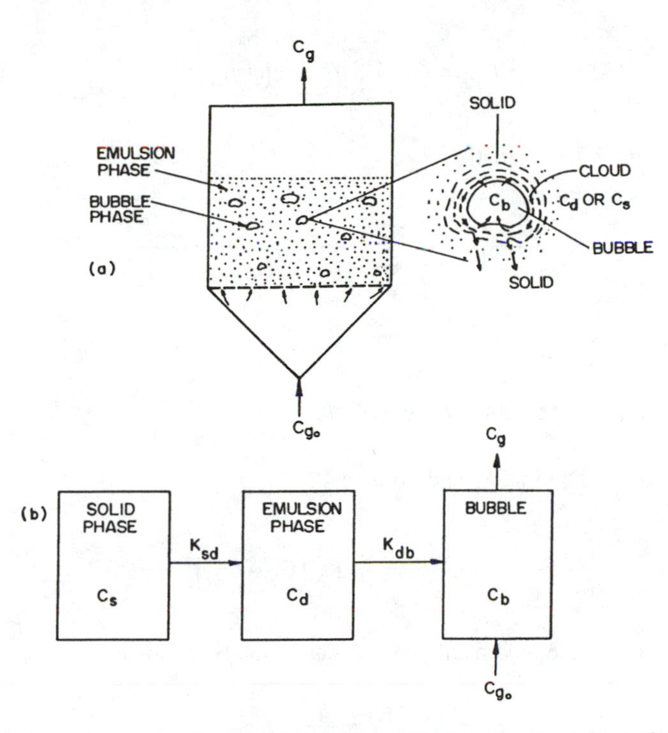

Figure 2.5 Schematic diagram for model definition: (a) fluid bed with various phases and (b) mass transport in three-phase model.

and Kobayashi[20] can be used in this case. The following is a model proposed by Verkooijen[19] for a fluidized bed where the bubbles are not surrounded by a cloud of gas–solid mixture.

c. Moisture at the Surface of the Particle (C_{sR})

For a wet solid particle, the moisture distribution inside the particle and the surface concentration can in principle be obtained from solution of the differential equation

$$\frac{\partial C_s}{\partial t} = \frac{D}{r^2}\frac{\partial}{\partial r}\left[\left(r^2 \frac{\partial C_s}{\partial r}\right)\right] \tag{2.6}$$

Solution of Equation 2.6 for the boundary conditions at $t = 0$ (i.e., $C_s = C_{sO}$ for $0 < r < R_p$) and $t > 0$ (i.e., $r = 0$, $\partial C_s/\partial r = 0$ and $r = R$, $C_s = C_{sR}$) will result in:

$$\frac{C_s - C_{sR}}{C_{sO} - C_{sR}} = \sum_{n-1}^{\alpha} \frac{6}{(n\pi)^2}\left[-(n\pi)^2 \frac{Dt}{R_p^2}\right] \tag{2.7}$$

For a perfectly mixed bed, the average moisture concentration (C_{sM}) that is the same as the exit concentration (C_{se}) can be obtained from the age distribution:

$$C_{sM} = C_{se} = \int_0^\infty C_s \exp\left(-t/t_m\right) dt \tag{2.8}$$

Using Equation 2.7 in Equation 2.8,

$$\frac{C_{se} - C_{sR}}{C_{sO} - C_{sR}} = \sum_{n-1}^{\alpha} \frac{6}{(n\pi)^2\left[(n\pi)^2 \alpha_p + 1\right]} \tag{2.9}$$

where $\alpha_p = Dt_m/R_p^2$.

d. *Mass Balance Across a Gas Bubble*

For a bubble of radius R_b and volume V_b,

$$V_b \frac{dC_b}{dt} = -a\, K_{bc}\left(C_b - C_{gR}\right) \tag{2.10}$$

where a is the surface area available for mass transport. Since

$$t = \frac{h}{U_b} \tag{2.11}$$

Equation 2.10, after integration for the boundary condition $h = 0$, $C_b = C_{go}$, results in:

$$\beta = \frac{C_{ge} - C_{gR}}{C_{go} - C_{gR}} = \exp\left(-\int_0^H \frac{a\, K_{bc}}{V_b U_b} dh\right) \tag{2.12}$$

From Equation 2.12, one obtains:

$$(C_{ge} - C_{go}) = (1 - \beta)\,(C_{gR} - C_{go}) \tag{2.13}$$

e. *Overall Mass Balance of a Fluid Bed Dryer*

If Q_s is the volumetric feed rate of solid and Q_g is the volumetric flow rate of gas, moisture lost by solid = moisture gained by gas gives:

$$Q_s\,(C_{sO} - C_{se}) = Q_g\,(C_{ge} - C_{go}) \tag{2.14}$$

Using Equation 2.13 in Equation 2.14, C_{se} can be predicted.

f. *Model Testing*

If the equilibrium moisture content of the solid is C_{seq}, then the ratio $(C_{sO} - C_{se})/(C_{sO} - C_{seq})$ is defined as the drying efficiency. The exponential term in Equation

2.12 is a function of the bubbling characteristics of a fluidized bed and has to be evaluated from basic hydrodynamic data. Verkooijen[19] tested his model for drying in a fluidized bed by experimentally measuring the drying efficiency and comparing it with his model prediction. His experiments were conducted on a 30-cm-ID fluidized bed provided with a distributor with 197 holes 2 mm in diameter. Silica gel with an initial moisture content C_{sO} = 88–145 g water per kilogram dry solid was dried using air at 50°C at velocities ranging from 5.5 to 11 cm/s. Bed height was 16 and 30 cm. His model prediction showed that for particles <120 μm, cloud resistance caused the drying efficiency to fall; for large particles, the drying efficiency predicted in the presence of cloud and by the hydrodynamic boundary layer model were the same. In other words, the model suggests that for particles of d_p <120 μm, mass transfer in the cloud phase must be considered. The model developed by Alebregtse[16] for drying in a fluidized bed is based on mass balance for solid phase, dense phase, and bubble phase. Simple two-phase theory is assumed to be obeyed. According to his model and experimental findings on the constant drying rate of wet salt (d_p = 40 μm) by air (T_{inlet} = 473 K), the following results were obtained:

1. The bed height above a certain limit has an inverse effect on the drying rate.
2. If the superficial gas velocity is maintained at low magnitudes, the drying gas at the exit can be well saturated.
3. The drying rate can be increased by distributing the gas uniformly and maintaining a small bubble size. Hence, a distributor with a low catchment area is recommended for this purpose.
4. The fluidized bed diameter has an insignificant effect at a high operating gas velocity.

g. Constraints

The above models do not take into account the volume increase of the gas inside the reactor due to the evaporation of moisture. Most basic data used for prediction of the drying rate in fluidized bed models were developed for dry particles that can be well fluidized. The fluidizing characteristics of wet particles are different from those of dry particles. Hence, the models should take this aspect into account. Furthermore, some of the data, such as sorption isotherm and diffusion coefficient for moisture inside the drying particle, are to be experimentally determined for use in model predictions. As previously mentioned, drying is a phenomenon that has complex heat and mass transport, and modeling this using a fluidized bed reactor increases the complexity further. Hence, there is room for further research on this tropic.

E. Vibro Fluidized Bed Dryer

The vibro fluidized bed was discussed when dealing with various types of dryers in the beginning of this section. Vibration in combination with fluidization enables the drying particle to fluidize smoothly. The gentle action of vibration helps the fluidization of fragile materials. Vibro fluidization is widely used for drying abrasive and heat-sensitive materials. The agglomerate which forms during the fluidization of a sticky material is kept in a mobile state, thereby enabling effective fluidization throughout the drying process. The amount of drying gas required to fluidize is reduced considerably. In a conventional fluidized bed, low gas velocity can fluidize fine particles, thereby reducing elutriation. The larger particles remain defluidized, causing partial fluidization of fine and partial defluidization of coarse or large particles. However, in a vibro fluidized bed, fines remain in a fluidized state with less gas, and coarse particles remain in a mobile state due to vibration. Drying of granitic particles ranging in size from 3 to 38 mm in a vibro fluidized bed was reported by Pye.[21] Wet particles near the feed are well distributed in the vibrated state, and even sticky or pasty materials in granulated or extruded form can be successfully processed in vibro fluidized beds. The gentle fluidization in a vibro fluidized bed creates an erosion-free environment for the fluidization vessel even when abrasive materials are handled.

1. Basics

The aerodynamics and thermal characteristics of vibrated fluidized beds were reviewed by Gupta and Mujumdar.[22] Studies on the drying of granular products in a vibro fluidized bed were reported by Strumillo and Pakowski.[23] The additional parameters in a vibro fluidized bed compared to conventional fluidized beds are those pertaining to the amplitude (a) of vibration and the vibration frequency (ω), which is applicable in general for angular motion. The product $a\omega^2$ is termed angular or vibrational acceleration and is used to characterize vibro fluidization. When the ratio $a\omega^2/g < 1$, the bed of solids is in contact with the distributor plate and the solids are kept away or levitated when the ratio is greater than unity. Hence, in vibro fluidization, solids can be brought into an incipient state of fluidization at a low drag force. The incipient velocity for a vibro fluidized bed (U_{mvf}) is smaller than the unvibrated bed velocity (U_{mf}). U_{mvf} can be determined by a conventional plot of bed pressure drop (ΔP_b) versus superficial fluid velocity. The difference between U_{mf} and U_{mvf} depends on the operating conditions. Details regarding the pressure drop–velocity curve were given as a map by Gupta and Mujumdar.[24] The difference between U_{mf} and U_{mvf} is small for deep beds vibrated at small amplitudes and is wide for shallow beds vibrated at high amplitudes. Beds operated between these two extremes correspond to a transition from fixed to fluidized bed. Gupta and Mujumdar[24] defined a new incipient vibro fluidized bed velocity (U_{mm}) for a visually well-mixed state and proposed a correlation:

$$\frac{U_{mm}}{U_{mf}} = 1.952 - 0.275\left(a\omega^2/g\right) - 0.686\left(a\omega^2/g\right)^2 \tag{2.15}$$

Equation 2.15 is valid for $a\omega^2/g < 1$. In general, it is observed that $U_{mm} > U_{mf} > U_{mvf}$. The bed pressure drop ($\Delta P_{b,mvf}$) at U_{mvf} was also presented by Gupta and Mujumdar[24] as:

$$\Delta P_{b,mvf}/\Delta P_{b,mf} = 1 - 0.0935\,(d_p/H)^{0.946}\,(a\omega^2/g)^{0.606}\,(\phi_v)^{1.657} \tag{2.16}$$

This relationship is valid for $25 < \omega < 40\ s^{-1}$, where ϕ_v is the equivalent volume shape factor. Equation 2.16 contains the bed height whereas Equation 2.15 does not. The difference in $\Delta P_{b,mvf}$ and $\Delta P_{b,mf}$ for $H > 0.05$ m was found to be negligible.[24]

Heat transfer in a vibro fluidized bed in general is enhanced, and as a result the heat transfer coefficient in a vibro fluidized bed is much higher even for gas velocities far below the minimum fluidization velocity. Studies by Yamazaki et al.[25] showed that the difference in the heat transfer coefficients of a vibrated and an unvibrated fluidized bed for smaller particles (≈114 μm) increases, and this difference shows a declining effect for larger particles. Heat transfer during constant rate drying can take place near the distributor, and hence the surface area available for heat transfer is lower than the total surface area of the bed. Hence, heat transfer coefficient calculations based on the total surface area of the particulate solid in the bed can result in lower values.[26] In a vibrated fluidized bed, the fraction of gas that passes through the emulsion phase is increased due to the fact that the number of bubbles and their size and frequency are reduced by vibration.[27] Because thermal equilibrium between the gas and the solid particles is achieved quickly, heat transfer between hot gas and solid in the emulsion phase is predominant.

2. *Vibro Inclined Fluidized Bed*

a. *Gas Velocity*

A vibro fluidized bed, in conjunction with an inclined distributor plate, can be used conveniently for drying and transportation of solids on a continuous basis. Such a class of vibro fluidized beds, called vibro inclined fluidized beds, was investigated in detail by Arai and Hasatani.[28] Their investigations mainly focused on the effect of vibration acceleration ($a\omega^2$) on the linear velocity of the solid particles, particle-to-gas heat transfer, and the drying of moist particles. The angle of inclination (θ) of the distributor plate ranged from 3 to 5°. Their investigation of transportation of solids by vibration revealed that particulate solids like sand require a certain minimum velocity to achieve solid transfer, and the linear velocity for transfer is enhanced by the intensity of the vibration acceleration. However, for particulate solids such as polystyrol, the vibration acceleration has no effect in enhancing the linear velocity. A correlation to predict the gas velocity ratio,

$$\left(U_g + U_g^*\right)/U_g$$

was porposed as:

$$\left(U_g + U_g^*\right)/U_g = \alpha\,\Gamma \tag{2.17}$$

where

$$\Gamma = \left(\frac{d_p^2 \rho_s U_g}{W}\right)^{0.3} \left(\frac{\rho_g U_g^2}{\left(\rho_s - \rho_g\right) d_p g}\right)^{-0.4} \left(\frac{a\omega^2}{g\sin\theta}\right)^{0.42} \tag{2.18}$$

$\alpha\Gamma = 1$ for $\Gamma < 3{,}5$, $\Gamma = 0.285$ for $3.5 < \Gamma < 15$, and U_g^* is the difference in gas velocity between the unvibrated and the vibrated inclined fluidized bed for the same linear velocity of the particulate solid in both systems.

b. Heat Transfer

The heat transfer rate between the particle and gas in a vibrated fluidized bed was predicted by the correlation[28]

$$Nu_p = 0.008 \ \mathrm{Re}_p^{0.53} \tag{2.19}$$

where

$$\mathrm{Re}_p = \left(U_g + U_g^*\right) d_p \ \rho_g / \mu_g \tag{2.20}$$

The mechanical vibration added vertically is found to enhance the heat transfer rate, and this would increase the drying rate. The distribution of moisture and temperature along the length of an inclined fluidized bed was predicted by a model which assumes the solid to be in a perfectly mixed state along the height of the bed and plug flow in its flow path. The gas flow is assumed to be in a plug flow state and the interparticle resistance for fine-sized particles is assumed to be negligible. This model[28] can fit well with experimental findings. However, the mathematical equations developed through the model must be solved by numerical methods. A vibro fluidized bed can give an enhanced drying rate uniformly as the linear velocity of the particle is evenly distributed and the velocity of the wet particles can be accelerated by mechanical vibration.

F. Spouted Bed Dryer

Drying in a spouted bed is suitable for coarse particles. The subject of spouted bed is entirely different from fluidization, and it has emerged as a separate topic since its introduction. A schematic representation of a spouted bed is presented in Figure 2.6. Spouted beds have several applications, such as in granulation; coating; drying of paste, solutions, and suspensions; and also for several other gas–solid reactions such as iron ore reduction. Initially, the spouted bed was developed for drying grains. The applications of the spouted reactor were reviewed by Mathur and Epstein,[29] who later published a book exclusively on spouted beds.[30] A spouted bed has either a conical vessel or a conical vessel in combination with a cylindrical upper portion. The spout emerging from the conical bottom penetrates the bed of particulate solids and induces good mixing in particular for coarse particles at lower pressure drops than in a fluidized bed. Particles that are sticky or have a wide size distribution can be processed well. Unfortunately, there is not much in the literature on the scaleup of spouted beds, and there seems to be no large commercial-scale unit in operation. A spouted bed essentially contains a central gas spout surrounded by an

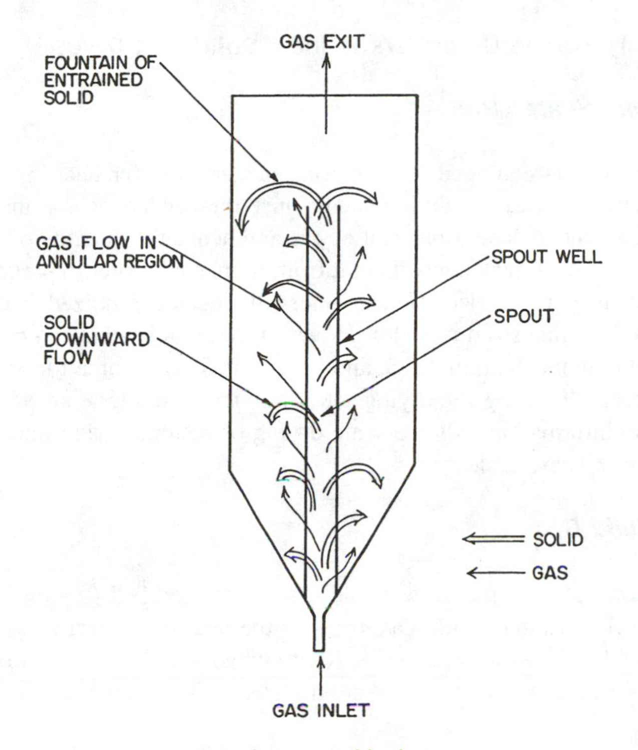

Figure 2.6 Schematic representation of a spouted bed.

annular bed of solids which are continuously drawn into the spout and entrained. The gas in the spout also enters into the annular bed of solids, thereby creating an environment conducive to good gas–solid contact. Spouted beds must be operated above a certain minimum velocity (U_{ms}), termed the minimum spouting velocity, which is similar to the minimum fluidization velocity. Unlike fluidization, a spouted bed has a limit in that there is a maximum spoutable bed height (H_{ms}) beyond which the spout ceases to exist. There are correlations available in the literature[29,30] to predict U_{ms} and H_{ms}. In a spouted bed, one of the important parameters is the spout diameter, which must be evaluated by the force balance in a spout and the surrounding annular region. A spouted bed with a central draught tube was reported[31] to promote better mixing at relatively low power requirements.

Most mass transfer data for spouted beds have evolved from experiments on drying. The downward flow rate of solids in a spout annulus is of a very high order of magnitude (≈10,000 lb/h), and the solid particles are dried in the annulus region while they travel downward. Mathur and Gishler[32] measured the air and solid temperature profile from the inlet and found the rise in the temperature of the particles to be only a few degrees due to their high flow rate and the fact that most drying occurs at the lower portion of the spout. Many useful applications of spouted dryers and their commercial applications on a continuous basis were reported by Romankov and Rashkovskaya.[33] This report is the compilation of the extensive work carried out in the Soviet Union. Spouted bed reactors have been tested successfully for the drying of pasty materials.

G. Internally Heated Dryer Versus Inert Solid Bed Dryer

1. Internally Heated Bed

Fluid bed dryers equipped with internal heaters transfer heat indirectly to the drying material; hence, the drying media just fluidize and carry the evaporated moisture. Thus, the total sensible heat of gas and hence the quantity of gas required are reduced. The heat transfer coefficient from immersed heater to particle increases with decreasing particle size. Hence, internally heated fluidized bed dryers are recommended for fine sized particles. However, internal heaters can be expected to raise the wet bulb temperature (T_{wb}), and hence the driving force for heat transfer is reduced, thereby lowering the drying rate of solids. The literature on this topic is scant, and no information on large-scale drying of material using internally heated fluid bed dryers is available.

2. Inert Solid Bed

A new concept is to dry the wet solids using an inert bed (Figure 2.7) of solids which, upon fluidization by hot gas, transfer the heat to the wet solid and dry the material quickly.[34,35] If the wet solid is light or fine and the inert solid is coarse or heavy, then the dryer can be operated continuously. The exit solids can be separated

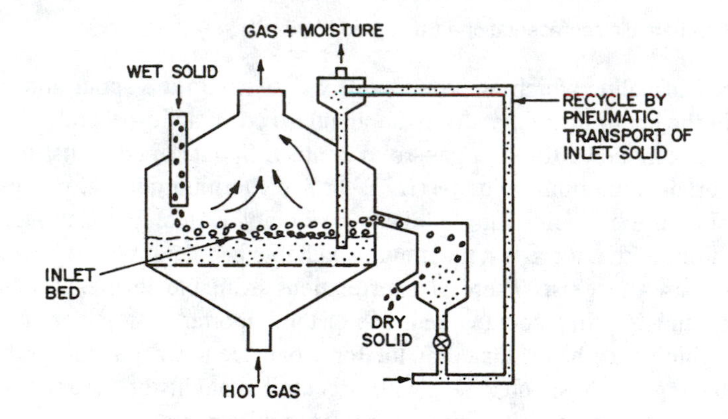

Figure 2.7 Schematic diagram of an inert bed (two-solid) fluid bed dryer.

by a simple cyclone and the inert solid can be recycled. Studies on drying of slurry such as moisture-laden NaCl were carried out[36] for drying on a bed of sand fluidized by hot air. The drying rate equation for such a two-solid dryer proposed by Mousa[36] is

$$R_{(g/min)} = \frac{1.187 \times 10^{-3} A\, G^{0.646}}{\lambda} (\Delta T)_{ln} \tag{2.21}$$

where A is the drying surface area, G is the air mass velocity, $(\Delta T)_{ln}$ is the log mean temperature difference between the drying feed material and the inert solid, and λ is the latent heat of evaporation. This drying rate was compared with the drying rate of through-circulation dryers. Drying by an inert solid bed of fluidized solids was found to be relatively fast.

3. Characteristics

a. Peformance

The design of fluidized bed dryers as well as coolers from a practical standpoint for project and production engineers and for applications in the chemical industry was discussed long ago.[37] In order to gain better insight into the performance of continuous fluidized bed dryers and also to develop the design principles, Vanecek et al.[38] carried out several experimental tests with 21 different materials on 12 continuous fluidized bed dryers of sizes ranging from pilot scale to full-size plant. The volume of drying section tested ranged from 0.0154 to 7.9 m^3, and the grid area used was from 0.0154 to 2.32 m^2. The materials tested were mainly inorganic and included minerals like ilmenite and inert beds like sand. The experimental data showed that the optimum velocity ratio, (U_{opt}/U_{mf}), is relatively lower for coarse particles than for fine particles. The term optimum velocity refers to the condition of good thermal efficiency and high rate of heat transfer. As a rule of thumb, the operating velocity for a fluidized bed dryer is suggested to vary with the square root of the particle diameter. For high gas velocities, perfect mixing tends to be achieved, and residence time distribution in such a situation corresponds to a perfect mixing condition.

b. Residence Time

For a tracer material whose amount is w in a bed of material with holdup W, the outlet concentration of the tracer as a function of the residence time ratio (t/τ) can be given by the equation

$$\log m = \log w/W - 0.4\, t/\tau \tag{2.22}$$

The tracer material distribution from the inlet of the dryer to the outlet can be approximated to follow the unidirectional diffusion of heat under steady state along an infinite slab, that is,

$$D_m \frac{d^2m}{dx^2} - \nu \frac{dm}{dt} = 0 \tag{2.23}$$

where ν is the mean velocity of the tracer material. The dimensionless parameter $\nu x/D_m$ is the ratio of the convection effect to diffusion caused by mixing. For a dryer with a circular diameter, the mean velocity (ν) can be taken as D/τ and the characteristic length as D. Based on the assumption that D_m remains unchanged for most dryers, the parameter D^2/τ is the measure for characterizing ideal mixing (i.e., for low values [10^{-3}–10^{-4} m²/s], mixing is close to the ideal situation). For a scaled dryer, if the value of D^2/τ is greater than 10^{-2}, then deviation from good mixing can be expected. The residence time for drying of materials will be different depending on their size. For example, fine particles that are elutriated have less residence time than coarse particles. Surprisingly, however, the final moisture content of the dried fine and coarse particles is nearly the same. This is attributed to the faster drying rate of fine particles even through their residence time is low.

c. Performance Assessment

The investigations of Vanecek et al.[38] resulted in the following very useful information:

1. The performance of different dryers can be compared based on the temperature drop of the drying gas provided the bed temperature is always kept above 100°C and the mass velocity is the same. A bed temperature of 100°C is advisable because the wet bulb temperature of ambient air has little significance when the dry bulb temperature of the hot gas is above 100°C.
2. The rate of evaporation per unit grid area is independent of dryer size; hence, a plot of this factor (W/a_{grid}) against gas temperature difference can be used to compare the performance of different dryers. Figure 2.8 shows such a plot for three different groups of particles: A for coarse, heavy, and moist (10%) particles; B for the same group A but for particle size ≈1 mm; and C for fine and light particles with less moisture (≈1%). Figure 2.8 is suggested as useful measure to compare different dryers.

H. Applications

1. Iron Ore Drying

Drying of iron ore concentrate in a large-scale fluidized bed was reported by Davis and Glazier.[39] Iron ore concentrate containing 3–4.5% moisture was dried mainly to avoid formation of lumps due to freezing of the moisture during winter and the subsequent associated problem of overland transport by train and unloading the concentrates from and reloading into boats. In order to avoid freezing, the concentrate was dried to a 1.5% moisture level. This job was accomplished in a fluidized bed dryer 3660 mm (12 ft) in diameter. A 4267-mm-diameter wind box below served to preheat the air by combustion of fuel oil and then to distribute the gas through a grid plate provided with 208 stainless-steel tuyeres. Each tuyere was originally made up of 5 rows of 12 holes, 6 mm each in diameter. The total free area for the gas flow through the distributor was 3.8% of the cross-section of the plate. The pressure drop across the tuyere for a gas flow rate of 1500 m³/h was 3 kPa, which corresponded to a distributor-to-bed pressure drop ratio of 0.45.

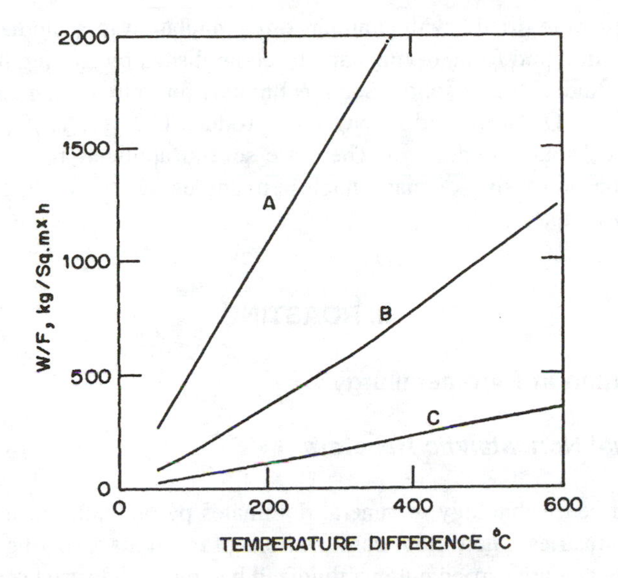

Figure 2.8 Performance of fluid bed dryers for different types of materials as a function of temperature drop on the fluidized bed: (A) coarse, heavy, and moist material; (B) same as Group A, d_p = 1 mm; (C) fine, light, and not very moist materials. (From Vanecek, V. et al., *Chem. Eng. Symp. Ser.*, 66(105), 243, 1970. With permission.)

An equal flow rate through all the tuyeres and smooth fluidization could be ensured at a distributor-to-bed pressure drop ratio of 0.45. The dryer was designed to process wet iron concentrate containing 3% moisture at a rate of 515 long tons per hour. The flow rate of air along with the combustion gas corresponded to a superficial gas velocity of 2 m/s which, along with the evaporated moisture, increased to 2.9 m/s. The residence time of the solid inside the bed ranged from 1.5 to 2 min, and the moisture content at the outlet was 0.1%. The required moisture content of 1.5% of the concentrate was obtained by blending the nearly bone dry concentrate with the wet concentrate. Heat transfer to solid in the fluidized bed dryer was reported to be 350,000–400,000 Btu/($h \cdot ft^3$ of bed volume). The dryers were reported to perform well, and several improvements to avoid erosion at the distributor were suggested. The replacement of tuyeres with many rows of holes by slotted openings in a plane may reduce erosion. Any external abrasion was further reduced by directing the gas jet from the tuyere slot between the tuyeres but not directly on the neighboring ones. Corrosion in metallic parts was observed due to SO_2, the source of which was the sulfur from fuel oil. The overall performance of this large-scale fluidized bed dryer was reported to be useful for economically exploring the technology in similar areas.

2. Miscellaneous Areas

Fluidized bed dryers were reported[40] to dehydrate alumina and remove 80% of the combined water of hydragillite ($Al_2O_3 \cdot 3H_2O$). The alumina thus obtained could readily react with ammonium fluoride to produce ammonium cryolite, which can subsequently be used to produce alumina by the electrolytic method. Zinc cakes from its pulps were dried[41] with simultaneous granulation in a fluidized bed dryer. When feed is in liquid form, drying can be accomplished by feeding the solution in a hot bed of fluidized inert solids. Such techniques for dehydration and calcination of solutions of $UO_2(NO_3)_2$ and $Al(NO_3)_3$ to produce UO_3 and Al_2O_3, respectively, were reported[42] four decades ago. There are several applications of fluidized bed reactors in the processing of many nuclear materials, all of which are dealt with separately in Chapter 3.

II. ROASTING

A. Fluidization in Pyrometallurgy

1. Industrial Noncatalytic Reactors

Fluidized bed technology in general dominates pyrometallurgical extraction in nonferrous industries. The various combinations of reactions involving noncatalytic-type reactions that are carried out in a fluidized bed on an industrial scale are shown in Figure 2.9. By and large, all these noncatalytic-type reactions are of a pyrometallurgical nature. The various areas of application of fluosolid reactors are depicted in Figure 2.10a, and their application for sulfide ores is depicted in Figure 2.10b. Fluid bed reactors in nonferrous pyrometallurgical industries came into commercial application in very large-scale operations probably in roasting. Sulfide concentrates of nonferrous metals such as zinc, copper, and gold and also sulfide ores of iron are roasted on a large scale using fluidized beds. Fluidized bed roasters came into major use in roasting of pyrite[43] or pyrrhotite mainly for the production of SO_2, which is subsequently used in the manufacture of sulfuric acid.

2. Early Fluid Bed Roasters

Fluid bed roasters were originally designed and operated on a large commercial scale by the Dorr Company in the United States and Badische Anilin und Soda Fabrik (BASF) in Germany. Experience on the Winkler generator,[44,45] used for the production of synthesis gas from coal, paved the way for roasting sulfide ores by BASF in Germany in 1943. A 30-ton/day (tpd) commercial unit to produce 36 tons of sulfuric acid was first commissioned in Ludwigshafen (Germany) in 1950, and seven years later a 200-ton pyrite capacity roaster was commissioned to produce 500 tpd of acid. The first fluosolid roaster of the Dorr Company went on line in 1952 at the Berlin, New Hampshire mill of the Brown Company.[46] The SO_2 obtained from the roaster was used to make sulfite cooking liquor. A 4.9-m-diameter, 1.5-m-deep fluidized bed reactor was fed with a slurry of 70–75% solid. During roasting at a temperature of 900°C, the water content of the slurry is instantly evaporated, and SO_2 by oxidation of sulfur and iron oxides from iron-bearing sulfides are formed. The highly exothermic roasting reaction requires close temperature control, which is accomplished either by spraying water (in the Dorr Company design) or by cooling with boiling water that passes through an immersed-type heat exchanger. Similar to

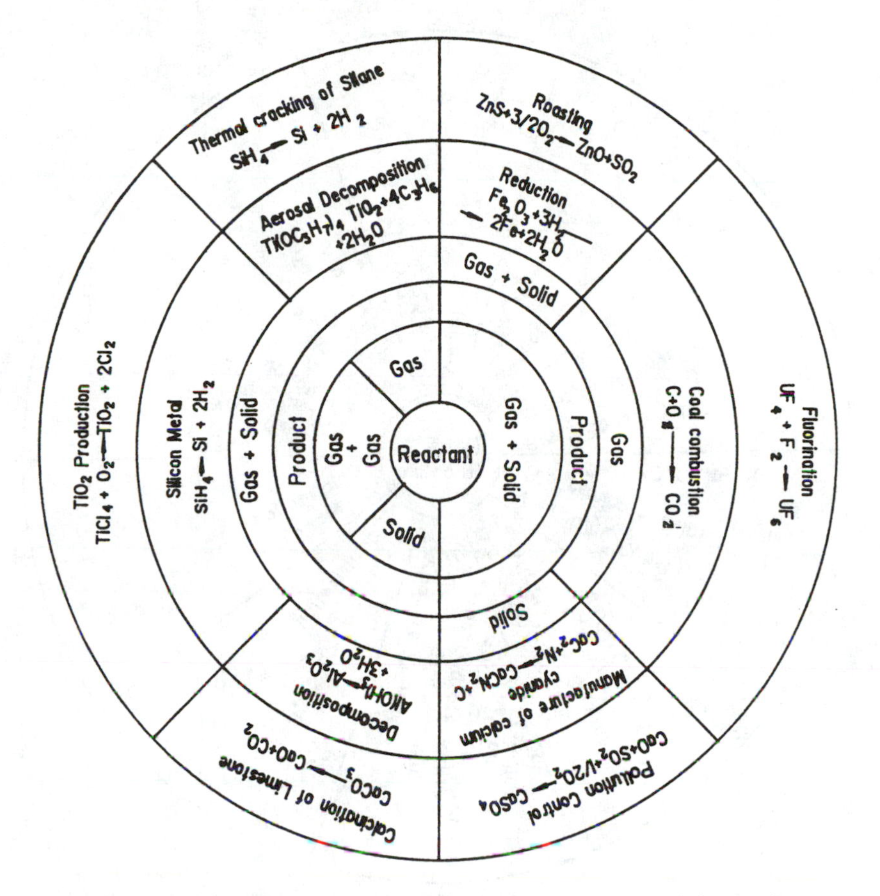

Figure 2.9 Application of fluidized bed for industrially important noncatalytic reactions.

the Brown Company, another fluid bed roaster 6.7 m in diameter and 1.8 m deep for zinc sulfite (150 tpd) was installed at Arvide, Quebec for the production of sulfuric acid at the rate of 100 tpd. Zinc oxide calcine from the roasters is usually subjected to an electrolytic route to recover zinc. Several fluidized bed roasters are now in operation throughout the world. The schematics of the roasters first conceived and operated by Dorr Company and BASF are presented in Figure 2.11a and b, respectively.

B. Zinc Blende Roasters

1. Commercial Plants

The first fluosolid roaster for zinc was put up in 1958 at Belen in Belgium by Viellie Montagne, the sole patent holder for the process, although the engineering design was patented by Lurgi of Germany. The fluidized bed acts as a thermal and chemical reaction inertia wheel which apparently equalizes wide differences in particle size. The development of a fluidized bed roaster for zinc concentrate by the General Chemical Division of Allied Chemical Corporation in the United States and

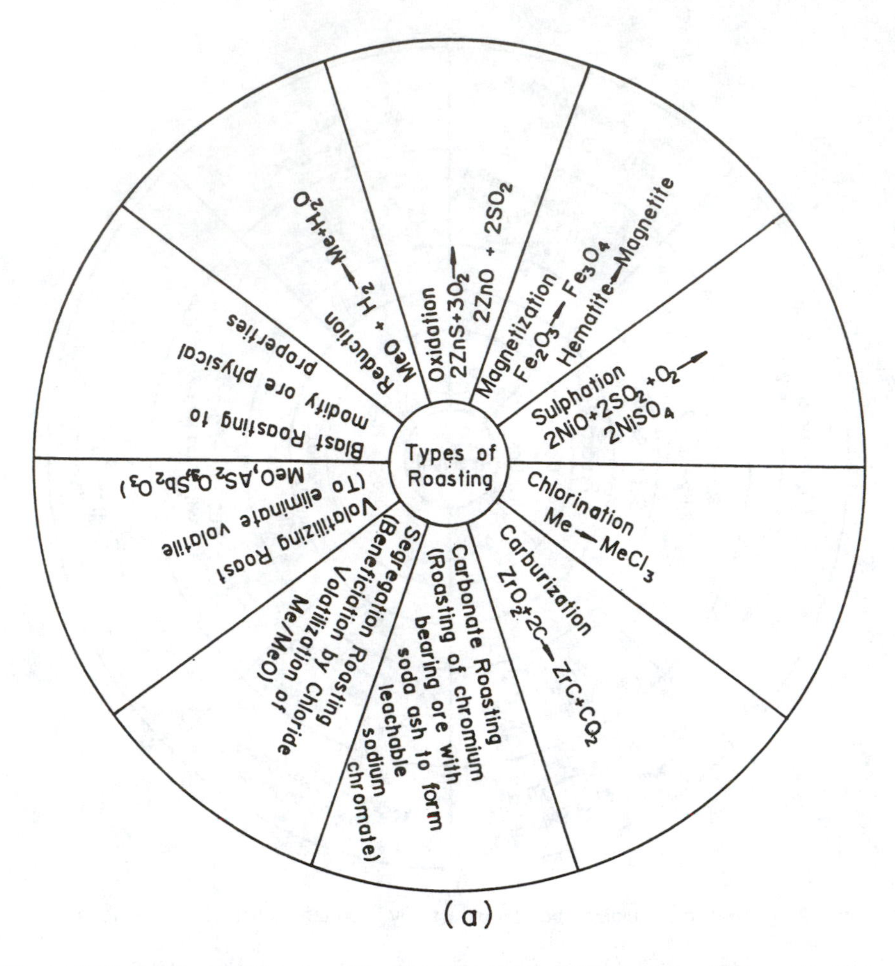

Figure 2.10 Fluidized bed roasters: (a) areas of application of fluosolid reactor for various types of roasting and (b) fluidized bed sulfide roasters.

the setup of a commercial-scale unit at Valleyfield, Canada were discussed in detail by Newmann and Lavine.[47] They projected the design details from the operation of a 152-mm-diameter, 1676-mm-high pilot-scale roaster to a commercial-scale 563-mm-diameter fluidized roaster. In the design of a roaster for optimum roasting, it is essential to control the temperature very closely so as to avoid zinc ferrite formation and to desulfurize efficiently, with a high oxidation rate. The retention time inside the reactor for the solid fines should be sufficient. Hence, a large suspension volume above the fluidized bed is usually recommended. The usual problems associated with large-scale roasters are due to mechanical devices such as screw or belt conveyers which feed the concentrate and discharge the hot calcine. For trouble-free filtration, a good calcine from a zinc roaster should have low sulfate and iron contents. Newman and Lavine[47] reported zinc recovery of 98.6% in their commercial zinc roasters. Zinc ferrite, which is similar to iron ferrite ($FeO \cdot Fe_2O_3$), is believed to be formed at high temperatures and prolonged contact time. Its formation can be

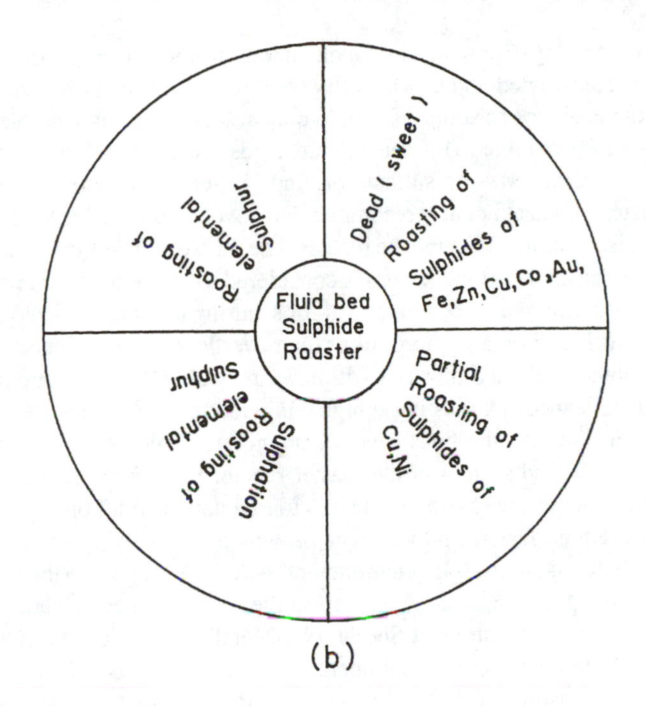

Figure 2.10b

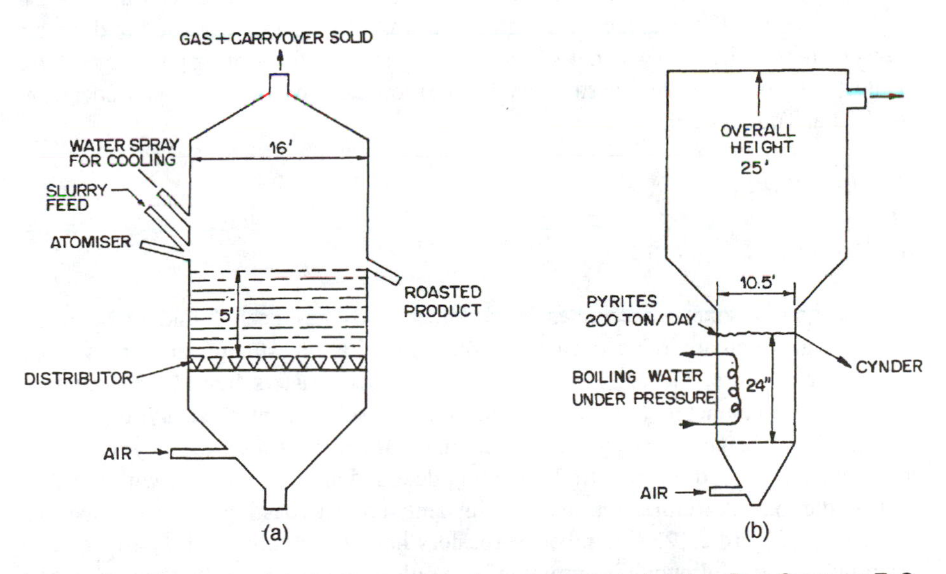

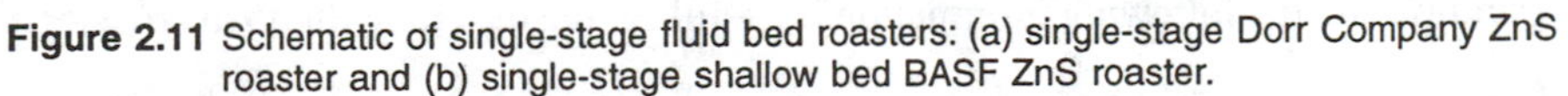
Figure 2.11 Schematic of single-stage fluid bed roasters: (a) single-stage Dorr Company ZnS roaster and (b) single-stage shallow bed BASF ZnS roaster.

suppressed by fine temperature control and a high SO_2 partial pressure. If the temperature is finely controlled, iron ferrite can be formed preferentially and a calcine with good leachability will be produced.

Sanki Engineering Company in Japan, under licence from Dorr–Oliver in the United States, constructed and operated fluosolid roasters ranging in size from 300 to 610 mm in diameter for roasting of sulfide concentrates such as zinc blende, copper zinc, pyrite, and pyrrhotite. The concentrate feeds were either dry or wet, and the roasting was of dead type or sulfate roasting. Based on experience with various fluosolid roasters, Okazaki et al.[48] recommended a wet slurry feed from the standpoint of good feed distribution all around the reactor. The ratio of carried-over particle calcine to the overflow (also termed space rate) is considered to be an important reactor design criterion. A low space rate is recommended for securing a stable fluidized bed. Carried over calcine was found to have more sulfur than overflow calcine. Hence, at low space rates, the combined calcined material will have low sulfate sulfur content. Carryover of calcine can be reduced by roasting at high temperatures (900–1100°C), which would produce fine calcines. As the fines have a tendency to agglomerate, carryover would be brought down. Good sealing of the roaster reactor is essential to avoid leakage of SO_2 from the roaster to the environment so as to maintain a pollution-free environment and also to prevent air leakage into the reactor, which is usually operated at a negative pressure mainly to maintain a safe environment. Any air leakage into the reactor would reduce SO_2 partial pressure, thereby increasing the sulfate content of the calcine. Fluid bed roasting of zinc concentrate at Sherbrooke Metallurgical Co. Ltd. (Ontario, Canada) was found[49] to be an economical operation. The Sherbrooke plant was originally designed for the roasting of 300 tpd of green zinc concentrates. The roaster, which normally operated at 1000–1030°C, also was tested successfully for lead–cadmium elimination by roasting at 1080–1100°C. Cadmium up to 90% and lead up to 92% were eliminated. The heat recovered through a waste heat boiler was used to drive air blowers, heat boiler feed water, dry green pellets, and also for heating purposes during winter. A suspension-type roaster[50] with a 350-ton capacity of ZnS began operation in 1962 at Trail, Canada.

2. Operation

a. Variables Selection

Roasting of zinc concentrates by fluidized beds has been found to be more economical[51] than any other roaster in terms of capital investments and operational costs. Furthermore, the calcines produced by these roasters have been found to possess good characteristics for leaching and the subsequent electrolytic process. The fluid bed behavior, as applied to the zinc roaster, was described by Themelis and Freeman[52] based on industrial operating data and also comparison with theoretical predictions. A fluidization diagram for zinc–copper roasting was developed as depicted in Figure 2.12. A number of roasters have been compared based on their operation and the following information is useful for the design of fluid zinc roasters:

1. The operating velocity ranges from 30 to 50 m/s, while the load factor is generally kept at 0.3 metric ton/h/m^2 of the grate area. One of Lurgi Chemie's largest zinc roasters, commissioned for Electrolytic Zinc Co. of Australia, Risdon, had a grate area of 123 m^2 and operated at a load factor of 0.27.

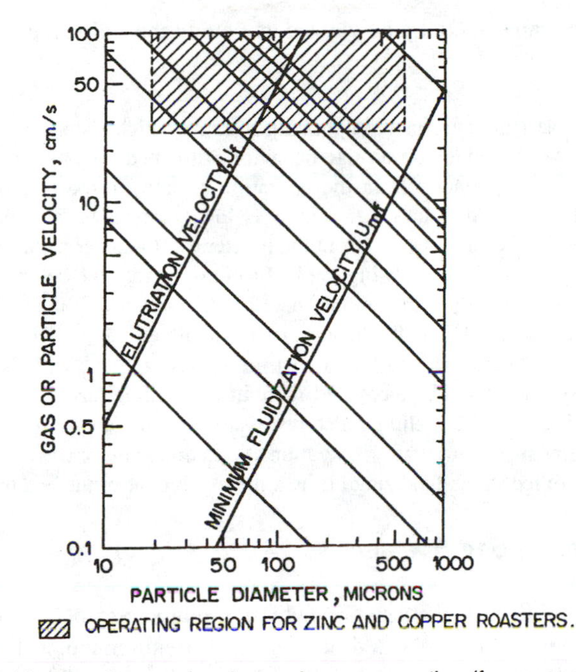

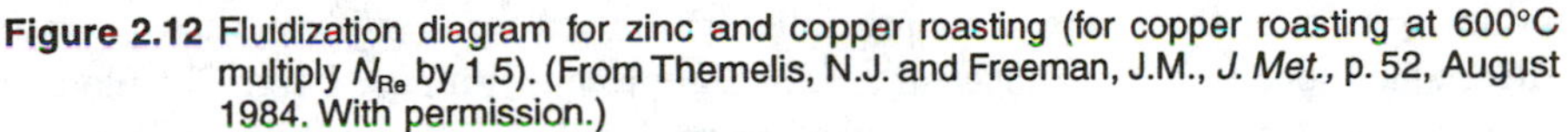

Figure 2.12 Fluidization diagram for zinc and copper roasting (for copper roasting at 600°C multiply N_{Re} by 1.5). (From Themelis, N.J. and Freeman, J.M., *J. Met.*, p. 52, August 1984. With permission.)

2. The particle residence time varies depending on the size distribution and falls in the range 0.4–12 h over a particle size range 22–315 μm.
3. The bed voidage is around 0.6, and 50% of the bed volume is estimated to be occupied by bubbles.
4. Bubble size is estimated to fall in the range 15–20 cm in diameter.
5. For a 20-cm bubble diameter, the superficial velocity corresponds to 99 cm/s.
6. The gas flow in the freeboard is within the turbulent regime (i.e., $4000 < Re < 100{,}000$).
7. The fluid bed temperature fluctuates within ±30°C and the heat transfer rate is not a controlling factor at the usual operating temperature as the sulfide particles ignite particles.

b. Constraints

A fluosolid roaster for zinc concentrate was first used in India at Debari in Udaipur by Hindustan Zinc Ltd. The details of this roaster (in operation since 1968) and the operational data were reported by Adhia.[53] The roaster, which has a 4850-mm bottom hearth diameter and an 8100-mm upper enlargement section and is fitted with 1848 special steel nozzles of 28-mm bore placed at 100 mm spacing, is used to roast 6.6 ton/m^2 of the hearth area. An endless belt drive (speed 72 km/h) feeds the concentrate, and a root blower (capacity 16,000 Nm^3/h) supplies air. The roaster is heated initially by oil firing up to 500°C and then with sulfur roasting up to 850°C. Thereafter, actual feeding of the concentrate and roasting are accomplished.

A zinc roaster can essentially be viewed as having the following three important constraints:

1. The first constraint concerns the operating regime, which should be between the particle carryover and the piston-like behavior of the bed. The interparticle forces which play an important role in the operating regime can be controlled by the addition of water. When the water content is low, particles become dry and light and are likely to be carried away from the roaster. On the other hand, wet particles become heavy and settle, allowing the bed to have piston-like behavior.
2. The second constraint pertains to the operating temperature; if the bed is allowed to operate above 1050°C, defluidization is inevitable due to particle sintering.
3. The third constraint relates to the control of sulfide–sulfur below 0.2%. This is essential, as zinc should be recovered without loss during leaching; furthermore, this low value of sulfide–sulfur is also necessary to maintain the roasting temperature, thereby sustaining roaster operation. Sometimes the oxygen concentration can be controlled in the fluidizing air as a means to control the bed temperature.

3. Turbulent Fluid Bed Roaster

A turbulent fluidized bed roaster in which a major portion of the material is carried over by the gas (and collected in post-reactor equipment such as waste heat boiler, cyclones, and electrostatic precipitator) is used for roasting pyrites, pyrrhotites, zinc blendes, copper concentrates, and cupronickel matte. Lurgi Chemie and Huttentechnik GmbH built and operated several turbulent-layer fluid bed roasters with capacities varying from 50 to 900 tpd; such roasters are automated and operated by a control center.

4. Design Data

Fluidized bed roasters have been in operation on a commercial scale for several decades. Yet there are still developments taking place on many frontiers in this field. More emphasis is now given to pollution abatement and instrumentation. The roasting reaction in a fluidized bed has been the subject of much research. Bradshaw[54] briefly described the roasting reaction, and the rate phenomena in many roasting processes were discussed by Sohn and Wadsworth.[55] Some important aspects of fluidized bed roasting of sulfides were discussed by Jully[56] and Pape.[57] In view of the varied design and ancillary parts of fluidized bed roasters, it is often important to know the details of the complete roasting systems; a schematic of one such system is depicted in Figure 2.13. The dimensions and operating parameters[52,58] of various commercial-scale zinc roasters and typical particle size distributions for feed and calcine are presented for comparative and design purposes, respectively, in Figure 2.14a and b. Commercial design details of roasters for pyrite, pyrrhotite, copper sulfide, copper–cobalt sulfide, and zinc sulfide, along with data on heats of reaction, were presented by Blair.[59]

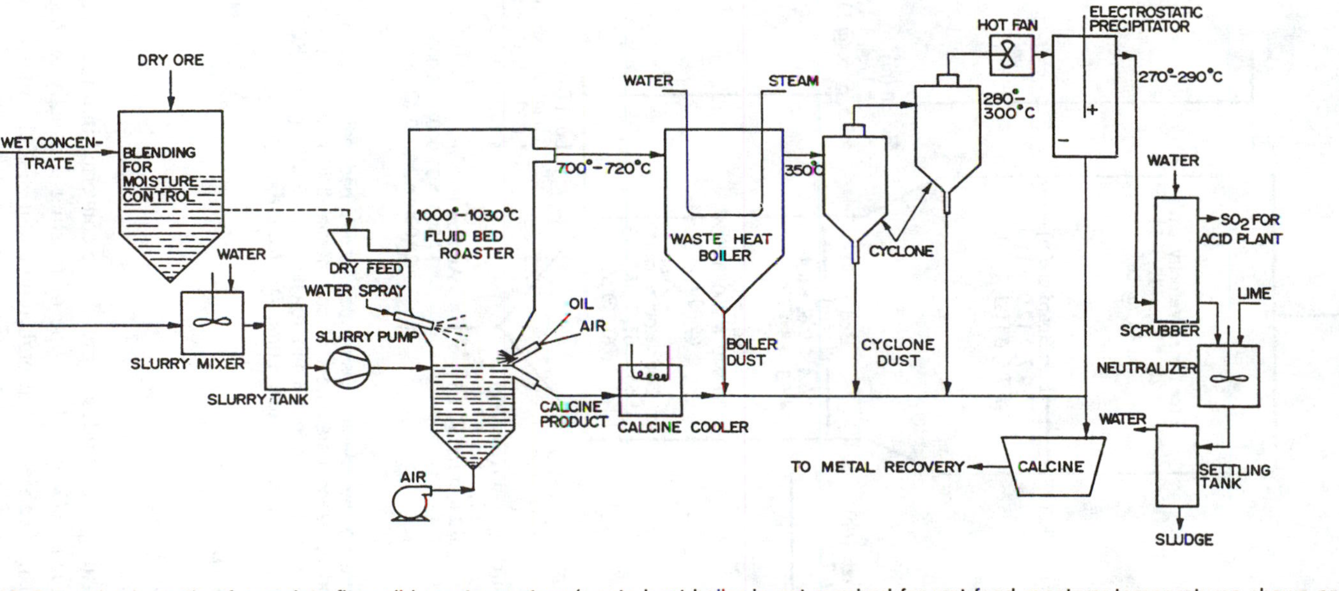

Figure 2.13 General schematic of complete fluosolid roaster system (waste heat boiler is not required for wet feed roasters; temperatures shown are for typical ZnS roaster[49]).

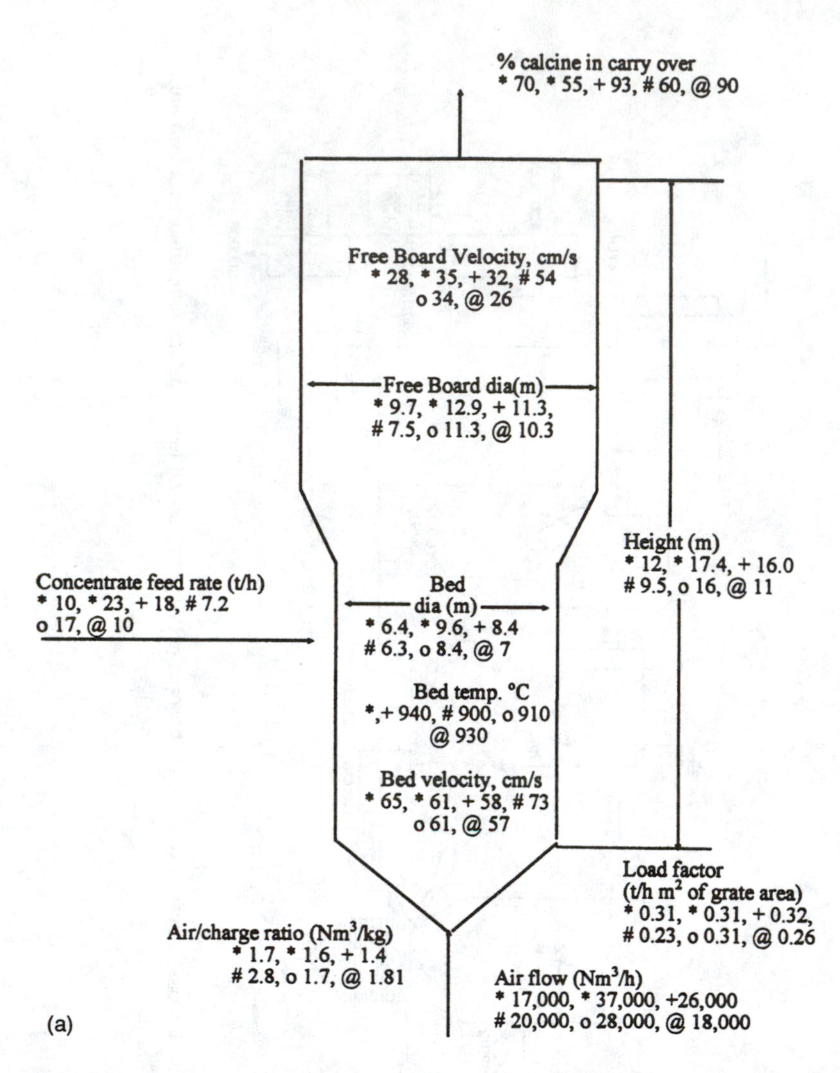

Figure 2.14 Dimensions and particle size distribution for fluid bed roasters: (a) typical dimensions and operating parameters of commercial fluid bed zinc roasters.[52,58] (b) Particle size distribution in a 6.4-in. roaster. * = CEZ, Lurgi; + = KCM, Lurgi; L = Trepca (Dorr–Oliver); o = Ruhr, Lurgi; @ = Norske, Lurgi.

C. Sulfation Roasters

1. Sulfation Principles

In oxidative roasting, the sulfides of metals are converted into their oxides, thus eliminating all sulfur as SO_2 gas. This type of roasting is usually termed dead or sweet roasting, and it is carried out at high temperatures. If the roasting is carried out at relatively low temperatures under controlled partial pressures of O_2 and SO_2, the sulfides of many or one of the metals can be converted into sulfates. The most

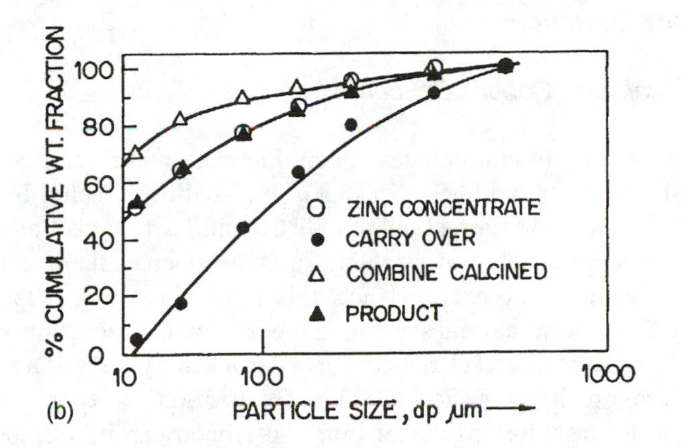

Figure 2.14 (continued)

general reaction for sulfation of many divalent metals such as copper, nickel, cobalt, zinc, etc. can be represented as:

$$MeS + 2O_2 \rightarrow MeSO_4 \tag{2.24}$$

$$2MeO + S2O_2 + O_2 \rightarrow 2MeSO_4 \tag{2.25}$$

The oxidation of SO_2 to SO_3 is a highly exothermic reaction, and hence it is favored at low temperatures. Sulfation, in principle, occurs when SO_3 reacts with a metal oxide. Sulfation is favored relatively at lower temperatures than dead roasting. Based on the thermodynamics of the sulfation reaction (Equation 2.25), the governing relationship for the partial pressures of O_2 and SO_2 is

$$\log P_{SO_2} = -1/2 \log K - 1/2 \log P_{O_2} \tag{2.26}$$

where K is the equilibrium constant for sulfation roasting. A predominance area diagram on log P_{SO_2} and P_{SO_3} is used as a key for sulfation. As an example, for nickel sulfate formation, the oxygen and SO_2 concentration should each be 3–10% at atmospheric pressure, and a roaster gas composition of 8% and 4% O_2 is most suitable for $CoSO_4$. By means of sulfation roasting, manganese from manganese ore and also from manganese sea nodules and copper and nickel from their sulfide concentrates that contain FeS can be sulfated. Similarly, cobalt oxide, galena (PbS), ZnS, and uranium shale can also be sulfated. The details of the above sulfation reaction were discussed by Sohn and Wadsworth.[55] Sulfation roasting, in principle, requires close control of temperature and also the partial pressures of O_2 and SO_2. This can be accomplished well in a fluidized bed reactor. Sulfation roasting for nonferrous metals was discussed by Stephens[60] as far back as four decades ago. The kinetics of sulfating pyrite cobalt concentrates in a fluidized bed and sulfating cobalt containing nickel and copper were investigated[61,62] in the 1960s.

2. Fluid Bed Sulfation

a. Iron, Nickel, and Copper Concentrates

Sulfation roasting of concentrates containing iron, nickel, and copper sulfate was studied by Fletcher and Shelef[63] using a fluidized bed reactor. In this type of sulfation, which has to be carried out at 680°C, sulfides of nickel and copper are preferentially converted into water-soluble sulfates, whereas the iron is converted to a stable water-insoluble oxide. Hence, this type of roasting helps directly in beneficiating the concentrate and separating the iron value from copper and nickel. Water-soluble sulfates of nickel and copper, after leaching the sulfated mass from the roaster, are usually subjected to solvent extraction to separate copper and nickel. The water-insoluble oxides of iron and silicon can be further treated by magnetic separation.

During sulfation, it is essential to control the temperature precisely, and this is possible in a fluidized bed reactor. In the event of large temperature variations, which are quite likely in conventional roasters such as shaft furnace, rotary kiln, and multiple hearth roasters, there can be some adverse reactions, such as the formation of water-insoluble ferrites of nickel. Fletcher and Hester[64] found that sulfation is hindered by the formation of impervious sulfate on the oxides of the metal. In order to eliminate this, Fletcher and Shelef[63] studied the effect of the addition of sulfates of alkali metals (i.e., lithium, sodium, cesium, rubidium, potassium) during sulfation roasting in a fluidized bed. The alkali metals are considered to enhance the sulfation unhindered as they form a thin liquid on each particle and dissolve the sulfate coat formed on the oxide particle. The alkali metal sulfates are also believed to decompose ferrites. For example, $Na_2S_2O_7$, which is formed by the reaction of Na_2SO_4 with SO_3, further reacts with nickel ferrites (NiO, Fe_2O_3) and decomposes the ferrites, forming the sulfates of nickel as well as sodium. It is essential to eliminate the ferrites, as they form a continuous series of solid solutions with Fe_3O_4, thereby rendering them inseparable. The fluidized, by virtue of its isothermal operating condition, is a potential reactor for sulfation roasting.

b. Cobaltiferrous Pyrite

A pyrite concentrate containing 0.35–0.40% cobalt was sulfation roasted in a 14-ft Dorrco fluosolid reactor at 630–675°C with a feed rate of 28–40 tpd in pilot-scale plants[65] at Akita, Japan. Addition of 0.3% Na_2SO_4 was found to enhance sulfation. The sulfation roasting of nickel-, copper-, and zinc-bearing cobaltiferrous pyrite concentrates as practiced by Outokumpu Oy, Finland, was reported by Palperi and Aaltonen.[66] The sulfation roasting was carried out in a fluidized bed reactor with a feed mixture of green concentrate. The calcine was obtained from dead roasting using a fluidized bed reactor. The fluidized bed reactor used was 7.5 m high and rectangular in cross-section with four compartments, each 16 m^2 in area (i.e., 4 m × 4 m). The fluidized charge height was 2–2.5 m and the reactor operated at 680°C. The temperature of the reactor was controlled by varying the feed and also by spraying water. Air was used for both reaction and fluidization.

For efficient sulfation in a fluidized bed reactor, it was found essential to distribute the air uniformly. The use of a distributor with a closely placed opening with high pressure drop was found necessary to achieve this condition. This corresponded to 20–30% of the sum of the pressure drop across the grate and the bed. The fluidized bed, divided into several compartments, can be used for reaction with different gas compositions at good temperature control for a longer retention time inside the reactor. A mean residence time of 25 h was found to be essential for complete sulfatization. The mean residence time and reactor volume could be brought down by 50% if the unreacted mass was separated magnetically and recycled. The fluidized bed reactor wall, in the course of time, was deposited with particulate lumps; when these fell into the fluidized charge, they caused defluidized zone formation. The occurrence of this deposit was prevented by building the reactor wall with a slight inward inclination. The sulfated charge from the fluidized bed was leached to recover cobalt, nickel, and zinc values from the pregnant solution. The residue obtained after leaching was rich in iron content.

c. Cupriferrous Iron Ore

A fluosolid sulfation roasting plant producing 15 tpd of cupriferrous iron ore was reported by Kwauk and Tai.[67] The fluidized bed reactor was divided into three sections. The sulfation was carried out under dense-phase fluidization at 500–550°C in a reactor with a 0.5-m bottom section and 3 m high. The preheated charge was fed through a valve positioned at the top of the disengaging zone (0.85 m ID and 3 m long). The sulfation was carried out using 6–7% SO_2 obtained separately from a pyrite roaster. A dilute-phase heat transfer section (0.85 m ID and 12 m long) above the disengaging section served as a preheater for the raw ore. The preheating section was provided with baffles, and preheating was accomplished by combustion of producer gas with air. The preheating reduced the combusiton time of pyrite to about one-fifth compared to that of autogeneous roasting carried out by mixing pyrite directly with ore. Sulfation roasting of chalcopyrite ($CuFeS_2$) concentrate in a fluidized bed was reported by Griffith et al.[68] This sulfation roasting was followed by leaching and electrowinning of copper in an economical way.

D. Magnetic Roasting

1. Importance

Magnetic roasting is an important step in beneficiating low-grade iron ores. By this roasting, nonmagnetic iron materials (mostly hematite [Fe_2O_3]) are converted into the magnetic form (Fe_3O_4). This is a reduction roasting, and the amount of oxygen removed by reducing the hematite to magnetite is merely one-ninth of the total oxygen. This type of roasting has a large requirement for thermal energy. However, the conversion of a nonmagnetic iron ore to a strongly magnetic material renders the low-grade iron ore suitable for easy beneficiation by magnetic methods. The reduction roasting of nonmagnetic iron material is followed by oxidation that yields a magnetic form of hematite (i.e., maghemite [γ-Fe_2O_3]).

2. Parameters

Magnetic roasting can be accomplished over the temperature range 500–690°C. The higher the temperature, the greater the reduction achieved in a short period. The reduction gas stream can have 5% H_2, 3.3% water vapor, and the rest nitrogen gas. Additions of alkaline metal oxides or alkaline earth oxides were not effective in reducing the induction time observed in a kinetic plot which showed a sigmoidal-type curve for percent reduction versus time plot. These results establish the characteristics of solid–solid transfunction. The many kinetic studies on magnetic roasting were reviewed concisely and lucidly by Khalafalla.[55]

The reduction of iron oxide directly to metal has been thoroughly studied, and this subject will be discussed later in this chapter. Although magnetic roasting dates back to the 19th century, not much has been reported on an industrial scale. However, this technique is known to be a very effective step for iron ore beneficiation when conventional methods such as froth flotation, gravity concentration, and direct magnetic separation are deemed to be unfit. Magnetizing roasting can be carried out in equipment such as a rotary kiln, shaft furnace, traveling grate roaster, and fluidized bed. Of these, the fluidized bed has distinct advantages due to its high rates of heat and mass transfer and nonmechanical agitation of the reacting species.

3. Reaction

The reduction of iron oxides by CO as well as by H_2 was reported by Narayanan and Subramanya.[69] The most general forms of the reduction reactions are

$$3Fe_2O_3 + x \rightarrow 2Fe_3O_4 + y \tag{2.27}$$

$$Fe_3O_4 + x \rightarrow 3FeO + y \tag{2.28}$$

$$FeO + x \rightarrow Fe + y \tag{2.29}$$

where x is 1 mol of CO (or H_2) and y is 1 mol of CO_2 (or H_2O).

Dorrco fluosolid roasters[70] were reported to have flexibility of fuel usage with the added advantage of using solids as well as gaseous reductants. Tomasicchio[71] described the magnetic reduction of iron ores by the fluosolid system by direct fuel injection. The performance of the first commercial Dorr–Oliver installation at the Motecatini Edison in Follonica, Italy was summarized and the operating requirements along with the basic dimensions of a plant to process 200 tpd of iron ore were presented.

4. Fluid Bed

a. IRSID Process

Magnetic roasting of iron ore by the IRSID (French Iron and Steel Institute) process was discussed by Boueraut and Toth.[72] The study was conducted jointly with the French Engineering Company CETIG to develop 1-metric ton/h pilot plant based on the results of a 200-kg/h fluid bed roaster. The study used a Lorraine (Europe) iron ore deposit, and magnetic roasting was chosen because of the high transportation cost from the mine to the steel plant and the high slag/pig iron ratio (i.e., 1.1:1.0) obtained in the blast furnace. After magnetic roasting, silicate and nonferrous gangue phases were eliminated by magnetic separation. As a thermal-intensive process, magnetic roasting needs to be carried out at a high thermal efficiency. Hence, the IRSID process used a fluid bed reactor in which heat was dissipated directly from an immersed gas burner. Direct heat transfer from the burner wall and the hot emerging gas into the fluidized bed was accomplished by this method. The sensible heat from the hot products and the exit gas were exchanged using a fluidized bed solid–solid heat exchanger. The exchanged heat was used to preheat the raw feed solid and the air to the burner. The reactor was operated at 650°C. Fuel oil and air consumption per metric ton of the dry ore were 21 kg and 210 Nm^3, respectively. The pressure drop recommended was 0.08 kg/m^2 across the distributor, and the roaster gas velocity used ranged from 15 m/s at the reactor bottom to 0.7 m/s in the main fluidized bed. In the IRSID process, careful attention was paid to recovery of heat from the fluid bed reactor as the reactor used for magnetic roasting should be thermally efficient, yielding a product with the best metallurgical properties. The heat consumption of the furnace varied from 130,000 to 270,000 kcal/ton of the dry ore depending on the mineralogical structure of the ore.

b. Two-Phase Reactor

A semiconveying two-phase fluid bed magnetizing roaster was reported.[73] This roaster consisted of three sections: (1) a 1.05-m dilute-phase top section for preheating the incoming feed as well as reducing gases (4% H_2 and CO), where the reducing atmosphere was maintained by combustion of producer gas; (2) a combustion zone at one-third the height from the bottom to burn the unreacted producer gas emerging from the reduction zone; and (3) a 0.825-m-ID dense-phase fluidized zone for reducing the fine ore to magnetic Fe_3O_4 using the product gas as the reductant.

The magnetizing reduction of iron ores was more recently reviewed in detail by Uwadiale,[74] and this review provides details on the process chemistry, the mechanism of reduction, the type of reductants, and the nature of the reduction products. In view of the high energy cost of magnetizing roasting and the desirable properties of roasted magnetic ore for easier separation, an efficient furnace is important, and in this regard fluid bed reactors have found a place in magnetic roasting.

E. Segregation Roasting

1. Cupriferrous Ore

Metals such as antimony, bismuth, cobalt, gold, lead, palladium, tin, and silver form volatile chlorides. An ore containing such metals can be subjected to segregation roasting (also called chlorometallization) to form volatile chlorides which, when reduced by carbonaceous matter, deposit as metal on the reductant surface. A fluidized bed reactor is the right choice for carrying out segregation roasting. Copper from cupriferrous ores can be extracted by heating the ore by direct coal injection in fluidized cupriferrous ores and then reacting this preheated charge with NaCl and powdered coal in a separate reactor where the copper content of the ore is transferred in gaseous form to the surfaces of coal particles and subsequently reduced to metallic copper. Copper metal can then be extracted from the copper-coated coal particles by ore dressing operations. A schematic of a fluidized bed segregation roaster is shown in Figure 2.15. A 250-tpd segregation pilot-plant roaster was reported by Kwauk.[73] In segregation roasting that uses NaCl, the silica in the ore combines with Na to form Na_2SiO_3, and the HCl vapor formed at 700–900°C is responsible for chlorinating the copper content. This type of roasting in a fluidized bed is useful for beneficiating copper and copper–nickel-oxidized ores.

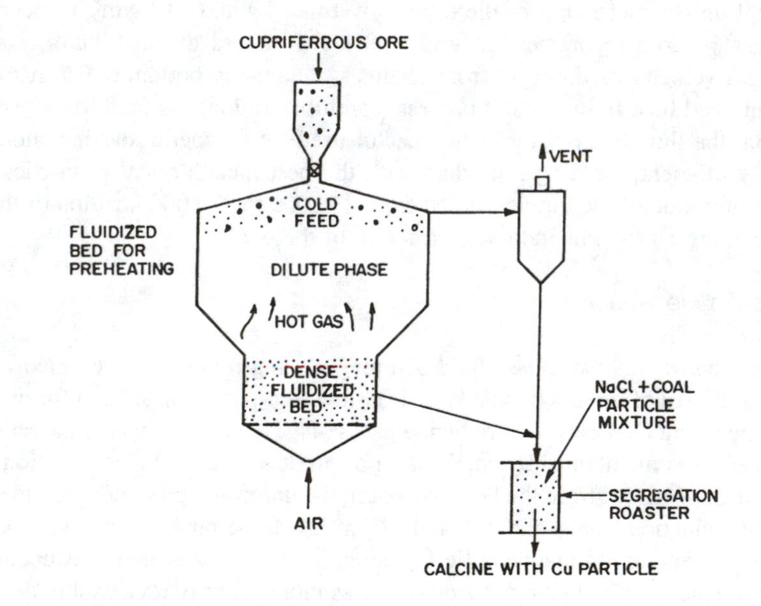

Figure 2.15 Fluid bed segregation roaster.

2. *Other Processes*

A copper segregation method called the TORCO process was reported[75,76] in which H_2 gas was used as the reducing agent. Chloridization during segregation roasting seemed to be the rate-controlling step. Brittan[77] developed a kinetic model for chloridization in a segregation process and applied it to fluid bed reactors. The TORCO process was proposed for many oxidic ores.[78] Studies on the segregation of nickel from its oxide and silicate ores by chlorination at 950°C in the presence of activated charcoal were reported by Pawlek et al.[79]

F. Fluid Bed Roasters for Miscellaneous Metal Sulfides

1. *Chalcocite, Copper, and Arsenic Concentrates*

The experience[80] of building and operating a fluosolid roaster for copper concentrates from an experimental setup 300 mm in diameter to a scaled-up large-scale 3660-mm-ID × 4880-mm-tall reactor at Copperhill by the Tennessee Copper Company in the early 1960s established the ability of fluosolid roasters to handle fine ore concentrates. A sand bed was employed to handle the fine concentrates. This sand bed could sustain the operation of the fluidized bed and also helped as the furnace fluxing. The roaster which operated at 590–650°C, was controlled by adjusting the feed rate. The purpose of the roaster was to increase the smelter capacity and to improve sulfur recovery. These objectives were well achieved.

Oxidizing roasting[81] of chalcocite (Cu_2S), carried out in a fluosolid roaster at 675–680°C, was reported to have optimum recovery of copper and cobalt. The Kennecott Copper Corporation[82] was successful in operating a fluid bed roaster at Ray Mines for copper concentrates. Fluidized bed roasting of arsenopyrites to obtain arsenic-free iron ore was discussed by Vian et al.[83] The roasting was proposed to be carried out in two stages. In the first stage, most of the arsenic would be removed by roasting in a fluidized bed with an atmosphere of air deficiency. In the second stage, sufficient air would be supplied to completely oxidize iron and also sulfur from the raw ore. Thermodynamic considerations identified the conditions under which arsenic would be fixed in calcine and oxidized iron. Roasting of arsenic pyrites in a fluidized bed to eliminate arsenic proved to be successful.

2. *Pyritic Gold Ore*

The roasting of pyritic gold ore to produce acid for extraction was carried out as early as four decades ago. The first gold-ore fluidized bed roaster[84] was built in 1947 by Cochenour William Gold Mines Ltd., in northwestern Ontario. The fluidized bed roaster was 2 m in diameter and 6 m tall. The aspect ratio was kept near unity. The reactor was provided with 120 cup-shaped orifices. A korundal ball 75 mm in diameter rested in each cup-shaped orifice, and this served to distribute the air and also prevented the bed materials from falling down into the air chamber. Roasting under a controlled air supply at 590°C could remove the arsenic and sulfur from the pyritic gold ore, and the calcine obtained was suitable for processing by cyanidation. The charge was originally fed into the bed by a screw feeder. Later, the charge was sprayed above the bed in the form of slurry. Gold ore was usually ground to 200 mesh to remove gangue and then subjected to froth flotation. The fluidized bed roaster was charged with such a fine floating concentrate and this proved to be viable. Processing of refractory gold ores and the roasting methods, including the application of circulating fluidized beds, were discussed by Fraser et al.[85]

3. Molybdenite and Cinnabar

Oxidative roasting of molybdenite (MoS_2) to MoO_3 is an important step in the extractive metallurgy of molybdenum. This roasting has to be carried out under close temperature control at 650°C. Above this temperature, MoO_3 will volatilize. As the roasting of MoS_2 is highly exothermic, close control of temperature can be accomplished in a fluidized bed reactor. Temperature control during fluidized bed roasting of MoS_2 was reported by Golant et al.[86] In order to achieve roasting of MoS_2 in two stages, namely, chlorination of the oxides followed by hydrogen reduction, a new fluidized bed known as a compartmented fluidized bed was proposed by Rudolph.[87] This new technique is discussed in Chapter 6.

Fluidized bed roasting of Egyptian molybdenite was reported by Doheim et al.[88] Their investigations showed that agglomeration-free roasting was possible with close temperature control, uniform air distribution, near-coarse average particle size distribution, and addition of inert bed material. Distributor plates were found to have a profound effect on the roasting process. Porous-plate-type distributors show the best performance. More information on the extractive metallurgy of molybdenum, the roasting of molybdenite, and the relevance of the fluidized bed for this duty was provided by Gupta.[89] A double-section (in a 2-m-diameter bottom × 3.6-m top) two-phase fluidized bed cinnabar roaster with a 300-tpd capacity to roast low-grade (0.06% Hg) cinnabar of crushed particle size as large as 12 mm was used[73] for mercury extraction. The roast temperature was kept at 800°C and the feed contained 9% coal.

G. Troubleshooting in Fluid Bed Roasters

1. Operation

Fluid bed roasting is always accompanied by operating and exit gas cleaning problems. When a fluidized bed zinc roaster is operated at higher temperatures (i.e., 1150°C), lead and cadmium content up to 80% can be removed.[90] However, this high temperature can lead to sintering of the bed materials, thus rendering the charge defluidized. At low operating velocities, there may be agglomeration of fine-sized particles. This can, however, be overcome with higher fluidization velocity. Coarse particles which do not have a tendency to agglomerate can create fine dust problems due to attrition. Sintering is generally due to hot spots caused by improper control of temperature for highly exothermic reactions such as those that take place in roasters and is also due to uneven gas–solid contacting in an improperly fluidized bed. For many exothermic reactions, good control of the bed temperature can be accomplished by mixing the charge with inert bed materials. For some kinds of roasting, such as the roasting of molybdenite, close temperature control is essential. The final residual sulfur content of the charge is controlled by: (1) bed temperature, (2) gas flow rate, (3) oxygen content of the gas, and (4) retention time of the charge in the bed. In the case of molybdenite roasting, temperatures above 580°C cause the bed to collapse due to sintering. As far as mechanical operations are concerned, charge feeding and product removal usually pose technological challenges. Wet solid spray, dry solid injection, and calcine removal with efficient heat recovery are all important from the standpoint of technoeconomic benefits.

2. *Exit Gas*

It is of utmost importance in any roaster to strictly control the dust and the fumes from the exit stream of the reactor. In today's atmosphere of strict pollution control laws and concern for workers' safety, fluid bed roasters should be able to completely eliminate all the dust and fumes. Conventional cyclones and electrostatic precipitators can do the job effectively. Dust loading of the gases[91] at the exit of the electrostatic precipitator is around 0.1 g/m^3. In the case of cinnabar roasters, where dust entrainment is severe, it is essential to use an efficient dust removal system. Most dust collection equipment for roasters works in hot corrosive environments, thereby increasing the problem of periodic shutdown and maintenance of the post-reactor systems. In some roasting operations, like molybdenite roasting, rhenium passes into the exit gas stream as Re_2O_7, and this source material must be collected from the dust. Circulating fluidized bed technology has come a long way in recent times. Collection of the entrained solids and recirculation still remain more an art than science and technology. Highly expanded fluid beds are applicable[92] for carrying out many exothermic and endothermic processes.

3. *Models*

Not much work has been reported on the modeling of fluidized bed roasters. Fukunaka et al.[93] proposed a model for oxidation of zinc sulfide in a fluidized bed. The model was developed based on two-phase theory and is applicable for a batch fluidized bed. This model could explain why most of the reactions occur in the emulsion phase with gas–film mass transfer control. Models for continuous fluidized bed operation are scant in the literature. Hence, design problems for many fluidized bed roasters remain unresolved due to lack of basic studies on large commercial-scale roasters.

III. CALCINATION

A. Definition

Calcination is a decomposition reaction by which chemically combined water of hydration and carbon dioxide are removed. It is an endothermic reaction. The term calcination is derived from the Latin word *calx*, which means chalk, and this was later used for earth and oxides. Although the oxidation of sulfides, dehydration, and thermal decomposition of carbonates are also calcination reactions, the last term is often called thermal decomposition of carbonates in the literature. Calcination is carried out for many commercially important materials such as dolomite (i.e., calcium carbonate), bauxite, magnesite, phosphate rock, etc. Calcination of limestone is a reaction of extractive metallurgical interest, as lime is used in steel as well as nonferrous industries. Calcination can be carried out in conventional furnaces such as a shaft kiln or rotary kiln. Because the heat requirements for these reactions are high, it is essential to conserve energy and carry out the calcination in an energy-saving reactor. As fluidized bed reactors are compact in size, fuel efficient, and operate isothermally at high heat and mass transfer rates, many calcinations are carried out on an industrial scale in fluidized bed reactors.

B. Fluid Bed Calcination

1. Limestone

The quality of lime is determined by the content of CaO in the calcined product. Excess burning can reduce the availability of CaO, and low-temperature operation can result in an unreacted ore. If the calcination is carried out near the theoretical dissociation temperature, CaO will not be lost by reacting with impurities such as iron, silicon, and alumina, which are usually present in limestone. The first multi-stage-type fluidized bed commercial reactor[94,95] for calcination of limestone was built in 1949 for the New England Lime Company. The reactor was 4 m ID and 14 m long and had five compartments. The reactor was used for three functions: (1) preheating the air which was also the fluidizing agent and the oxygen source for burning the fuel oil, (2) carrying out the calcination reaction which could be accomplished by the heat supplied by the combustion of sprayed fuel oil over the surface of the limestone, and (3) preheating the incoming limestone feed by the hot outgoing combustion product gas. The five-compartment concept was found to be useful for carrying out calcination economically, conserving the fuel oil. The calcination of limestone was accomplished at 1000°C with a heat input of 42.9 kcal/mol of limestone. Limestone conversion of 96.8% was achieved. A relatively large amount of unconverted fines, amounting to 14%, was entrained, and this was found to be responsible for lowering the overall conversion of limestone calcination. The particle size of the raw material fed into the top of the multicompartment calciner was in the range 6–65 mesh. In order to calcine calcium carbonate fines (<50 μm), a new technique known as pelletization[96] was developed. In this technique, soda ash or caustic soda is mixed along with the feed. The calcined material is coated over the fines as a sticky mass, thereby allowing the fines in the reactor to grow by adhesion or pelletization. The growth rate of the particles inside the reactor can be controlled by controlling the feed rate of the material into the reactor.

Limestone calcined in a fluidized bed reactor usually has improved properties; hence, many commercial fluidized bed limestone calciners came into existence as early as three decades ago, and reports[97,98] on these are available. A multistage fluidized bed calciner that produced a very active quicklime was reported by Van Thoor.[99] Details and the kinetics of limestone calcination can be found in the literature.[100] The decomposition of limestone is assumed to start from the surface and then proceed toward the center. The reaction is believed to occur in a thin layer between the unreacted limestone and the product. Although this type of theory is accepted, many conflicting issues crop up in terms of the controlling mechanism. Calcination can proceed if the latent heat of decomposition is available. The heat flux and hence the rate of calcination are determined from knowledge of the sample geometry and external conditions using Fourier's law of heat conduction.

2. *Cement, Bauxite, and Phosphate Rock*

A process for the production of cement clinker,[101] named after the inventor, Pyzel, was developed in 1944. It incorporated a fluidized bed reactor with a 2.5-m ID and operated at 1300°C. The reactor, when operated in a single stage, lost heat through the exit gas, and the product obtained was of the order of 1050–1180 kcal/kg of clinker. Hence, improved heat recovery systems were deemed essential. Mitsubishi developed a new suspension[102] preheating system for cement clinker production and incorporated a fluidized bed limestone calciner. Fine lime powders were produced by spraying lime sludge over the carrier bed of agglomerated lime particles. The slurry, after being sprayed on the bed, decomposed in a short time to fine lime particles, and the product was collected after cooling rapidly.

A fluidized bed calcination process for converting low-grade bauxite-containing selenium oxide was reported[103] to be a viable technical route to produce aluminum sulfate after treating the calcined product with H_2SO_4. The calcination of low-grade bauxite with 1:1 soda ash at temperatures around 700–1400°C in a fluidized bed reactor can produce water-soluble sodium aluminate which, upon leaching and treating with sulfuric acid, can yield aluminum sulfate. Calcination of low-quality phosphate rock from the western United States was reported by Priestly.[97] The phosphate rock contained 3.5% hydrocarbon and could generate much of the heat required for the calcination carried out at 760°C in a 6-m-diameter three-compartment Dorr–Oliver fluosolid calciner. During the calcination of phosphate rock, any limestone contained in it was also calcined, thereby yielding useful lime which was leached out easily. The calcined phosphate rock was cooled by direct water spray over a separate fluidized bed, thus bringing down the charge temperature from 538 to 121°C. A continuously operated pilot-scale fluidized bed reactor for the decomposition of ferrous sulfate at 700°C to produce a pigment containing 96% Fe_2O_3 with a mean pigment size of about 1 mm was reported by Fenyi.[103]

3. *Aluminum Trihydrate*

a. Circulating Fluidized Bed

Vereinigte Aluminium Werke (VAW) and Lurgi developed a highly expanded fluidized bed for processing fine-grained solid particles. The reactor was operated at a high velocity, which increased the capacity significantly. Reh[92] described highly expanded fluid beds for application in industrially important exothermic and endothermic processes and also gave an account of the developments carried out by VAW/Lurgi for the calcination of $Al(OH)_3$. In a highly expanded bed, the product is collected mostly from the exit gas. VAW/Lurgi operated a pilot plant for several years to calcine aluminum trihydrate. The capacity of the pilot plant was 28 tpd, which was later scaled up to 560 tpd.

b. Operation

The circulating fluidized bed for calcining $Al(OH)_3$ was fluidized by air that was divided into primary and secondary supplies. The primary air was preheated by passing through suspended pipe coils immersed inside the four-stage fluidized bed of the hot product, alumina (200°C), and then was used to fluidize the $Al(OH)_3$ for calcination. The feed, $Al(OH)_3$, was dewatered and preheated using waste heat. Venturi-type highly expanded fluid beds were employed for this duty. Venturi-type conical fluidized beds are grateless and offer uniform temperature distribution and relatively lower pressure drops than conventional fluidized beds. The heat for the calcination of $Al(OH)_3$ was supplied by direct oil burning inside the fluidized bed of high solid concentration. The secondary air supplied through the hot four-stage fluidized bed of hot alumina helped to effect complete combustion. The combustion was near stoichiometric without any soot formation or superheating. The CO content at the outlet of the reactor and the O_2 content in the flue gas were 0.5 and 1%, respectively. The operating gas velocity in the pilot plant was 3 m/s. The reactor was 100 mm ID and 800 mm in height. The mean solid concentration was 16 kg/m^3 of the furnace volume. A schematic diagram of a typical circulating fluidized bed used for calcination is shown in Figure 2.16. Based on the experience with a 28-tpd pilot plant, a 560-tpd-capacity calcination plant was commissioned. A 560-tpd-capacity circulating-type fluidized bed calciner is reported to require merely two-thirds the inside diameter and one-third the grate area of a conventional multistage fluidized bed calciner (6700 mm in diameter) with a capacity of 280 tpd.

4. Alumina

Schmidt et al.[104] gave a complete process description of the methods and means of producing alumina of different quality using a fluidized bed calciner. The fluidized bed process was reported to offer optimal performance to produce alumina of the desired quality. The two frequently used types of alumina are (1) fine-grained, high-calcined floury type and (2) large-grained, low-calcined sandy type. The VAW/Lurgi process for calcining alumina can optimally adopt the changeover from the production of floury-type to sandy-type alumina. Investigations carried out to minimize particle attrition and elutriation improved the process to meet pollution control requirements. In the new Toth[105] process for the production of alumina, drying and calcination are accomplished in a fluidized bed. The Toth process adopts fluidized bed chlorination of clay followed by the reduction of aluminum chloride by manganese metal pellets. The by-product, $MnCl_2$, is dechlorinated in a fluidized bed for recycling Cl_2 and Mn.

5. Waste Calcination

a. Chloride Waste

Chlorination of nonferrous ores and minerals usually ends up with the generation of highly corrosive pollutants such as ferrous or ferric chloride. The loss of

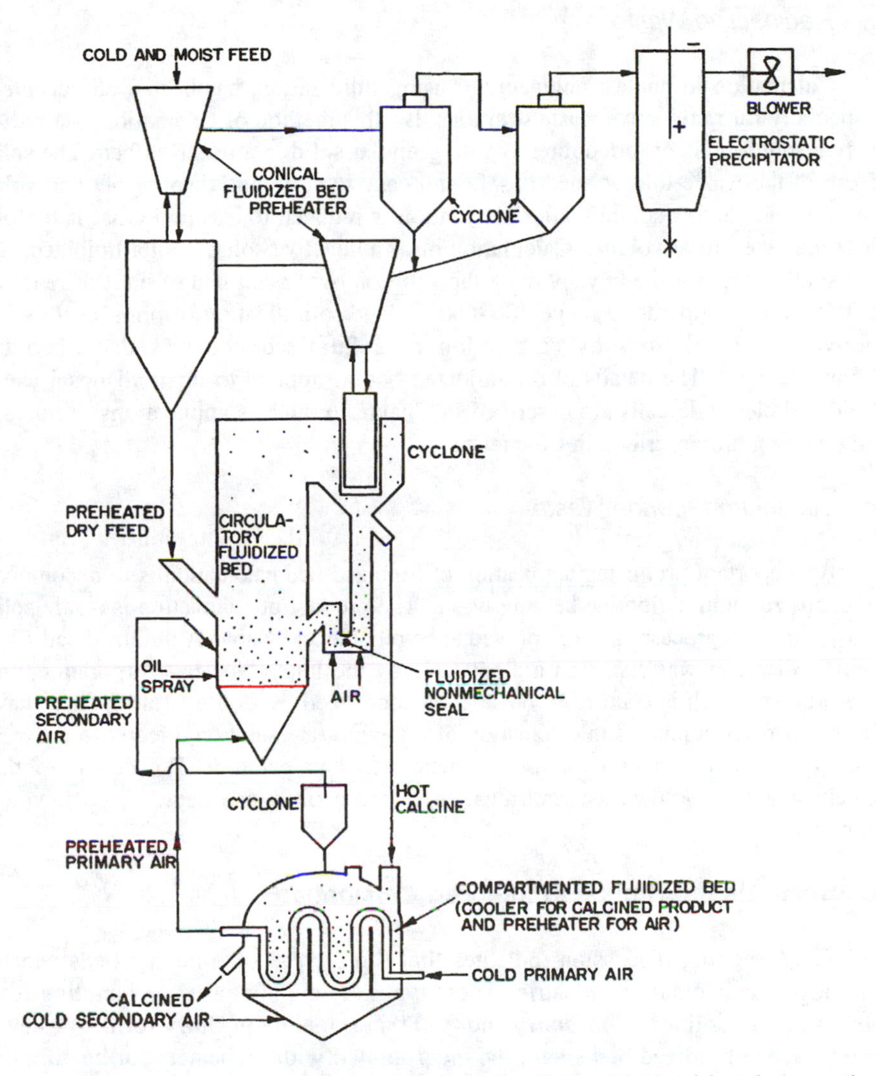

Figure 2.16 Typical schematic diagram of a circulating fluidized bed calciner incorporating waste heat and hot calcine heat recovery system.

chlorine and the environment problem can be overcome by calcining the ferric chloride[106-108] over a bed of fluidized iron oxide. The calcined product is constituted of nonpolluting iron oxide and the chlorine obtained can be recycled for chlorination reactions. Recovery of chlorine and iron oxide by dechlorination of $FeCl_3$ using a fluidized bed of Fe_2O_3 was discussed by Paige et al.[109] For smooth operation and better dechlorination in a fluidized bed, a feed composition of 75% $FeCl_3$ and 25% Fe_2O_3 was recommended. An exit chlorine concentration >80 wt% (for feed $FeCl_3$ of bulk density <13 g/cc) and Fe_2O_3 (after washing) calcine of 70 wt% suitable for blast furnace feed can be obtained from a fluidized calciner.

b. Radioactive Waste

Calcination in nuclear engineering, using fluidization, has been well accepted, especially for radioactive waste solutions. By this method of calcination, the radioactive solution is converted directly into granular solid in a fluidized bed. The solid form of the waste thus formed is safe and easy to store and dispose of. The voluminous radioactive solution after calcination is reduced to a volume that is tenfold less than the initial volume. Calcination of a radioactive solution in a fluidized bed is usually accomplished by spraying the solution over a hot bed of inert or reactive solids in the temperature range 400–600°C. The solidification of high-level radioactive waste solutions by calcination in a fluidized bed was described by Schneider.[110,111] The details of the fluidized bed as applied to waste disposal using fluidized electrode cells are described in Chapter 6, and its applications in nuclear engineering are described in Chapter 5.

c. Zirconium Fluoride Waste

An important promising application of fluidized bed calcination is to decompose aqueous zirconium–fluoride-bearing waste. The product of calcination is a safe solid waste, and this process has been proved to be reliable and viable. A fluidized bed 1200 mm in diameter was tested on a plant scale by the Idaho Nuclear Corporation, and the process which is claimed to be safe was described by Lohse et al.[112] There have been other developments in calcination of uranyl nitrate solution directly to arrive at useful oxides of uranium for use in nuclear fuel preparation. The details of this calcination (also known as denitrification) are described in Chapter 3 on uranium extraction.

C. Some Useful Hints on Fluid Bed Calcination

The foregoing discussion indicates that three types of fluidized beds can be employed in calcination industries These types are mainly based on handling feed that is coarse or fine but not slurry and feed that is mainly in slurry form. Whatever the type, the fluidized bed should be incorporated with preheaters for heating the fluidizing gas and the charge to the calciner in order to maintain an optimum thermal balance. A fluidized bed preheater (or cooler for hot calcine) for the fluidizing gas as well as the feed charge is an ideal choice. This method of recuperating the heat in a circulating fluidized bed is depicted in Figure 2.16. In the case of a conventional (i.e., noncirculating-type) fluidized bed, the right choice is a multistage fluidized bed with three zones: a top zone for preheating the incoming solid, a central zone for the calcination reaction, and a bottom zone for preheating the air by the hot calcine. In the recent past, the commercial potential of circulating fluidized beds has been established, and many international firms are now available to execute turnkey projects using circulating-type fluidized bed calciners.

IV. DIRECT REDUCTION

A. Significance

Direct reduction has gained importance over the years in the production of metal powders that can be used on a moderate scale of operation to prepare the feed directly in powder metallurgical applications. Direct reduction of iron ore in ferrous industries is considered an alternative to the blast furnace, when sponge iron is produced for mini steel plants. For centuries, the blast furnace has played a predominant role in large-scale iron making. However, it suffers from some drawbacks such as the feed in the form of sinters or pellets, high-grade coking coal, and large-scale infrastructure. A blast furnace is reported to be economical when iron making is of the order of 3 million tpa. In such a large-scale operation, the obvious constraints are limited flexibility in operation and choice of materials. For small-scale units (250,000–500,000 tpa), an alternative route without using a blast furnace is preferred. If it can accept feed iron ores as fines and coal instead of metallurgical-grade coke, this route would be technically and commercially suitable. Fluidized bed technology for direct reduction of iron ores came into use as such an alternative. With recent research into an entirely new route for iron making, direct smelting technology is gaining importance. The fluidized bed is used in both direct reduction and direct smelting technology. Flexibility in operation and the ability to maintain a clean environment have prompted many commercial giants in the iron and steel industries to adopt this technology in future developments and expansion. A detailed review of iron making was presented by Wright et al.[113] The review focuses mainly on the latest upcoming new direct smelting technology with commercial potential. This technology is now a step ahead of direct reduction. However, direct reduction has its own merits in the sense that the end product is pure and useful for iron-making and powder metallurgical applications. The application of the fluidized bed in various new routes of iron making (i.e., direct reduction and direct smelting) is schematically represented in Figure 2.17. Next, we will discuss the various commercial routes that have been developed for the direct reduction of iron ores using fluidized beds.

B. Iron Ore Reduction

1. *Advent of the Fluid Bed*

Various applications of fluidization in ferrous industries were brought out in a review by Doheim.[114] The fluidized bed reactor, as applied to direct reduction, was also discussed briefly. The three major processes developed for direct reduction of iron ore fines using the fluidized bed reactor are the (1) H–iron process, (2) fluidized iron ore reduction process, and (3) Nu–iron process. In all these direct reduction processes, the fluidizing gas as well as the reducing agents are the same. The purpose of these processes is to produce a quality product that has particles of high density, low porosity, and good thermal conductivity with uniform chemical and size composition. It is essential to ensure that the reduced iron ore particles have these

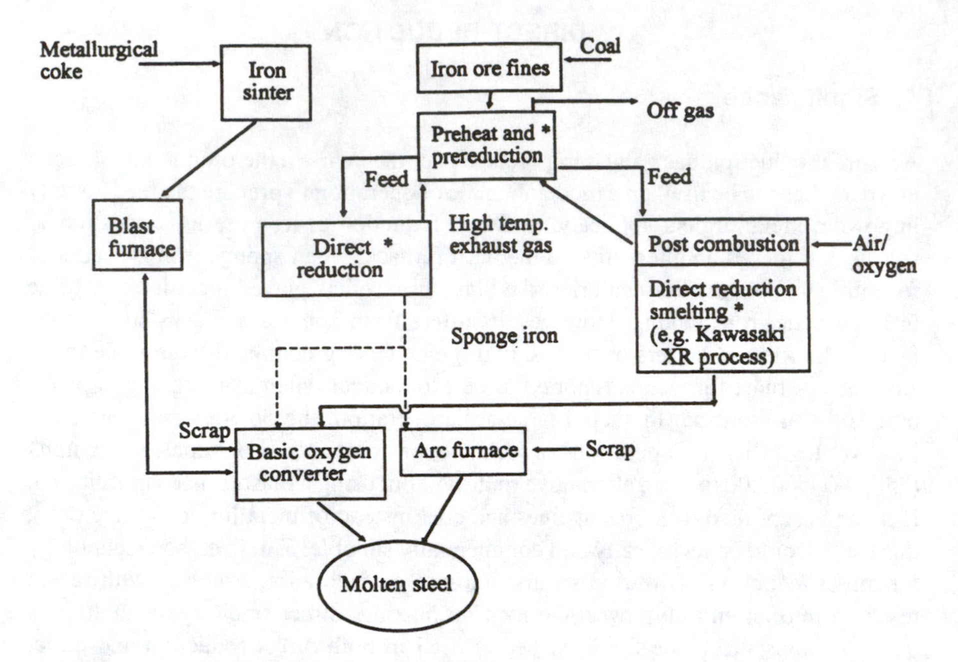

Figure 2.17 Fluidized beds in iron-making process.

properties so as to prepare a reliable quality feed for the continuous iron-making furnace. For example, melting is easy with high-thermal-conductivity particles, and reoxidation is less with high-density or nonporous particles. A prereduced iron ore, when used in a steel-making electric furnace in place of iron scrap, can improve productivity by 18% and reduce consumption of electrodes and oxygen by 22 and 40%, respectively. The three processes for the direct reduction of iron ore using a fluidized bed developed by various research groups are outlined briefly in the following sections.

2. Fluid Bed Processes

a. H–Iron Process

A multistage fluidized reactor to reduce pure iron oxide powder with hydrogen gas was developed jointly by Hydrocarbon Research and Bethlehem Steel. The reactor operated at 450–500°C and 46 atm pressure. High pressure was required to operate the reactor at temperatures suitable for producing nonsticky and unsintered iron powders and also for increasing the reaction rate. The conversion of iron oxide discharged at the lowest stage was 98%, while the conversions achieved in the middle and upper stages were 87 and 47%, respectively. Hydrogen utilization was only 5%, thus necessitating the recirculation of the exit gas after drying. The highly pyrophoric iron product was treated with N_2 at 810–870°C before storing. This product has applications in the powder metallurgical and briquetting industries. For details on this process, refer to the publication by Labine.[115] The process was tested on 50-tpa (by Alan Wood Steel Co.) and 100-tpa capacities (by Bethlehem Steel). The typical dimensions of the 50-tpa pilot-plant fluidized bed reactor were 1.7 m OD and 29 m height. Approximately 0.051–0.056 ton of H_2 and 0.25 ton of oxygen were required to process 1.4 tons of high-grade magnetite to arrive at 1 ton of iron by this H–iron process. A schematic of the H–iron process is presented in Figure 2.18a.

b. Fluidized Iron Ore Reduction

This process of reducing high-grade iron ore was developed by Exxon.[116] The ore fines, after preparation and preheating, were reduced sequentially in three reactors using a reducing gas mixture, CO–H_2, obtained by steam reforming of natural gas followed by a shift reaction. The reduction was carried out at 800°C and the product obtained was 25–45% passing through 325 mesh. The metallization achieved was 90–95%. The technical feasibility of this process was first demonstrated on a 5-tpd pilot plant, and then a continuous 300-tpd plant was built for Imperial Oil Refinery in Nova Scotia and operated in late 1965. The only commercial FIOR plant built by Davy McKee under licence from Exxon has been in operation in Venezuela since 1980 at a scheduled production level of 1000 tpd of iron briquettes. A schematic of the FIOR process is shown in Figure 2.18b.

c. Nu–Iron Process

The Nu–iron fluidized bed process,[117-119] also known as the HIB process, was developed by U.S. Steel Corporation. After drying and preheating in a fluidized bed, ore with a particle size of –10 mesh (–1.67 mm) was reduced in the temperature range 700–750°C by hydrogen gas in a two-stage fluidized bed, yielding a product with 86.5% metal. U.S. Steel[120] is reported to have operated an HIB process to produce 75% reduced iron ore briquettes intended for use in a blast furnace. This process was later modified to produce heavily reduced briquettes for steel making, and the results showed that the steel produced was of satisfactory quality. The exit gas leaving the reactor was used as fuel in the reformer furnace. A simplified flow diagram of the Nu–iron process is shown in Figure 2.18c.

d. Other Reduction Processes

Fluosolid reactors were reported[71] to be used for the direct reduction of iron ore by injection of fuel oil on the hot bed. Such a process on a commercial scale was first used at the Montecatini plant in Follonica, Italy, for reducing hematitic pyrite cinder to magnetite at 530°C on a scale of 430 tpd. Another process[121] developed in Italy for reduction using H_2 at 650–700°C in a fluidized bed was reported to have produced sponge iron at quantities of 20 tpd/m^3 of the reactor. The process was designed to use nuclear energy to superheat the reducing hydrogen gas and also for steam reforming. A fluidized bed preheater[122] for iron ore was used in a pilot plant that could produce high-purity molten iron in a furnace where melting and reduction were accomplished. As mini steel mills require a smaller and efficient direct reduction process, a technique involving two interconnected fluidized bed reactors,[123]

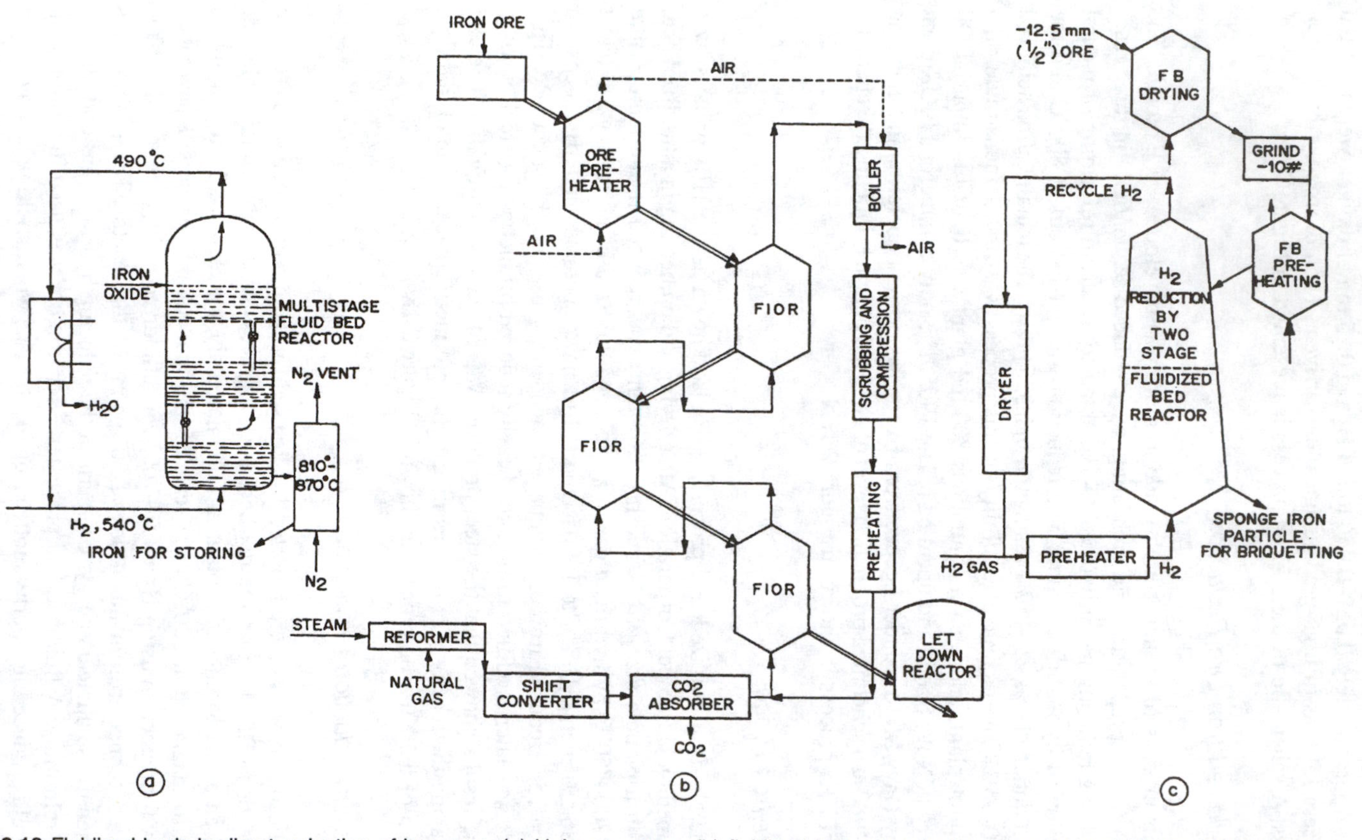

Figure 2.18 Fluidized beds in direct reduction of iron ores: (a) H–iron process, (b) fluidized iron ore reduction process, and (c) Nu–iron process.

where reduction is carried out in one vessel and the combustion of coke in another, was developed by Kawasaki Steel. The heat generated in the combustion chamber by this technique was directly transferred by the circulating solid and the products of combustion were used as the reducing agent.

3. Reaction Aspects in Direct Reduction

a. Reduction

As the feed material for direct reduction is comprised of iron ore fines, which is the most prevalent form of iron ore worldwide, and the reductant is a gas, direct reduction using a fluidized bed is a readily acceptable process in countries where coke or coal resources are scant but natural gas is abundant. Even the most modern direct reduction smelting requires coal or electrical energy. In order to optimize the use of these forms of heat input, prereduction is recommended using a fluidized bed reactor. This aspect will be discussed later, but let us now consider the reaction and the resources for reductants in direct reduction as applied to iron making.

For reduction of iron oxide by CO (or H_2):

$$\text{Hematite–magnetite} \quad 3Fe_2O_3 + CO \text{ (or } H_2) \rightarrow 2Fe_3O_4 + CO_2 \text{ (or } H_2O) \quad (2.30)$$

$$\text{Magnetite–wustite} \quad Fe_3O_4 + CO \text{ (or } H_2) \rightarrow 3FeO + CO_2 \text{ (or } H_2O) \quad (2.31)$$

$$\text{Wustite–iron} \quad FeO + CO \text{ (or } H_2) \rightarrow Fe + CO_2 \text{ (or } H_2O) \quad (2.32)$$

The reduction proceeds depending on the temperature and the partial pressure ratio, CO/CO_2 (or H_2/H_2O). A reduction equilibrium diagram[124] (shown in Figure 2.19) is helpful in assessing reduction from higher oxides to lower oxides. In general, higher oxides can be reduced to wustite (FeO) at temperatures above 600°C for the partial pressure ratio CO/CO_2 (or H_2/H_2O) at 1:1.

b. Reductants

Generation of the gaseous reductants is governed by the following three types of reactions.

1. Water gas-shift reaction:

$$CO + H_2O \leftrightarrow CO_2 + H_2 \quad (2.33)$$

2. Boudouard reaction:

$$2CO \leftrightarrow C + CO_2 \quad (2.34)$$

3. Gasification:

$$C + H_2O \leftrightarrow CO + H_2 \quad (2.35)$$

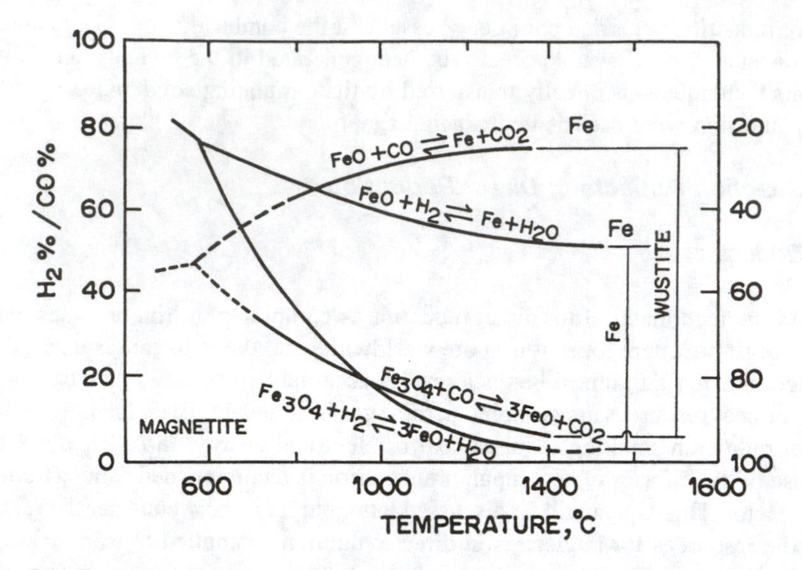

Figure 2.19 Reduction equilibrium diagram for oxides of iron. (From Stephens, F.M., Jr., *J. Met.*, 5, 780, 1953. With permission.)

The reduction of hematite to wustite was investigated by Doherty et al.[125] with CO–CO_2 and H_2–H_2O using particles in the size range 180–250 µm in a laboratory fluidized bed reactor, and it was shown that the reaction was efficient even for gas flow rates of 7–15 U_{mf}. Uniform internal reduction of magnetite to wustite was observed. Off-gas analysis showed the importance of the water-shift reaction within the pores of fluidized bed particles. The reducing gases should be rich in CO and H_2 and less so in CO_2 and H_2O, and they should be available at temperatures around 1000°C. Although oil and neutral gases are the sources for gaseous reductants, it was reported[114] that the reactant gas can be produced cheaply using nuclear heat by the gasification of brown coal in a fluidized bed reactor. A solid reductant such as coal, which has less volatiles and is rich in carbon, can be produced by carburization of coal at high temperatures[126] (750–1200°C) in a fluidized bed. Reduction of metal oxides by carbon is in principle a gas–solid reaction due to the intermediate gaseous product CO. The details of the gas–solid reaction pertaining to this topic were dealt with clearly by Szekely et al.[127]

C. Troubleshooting in Fluosolid Reduction

1. Defluidization

a. Nodule Formation

During reduction by hydrogen gas iron ore has a tendency to defluidize. The defluidization is generally attributed to temperatures above 704°C at higher levels of conversion amounting over 90%. A possible reason is the formation of nodules which, upon impingement on the surface of other particles, form microwelds, resulting in agglomeration and defluidization.

b. Temperature Effect

Agarwal and Davis[128] studied the dynamics of fluidization of iron ore and investigated the aspect of defluidization. They found that defluidization of reduced iron ore was pronounced when a high degree of reduction was approached. For temperatures below 620°C, there was no defluidization. In the temperature range 620–730°C, defluidization appeared to occur due to a sticky mass, and this was verified after cooling and testing the bed material which contained fritted material. Defluidization caused a reduction in the conversion, and this could be overcome by an increased fluidization velocity. The effect of inserting internal baffles was found to be negligible in eliminating defluidization. High-temperature operation for an endothermic reaction of this kind requires an enormous heat supply; this renders the high-temperature operation uneconomical. Gransden et al.[129] studied the defluidization of iron ore and discussed various means of eliminating defluidization. Although suggestions such as covering the surface of iron particles with carbon or calcium oxide, controlling the particle size distribution either by withdrawing fine or adding coarse particles, and using preused ore for high-temperature reduction have been put forward to overcome the problem of defluidization, the success rate is not the same in all cases. The mechanism of defluidization still seems to be somewhat elusive. If the increased tendency to sinter at temperatures above 710°C is regarded as the cause of defluidization, it is overruled by the results of prolonged fluidization of the prereduced iron ore.

c. Mechanism

Gransden et al.[129] proposed a mechanism based on microscopic study and examination of the quality of fluidization. According to them, the fluidization power of H_2 gas toward the end of the reduction is weakened due to unavailability of water vapor which, when formed during reduction, takes part in fluidizing the bed along with hydrogen gas. It should be mentioned that water vapor is a more powerful fluidization agent than H_2 gas. Hence, the absence of water vapor toward the end of reduction causes defluidization of the bed. Another explanation is that iron nucleates on the surface at temperatures above 710°C and grows as spikes or nodules, degrading the bed to a state of defluidization. If spikes are formed within the porous structure of the wustite at temperatures below 710°C, they do not interfere with the fluidization behavior of the bed. The exact picture of defluidization remains unsolved as the above mechanisms are not able to satisfactorily explain all the observations made on the defluidization phenomenon.

2. Feed Preheating

Iron ore reduction is highly endothermic, and hence drying and preheating of the feed material to the reactor are essential. Preheating can be done in three ways: (1) by using an external heat exchanger, (2) by heating the incoming fluidizing gas below the distribution chamber by combustion of oil, and (3) by direct oil or gas injection into the bed. Submerged heating by direct oil or gas injection into the bed has proved to be more efficient than the other two methods. While heating below the distributor plate, the distributor plate can malfunction due to high thermal stresses and expansion of the construction material of the grid plate. Two equilibrium stages are encountered during reduction of the iron ore. The equilibrium constant for reduction with hydrogen for the higher oxides (ferric oxide or ferroferric oxide) is 1.2 and is 0.42 for the lower oxides. This implies that the ferrous oxides should be reduced with fresh hydrogen at the lower stage and the incoming higher oxides with partially consumed hydrogen. This necessitates the use of a two-stage fluidized bed, as in the Nu–iron process. If a single fluidized bed is used, reduction depends on the particle residence time and the gas–solid contact. For example, in a 457-mm-diameter fluidized bed for iron ore reduction, a bed height of 4268 mm can give up to 75% equilibrium conversion. Beyond this height, the gas–solid contact becomes poorer due to slugging and gas bypassing. Furthermore, the particle residence time in a single stage is low compared to a double stage. All these indicate the necessity of using a two-stage fluidized bed for iron ore reduction.

3. *Particle Carryover*

Particle carryover from one stage to the other must be eliminated. If the metallic iron reduced from the first stage is elutriated or entrained into the upper stage, then the reduced charge will be reoxidized. This warrants the provision of enough entrainment space or equilibrium disengaging height along with internal cyclones. As a rule of thumb, the cyclone inlet should be located 300 mm above the expanded height of the fluidized bed for a 300-mm-diameter bed. The dipleg of the cyclone must be immersed well into the fluidized bed to provide enough sealing.

4. *Sulfur Control*

Sulfur is an undesirable impurity in steel making. In reducing operations, sulfur will not be removed. Any sulfur remaining in the ore or in the reducing gas will then be taken up readily by the nascent iron, which has great affinity for sulfur. As the final sulfur content of the iron should not be more than 0.03%, it is essential to control sulfur in the reducing gas to 20 ppm and in the feed ore to no more than 0.02%. However, many ores contain sulfur in the range 0.03–0.05%, and this is increased substantially after reduction. Hence, it is essential to remove sulfur in a preheating stage and also to use a fuel free of sulfur.

5. *Carbon Reductant*

Reduction of iron ore with carbon particles has not been reported on a large-scale operation. Carbon reduction of iron oxide sludge[130] in a 45-mm-ID laboratory fluidized bed was reported. The charge to the fluidized bed consisted of pelletized particles of iron and carbon, and it was fluidized by N_2 gas and recycled off-gas from the reactor. Although the reduction was reported to be successful, large-scale operation was not declared based on this.

D. Fluidization in Modern Iron Making

1. Novel Processes

Fluidized bed reactors have potential application in new iron-making technologies.[113,131] Direct reduction using iron bath smelting or Hi-plas (Plasma Smelting) has emerged as a promising iron-making method to substitute for the blast furnace in the future. The underlying principle of this technique is based on the fact that molten iron has great affinity for carbon, and the molten iron bath with dissolved carbon, in the temperature range 1300–1600°C, is a powerful reducing medium. Hence, air or O_2, when injected into the bath, generates CO gas, and when carbon or coal is charged they are dissolved, resulting in a gasification process with an iron bath. If this iron bath is charged with iron ore, they are readily reduced to iron in the molten state. The gas mixture generated from the iron bath emerges at a high temperature and has a high reducing potential due to the high content of CO and H_2. The heat supplied in direct smelting is by combustion of coal with O_2 or air in the molten iron bath or by electrical energy in the case of plasma smelting. In order to optimize the heat input, the off-gas emerging from the bath can be partially combusted to release heat energy to the smelting and partially used to reduce the iron ore feed. Fluidized bed direct reduction has proven to be a potential and viable method. The direct reduction of iron ore can be carried out with smelting furnace off-gases either after partial combustion or without any combustion. One such process which uses off-gases directly for reduction of the incoming charge using a shaft-type reactor is the COREX[132] process, operating on a commercial scale at 300,000 tpa capacity in Pretoria, South Africa since 1988. In another process of iron bath smelting, known as the NKK (Japan) process, the off-gas after combustion is used to reduce the iron ore in a fluidized bed. A 5-ton capacity pilot plant built in Fukuyama Works (Japan) is believed to have paved the way for commissioning a 500-tpd plant at Keihin Works, Japan.

2. Two-Stage Process

In a two-stage process, where post-combustion and prereduction are used in iron bath smelting, the rate of coal consumption depends on the degree of post-combustion as well as prereduction. As post-combustion is increased, coal consumption falls, and it again falls as the percent prereduction is increased. However, the percent post-combustion cannot be increased if the percent prereduction is required to be carried out at a higher level. Wright et al.[113] gave a detailed plot of the coal consumption rate versus percent post-combustion as function of percent prereduction. It was also shown that any degree of post-combustion above 30% limits the reduction to below 29.6%. If air instead of O_2 is used for post-combustion, the reducing potential of the off-gas leaving the iron bath is lowered due to dilution by nitrogen gas. No commercial plant seems to be operating based on this two-stage process in which fluidized bed prereduction plays a key role. Research is being carried out worldwide in a competitive manner by many commercial giants in an effort to complement such a promising new iron-making technology.

3. *Flue Dust Control*

In steel-making industries, depending on the type of steel (e.g., carbon steel and stainless steel), the furnace off-gas contains huge quantities of dusts which are objectionable from the standpoint of environmental pollution. Irrespective of the type of steel-making furnace, dust emission is inevitable. It is estimated that roughly 10–20 kg of dust is emitted for every metric ton of steel produced. The dust emitted from a carbon steel-making furnace is rich in zinc and lead, while alloying elements such as chromium, nickel, manganese, etc. appear in the dust in a stainless-steel-making furnace. In addition, iron dusts are predominant in both steel-making furnaces. Dumping and disposal of waste after chemical treatment pose problems to the environment. In this context, an attempt is made to recycle some of the metal content of the dust after some processing by using efficient post-furnace off-gas treatment systems. Nyirenda[133] reviewed the processing of steel-making dust and pointed out the use of the fluidized bed in this application. Direct reduction of flue dust with coal is a promising method. By this technique, the iron oxide content is reduced to metal, while zinc, lead, chloride, and sulfur are volatilized. The zinc and lead from the reactor off-gas can be recovered and sold and the iron recycled. Development of a circulating fluidized bed[134,135] for treating flue dust is under way and the commercial operation of such a unit is expected in the near future.

E. Direct Reduction in Nonferrous Industries

1. *Metal Powder Production*

Metal powders of nonferrous metals such as copper, nickel, and cobalt can be produced by the direct reduction of their compounds, which need not necessarily be in the oxide form. Pure metal powders, starting with ingot material, are prepared by atomizing the molten ingot in air and subsequently reducing the oxide powder with a gaseous reductant such as hydrogen. For example, copper powders can be produced by atomization/electrolysis, by reduction of their sulfate solution using H_2, and by solid-state reduction. The characteristics of the powders produced by each of these methods are different, and none of the above processes, except solid-state reduction, can directly and continuously yield metal powders.

2. *Fluid Bed Reduction*

a. *Copper and Nickel Powders*

The fluidized bed can be successfully employed in solid-state reduction. Use of the fluidized bed for reduction of metal oxides[136] was tested and patented as early as four decades ago. The starting solid materials for reduction can be chlorides, sulfides, or sulfates of the metals. The kinetics of hydrogen reduction of copper sulfate[137] showed that copper metal can be obtained at 300°C, and elevated temperatures (up to 606°C) can increase the reduction rate and the conversion. The possibility of obtaining nickel metal powders directly by reducing nickel sulfide (Ni_3S_2)

by hydrogen and the kinetics of reduction in the temperature range 475–360°C were reported by Chida and Ford.[138] The reduction of nickel chloride in the temperature range 260–515°C was reported by Williams et al.[139] who proposed an autocatalytic model. In principle, it can be inferred from the foregoing that all the above reduction reactions can be accomplished in a fluidized bed reactor. Toor et al.[140] produced copper powder by reducing copper sulfate ($CuSO_4$) by hydrogen in a fluidized bed. The copper powder thus obtained, after compaction, was found to have good green density and good strength, but the flow characteristics were rather poor. The properties of powders produced by the fluidized bed technique were found to be different from those of powders produced by conventional methods. The particle size distribution was bimodal and particle sizes were very irregular. The final particle size distribution can be controlled by properly selecting the starting material. Fine powders can be produced by simply grinding the starting copper sulfate crystals. This method of producing copper powder can eliminate the electrolytic route, which is a slow and batch process and which requires washing and drying operations.

b. Nickel and Titanium Powders

Reduction of nickel oxide[141] in a fluidized bed was tested long ago. Thermogravimetric investigations[142] on the reduction of $NiSO_4$ by hydrogen in the temperature range 450–650°C showed that the product contained Ni_3S_2 and Ni. Dynamic thermogravimetric runs showed that the reduction proceeded in two stages. In the first stage, Ni_3S_2 formed at 360°C, and in the second stage, Ni was formed with a starting temperature of 520°C. The fluidized bed reactor was used to prepare powder from $NiSO_4$ by hydrogen reduction in two stages. The second-stage reduction, which was carried out in the temperature range 520–600°C, showed that nickel powders with a low sulfur content could be produced. The second-stage reduction followed the pseudo zero-order reaction. The possibility of producing titanium metal powder by the reduction of titanium tetrabromide in a fluidized bed was reported by Coffer.[143] This method, although it may be more economical than reducing the halide of this metal by sodium or magnesium, did not come into the limelight. Any nitrogen present as an impurity in the reducing gas (i.e., hydrogen) can easily combine with the freshly reduced titanium at the reducing temperature (i.e., 1400°C). This method was explored in 1959 and was not pursued because of insufficient knowledge on fluidization available at that time. No work seems to have been carried out to date based on this technique. The decomposition of iron carbonyl[144] over a fluidized bed of iron powders to produce high-purity iron was known long ago. This method of carbonyl decomposition is a promising route for separating iron from laterite-containing nickel. The carbonyls of iron and nickel can be easily separated by distillation and the separated chlorides can be independently reduced to their respective pure metals.

c. Molybdenum Powder

Fluidized bed reactors were reported to be useful in preparing molybdenum powder. Molybdenum powder can be produced by various methods, such as metallothermic and nonmetallic reduction. For details of these reduction processes, refer to a monograph[89] on molybdenum. Hydrogen can be used to reduce compounds of molybdenum such as its oxides, chlorides, and sulfides. The fluidized bed reactor was employed[145] for the production of metallic Mo powder by reducing MoO_3 to Mo by hydrogen in a continuous manner. The fluidized bed was initially fed with molybdenum powder and was fluidized by preheated (at 600°C) hydrogen. Heating of the bed was accomplished by an external furnace. Once a reaction temperature of 955°C was reached, MoO_3 was charged into the reactor by a screw feeder. The metal reduced was collected from a weir-type overflow discharge tube. A hydrogen flow equal to 19 times the stoichiometric amount and an oxide feed rate of 400 g/hr were maintained to achieve 99% conversion of the oxide to the metal. It should be noted that the reduction of MoO_3 to Mo should be carried out in two stages. The reduction of MoO_3 by hydrogen is highly exothermic and it is necessary to control the temperature in order to avoid vaporization of MoO_3 (the vapor pressure ranges from 7.08×10^{-5} to 208.93 mmHg at 500–1000°C). Hence, MoO_3 should be reduced first to MoO_2 at 500–600°C and then to Mo at 600–1000°C. Sathiyamoorthy et al.[146] carried out hydrogen reduction of ammonium molybdate in a fluidized bed in two stages to prepare molybdenum metal powder. Detailed kinetic studies on the second-stage reduction were carried out in the temperature range 900–1050°C using hydrogen concentrations of 50–100% in the gas stream of argon. The rate equation and the activation energy were determined. The advantages of the fluidized bed, starting with the decomposition of ammonium molybdate in air and reducing the resultant oxides successfully in two stages to molybdenum metal powder, were thus established.

d. Recommendations

Metals prepared by hydrogen reduction of their chlorides are niobium, silicon, tantalum, and vanadium. Tungsten is prepared by the reduction of its fluoride by hydrogen. The details of the reaction involving the reduction of halides by hydrogen are reported in the literature.[147] The reduction of all these commercially important nonferrous metals can be carried out using the fluidized bed. To date, however, little seems to have been done on this aspect.

V. FLUID BED HALOGENATION

A. Fluidization in Halide Metallurgy

1. Introduction

Halogenation is an important process in halide metallurgy. Metals can be produced by chemical reduction of their halides, oxides, or sulfides using either metallic reductants such as sodium, magnesium, silicon, aluminum, and calcium or by nonmetallic reductants like C, CO, and H_2. Chemical reduction of metallic compounds by nonmetallic compounds can be successfully carried out in a fluidized bed reactor. Metals are also produced by electrowinning. Fluidized cathode techniques are useful and are applied in this area. This aspect will be discussed in Chapter 6 in the treatment of fluidized electrodes. Metals are also produced by thermal decomposition of their carbonyls. Nickel powder production by the decomposition of its carbonyl is a typical example and the fluidized bed can be employed here. Applications of fluidization in metal production are shown in Figure 2.20a.

2. Chlorination and Fluidization

Not much is reported on this topic in the literature. However, we will take a brief look at certain processes which are widely accepted and practiced in metallurgical industries. One branch of metallurgy that has emerged as a separate topic is chloride (halide) metallurgy. This subject has been amply dealt with in the literature.[148-150] Chlorination processes employed in chloride metallurgy are classified as in Figure 2.20b. In principle, two-phase (i.e., gas–solid) fluidization can be employed for chlorination of metals, metal oxides, and ore concentrates by chlorinating agents (which also constitute the fluidizing medium) such as Cl_2, HCl, and some metal chlorides. Where a metal chloride is used for chlorination, an inert gas can be used as a fluidizing agent. Chlorination is usually an exothermic reaction. In such cases, using a bed diluent or dilution of the fluidizing gas with an inert gas can control the reaction temperature. Fluidization with dry chlorinating agents is a gas–solid reaction and falls into the category of anyhydrous chlorination. If a third phase (i.e., liquid) is also involved, then the fluidization is of the three-phase type and belongs to the category of aqueous chlorination. Aqueous chlorination is mainly confined to base metal sulfides. Although the title of this section refers to halogenation, we are constrained to write more on chlorination because of its importance and the large volume of literature concentrated on this subject. The terms chlorination and chloridization are often encountered in chloride metallurgy. The former is carried out using elemental Cl_2 and the latter with hydrogen chloride or metal chlorides.

3. Chloride Metallurgy

Chlorination is particularly popular in nonferrous metals extraction. This important technology faced difficulties in the past due to the corrosion problems associated with the high-reaction-temperature environments; this made the construction material for the reactor a main handicap for commercial-scale operation. With the advent of new construction materials and fluidization engineering, added to cost-effective chlorinating agents such as Cl_2, NaCl, $CaCl_2$, etc., chloride metallurgy has emerged as a prime route in the processing of base, rare, refractory, and reactive metals. The main reasons that make chloride metallurgy attractive are the many advantageous properties of chlorides such as low melting point and low vapor pressure, which make the chlorides of metals easily separable from their concentrates or low-grade ones. Additionally, chlorides are highly soluble in water, easily oxidizable, and readily reducible by H_2. Thus, chlorination has become more attractive than conventional hydro- and pyrobeneficiation of ores. Furthermore, many chloride by-products can be regenerated and recycled, thus rendering the process pollution free. It is worth mentioning that sulfur in nonpolluting form is freed in chlorination of sulfides. Chlorination is also applied in refining metals such as aluminum. The more

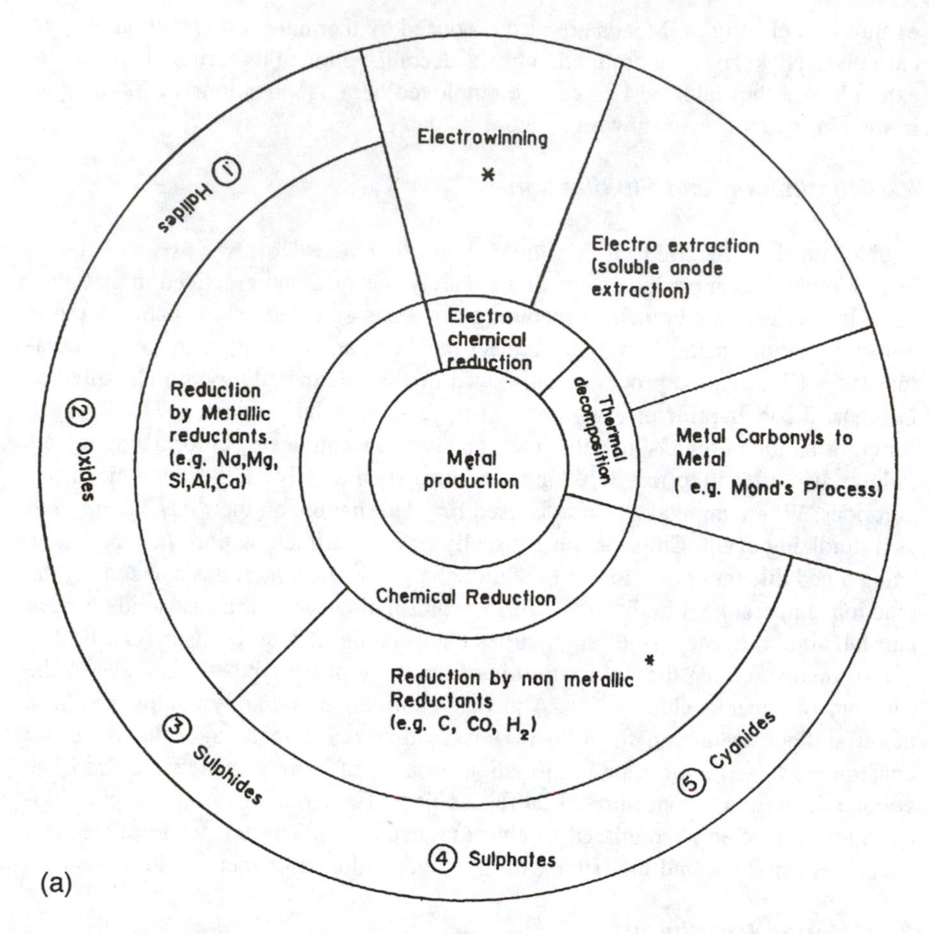

Figure 2.20 Fluidization in metal production and chloride metallurgy: (a) fluidization in metal production and (b) areas of application of fluidization in chloride metallurgy.

volatile impurities are preferentially removed during refining, resulting in pure metal. The following discussion focused more on chlorination than fluorination. As fluorination is more relevant to nuclear engineering applications, it is discussed separately in Chapter 3.

B. Fluid Bed Chlorination

1. Metal Oxide Chlorination

Metal oxides mixed with carbon are chlorinated more frequently in fluidized bed chlorination. A mixture of CO–Cl_2 is often also used for chlorination. If HCl is the chlorinating agent, the products generated are oxychloride types, and high temperatures (1000–1200°C) are often required for this reaction. In certain specific cases, raw materials containing CaO can be chlorinated using a mixture of SO_2 and Cl_2 so as to chlorinate the metal values and convert the CaO to $CaSO_4$. In the absence of

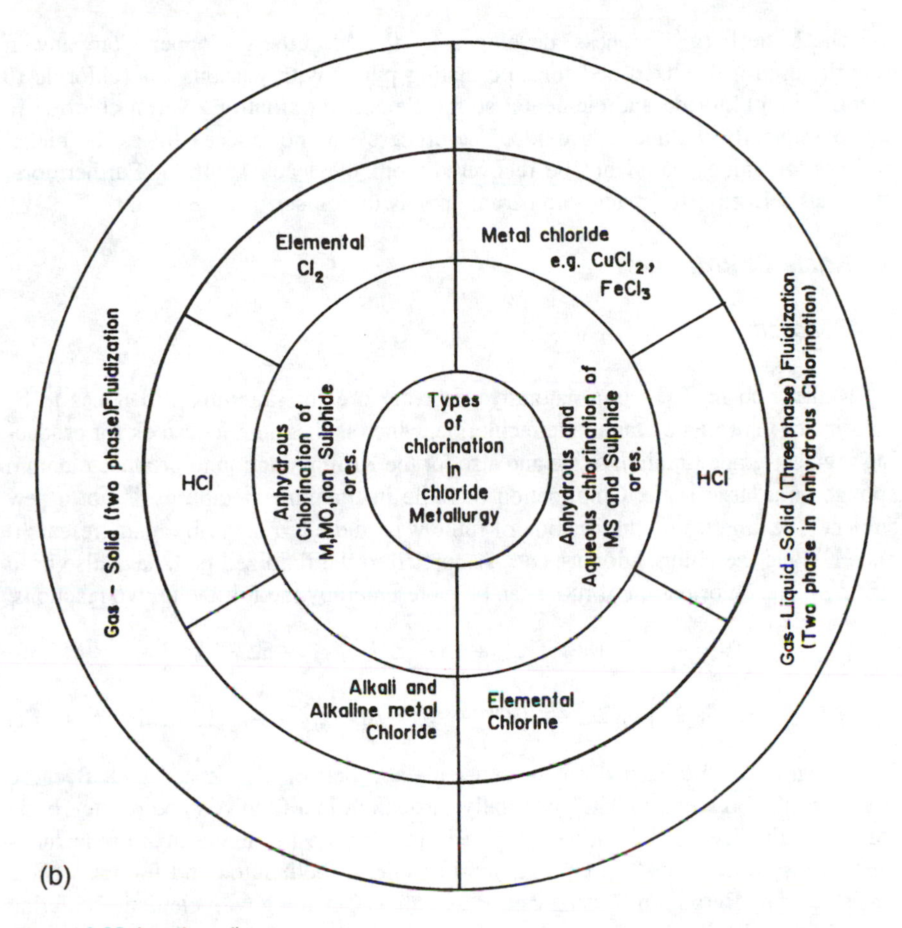

(b)

Figure 2.20 (continued)

SO_2, $CaCl_2$, which is a hygroscopic and water-soluble compound, will be a resultant product. Although chlorination using metal chlorides can be carried out in a fluidized bed by mixing with an inert fluidizing gas, not much has been reported on this in the literature. Selective chlorination, which is practiced in ilmenite beneficiation, has been extensively studied by various groups of investigators. This aspect will be discussed later in this section. When chlorination of a metal oxide is carried out without carbon, the forward reaction will be favored if the O_2 liberated is removed from the reaction site. If carbon is also used, the oxygen is removed as CO_2 or CO, depending on the temperature of the reaction. Chlorination of metal oxides can be accomplished after converting them into their carbides in an electric furnace and then reacting them in a fluidized bed. This is a two-step process. In the first step, the carbide is formed at a high temperature. In the next step, the carbide is chlorinated. Carbide chlorination is highly exothermic and thus requires close temperature control. The fluidized bed is ideal for this type of reaction. Some basic aspects of chlorination of metal oxides can be found elsewhere in the literature[147] on extractive metallurgy.

The Mitterberg[151] process, developed by the Mitterberg Copper Company in Austria, used a fluidized bed for chlorinating pyrite with gaseous iron chloride to produce iron chloride and elemental sulfur. Selective oxidation of iron chloride in air subsequently yielded iron oxide. The process was not successful as the nickel and copper values could not be recovered from the leach solution. Furthermore, materials selection for reactor construction was then a serious constraint.

2. Rutile Chlorination

a. Reaction

Rutile, rich in TiO_2, is a naturally occurring ore for titanium, and it has to be chlorinated to produce titanium tetrachloride, which is the main feedstock for producing pigment-grade titanium oxide and also for the Kroll reduction to produce titanium sponge on a large scale. Chlorination of rutile in titanium metallurgy is not a new subject. The kinetics of chlorination of rutile was studied in detail by many researchers,[152-155] and the chlorination aspects, as applied to the fluidized bed, have also been reported. The chlorination of rutile can be represented by the following two reactions:

$$TiO_2 + 2Cl_2 + \leftrightarrow TiCl_4 + CO_2, \Delta H_{1200\,K} = -52.25 \text{ kcal} \tag{2.36}$$

$$TiO_2 + 2Cl_2 + 2C \leftrightarrow TiCl_4 + 2CO, \Delta H_{1200\,K} = -11.15 \text{ kcal} \tag{2.37}$$

The reaction can be carried out in either a static bed or a fluidized bed. Because commercial production of $TiCl_4$ is usually carried out in a fluidized bed reactor, basic studies on fluidized bed chlorination have been of interest to many titanium technologists for several years. The chlorination of rutile, in both static and fluidized beds, was studied by Bergholm[152] using carbon as well as CO along with elemental chlorine gas. The chlorination of rutile with a mixture of CO and Cl_2 for rutile is given by:

$$TiO_2 + 2Cl_2 + 2CO \leftrightarrow TiCl_4 + 2CO_2 \tag{2.38}$$

The reaction represented by Equation 2.38 is slower than the reaction represented by Equation 2.36 below 1000°C; the reaction velocity is proportional to the CO concentration and independent of the Cl_2 concentration. The converse is true for the reaction given by Equation 2.36. When chlorination is carried out with insufficient or low-reactivity carbon below 1000°C, the carbon is converted mainly to CO_2. Bergholm[152] pointed out that carbon with rutile in a fluidized bed is brought into contact within 200 μm instead of direct contact as in rutile–carbon pellets.

b. Mechanism

It was first believed that an intermediate compound such as phosgene is responsible for the chlorination of rutile in the presence of carbon particles. Phosgene decomposed progressively with temperature, and the decomposition was nearly complete at 1000°C, but there was no sign of reduction in the reaction rate. Hence, it was postulated that phosgene was not the intermediate responsible for chlorination.

The reaction was believed to be analogous to the reaction of H_2 and Cl_2 or O_3 and Cl_2 mixture, where the Cl_2 atoms and the COCl radicals played important roles. The fluidized bed, due to its backmixing characteristics, requires a reasonably good Cl_2 concentration to achieve high conversion. Gas diffusion in a fluidized bed is not a limiting factor. Experimental showed that the reaction rate of a solid in dense pellets of a fine grain size is faster with Cl_2 than with a loosely packed mixture of low-depth rutile–carbon. At temperatures above 1000°C, chlorination with a CO mixture is rapid and is independent of Cl_2 concentration. Hence, the fluidized bed is recommended for this high-temperature chlorination.

c. *Parameters*

Studies carried out by Vijay et al.[156] on the chlorination of natural rutile mixed with coke showed some interesting results. The conversion of rutile increased with an increase in the flow rate. The increase in gas flow resulted in the formation of gas bubbles that escaped quickly without participating in the reaction. Hence, an increased degree of bed turbulence and good mixing are required for a high degree of conversion. The rutile-to-coke ratio at 900°C was found to play an important role in the reaction. As the ratio fell, the rutile reaction rate also fell, and a maximum was observed at a rutile-to-coke ratio of 0.25; this was followed by a minimum corresponding to a rutile-to-coke ratio of 0.33, and a change in the reaction mechanism to something other than C–Cl_2 (like CO–Cl_2) was presumed. The reaction of chlorine with rutile was shown to follow the shrinking core model with the surface reaction as the controlling factor. In fluidized bed chlorination with coke, 3 mol of gaseous product would be produced for every 2 mol of Cl_2 reacted, causing expansion of the gas fluidized bed. When the bed height was increased, the utilization of Cl_2 gas increased up to 100%.

d. *Models*

The chlorination of rutile with coke in a fluidized bed was modeled by Youn and Kyun[157] to predict the conversion of chlorine, the particle size distribution, the product gas composition, and the reaction time. The details and the types of the various models applicable to fluidized bed reactors are presented in Chapter 5. Youn and Kyun[157] used a bubble assemblage model[158] to predict the gas exchange from the bubbles to the dense emulsion phase in a fluidized bed. The overall conversion of Cl_2 was predicted, neglecting the reaction in the freeboard. The model was tested using the experimental data of a pilot-plant chlorinator and the model prediction was quite satisfactory. The use of a fluidized bed section in a vertical shaft kiln and also an electrothermally heated bed for the chloridization of rutile to produce $TiCl_4$ was discussed by Barkdale.[159]

3. *Ilmenite Chlorination*

a. *Chlorination for Beneficiation*

Ilmenite, a ferrous titanate ore, is an alternative raw material used in place of naturally occurring rutile in the titanium metal and pigment industries. In view of the limited supply of natural rutile and the abundant availability of ilmenite, many methods have emerged to beneficiate ilmenite. The iron content of the ilmenite is usually removed by the well-known, commercially accepted sulfate and chloride routes. The sulfate route is an aqueous process and the chloride route is nonaqueous. In the chloride route, ilmenite is chlorinated either totally or selectively for the removal of iron. The titaniferous slag, obtained after the removal of iron by electromelting, is also chlorinated to remove the last traces of iron and impurities.

b. Direct Chlorination

Chlorination can be accomplished by a moving bed, a static bed, and a fluidized bed. Chlorination of ilmenite by the fluidized bed is now well established. Doraiswamy et al.[160] reported chlorination of ilmenite in a fluidized bed, and Perkins et al.[161] reported chlorination of ilmenite and slag. The work of Perkins et al.[161] was mainly an evaluation study carried out in a 150-mm and 2-m silica column fluidized bed operated at 1100°C using an inert bed of silica that acted as a diluent for impurities. Impurities such as chlorides of calcium and magnesium remained in the liquid state and created a defluidized mass, thereby necessitating removal of part of the bed. This removal is essential if the ore contains impurities such as calcium and magnesium. Heating of the bed was initially accomplished with natural gas burners. During shutdown, the bed was kept fluidized with helium to flush the unreacted chlorine and also to prevent its consumption of unreacted Cl_2. Chlorine utilization of over 90% was reported for most of the titaniferous slags and Australian rutile. Titanium extracted amounted to 90–95%. Ilmenite chlorination in fluidized bed reactors was studied and discussed by several research workers.[162,163] Dooley[164] assessed the production of titanium metal and discussed the use of fluidized bed chlorination of ilmenite (or rutile).

In direct chlorination of ilmenite, most impurities are also chlorinated, and hence the chlorine consumption is very high. Furthermore, the chlorides have to be separated by sublimation to remove $FeCl_3$ and to recover $TiCl_4$. If, on the other hand, iron can be chlorinated selectively, both energy and chlorine consumption can be reduced tremendously. The chlorinated bed, after removal of iron, contains a TiO_2-rich material which is left in a porous and highly reactive state for further reaction.

c. Selective Chlorination

Basic investigations on the kinetics of selective chlorination of ilmenite in a fluidized bed were reported by Rhee and Sohn,[165] who also presented modeling[166] of the fluidized bed for the selective chlorination of ilmenite ore by a $CO–Cl_2$ mixture. The kinetics in the temperature range 923–1123 K were represented by the pore-blocking rate law and the activation energy was determined to be 37.2 kJ/mol. The partial pressure of CO was found to influence chlorination more strongly than the partial pressure of chlorine. Experiments were conducted in a shallow fluidized bed in order to minimize heat and mass transfer rates and also to eliminate side reactions of the chlorides. For selective chlorination, the partial pressure of oxygen had to be controlled in the range 10^{-8}–10^{-3} atm when the chorine partial pressure was kept in the range 10^{-2}–1 atm. The reaction mechanism for selective chlorination can be represented as:

$$6FeTiO_3 + Cl_2 + 2(FeCl_3) \rightarrow 4FeCl_2 + 2Fe_2O_3 + 6TiO_2 \quad (2.39)$$

$$2Fe_2O_3 + 3Cl_2 + 2(FeCl_3) \rightarrow 6FeCl_2 + 3O_2 \quad (2.40)$$

$$FeCl_2 + {}^1/_2CL_2 \rightarrow FeCl_3 \quad (2.41)$$

$$O_2 + 2CO \rightarrow 2CO_2 \quad (2.42)$$

Reactions 2.39 and 2.41 were confirmed to be fast by experiments. Reaction 2.42 was fast in the temperature range 878–1073 K. Hence, Reaction 2.40 can be regarded as the overall rate-controlling step. At temperatures below 773 K, $FeCl_3$ formed FeOCl, which decomposed above 773 K. Unstable compounds ($FeCl_3 \cdot nFe_2O_3$) formed above 773 K and also served as the effective chlorinating agent. Hence, temperatures above 773 K appeared to be suitable for selective chlorination.

d. Models

The two-phase model for the fluidized bed was modified, and two models were proposed for application in selective chlorination. In the first model, the bubble and emulsion phase were considered as a separate continuum with mass exchange between them; the second model was based on compartmentalization of the fluidized bed into a network of perfectly mixed reactors. These models were tested to predict the effect of various variables such as superficial velocity, exchange between the phases, and reaction rate. The compartmentalization method was found to give satisfactory results.

e. Some Highlights of Beneficiation

Chlorination of ilmenite in a fluidized bed reactor using solid carbon is more attractive than using CO gas, because carbochlorination is faster than chlorination by a CO–Cl_2 mixture. This was established[154] in rutile chlorination with a mixture of solid carbon-rutile. The literature on carbochlorination of titaniferrous magnetite is scant, even though a good number of studies have been reported on selective chlorination of low-grade ilmenite ores. Rhee and Sohn[167] investigated the selective chlorination of iron in titaniferrous magnetite ore in a fluidized bed and examined the effects of the chlorination temperature, the partial pressure of chlorine, the particle size of the ore and carbon, and the quantity of carbon in the ore. The chlorination rate with respect to chlorine concentration is first order, and the best results on selective chlorination of iron are obtained in the temperature range 900–1000 K, with 33 wt% of the carbon having an ore-to-carbon particle size ratio of 0.5. The law of additive times for chemical reaction and pore diffusion mixed control could adequately explain the experimental results. Ilmenite beneficiation by a method other than chlorination was carried out by reducing[168,169] the iron content by hydrogen gas, followed by smelting or leaching to remove iron so as to obtain a TiO_2-rich ore. This route seems to be more attractive than the chloride route from the standpoint of pollution control and corrosion problems.

4. Chlorination of Zirconium-Bearing Materials

a. Chlorination in Zirconium Metallurgy

Chlorination of zirconium-bearing materials is an important unit process in the metallurgy of zirconium. Zirconium metal is produced by Kroll reduction of the tetrachloride of zirconium. Chlorination of zirconium-bearing materials or impure chloride materials plays an important role in the process flowsheet of zirconium extraction. Chlorination was carried out in the past in either shaft or packed bed chlorinators. The feed materials for the chlorinator can consist of oxides as well as carbides. Zircon, which is the main ore of zirconium, is a silicate material, and its chlorination can be accomplished at higher temperatures.

b. General Studies

Fluidized bed chlorination of zirconium-bearing materials was investigated by Spink et al.[170] with four types of feed: zirconium oxides, zirconium carbides, fused oxides, and zircon. The carbide of zirconium was first prepared in an electric furnace and then was chlorinated. Chlorination of the carbide is a highly exothermic reaction and the reaction temperature is around 500–600°C. The use of a fluidized reactor for this purpose is appropriate because heat dissipation can be controlled more efficiently than in any other reactor. The chlorination can be carried out in a conventional fluidized bed reactor, heated by an externally wound electrical resistance heating coil. The chlorination of oxides or fused oxides and zircon present in electrically conducting carbon can be carried out by using a resistively heated fluidized bed reactor in the temperature range 1000–1200°C. Coke particles of +65 –100 mesh and 100-mesh zircon were used in a resistive fluidized bed. Spink et al.[170] used a 100-mm × 125-mm cross-section and 610-mm-long graphite resistive-type reactor heated by a 220-V AC supply and controlled by a 10-kW variable transformer. The details of resistively heated reactors, also known as an electrothermal fluidized bed reactor, are dealt with separately in Chapter 4. Because chlorination of the oxides of zircon can be carried out directly in a fluidized bed without producing a separate intermediate carbide, direct chlorination is now extensively used in large commercial-scale operations. The chlorination of oxides as well as zircon is highly endothermic and thus requires continuous heat input. A resistively heated bed has been found to be suitable, especially for chlorination at high temperatures. Because the reaction is fast at high temperatures, elutriation of fine particles in a resistively heated fluidized bed is minimized. The conversions of ZrC, ZrO_2, fused zirconia, and zircon achieved in a fluidized bed chlorinator[170] are, respectively, 81.5% (at 540°C), 94% (at 1200°C), 94% (at 1200°C), and 74% (at 1200°C).

c. Chlorination of Nuclear-Grade Zirconium Dioxide

Fluidized bed chlorination of nuclear-grade ZrO_2 with a gaseous mixture of CO and Cl_2 was reported by Sehra.[171] To test the feasibility of the reaction experimentally, this type of chlorination was done on a laboratory scale with the goal of replacing a vertical shaft chlorinator which used a briquetted ZrO_2 and carbon mixture. Because the shaft chlorinator malfunctioned due to the channeling caused by briquette disintegration, fines filling, and choking, a fluidized bed chlorinator was considered an appropriate alternative. In a laboratory-scale fluidized bed reactor with a 252-mm-diameter × 600-mm-long cylindrical column, chlorination of ZrO_2 with an average particle size of 120 μm at 800°C with 1:1 ratio of a CO and Cl_2 mixture amounting to 400 cc/min resulted in chlorine utilization of 72.93% at a reaction rate of 3.42 g/hr/cm^2. Chlorination of nuclear-grade ZrO_2 in a fluidized bed is preferred to a vertical shaft or packed bed chlorinator due to the many well-known advantages of the fluidized bed and elimination of the intermediate briquetting step for pellet feed preparation.

d. Chlorination Results

The chlorination kinetics of zirconium oxide in the presence of carbon was studied by Biceroulu and Gauvin[172] over the temperature range 1400–1950 K in a radio-frequency chlorine plasma tail flame. The investigations were concerned with the influence of temperature, chlorine concentration, and pellet carbon content. At temperatures above 1700 K, the reaction was determined to be controlled by ash diffusion. High reaction temperatures (>1700 K) offset the chlorination rate when the carbon content was less. A fluidized bed reactor was suggested to be the best choice to eliminate ash diffusion at reaction temperatures greater than 1700 K. Chemical reaction was the rate-controlling step when the reaction was carried out below 1700 K. A high carbon content is essential to maintain a good chlorination rate. Rate expressions for the chemical reaction were presented in terms of temperature, chlorine concentration, and initial weight fraction of carbon. The activation energy determined for the chemical reaction control regime was 93.325 kJ/mol. These studies brought out the importance of carrying out chlorination of ZrO_2 at high temperatures and the application of high-temperature reactors such as electrothermal fluidized bed and plasma flame reactors for this purpose.

e. Direct Chlorination of Zircon

In the preceding section, we dealt with the chlorination of ZrO_2. With the advent of nuclear energy, the importance of Zr metal and the demand for supply zirconium-bearing minerals have increased. Zirconium occurs in the ores of baddeleyite (ZrO_2), zircon ($ZrSiO_4$), or zirkite ($ZO_2 \cdot ZrSiO_4$). Among these ores, zircon, which theoretically contains 67.2% ZrO_2 and 32.8% SiO_2, is the most abundant zirconium-bearing ore. For the production of Zr metal, the silicate ore (i.e., zircon) has to be opened. Zircon is chemically stable even in hydrofluoric acid at all concentrations and temperatures. The breaking of zircon is conventionally carried out by alkali fusion at 650–750°C or by direct chlorination at 1100–1200°C. Because chlorination is an essential and unavoidable part of zirconium metallurgy, there have been continued efforts to develop a new reactor that can work at temperatures from 1100 to 1200 K. A fluidized bed reactor is the right choice in this regard, but heating the charge inside the reactor is a problem. Heat can be generated by combustion of carbon/coke in the reactor. However, this would dilute the off-gas containing $ZrCl_4$, posing a condensation problem. Induction heating in a large-diameter reactor is not practicable. External heating through the wall requires a highly heat-conductive and corrosion-resistant material. Hence, the final choice is an electrothermal fluidized bed reactor in which heat is generated by the bed resistance precisely at the reaction site. The chlorination kinetics of zircon in a static bed and in an electrothermal fluidized bed reactor were studied by Manieh and Spink.[173] The kinetic data generated in static boat experiments on chlorination were used to fix the operating parameters and also to compare the chlorination behavior of an electrothermal fluidized bed. Zircon mixed with petroleum coke was chlorinated and the chlorination rate was found to increase with Cl_2 gas concentration, temperature, and carbon-to-zircon ratio. The chlorination rate was found to be proportional to the power 0.32 of the chlorine concentration and 0.15 of the carbon-to-zircon ratio. The activation energy was found to be 4.01 kcal/g · mol. Chlorination was conducted in an electrothermal fluidized bed (50 mm in diameter × 400 mm tall with a 75-mm-diameter × 150-mm-tall disengaging section) with a charge of zircon (–100#) and coke (–20 +65 mesh) mixed at a carbon-to-zircon ratio of 12:1: Heating was carried out with a 600-A × 40-V power supply using a 25-kVA transformer. The effects of gas flow rate, bed height, and chlorine concentration were investigated at a temperature of 1015°C. The best chlorine conversion of 51.1% was attained with a flow rate of 0.061 g · mol/min, starting with an initial bed height of 600 mm. A minimum bed height of 250 mm to bring the temperature of Cl_2 to the reaction temperature and dense graphite material as the reactor construction material were suggested, and the electrothermal fluidized bed was recommended as the only choice for chlorination of zircon at high temperatures. Kinetics and mechanisms of zircon chlorination studied by Sparling and Glastonbury[174] showed that chlorination progresses with the formation of an intermediate metal oxychloride. The surface reaction is fast, and hence it is not the rate-controlling step. In a carbon-deficient system, the rate-controlling step is the diffusion of $MeOCl_2$ through the boundary layer to the carbon surface, whereas in a zircon-deficient system diffusion of $MeOCl_2$ away from the zircon surface controls the reaction in the temperature range 900–1200°C. A detailed chlorination study carried out later by Manieh et al.[175] established that the reaction is of zero order with respect to chlorine concentration. The chlorine is presumed to be strongly absorbed on the solid phase and the reaction is supposed to proceed subsequently. The chlorine concentration in the gas phase has no effect on the reaction rate. The reaction rate is controlled by surface reaction or adsorption. The activation energy was found to be 10.55 kcal/g · mol.

5. *Columbite Ore and Molysulfide Chlorination*

a. *Columbite Chlorination*

Columbite is a columbium (niobium)–tantalum-bearing mineral. Niobium and tantalum have very similar chemical properties. Hence, they can be chlorinated and separated from columbite ore, as these metal chlorides are volatile. A laboratory-scale Vycor (96% silica glass produced by Corning) fluidized bed reactor was tested on a continuous-run basis by Oslen and Block.[176] They achieved complete chlorination of the niobium and tantalum values in the mineral without formation of intermediate oxychlorides. The fluidized bed was charged with 50% columbite mineral and 50% metallurgical coke and was fluidized using elemental chlorine at 500°C. Any oxychloride formed was rechlorinated at 600°C in the top section of the reactor, where a bed of charcoal filter helped the rechlorination and also eliminated the carryover of mineral dust along with the product vapors. The chlorination rate of the mineral was found to be independent of chlorine concentration and proportional to the power 0.13 of the mineral concentration with a reaction rate constant of 0.0522 g. mineral/cc/min. A relation for the fraction of mineral (X) converted to chloride was proposed as a function of the mineral residence time (τ, in minutes), as:

$$\frac{X}{(1-X)^{0.13}} = 0.0092\ \tau \tag{2.43}$$

A fluidized bed reactor was reported[177] for chlorinating tantalum oxide concentrate and removal of the tantalum values as chloride vapor, leaving behind the manganese content of the concentrate in the bed. Pure tantalum chloride was obtained after distillation of the crude chlorides.

b. *Molysulfide Chlorination*

Nair et al.[178] chlorinated commercial-grade molybdenite in a fluidized bed reactor using a mixture of Cl_2, N_2, and O_2 in the proportion of 2:5:23. Molybdenum recovery of 99% and chlorine utilization efficiency of 84% were reported to be achieved at 300°C. The reaction appeared to be of the first order with respect to particle size, and the overall reaction was controlled by chemical reaction. A fluidized bed reactor used for the chlorination of low-grade Indian molybdenite concentrate was reported by Nair et al.[179] Low-grade Indian molybdenite received from the froth flotation cells can be upgraded by the conventional hydro- and pyrometallurgical route. However, the product obtained can be used as a steel additive but not for pure Mo metal production. An alternative way is to chlorinate the low-grade Mo concentrate in the presence of O_2 to generate the more volatile oxychloride of Mo (at 275°C), leaving behind the Cu, Ni, and Fe values as a residue in the bed. Molybdenum recovery of 99% was obtained with 75-µm concentrate particles for an aspect ratio of 2 in a 40-mm-ID Pyrex glass column reactor. An N_2–O_2–Cl_2 mixture of 13:5:2 composition was used for 25 min at a rate of 25 lpm at 275°C during oxychlorination for optimum performance. In the absence of chlorination, (i.e., during start-up and the cooling stages), N_2 gas alone was used just to keep the bed material in a state of fluidization. Based on the oxychlorination technique, a completely new flowsheet was proposed for processing low-grade molybdenum concentrate. As the oxychlorination was exothermic, a bed diluent like alumina (Al_2O_3, ρ_s = 2.64 g/cc, d_p = 80 μm) was mixed up to 40% by weight. The addition of a bed diluent for a continuous process was deemed necessary for close control of the bed temperature. The fluidized bed reactor contained residue chlorides of Ni, Fe, and Cu after oxychlorination of Mo, and the residue was roasted with O_2 in the same reactor, mixing NaCl and converting the $FeCl_3$ portion into iron oxides. After leaching, the roasted residue yielded chloride solutions of Ni and Cu which could be separated by solvent extraction. Nair et al.[180] conducted studies on the oxychlorination kinetics in a fluidized bed reactor over a bed temperature range of 250–250°C and a particle size range of 75–200 μm. The reaction rates were determined over these ranges of variables and the specific reaction rate constant was evaluated. As these authors determined earlier in their experiments on commercial-grade molybdenite,[178] the reaction was of the first order with respect to particle size, and overall oxychlorination was found to be controlled by chemical reaction.

6. Chlorination in Silicon Metal Production

a. Chlorosilane

Crystalline silicon metal in its pure and dense form, when produced at a cheap price, is commercially competitive, and demand for it in the electronics and photovoltaic industries is ever-increasing. Metallurgical-grade silicon or the chlorides of silicon, obtained as by-products from zirconium or other industries, is the usual starting material to produce pure silicon metal via chlorination and subsequent reduction. Chlorosilanes were produced[181] by the reaction of gaseous CH_3Cl with silicon powder using a copper catalyst in a fluidized bed reactor at temperatures in the range 250–450°C at a pressure of 6 atm. In order to have a longer residence time, a low gas velocity was used, incorporating a mechanical agitator. The fluidized bed has been identified as a potential reactor and has been used in several steps in the process flowsheets of pure silicon metal production.

b. Fluidization in Silicon Metal Production

The various steps in producing pure silicon metal and the use of fluidized bed reactors are shown in Figure 2.21. Pure silicon metal is obtained by decomposition[182,183] of silane (SiH_4) or by hydrogen reduction of trichlorosilane[184] in a fluidized bed seeded with pure silicon particles. A radiantly heated, tapered fluidized bed was used[185] for the decomposition of SiH_4 to Si and H_2, as this offered ease of operation compared to conventional externally or internally heated fluidized beds,

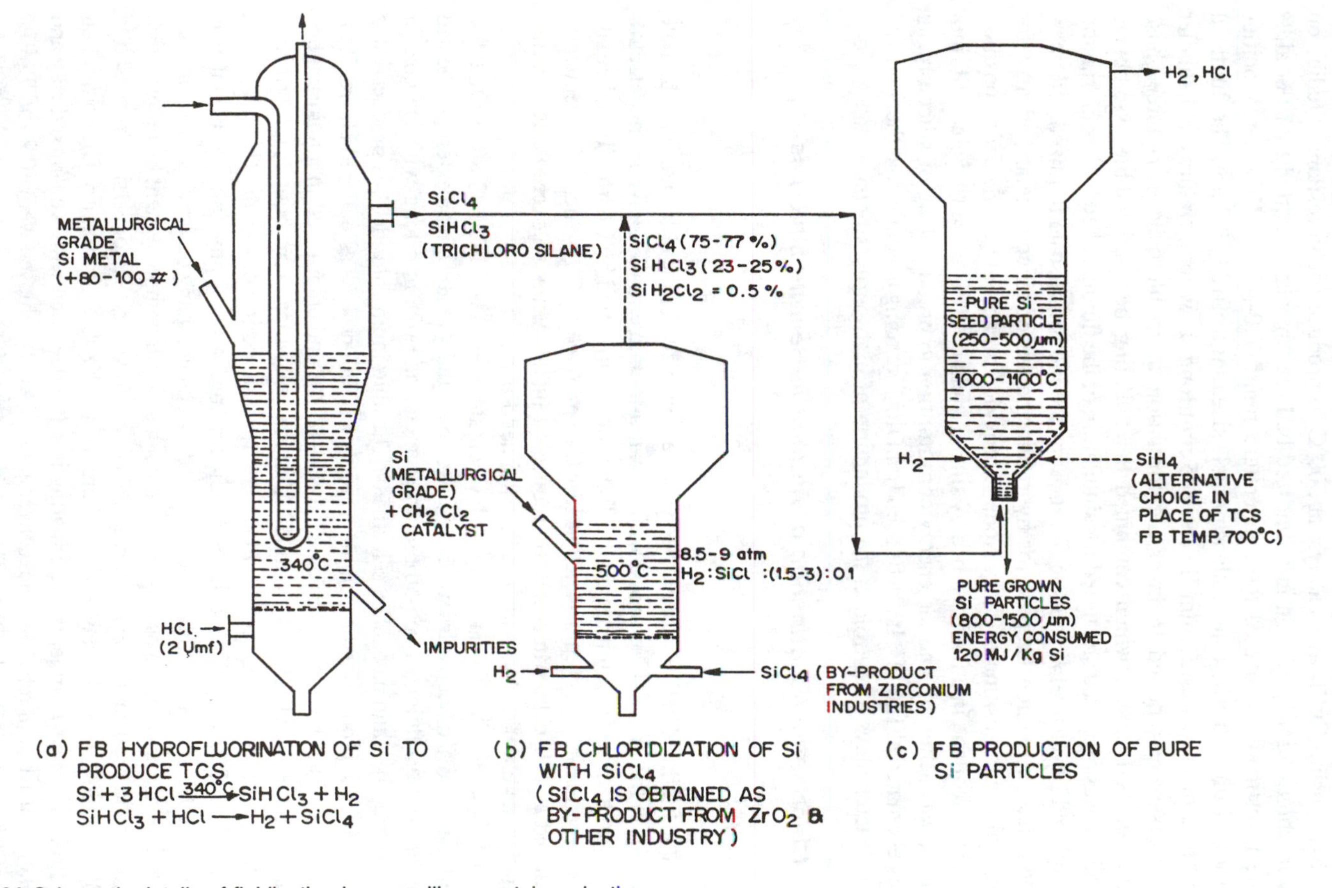

Figure 2.21 Schematic details of fluidization in pure silicon metal production.

which have the problem of silane condensation on hot heating surfaces. Industrial-grade silicon metal is hydrochlorinated in a fluidized bed at 360°C or chloridized by $SiCl_4$ using $CuCl_2$ as a catalyst at 500°C to produce trichlorosilane. Studies on hydrochlorination of silicon in a fluidized bed reactor were reported by Li et al.[186] The investigations covered the temperature range 340–400°C, HCl gas velocities varying from 2 to 5 U_{mf}, and a silane particle size range of 124–297 μm. The fluidized bed reactor used was 25 mm ID and 375 mm long. Low temperatures and smaller particles were reported to favor the formation of trichlorosilane. A three-phase fluidized bed model, which combined the bubbling bed and bubble assemblage models, was proposed and tested for simulation of the fluidized bed hydrochlorinator. Kunii and Levenspiel[187] cited a reference for the industrial-scale fluidized bed operation used by Union Carbide for hydrofluorination of silicon metal and also for thermal decomposition of SiH_4. Noda[184] provided an account of the decomposition of trichlorosilane in a fluidized bed. A silicon yield of 20%, close to the equilibrium value, was achieved, and the energy consumption to produce 1 kg of silicon metal was estimated at 120 MJ. It can be seen from the foregoing accounts that the fluidized bed reactor plays a key role in the production of high-purity silicon metal.

7. *Chlorination/Fluorination of Aluminum-Bearing Materials*

a. *Toth Process*

The fluidized bed reactor has been used extensively for the chlorination of bauxite or clay in the Toth[105,188] process, which is an alternative to the conventional electrolytic production of aluminum. In the Toth process, invented by Charles Toth, bauxite or clay, after drying, is chlorinated at 125°C to produce Al_2Cl_6, which is subsequently reduced at 220°C by Mn metal to produce Al metal. Mn is recovered after dechlorinating its chloride by oxidation in a fluidized bed.

Chlorination studies on Georgia clay containing 39.1% Al_2O_3, 43.8% SiO_2, 0.9% Fe_2O, and 2.6% other oxides were conducted by Ujhidy et al.[189] in an externally heated (by electricity) fluidized bed 50 mm in diameter and 625 mm long with the goal of recovering the alumina content selectively. The aluminum chloride thus produced was meant for the production of aluminum using the Toth process as outlined above. Prior to chlorination, the clay was calcined at 800–850°C and chlorinated in a fluidized bed reactor with a $CO–Cl_2$ gas mixture. Alumina and silica were equally chlorinated, achieving 40–80% Al_2O_3 conversion over a reaction period of 4–8 hr. The conversion attained in 2 hr at 800°C with a $CO–Cl_2$ mixture without any solid reductant was merely 25%, and this could not be increased by changing any other reaction parameter or operating parameter. The conversion did not show any improvement by chlorination with a solid reductant such as brown coal coke. However, the conversion increased to 40% by chlorinating the clay–coke mixture with a $CO–Cl_2$ gas mixture. The conversion could be increased further by chlorinating in a fluidized bed with well-mixed clay and coke. In just 1 hr of reaction time, conversions of Al_2O_3 achieved were 80.2 and 79.9%, respectively, at 850 and 930°C. The chlorination of SiO_2 was suppressed by 50% by adding 5% sodium chloride. Thus, a fluidized bed proved to be a potential reactor in chlorinating Georgia clay.

b. Aluminum Trifluoride

Aluminum trifluoride (AlF_3), which is used in the aluminum electrolytic cell, is produced by reacting calcined aluminum trihydrate, $Al(OH)_3$, with hydrofluoric acid vapor. The reaction is accomplished in a fluidized-bed-type reactor. A three-stage fluidized bed reactor for the calcination of $Al(OH)_3$ and the reaction of $Al(OH)_3$ with HF at 595°C was reported[190] by Kaiser Aluminum Corporation in the United States for the commercial production of AlF_3. The Montedison/Lurgi process[92] for the production of AlF_3 used a highly expanded circulating-type fluidized bed reactor, and a 28-tpd plant at Montedison came into operation in 1958 in Porto Marghera, Italy. Dried alumina was hydrofluorinated using 98% hydrofluoric acid to achieve 92–94% pure AlF_3. The fluidized bed reactor heated by methane combustion had two zones: (1) a dense bottom zone where the trihydrate of aluminum reacted with evaporated and superheated (70°C) hydrofluoric acid vapor and (2) an expanded top zone where a flue gas velocity of 120–152 cm/s facilitated gas circulation at 537°C in the first cyclone of the two in series. The waste gases, after secondary cleaning, were cooled to 70–77°C by adiabatic absorption and 20–23% purity HF was recovered.

8. Selective Chlorination for Nickel and Cobalt Recovery

a. Principles

Heertjes and Jessurun[191] demonstrated the selective chlorination of nickel and cobalt in lateritic iron ore in a fluidized bed. The underlying principle regarding this selective chlorination lies in the fact that when lateritic iron ore that contains hydrated nickel and cobalt oxides is chloridized with a mixture of steam and HCl, water-soluble chlorides of nickel and cobalt are formed, leaving behind water-insoluble Fe_2O_3 in the ore. A fluidized bed reactor was found to be ideally suited for this purpose.

b. Chlorination Studies

New Guinea laterite ore containing nickel, cobalt, and iron in the proportion 6:1:200, after subjected to chloridization, was analyzed and found to contain the above elements in the proportion 27:6:1, thereby establishing the high enrichment of nickel and cobalt values with respect to iron in the chloridized and leached product. Experimental investigations on a batch fluidized bed 50 mm in diameter, fitted with a titanium or Monel sieve plate with 0.4-mm openings, were carried out with the fines fractions ($d_p < 100$ μm) obtained after wet sieving and drying of laterite ore. Variables such as preroasting temperature, reaction time, fluidization gas velocity, and bed height in a fluidized bed reactor were investigated to arrive at an optimum condition of selective chlorination. The recovery of nickel by and large was poor if the chlorination was carried out with preroasted ore. Hence, a simple drying operation of the ore at 100–110°C before chlorination was recommended. Three compounds of nickel and two compounds of cobalt were found to exist in laterite ore, as inferred from the recovery of these elements at various temperatures. One compound of nickel remained practically unattacked by HCl, the second compound reacted at 400°C when heated for a long time, and the third compound was reactive during preheating as well as during chloridization. One of the compounds of cobalt was not affected by the preroasting temperature. Although the reaction rate increased with higher HCl concentration, the fluidized granules became sticky at high HCl partial pressures. An HCl-to-H_2O molar ratio of 45:55, a fluidizing gas velocity two times the incipient value, and a bed height equal to the reactor diameter were recommended for optimum recovery of nickel and cobalt values by selective chloridization of lateritic iron ores in a batch fluidized bed reactor.

NOMENCLATURE

a	interfacial surface area (m^2)
a'	half amplitude of vibration (m)
C	moisture content in the solid (kg/kg)
C_0	moisture content of bubble gas at height $h = 0$, $C_b = C_{g0}$ (kg moisture/kg dry gas)
C_b	moisture content in the bubble (kg/m^3)
C_{ge}	moisture content of bubble at the exit (kg moisture/m^3 dry gas)
C_{go}	moisture content of bubble at height $h = 0$ (kg moisture/m^3 dry gas)
C_{gR}	moisture content of gas at particle radius $r = R$ (kg moisture/kg dry gas) (at bubble radius R in Equations 2.10 and 2.12)
C_s	moisture content at the surface at time $t = 0$ (kg/kg dry solid)
C_{se}	moisture content of gas at the exit (kg moisture/kg dry gas)
C_{sR}	moisture content of solid at $r = R$ at the exit (kg moisture/kg dry solid)
C_{sM}	average moisture content of drying solid (kg moisture/kg dry solid)
C_{so}	moisture content at the surface (kg/kg dry solid)
D	diffusivity in particle (m^2/s)
D_m	mixing coefficient (Equation 2.23)
d_p	particle size (weight mean size in Equation 2.16) (m)
G	air mass velocity (kg/m^2)
H	bed height (m)
h	height inside the bed above the distributor but less than H (m)
K	mass transfer coefficient (mol/m^2 · atm · s)
K_{bc}	bubble-to-cloud mass transfer coefficient (m/s)
K_{eq}	equilibrium constant (–)
m	mass fraction of tracer material (–)
N	vapor or moisture flux (mol/m^2 · s)
P_b	partial pressure at bulk mean temperature (atm)
P_{SO2}	partial pressure of sulfur dioxide (atm)
P_{wb}	partial pressure at wet bulb temperature (atm)
Q_g	volumetric flow rate of gas (m^3/s)
Q_s	volumetric feed rate of solid (m^3/s)

R	drying rate (kg/s)
r	radial position in the particle (m)
Re_p	Reynolds number (Equation 2.20) (–)
R_p	radius of particle (m)
t	residence time(s)
T_b	bulk mean temperature (K)
t_m	mean residence time(s)
T_{wb}	wet bulb temperature (K)
U_b	bubble rise velocity (m/s)
U_g	gas velocity (m/s)
U_b^*	difference in the gas velocity between the unvibrated and vibrated inclined fluidized bed for the same linear velocity of particulate solid (m/s)
U_{mf}	minimum fluidization velocity in unvibrated fluidized bed (m/s)
U_{mm}	minimum superficial gas velocity at which mixing is observed in vibro fluidized bed (m/s)
U_{mvf}	minimum fluidization velocity in vibro fluidized bed (m/s)
v	mean velolcity of tracer (m/s)
V_b	bubble volume (m^3)
X	moisture content (kg/kg dry solid)
x	distance (m/s)
X_e	equilibrium moisture content (kg/kg dry solid)
X_f	fractional conversion (–)
X_m	mean moisture content of solid (kg/kg dry solid)
W	bed holdup (kg)
w	tracer holdup (kg)

Greek Symbols

α_p	radial dispersion number (–)
α_H	axial dispersion number, Dt_m/H^2
$\Delta p_{b,mvf}$	pressure drop at U_{mvf} (Pa)
$\Delta p_{b,mf}$	pressure drop at U_{mf} (Pa)
θ	dimensionless particle residence time, t/t_m
λ	latent heat of vaporization (kJ/kg)
μ_g	viscosity of gas (Pa · s)
ρ_p	particle density (kg/m^3)
ϕ_v	particle volume equivalent of shape factor (–)
ω	angular frequency (l/s)

REFERENCES

1. Romankov, P.G., Drying, in *Fluidization,* Davidson, J.F. and Harrison, D., Eds., Academic Press, London, 1971, chap. 12.
2. Kunii, D. and Levenspiel, O., *Fluidization Engineering*, Butterworth-Heinemann, Boston, 1991, 21.

3. Reay, D. and Baker, C.G.J., Drying, in *Fluidization,* Davidson, J.F., Clift, R., and Harrison, D., Eds., Academic Press, London, 1985, chap. 16.
4. Jobes, C.W., Fluidized crystal dryer pays off, *Chem. Eng.*, 61, 166, 1954.
5. Jensen, A.S., Drying under pressure, in *Fluidization*, Ostergaard, V.K. and Sorensen, A., Eds., Engineering Foundation, New York, 1986, 651.
6. Kjaegaard, O., Mortensen, H.B., and Mortensen, S., Designing fluid bed dryers for low energy operations, in *Fluidization,* Ostergaard, V.K. and Sorensen, A., Eds., Engineering Foundation, New York, 1986, 659.
7. Liu, D. and Kwauk, M., Pneumatically controlled multistage fluidized beds, in Third Int. Conf. on Fluidization, Henniker, NH, 1980, 485.
8. Liu, D., Li, X., and Kwauk, M., Pneumatically controlled multistage fluidized bed and its application in solvent recovery from a plant waste gas, *Sel. Pap. J. Chem. Ind. Eng.*, 1, 1, 1981.
9. Kwauk, M., Particulate fluidization: an overview, in *Advances in Chemical Engineering*, Vol. 17, Wei, J., Anderson, J.L., Bischoff, K.B., and Seinfeld, J.M., Eds., Academic Press, New York, 1992, chap. 2.
10. Jones, P.S. and Leung, L.S., Downflow of solids through pipes and valves, in *Fluidization,* Davidson, J.F., Clift, R., and Harrison, D., Eds., Academic Press, London, 1985, chap. 8.
11. Mustafa, I., Ph.D. thesis, Imperial College of Science and Technology, University of London, 1977.
12. Reay, D. and Allen, R.W.K., in Proc. Research Session on Solids Drying, Inst. Chem. Eng. Jubilee Symp., London, 1982.
13. Vanecek, V., Picka, J., and Najmer, S., Some basic information on the drying of granulated NPK fertilizers, *Int. Chem. Eng.*, 4(1), 93, 1964.
14. Rosensweig, R.E., Fluidization, *Science*, 204, 57, 1979.
15. Geuzens, P.L. and Thoenes, D., Axial and radial gas mixing in a magnetically stabilized fluidized bed, in *Heat and Mass Transfer in Fixed and Fluidized Beds*, Van Swaaij, W.P.M. and Afgan, N.H., Eds., Hemisphere Publishing, Washington, D.C., 1986, 697.
16. Alebregtse, J.B., Fluidized bed drying: a mathematical model for hydrodynamics and mass transfer, in *Heat and Mass Transfer in Fixed and Fluidized Beds*, Van Swaaij, W.P.M. and Afgan, N.H., Eds., Hemisphere Publishing, Washington, D.C., 1986, 511.
17. Palanez, B., A mathematical model for continuous fluidized bed drying, *Chem. Eng. Sci.*, 38, 1045, 1983.
18. Hoebink, J.H.B.J. and Rietema, K., Drying granular solids in a fluidized bed. II. The influence of diffusion limitations on the gas–solid contacting around bubbles, *Chem. Eng. Sci.*, 35, 2257, 1980.
19. Verkooijen, A.H.M., Fluidized bed drying of fine particles with internal diffusion limitation, in *Fluidization,* Ostergaard, V.K. and Sorensen, A., Eds., Engineering Foundation, New York, 1986, 643.
20. Chiba, T. and Kobayashi, H., Gas exchange between the bubble and emulsion phases in gas–solid fluidized beds, *Chem. Eng. Sci.*, 25, 1375, 1970.
21. Pye, E.W., Vibration shakes up thinking on fluid bed drying, *Process Eng. (London)*, March 1974, 61.
22. Gupta, R. and Mujumdar, A.S., Aerodynamics and thermal characteristics of vibrated fluid bed — a review, *Drying*, 80(1), 141, 1980.
23. Strumillo, C. and Pakowski, Z., Drying of granular products in vibrofluidized beds, *Drying*, 80(1), 211, 1980.

24. Gupta, R. and Mujumdar, A.S., Aerodynamics of a vibrated fluid bed, *Can. J. Chem. Eng.*, 58, 332, 1980.
25. Yamazaki, R., Kanagawa, Y., and Jimbo, G., Heat transfer in a vibrofluidized bed. Effect of pulsated gas flow, *J. Chem. Eng. Jpn.*, 7, 373, 1974.
26. Pakowski, Z. and Strumillo, C., *Drying*, 80(2), 208, 1980.
27. Bratu, E.A. and Jinescu, G.I., Heat transfer in vibrated fluidized layers, *Rev. Roum. Chim.*, 17, 49, 1972.
28. Arai, N. and Hasatani, M. Heat and mass transfer in vibro inclined fluidized bed, in *Fluidization, 85*, Kwauk, M. and Kunii, D., Eds., Science Press, Beijing, 1985, 445.
29. Mathur, K.B. and Epstein, N., Developments in spouted bed technology, *Can. J. Chem. Eng.*, 52, 129, 1974.
30. Mathur, K.B. and Epstein, N., *Spouted Bed*, Academic Press, New York, 1974.
31. Schwedes, J. and Otterbach, J., *Verfahrenstechnik (Mainz)*, 8(2), 42, 1974.
32. Mathur, K.B. and Gishler, P.E., A study of the application of the spouted bed technique to wheat drying, *J. Appl. Chem.*, 5, 624, 1955.
33. Romankov, P.G. and Rashkovskaya, N.B., *Drying in Suspended State*, 2nd ed., Chemistry Publishing House, Leningrad, 1968.
34. Schmidt, A., Continuous drying of solutions, *Chem. Zentralbl.*, 126, 3219, 1955.
35. Frantz, J.E., Ph.D. thesis, Louisiana State University, 1958.
36. Mousa, A.H.N., Drying by fluidized bed that contains inert particles, *Int. J. Miner. Process.*, 6, 155, 1979.
37. Wormald, D. and Burnell, E.M.W., Design of fluidized beds driers and coolers for the chemical industry, *Br. Chem. Eng.*, 16(4/5), 376, 1971.
38. Vanecek, V., Markvart, M., Drabohlav, R., and Hummel, R.L., Experimental evidence on operation of continuous fluidized bed driers, *Chem. Eng. Symp. Ser.*, 66(105), 243, 1970.
39. Davis, W.L., Jr. and Glazier, W., Large scale fluidized bed drying of iron ore concentrate, *AIChE Symp. Ser.*, 70(141), 137, 1974.
40. Anon., Aluminum fluoride from wet process phosphoric acid wastes, *Br. Chem. Eng. Process Technol.*, 17(7/8), 609, 1972.
41. Ogienko, A.S., Drying zinc cakes from pulp in a fluidized bed with simultaneous granulation of the products, *Sov. J. Nonferrous Met.*, 10, 39, 1965.
42. Jonke, A.A., Petkus, E.J., Loeding, I.W., and Lavrovski, S., The use of fluidized beds for the continuous drying and calcination of dissolved nitrate salts, *Nucl. Sci. Eng.*, 2, 303, 1957.
43. Thompson, R.B. and MaeAskill, D., Anacond's Yerington plant demonstrates economical recovery of sulphur flow low grade ore, *Chem. Eng. Prog.*, 51, 369, 1955.
44. Newman, L.L., Oxygen in the production of hydrogen or synthesis gas, *Ind. Eng. Chem.*, 40, 559, 1948.
45. Winkler, F., U.S. Patent 1,687,118, October 9, 1943.
46. Anony., SO_2 by fluidization, *Chem. Eng.*, 60, 238, 1953.
47. Newman, R.I. and Lavine, A.J., Development of a fluid bed roaster for zinc concentrates in pyrometallurgical processes in nonferrous metallurgy, in *Metallurgical Society Conferences*, Vol. 39, Anderson, J.N. and Queneau, P.E., Eds., Gordon and Breach, New York, 1967, 1.
48. Okazaki, M., Nakane, Y., and Noguchi, H., Roasting of zinc concentrate by fluo solids systems as practiced in Japan at Onahama Plant of Toho Zinc Company Ltd., in *Metallurgical Society Conferences*, Vol. 39, Anderson, J.N. and Queneau, P.E., Eds., Gordon and Breach, New York, 1967, 19.

49. Coolbaugh, E.W. and Neider, R.F., Fluid column roasting at Sherbrooke Metallurgical Co. Ltd., Portland, Ontario, in *Metallurgical Society Conferences*, Vol. 39, Anderson, J.N. and Queneau, P.E., Eds., Gordon and Breach, New York, 1967, 45.
50. Reid, J.H., Operation of a 350 ton per day suspension roaster at Trail, British Columbia, in *Metallurgical Society Conferences*, Vol. 39, Anderson, J.N. and Queneau, P.E., Eds., Gordon and Breach, New York, 1967, 69.
51. Stankovie, G., Fluosolid roaster and acid plant at Trepca lead and zinc mines and refineries, in Proc. World Symp. on the Min. Met. Pb and Zn, Vol. 2, 1970, chap. 4.
52. Themelis, N.J. and Freeman, G.M., Fluid bed behaviour in zinc roasters, *J. Met.*, p. 52, August 1984.
53. Adhia, J.D., First fluosolid roaster in India — design and operation, in *Recent Developments in Nonferrous Metals Technology,* Vol. III, Proc. Symp. Nonferrous Metals Technology, Maynar, J.E. and Gupta, R.K., Eds., National Metallurgical Lab., Jamshedpur, 1968, 49.
54. Bradshaw, A.V., Rate controlling factors in gas/solid reactions of metallurgical interest, *Trans. Inst. Min. Met.*, 79, C281, 1970.
55. Khalafalla, S.E., in *Rate Processes of Extractive Metallurgy,* Sohn, N.Y. and Wadsworth, M.E., Eds., Plenum Press, New York, 1979, chap. 4.
56. Jully, J.W., Practical operating aspects of fluid bed roasting, in Symp. on Fluidization in Theory and Practice, Intn. of Chem. Eng., Austr., Chem. Inst., Melbourne, 1967.
57. Pape, H., Roasting of sulphide ores, in Symp. on Fluidization in Theory and Practice, Intn. of Chem. Eng. Austr., Chem. Inst. Melbourne, 1967.
58. Mathews, O., *Trans. Can. Inst. Min. Metall.*, 52, 97, 1949; cited in Reference 52.
59. Blair, J.C., Fluid bed roasting, in *SME Mineral Processing Handbook*, Vol. 7, Weiss, N.L., Ed., Society of Mining Engineers of the American Institute of Mining, Metallurgical and Petroleum Engineers, New York, 1988, 12.
60. Stephens, F.M., Jr., The fluidized bed sulphate roasting of nonferrous metals, *Chem. Eng. Progr.*, 49(9), 455, 1953.
61. Margulis, E.V. and Cherednik, I.M., Kinetics of sulphating pyrite cobalt concentrate in a fluidized bed, *Tsvetn. Met.*, 18(8), 69, 1967.
62. Chi, Tsain, S. and Smirnov, V.I., Investigating fluo-solid sulphating roasting of the cobalt mattes of nickel and copper plants, *Tsvetn. Met.,* 1, 37, 1960.
63. Fletcher, A.W. and Shelef, M., A study of the sulphation of a concentrate containing iron, nickel and copper sulphides, *Trans. AIME*, 230, 1721, 1964.
64. Fletcher, A.W. and Hester, K.D., *Trans. AIME*, 229, 282, 1964.
65. Sakano, T. and Nagano, T., Cobalt recovery from cobalt bearing pyrite concentrate, unit process, *Hydrometallurgy,* 24, 770, 1964.
66. Palperi, M. and Aaltonen, O., Sulphatizing roasting and leaching of cobalt ores at Outokumpu Oy, *J. Met.*, p. 34, February 1971.
67. Kwauk, M. and Tai, D.W., Transport processes in dilute phase fluidization as applied to chemical metallurgy (in Chinese, with English abstract), *Acta Metall. Sin.*, 7, 391, 1964.
68. Griffith, W. et al., Development of the roast leach process for Lekeshore, *J. Met.*, 27(2), 17, 1975.
69. Narayanan, P.I.A. and Subramanya, G.V., Laboratory studies on the beneficiation of some furruginous manganese ores of India for the production of ferromanganese, *Trans. Indian Inst. Met.*, 66, 49, 1956.
70. Priestly, R.J., Upgrading iron ore by fluidized magnetic conversion, *Blast Furn. Steel Plant*, 46, 303, 1958.

71. Tomasicchio, G., Magnetic reduction of iron ores by the fluosolids systems, in *Proc. Int. Symp. on Fluidization*, Drikenburg, A.A.H., Ed., Netherlands University Press, Amsterdam, 1967, 725.
72. Boueraut, M.M. and Toth, I., IRSID process of fluid bed magnetic roasting, *Chem. Eng. Prog. Symp. Ser.*, 62(62), 15, 1966.
73. Kwauk, M., Particulate fluidization: an overview, in *Advances in Chemical Engineering*, Vol. 17, Anderson, J.L., Bischoff, K.B., and Seinfeld, J.M., Eds., Academic Press, New York, 1992, 311.
74. Uwadiale, G.G.O.O., Magnetizing reduction of iron ores, *Miner. Process. Extract. Metall. Rev.*, 11, 1, 1992.
75. Pinkey, E.T. and Plint, N., Treatment of refractory copper ores by the segregation process, *Trans. Inst. Min. Met.*, 76, C114, 1967.
76. Anon., Copper TORCO segregation process, *Br. Chem. Eng.*, 14(3), 264, 1969.
77. Brittan, M.I., Fluidized bed copper chloridizing, *Trans. Inst. Min. Metall.*, 80, C262, 1971.
78. Jackson, D.V. and Taylor, R.F., *High Temperature Chemical Reaction Engineering: Solids Conversion Processes*, Inst. Chem. Eng., London, 1971, chap. 3.
79. Pawlek, F.E. et al., Chlorination of garnertite, presented at the 100th AIME Annu. Meet. Met. Soc., New York, 1970.
80. Blair, J.C., Fluosolid roasting of copper concentrates at Copperhill, in *Nonferrous Metallurgy and Metallurgical Society Conferences*, Vol. 39, Anderson, J.N. and Queneau, P.E., Eds., Gordon and Breach, New York, 1967, 55.
81. Thoumsin, F.J. and Coussemant, R., Fluid bed roasting reactions of Cu and Co sulphide concentrates, *J. Met.*, 16(10), 831, 1964.
82. Nebeker, J.S. et al., Operation of the fluosolids roaster at Kennecotts, paper presented at the 99th AIME Annu. Meet. Met. Soc., Denver, 1969.
83. Vian, A., Iriarte, C., and Romero, A., Fluidized roasting of arsenopyrites, *Ind. Eng. Chem. Process Des. Dev.*, 2(3), 214, 1963.
84. Mathews, O., *Trans. Can. Inst. Min. Metall.*, L11, 97, 1949.
85. Fraser, K.S., Walton, R.H., and Wells, J.A., Processing of refractory gold ores, *Miner. Eng.*, 4(7–11), 1029, 1991.
86. Golant, A.I., Korneeva, S.G., and Stepanov, A.V., Quality and effectiveness of controlling the temperature of the fluidized bed roasting of molybdenite concentrates, *Tsvetn. Met. Mosk.*, 43(3), 45, 1970.
87. Rudolph, V., Selection and application of gas fluid bed reactors in the nonferrous metals industry, *Miner. Process. Extract. Metall. Rev.*, 10(1–4), 87, 1992.
88. Doheim, M.A., Abdel-Wahab, M.Z., and Rassoul, S.A., Fluidized roasting of molybdenite, *Trans. Inst. Min. Metall.*, 84, CIII, 1975.
89. Gupta, C.K., *Extractive Metallurgy of Molybdenum,* CRC Press, Boca Raton, FL, 1992, chap. 4.
90. Taverner, M.G., Experimental Fluid Bed Roasting of Zinc Concentrate Pellets, Met. Statement No. 65, Sulphide Corp. Pty. Ltd., Cockle Creek Works, 1967.
91. Gordon, G. and Peisakhov, I., *Dust Collection and Gas Cleaning,* Vol. 4, Mir Publishers, Moscow, 1972.
92. Reh, L., Fluidized bed processing, *Chem. Eng. Prog.*, 67(2), 58, 1971.
93. Fukunaka, T., Monta, T., Asaki, Z., and Kondo, Y., Oxidation of zinc sulphide in a fluidized bed, *Metall. Trans. B*, 7B, 307, 1976.
94. Ellis, J.N., *Chem. Eng. Min. Rev.*, 41, 295, 1949.
95. Kite, R.P. and Roberts, E.J., Fluidisation in non-catalytic operations, *Chem. Eng.*, 54, 113, 1947.

96. Thomson, R.B., in *Fluidization*, Othmer, D.F., Ed., Reinhold, New York, 1956, 212.
97. Priestly, R.J., High temperature reactions in a fluidized bed, in *Proc. Int. Symp. on Fluidization*, Drinkenburg, A.A.H., Ed., Netherlands University Press, Amsterdam, 1967, 701.
98. Geldart, D., Gas–solid reactions in industrial fluidized beds, *Chem. Ind.*, 13, 41, 1968.
99. Van Thoor, T.J.W., in *Materials and Technology Encyclopaedia,* Vol. 2, de Bussy, J.H., Ed., Longman, London, 1971, chap. 2.
100. Turkdogan, E.T. and Sohn, H.Y., Calcination, in *Rate Process of Extractive Metallurgy,* Sohn, H.Y. and Wadsworth, M.E., Eds., Plenum Press, New York, 1979, chap. 4.3.
101. Kunii, D. and Levenspiel, O., *Fluidization Engineering*, John Wiley & Sons and Toppan Company Ltd., Tokyo, 1969, 56.
102. Kunii, D. and Levenspiel, O., *Fluidization Engineering*, John Butterworth-Heinemann, Boston, 1991, 50.
103. Fenyi, G., Kaldi, P., and de Jonge, J., The application of gas–solid fluidization in high temperature processes, in *Proc. Int. Symp. on Fluidization*, Drinkenburg, A.A.H., Netherlands University Press, Amsterdam, 1967.
104. Schmidt, H.W., Beisswenger, H., and Kampf, F., Flexibility of the fluid bed calciner process in view of changing demands in the alumina market, *J. Met.*, 32(2), 31, 1980.
105. Anon., Aluminum from clay chlorination, *Proc. Technol. Int.*, 18(11), 417, 1973.
106. Sawyer, R.H., U.S. Patent 2,642,339, 1948.
107. Hughes, W.H. and Arkless, K., British Patent 992,317, 1965.
108. Henderson, A.W., Campbell, T.T., and Block, F.E., Dechlorination of ferric chloride with oxygen, *Metall. Trans.*, 3, 2579, 1972.
109. Paige, J.I., Robidart, G.B., Harris, H.M., and Campbell, T.T., Recovery of chlorine and iron ioxide from ferric chloride, *J. Met.*, 27, 12, 1975.
110. Schneider, K.J., Solidification of radioactive wastes, *Chem. Eng. Prog.*, 66(2), 35, 1970.
111. Schneider, K.J., Solidification and disposal of high level radioactive wastes in the United States, *React. Technol.*, 13(4), 387, 1970/1971.
112. Lohse, G.E. et al., Calcination of zirconium fluoride wastes, in Proc. Symp. on Fluidization Tripartite, Chem. Eng. Conf., Montreal, Canada, 1968, 75.
113. Wright, J.K., Taylor, I.F., and Philip, D.K., A review of progress of the development of new iron making technologies, *Miner. Eng.*, 4(7–11), 983, 1991.
114. Doheim, M.A., Fluidization in the ferrous industry, *Iron Steel Int.*, 47(5), 405, 1974.
115. Labine, R.A., *Chem. Eng.*, p. 96, February 1960.
116. Brown, J.W., Campbell, D.L., Saxton, A.L., and Carr, J.W., Jr., FIOR — the Esso fluid iron ore direct reduction process, *J. Met.*, 18(2), 237, 1966.
117. Reed, T.F., Agarwal, J.C., and Shipley, E.H., Nu–iron: a fluidized bed reduction process, *J. Met.*, 12, 317, 1960.
118. Agarwal, J.C., Davis, W.L., Jr., and King, D.T., Fluidized bed coal dryer, *Chem. Eng. Prog.*, 58(11), November 1962.
119. Cronan, C.S., Nu–iron: good process, *Chem. Eng.*, 67, 64, 1960.
120. Pitt, R.S., Cross country repeater mill for district iron and steel, *Iron Steel Int.*, 46(3), 242, 1973.
121. Wenzel, W., Iron making with the aid of nuclear energy, *J. Egypt. Soc. Eng.*, 11(1), 11, 1972.
122. Anon., Continuous iron process nears pilot plant, *Chem. Eng. News*, 45(33), 46, 1967.
123. Kunii, D. and Levenspiel, O., *Fluidization Engineering*, Butterworth-Heinemann, Boston, 1991, 57.

124. Stephens, F.M., Jr., Langston, B., and Richardson, A.C., The reduction–oxidation process for the treatment of taconites, *J. Met.*, 5, 780, 1953.
125. Doherty, R.D., Hutchings, K.M., Smith, J.D., and Yoruk, S., The reduction of hematite to wustite in a laboratory fluidized bed, *Metall. Trans. B*, 16, 425, 1985.
126. Bowling, K. McG. and Waters, P.L., *Br. Chem. Eng.*, 13(8), 1127, 1968.
127. Szekely, J., Evan, J.W., and Sohn, H.Y., *Gas–Solid Reactions*, Academic Press, New York, 1976, chap. 5.
128. Agarwal, J.C. and Davis W.L., Jr., The dynamics of fluidization of iron and its ores, *Chem. Eng. Prog. Symp. Ser.*, 62(67), 101, 1966.
129. Gransden, J.F., Sheaby, J.S., and Bergougnou, M.A., An investigation of defluidization of iron ore during reduction by hydrogen in a fluidized bed, *Chem. Eng. Prog. Symp. Ser.*, 66(105), 209, 1970.
130. Espiridion, G.V., Henry, D., and Karl, C.D., Carbon reduction of iron oxide waste materials to reactive sponge iron in a fluidized bed reactor, *Chem. Eng. Prog. Symp. Ser.*, 66(105), 221, 1970.
131. Naden, D., Wrightman, G., and Jaworski, H.K., Fluid ore reduction technology for direct smelting processes, in Proc. 46th Iron Making Conf., AIME, Pittsburgh, March 1987, 409.
132. Steffen, R., The COREX process: first operating results in hot metal making, *Met. Plant Tech.*, 2, 20, 1990.
133. Nyirenda, R.L., The processing of steel making flue dust: a review, *Miner. Eng.*, 4(7–11), 1003, 1991.
134. Anon., Finding Value in Waste, Metal Price Report, Scrap Supplement, October 23, 1990, 6.
135. Mackey, T.S. and Prengaman, R.D., in *Lead and Zinc, 90*, TMS, Warrendale, 1990, 453.
136. McGrath, H.G. and Rubin, L.C., Fluidized Reduction of Metal Oxides, U.S. Patent 2,61,765, March 9, 1954.
137. Duvall, F.E.W. and Sohn, H.Y., Kinetics of hydrogen reduction of porous copper sulphate pellets, *Trans. Inst. Min. Metall.*, 92C, 166, 1983.
138. Chida, T. and Ford, J.D., Kinetics of hydrogen reduction of nickel sulfide, *Can, J. Chem. Eng.*, 55, 313, 1977.
139. Williams, D.T., El-Rahaiby, S.K., and Rao, Y.K., Kinetics of reduction of nickel chloride with hydrogen, *Metall. Trans. B*, 12B, 161, 1981.
140. Toor, T.P.S., Aangers, R., and Kaliaguines, S., Copper powders produced by fluidized bed reduction of $CuSO_4$ with hydrogen, *Can. Metall. Q.*, 21(1), 79, 1982.
141. Kivniek, A. and Hixson, N., Reduction of nickel oxide in a fluidized bed, *Chem. Eng. Prog.*, 48, 394, 1952.
142. Swaminathan, K., Sreedharan, O.M., Sathiyamoorthy, D., and Bose, D.K., Thermogravimetric and fluidized bed studies on hydrogen reduction of $NiSO_4$, *Trans. Indian Inst. Met.*, 39(6), 637, 1986.
143. Coffer, L.W., Extractive metallurgy techniques possibly applicable in India, in *Proc. Symp. Nonferrous Metals Technology*, Vol. 2, Mannar, J.E. and Gupta, P.K., Eds., National Metallurgical Laboratory, Jamshedpur, India, 1968, 238.
144. Lewis, C. and Stephens, L.W., Iron and nickel by carbonyl treatment, *J. Met.*, p. 419, June 1958.
145. Michael, A.B. and Hanway, J.E., Jr., Hydrogen reduction of molybdic oxide by a fluidized bed reactor, *J. Met.*, 16, 881, 1964.
146. Sathiyamoorthy, D., Shetty, S.M., and Bose, D.K., Hydrogen reduction of ammonium molybdate in fluidized bed, *Trans. Indian Inst. Met.*, 6, 529, 1986.

147. Habashi, F., *Principles of Extractive Metallurgy,* Gordon and Breach, New York, 1986, 350.
148. Altewkar, V.A., Developments in chlorine metallurgy, in Int. Conf. on Advances in Chem. Met., Preprints Vol. 1, BARC, Dept. Atomic Energy and IIM, Bombay, January 3–6, 1979, 4/1.
149. Mukherjee, T.K. and Gupta, C.K., Base metal processing by chlorination, *Miner. Process. Extract. Metall. Rev.*, 1, 111, 1983.
150. Jena, P.K. and Brocchi, E.A., Halide metallurgy of refractory metals, *Miner. Process. Extract. Metall. Rev.*, 10(1–4), 29, 1992.
151. Hohn, H., Jangg, G., Putz, L., and Schmiedl, E., in Proc. Int. Mineral Dressing Conf., Stockholm, 1957, 683.
152. Bergholm, A., Chlorination of rutile, *Trans. Metall. Soc. AIME*, 221, 1121, 1961.
153. Dunn, W.E., Jr., High temperature chlorination of titanium bearing minerals, IV, *Metall. Trans. B*, 10B, 271, 1971.
154. Morris, A.J. and Jensen, R.F., Fluidized bed chlorination of Australian rutile, *Metall. Trans. B*, 7B, 89, 1976.
155. Barin, I. and Schulen, W., On the kinetics of the chlorination of titanium dioxide in the presence of solid carbon, *Metall. Trans. B*, 11B, 199, 1980.
156. Vijay, P.L., Subramanian, C., and Rao, Ch.S., Chlorination of rutile in fluidized beds, *Trans. Indian Inst. Met.*, 29(5), 355, 1976.
157. Young, In-Jn and Kyun, Y.P., Modelling fluidized bed chlorination of rutile, *Metall. Trans. B*, 20B, 959, 1989.
158. Kato, K. and Wen, C.Y., Bubble assemblage model for fluidized bed catalytic reactors, *Chem. Eng. Sci.*, 24, 1351, 1969.
159. Barkdale, J., *Titanium*, 2nd ed., Poland Press, New York, 1966.
160. Doraiswamy, L.K., Bijawat, H.C. and Kunte, M.V., Chlorination of ilmenite in a fluidized bed, *Chem. Eng. Prog.*, 55(10), 80, 1959.
161. Perkins, E.C., Dolezal, H., Taylor, D.M., and Lang, R.S., Fluidized Bed Chlorination of Titaniferrous Slags and Ores, Bureau of Mines Report 6317, 1963, 13.
162. Doheim, M.A., Raaouf, A.A., and Rassoul, S.A., Chlorination of high titanium slag in fluidized beds, *Trans. Jpn. Inst. Met.*, 14(16), 483, 1973.
163. Othmer, D.F. and Nowak, R., TiO_2 and Ti metal through the chlorination of ilmenite, *Chem. Eng. Progr.*, 69(6), 113, 1973.
164. Dooley, J., Titanium production, ilmenite vs. rutile, *J. Organomet. Chem.*, 27, 8, 1975.
165. Rhee, K.I. and Sohn, H.Y., The selective chlorination of iron from ilmenite ore by $CO–Cl_2$ mixtures. I. Intrinsic kinetics, *Metall. Trans. B*, 21B, 321, 1990.
166. Rhee, K.I. and Sohn, H.Y., The selective chlorination of iron from ilmenite ore by $CO–Cl_2$ mixtures. II. Mathematical modelling of the fluidized bed process, *Metall. Trans.*, 21B, 331, 1990.
167. Rhee, K.I. and Sohn, H.Y., The selective carbochlorination of iron from titaniferrous magnetite ore in a fluidized bed, *Metall. Trans. B*, 21B, 341, 1990.
168. Volk, W. and Stotler, H.H., Hydrogen reduction of ilmenite ores in fluidized beds, *J. Met.*, 22(1), 50, 1970.
169. Vijay, P.L., Ramani, V., and Sathiyamoorthy, D., Peroxidation and hydrogen reduction of ilmenite in a fluidized bed, *Metall. Trans. B*, 27B, 731, 1996.
170. Spink, D.R., Cookston, J.W., and Hanway, J.E., Jr., The fluidized bed chlorination of zirconium bearing materials, presented in the Annual Meeting of American Institute of Mining, Metallurgical and Petroleum Engineers, New York, February 19, 1964.
171. Sehra, J.C., Fluidized bed chlorination of nuclear grade ZrO_2 with CO and Cl_2, *Trans. Indian Inst. Med.*, 27(2), 93, 1974.

172. Bicerolu, O. and Gauvin, W.H., The chlorination of zirconium dioxide in the presence of carbon, *Can. J. Chem. Eng.*, 58(3), 357, 1980.
173. Manieh, A.A. and Spink, D.R., Chlorination of zircon sand, *Can. Metall. Q.*, 12(3), 331, 1973.
174. Sparling, D.W. and Glastonbury, J.R., Kinetics and mechanism of zircon chlorination, in Aust. Inst. Min. Met. Conf., Western Australia, May 1973, 455.
175. Manieh, A.A., Scott, D.S., and Spink, D.R., Electrothermal fluidized bed chlorination of zircon, *Can. J. Chem. Eng.*, 52, 507, 1974.
176. Oslen, R.S. and Block, F.E., The chlorination of columbite in a fluidized bed reactor, *Chem. Eng. Prog. Symp. Ser.*, 66(105), 225, 1970.
177. Glaussen, W.E. and Todd, F.A., Economic and technical considerations in chlorine metallurgy, presented at the 103rd AIME Annu. Meeting Met. Soc., Dallas, 1974.
178. Nair, K.U., Sathiyamoorthy, D., Bose, D.K., Sundaresan, M., and Gupta, C.K., Chlorination of commercial molybdenite concentrate in a fluidized bed reactor, *Metall. Trans. B*, 18B, 445, 1987.
179. Nair, K.U., Sathiyamoorthy, D., Bose, D.K., and Gupta, C.K., Processing of low grade Indian molybdenite concentrate by chlorination in a fluidized bed reactor, *Int. J. Miner. Process.*, 23, 171, 1988.
180. Nair, K.U., Sathiyamoorthy, D., Bose, D.K., and Gupta, C.K., Studies on oxychlorination of MoS_2 in a fluid bed reactor, *Metall. Trans. B*, 19B, 669, 1988.
181. Nagata, S., Matsuma, T., Hashimoto, N., and Hase, H., *Chem. Eng. (Japan)*, 16, 301, 1952.
182. Rohatgi, N. and Hsu, G., Report from jet propulsion laboratories, California Institute of Technology, October 1983.
183. Christen, E., Flat Plate Solar Array Project, 10 Years of Progress, Jet Propulsion Lab., California Institute of Technology, (for DOE and NASA), October 1985.
184. Noda, T., in Flat Plate Solar Array Project, Workshop on Low Cost Polysilicon for Terrestrial Photovoltaic Solar Applications, Jet Propulsion Labs, Las Vegas, October 1985.
185. Levenspiel, O., Larson, M.B., Zhang, G.T., and Ouyang, F., Preliminary study of a radiantly heated bed for the production of high purity silicon, *AIChE Symp. Ser.,* 80(241), 87, 1984.
186. Li, K.Y., Peng, S.H., and Ho, T.E., Hydrochlorination of silicon in a fluidized bed reactor, *AIChE Symp. Ser.,* 84(262), 114, 1988.
187. Kunii, D. and Levenspiel, O., *Fluidization Engineering,* Butterworth-Heinemann, Boston, 1991, chap. 2.
188. Peacey, J.G. and Davenport, W.G., Evaluation of alternative methods of aluminium production, *J. Metal.*, 26(7), 25, 1974.
189. Ujhidy, A., Szepvolgyi, J., and Borlai, O., Applications of fluidized bed reactor to chlorination at high temperatures, in *Fluidization Technology*, Vol. 11, Keairns, D.L., Ed., Hemisphere Publishing, Washington, D.C., 1976, chap. 4.
190. Anon., *Chem. Eng.*, 72(2), 92, 1965.
191. Heertjes, P.M. and Jessurun, P.M., Recovery of nickel and cobalt from lateritic iron ores in a batch fluidized bed, in *Proc. Int. Symp. on Fluidization*, Drinkenburg, A.A.H., Ed., Netherlands University Press, Amsterdam, 1967, 688.

CHAPTER 3

Fluidization in Nuclear Engineering

I. LEACHING

A. Fluidized Bed Leaching

1. General Description

There are a variety of liquid–solid contacting devices that do not use any mechanical means for agitation to achieve better liquid–solid contact. Slater[1] reviewed the various types of continuous countercurrent contactors for liquids and particulate solids and stressed that a contactor should necessarily have a high processing capacity even at the cost of a slightly lower efficiency. The areas of importance for continuous contactors include water treatment, uranium extraction from ore slurries, and drying of petroleum fractions with alumina and silica gel. Fluidized bed leaching equipment occupies less space and does not require mechanical agitation. Its operation is fully hydraulic and automatic. Its continuous countercurrent operation with a low liquid-to-solid ratio renders complete leaching of solid. Fluidized beds with stages or pulsation can be employed for efficient liquid–solid contacting. Pulsed fluidized beds have the advantage of maintaining the same liquid rate as moving beds but at a reduced pressure drop. Multistaging is essential to eliminate the effect of backmixing. A perforated plate or conical plate with a high head of solid in the downcomer would eliminate liquid bypassing and also axial mixing. The solids residence time in a fluidized bed column can be made independent of flow ratios, and thus leaching can be effected to the fullest extent. The high axial mixing of a fluidized bed contactor can be offset by the advantage of high flow rates of both phases with the possibility of processing turbid fluids. A low specific volume compared to a fixed bed to perform the same function is an added advantage.

An account of the pulsed fluidized bed for contacting granular solids was given in a review by Rickles.[2] Several patents exist on this topic. Work carried out in China was summarized by Kwauk.[3]

2. Uses

In principle, fluidized bed leaching equipment is useful to leach out the soluble content of the ore or the roasted product from the inert solid matrix, leaving the residual solids and continuously bringing out the leached matter into the solution for further processing. Fluidized bed leaching was reported[3] for leaching the copper values after the sulfation roasting of cupriferous iron ores, for leaching out the alumina by an alkaline solution from calcined bauxite, and also for leaching the ferrous components from pyrite cinders after chloridizing roasting. Fluidized beds have been used for washing pigments, and fluidized bed washing columns have been discussed by Yakubovoch et al.[4]

The use of fluidized bed leaching with regard to zinc calcines[5,6] has also been reported. Fluidized bed leaching and washing are also used (1) to remove wax from certain coals by leaching with suitable organic solvents, (2) to prepare activated carbon from crude char by acid leaching followed by water washing, and (3) in supercritical extractions.

3. Leaching Equipment

A schematic diagram of leaching equipment is shown in Figure 3.1. The slurry is introduced from the top countercurrent to the rising stream of the leaching liquid which is fed and distributed at a suitable location above the underflow discharge. Above this feed point, leaching action is initiated by the fluidization of the solid particles by the upflow of the leachant liquid. The fluidization zone essentially consists of two parts: a dense phase and a lean phase with a clear interface between them. Above the fluidization zone is a settling zone where the clear decanted solution, rich in the leached matter, moves away as overflow.

B. Design Aspects of Leaching Column

For leaching solid particles whose downward mass flow rate is S and the flow rate of the upflowing liquid is L, the volume of the reactor (V) can be determined from knowledge of the particle residence time (θ).

Thus, if Z is the column height, the residence time (θ) for the column of cross-sectional area A is

$$\theta = \frac{AZ(1-\epsilon)\rho_s}{S} = \frac{Z(1-\epsilon)}{(S/A\rho_s)} = \frac{Z(1-\epsilon)}{U_s} \tag{3.1}$$

In the above equation, the mass flow rate of liquid (L) does not appear. In a fluidized bed there exists a relationship between L and S as a function of the particle

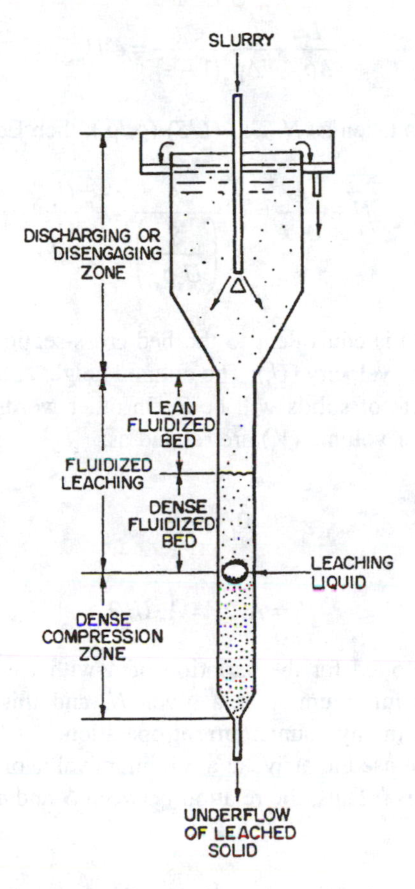

Figure 3.1 Fluidized bed leaching equipment.

terminal velocity (U_t) and the bed voidage (ϵ). This relationship can be deduced from the Richardson–Zaki equation, that is,

$$U = U_t\, \epsilon^n \tag{3.2}$$

where U is the superficial velocity of the liquid when the solid is not set in motion. For simultaneous motion of fluid and solid, one has:

$$U = U_{rf} \cdot \epsilon \tag{3.3}$$

where

$$U_{rf} = \frac{U_f}{\epsilon} + \frac{U_s}{1-\epsilon} \tag{3.4}$$

for a countercurrent flow. Using Equations 3.3 and 3.4 to eliminate U in Equation 3.2 and expressing U_f and U_s in terms of their respective mass flow rates L and S,

$$\frac{L}{A\rho_l}+\frac{S\epsilon}{A\rho_s(1-\epsilon)}=\epsilon^n U_t \tag{3.5}$$

If the ratio U_f/U_s is taken as N, i.e., (L/S) (ρ_s/ρ_f), then Equation 3.5 becomes

$$N+\frac{\epsilon}{1-\epsilon}=\epsilon^n\frac{A}{\left(\dfrac{S}{U_t\rho_s}\right)}=\epsilon^n\frac{U_t}{U_s} \tag{3.6}$$

where the ratio $(S/U_t\rho_s)$ is equivalent to the bed cross-section (A_t) when solids are moving at their terminal velocity (U_t). The column height at this velocity to process the same mass flow rate of solids will be Z_t. In other words, the solids residence time (θ_t) and the reactor volume (V_t) are related as:

$$\theta_t=\frac{Z_t}{U_t} \tag{3.7}$$

$$V_t=A_t\,Z_t=A_t\,U_t\,\theta_t \tag{3.8}$$

Equation 3.6 can be plotted for the variation of A with voidage (ϵ) at a given N. Such plots[7-9] reveal a minimum A at a given N, and this minimum is said to correspond to flooding in any countercurrent operation.

Equation 3.6 can be used to arrive at a minimum value of the ratio $A/A_t = \bar{A}$ for any given N or vice versa. Thus, the relation between N and ϵ when $dA/d\epsilon = 0$ (i.e., at minimum $\bar{A}$) is

$$N=\frac{\epsilon[1-n(1-\epsilon)]}{n(1-\epsilon)^2} \tag{3.9}$$

Substituting N from Equation 3.9 in Equation 3.6, the minimum cross-section ratio ($\bar{A}_{\min}$) is obtained as:

$$\frac{A_{\min}}{A_t}=\bar{A}_{\min}=\frac{1}{n(1-\epsilon)^2\epsilon^{n-1}} \tag{3.10}$$

This ratio is also same as the reciprocal of

$$\bar{U}_s=\left(\frac{U_s}{U_t}\right)_{\bar{A}_{\min}} \tag{3.11}$$

Substituting this U_s in Equation 3.1 for the desired residence time, the reactor height can be evaluated when the reactor is to have the minimum cross-section. For any other U_s for a known N, using Equations 3.6 and 3.1, the reactor height can be evaluated for any desired residence time.

Feed particles to the leaching or washing equipment usually are polydisperse in nature and are seldom monosized. Hence, the leaching equipment should be designed in such a way that the finest particles would not be elutriated and at the same time the largest ones would not be left unleached at the cost of a low velocity which is otherwise required for arresting the elutriation of fine particles. Kwauk[3] showed that a dimensionless parameter $(U_{t,b}/U_{t,s})(\epsilon^{n_b-n_s})$ (where the subscripts b and s are, respectively, for big and small particles) has to be taken for the correcton of $\bar{A}$ (see Equation 3.10) determined on the basis of monosized particles. When this dimensionless number is below 0.2, the volumetric utilization of leaching equipment is drastically reduced. In that case, selection of staged fluidized leaching is advised.

C. Staging

1. Staged Fluidization

a. Leaching

Industrial-scale staged fluidized leaching (SFL) was developed[10] for washing uranium ore. In SFL, the volume of the leaching or washing apparatus is reduced by an order of two. A staged fluidized bed operates in series, with its top end connected with a single entry for the feed just above the first column. The first column is fed with the leaching liquid at a velocity sufficient for the coarse or heavy particles to descend slowly. Thus the first column serves as a long residence time leacher for coarse particles and the leachable matter from the solid particles is carried up by the liquid along with fine solid particles. The leaching liquid in the subsequent columns of an SFL is fed at the bottom at a flow velocity appropriate to allow the fine particles to settle down after a longer time. The last column in a SFL will have the finest particles settling down and discharging at the bottom. In SFL, all particles, including coarse ones, are subjected to leaching for sufficient time to recover the soluble matter effectively. This is not possible in single-column leaching as the major fractions of fines are elutriated and thus left unleached. A schematic diagram of SFL equipment is shown in Figure 3.2a. The fluidized bed can also be used for washing purposes. Kwauk and Wang[11] gave an account of stripping nickel from laterite fines by liquor ammonia in a 1-m-diameter, 14.5-m-long fluidized bed washer and a fluidized bed leacher with a disengaging section 1.6 m in diameter.

The nickel concentration in a fluidized bed leaching column remained constant from the top to the interface which separated the dilute and the dense phases of the fluidized bed. At the interface, the nickel concentration reduced sharply and then remained constant in the dense fluidized phase. The pulp density of the solids exhibited similar trends but in the opposite sense. The reason for a sharp change in the nickel concentration and the pulp density at the interface was attributed to lack of backmixing, which was predominant above and below the interface.

b. Multistaging

The existence of a concentration gradient between the dilute- and the dense-phase regions of a fluidized bed can be exploited by making a multistage fluidized bed and

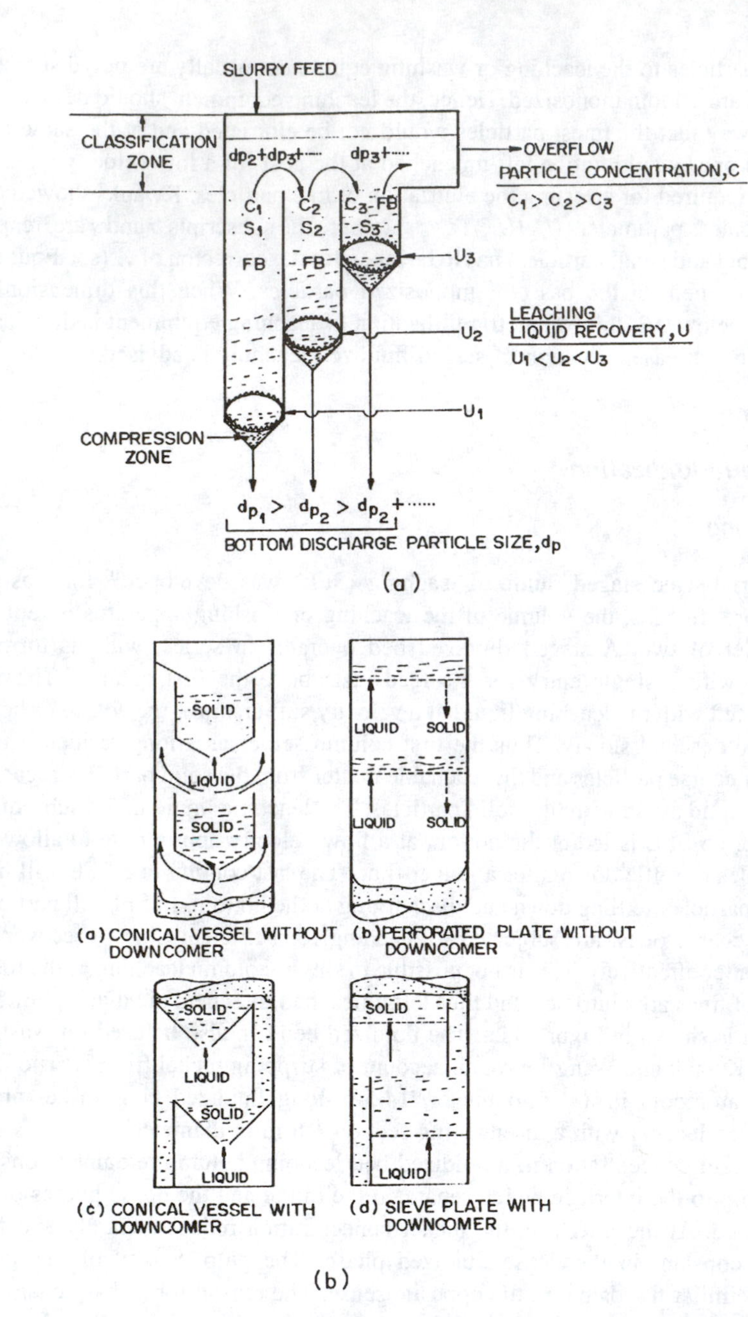

Figure 3.2 Staged and multistaged fluidized bed leaching equipment. (a) Staged fluidization for leaching. (b) Multistaging for fluidized bed leaching. FB = fluidized bed.

allowing the last stage to process the two nonmixing zones. Multistaging can be done in several ways, as shown in Figure 3.2b. There are some hydrodynamic problems in the movement of the solid and the liquid phases. Fine solids along with liquid make slurry, and they flow down between the stages easily. For coarse particles, this is not so. In order to avoid bypassing the leaching fluid through the downcomer, the sealing due to the liquids–solid bed should be adequate. In a conical-type multistage fluidized bed, the upflowing fluid keeps the bed of solid in a fluidized state in each stage.

In the event of excessive solid buildup, the pressure drop across the bed due to the fluidizing fluid will increase, and this may ultimately lead to defluidization or choking. In order to avoid this situation, a liquid bypass system, based on the principle of the autosiphon, is advised. The technical details of two types of autosiphons, open and gas sealed, were given by Kwauk.[3] These autosiphons operate when the pressure drop across each stage is increased beyond a certain level and allow the liquid to bypass from one stage below to another stage above, thereby allowing the defluidized solids to flow down through the conical bottom, unrestricted by the opposing fluid flow resistance.

II. URANIUM EXTRACTION

A. Fluidization in Nuclear Fuel Cycle

The application of fluidization in nuclear engineering is well recognized[12] from the mining of uranium ore to its disposal after burnup in a nuclear reactor. Fluidization as a useful unit operation is accepted in uranium extraction, nuclear fuel fabrication, reprocessing of fuel, and waste disposal. Efforts to make a fluidized bed nuclear reactor took place two decades ago. The only area where the application of the fluidization technique has not yet been adopted in the nuclear fuel cycle is isotope enrichment. The areas of application of fluidization in the nuclear fuel cycle are depicted in Figure 3.3. The details of the various processes in uranium extraction will be discussed next.

As discussed in the preceding section, a fluidized bed can serve as leaching as well as washing equipment. Conventionally, mined uranium ore is subjected to acid leaching[13] to dissolve the uranium values in a mechanically agitated vessel, followed by solvent extraction. The leaching operation is also carried out in an agitated three-phase tall reaction column. This type of reactor represents a form of three-phase fluidization where the acidic liquid can be either stationary or continuous. Three-phase fluidized beds are now extensively used as bioreactors. The application of fluidization in aerobic and anaerobic processes of wastewater treatment is well established. Conventional heap leaching of the ore can be carried out in an enhanced manner by using a fluidized bed bioreactor. Fluidized bed bioreactors are discussed in Chapter 6, and the application of fluidization for the extraction of uranium just after its leaching from the ore is discussed next.

B. Fluid Bed Denitration

1. Thermal Decomposition of Uranyl Nitrate

Uranyl nitrate solution (with six water molecules) is obtained after leaching of the uranium ore and solvent extraction. Uranyl nitrate hexahydrate (UNH) is a

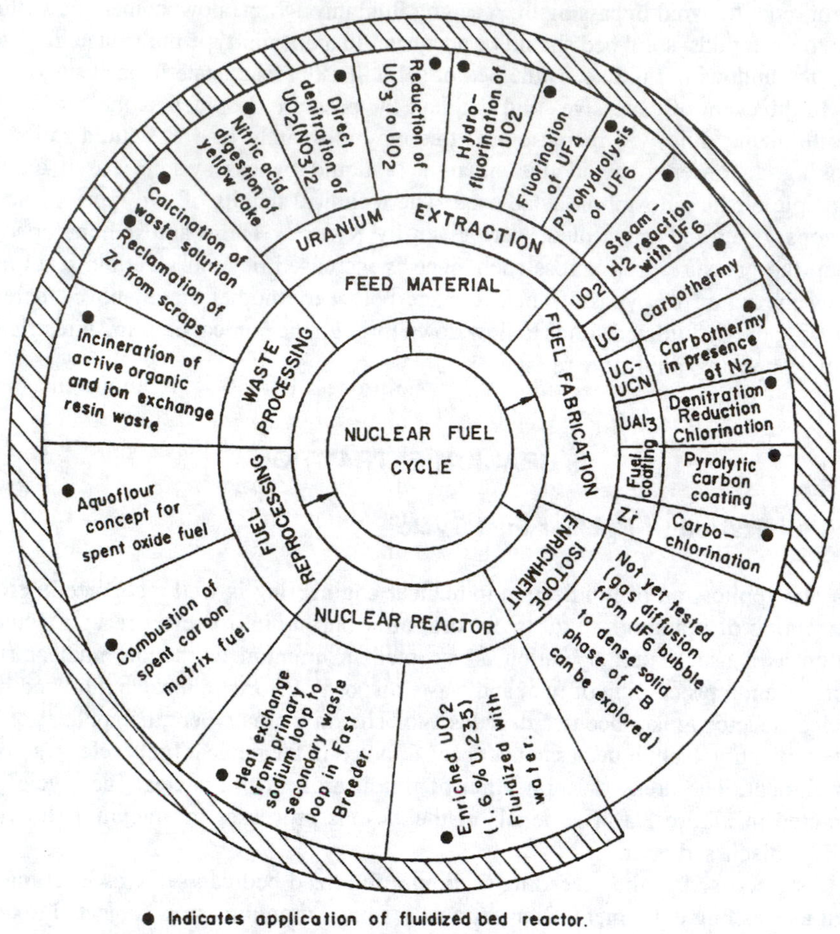

Figure 3.3 Fluidized in nuclear fuel cycle.

thin liquid (freezing point 78°C, boiling point 124°C, density 2.47 g/cc, viscosity 1.55 cP at 100°C). Its decomposition begins at 60°C. $UO_2(NO_3)_2$ decomposes at 200°C. The oxide of uranium, U_3O_8, is obtained[14] after treating this nitrate solution with ammonia and subsequently calcining the resultant ammonium diuranate (ADU) at 350°C. This aqueous process involves several steps, such as neutralization of the nitrate solution with ammonia, precipitation, filtration, drying, and calcination. Because of these various steps, the overall efficiency is reduced. Furthermore, the process is not environmentally safe. The effluent generated by this route cannot be let out directly into the waste stream. Hawthorn et al.[15] pointed out that the uranium oxides derived by the ADU route are amorphous and very fine and cannot be successfully fluidized later to obtain UF_6 by fluorination of UF_4 in a fluidized bed reactor. An economical and efficient way is to decompose the uranyl nitrate solution thermally and obtain the oxides of uranium in a single step.[12,16-23] Thermal decomposition of UNH is conveniently carried out in a fluidized bed reactor. The decomposition reaction is accomplished by spraying uranium nitrate solution over a hot fluidized bed of UO_3. The decomposition reaction is

$$UO_2(NO_3)_2 \cdot 6H_2O \xrightarrow{300°C} UO_3 + xN_2O + yNO_2 + zO_2 + 6H_2O \quad (3.12)$$

The reaction is endothermic, requiring 144,100 kcal/kmol of uranium at 230°C. The heat required for this reaction is usually supplied by an externally heated electrical heater or by internal heating using a fused salt heat exchanger or an immersed electrical heater. The fluidizing gas is air. The bed, to start with, contains UO_3. As the decomposition proceeds, UO_3 particles are built up gradually and discharged by an overflow outlet. Thus the process is made continuous and there is no consumption of chemicals in this nonaqueous route. This is a single-step process, and hence the efficiency is high. The decomposition reaction does not require any chemical, and the products of the reaction are solid uranium oxide and a gaseous mixture of NO, NO_2, water vapor, and air. The temperature of calcination[23] is a sensitive factor in maintaining the activity of the uranium oxide. The higher the temperature the lower the activity of the calcined product. Hence, mixing sulfuric acid with the feed to maintain 0.15–0.20% sulfur in the calcined product that will have high activity for subsequent operations is recommended. Uranyl nitrate solution, evaporated to 70–100% UNH, has been successfully calcined in a fluidized bed with a continuous discharge of the calcined produce by either bottom discharge or top overflow discharge.

2. *Fluidized Bed*

a. *Calcination*

Otero et al.[20] described the fluidized bed calcination of uranyl nitrate solution in Nuclear Energy Council, Madrid (Spain). The fluidized bed reactor used was made up of stainless steel. It had three sections: (1) a bottom section 14 cm in diameter with an internal immersed electrical heater and a bubble-cap-type distributor, (2) a 20-cm-diameter central part provided with a nozzle for spraying UNH and a 25-mm overflow tube to collect the calcined UO_3 product, and (3) a 40-cm upper part fitted with stainless-steel filters. Preheated UNH was sprayed through the nozzle inside the bed. The process of calcination was visualized to occur in five steps: (1) formation of U_3O_8 particles by drying and decomposition of these particles, (2) growth of the existing particles due to coalescence promoted by liquid, (3) agglomeration of wet particles, (4) abrasion of agglomerated particles due to collisions, and (5) fines generation due to attrition. The important parts of the fluidized calciner were the heater, the nozzler, and the filters. The electrical heater was comprised of six units with a total power of 27 kW and a provision for overloading up to 4.5 kW per unit. The inner heater was made up of kanthal, embedded in arc-melted magnesia and cladded by magnesia.

b. Caking

The minimum fluidization velocity to fluidize a bed of UO_3 with air at room temperature and ambient pressure was determined by the equation

$$U_{mf}(\text{cm/s}) = 166\ D_p^{1.75} \tag{3.13}$$

where the particle size (D_p) is expressed in millimeters.

During operation of the reactor, cake formation on the nozzle and the wall was a practical problem as this cake has a tendency to spread throughout the reactor. The angle of the spray nozzle was an influencing factor for cake formation on the cone, and cake formation on the wall was due to the high penetrating power of the spray jet emerging from the nozzle. The theory of cake formation was developed by Otero and Garcia,[21] and an equation to predict the cake weight as a function of the operating variables was proposed. Cake formation in the spray nozzle could be reduced by protruding the liquid conduit which was coaxially surrounded by an annular-flow air nozzle conduit. Gonzalez and Otero[24] proposed a theory for prediction of the average particle diameter of UO_3 formed in a fluidized bed. Attrition occurred during the initial stage of UO_3 formation and not after calcination. The UO_3 product obtained after calcination was reddish-orange in color, spherical in shape, and had a fracture angle of 30–35°. Particle growth in a fluidized calciner was found to occur in concentric layers of UO_3. Iron contamination of the product from the reactor material was below 5 ppm. The nitrate and the moisture content of the material were a function of the operating temperature: the higher the temperature, the lower the moisture content of the product.

c. Denitration Unit

Legler[12] reported that a fluidized bed denitrator was built by the Atomic Energy Commission at Weldon Springs, Missouri, and operated by Mallinckrodt Chemical Works. This plant could not sustain production, as the product UO_3 had less reactivity for subsequent hydrofluorination. It was ascertained that a fluidized bed was not responsible for this low reactivity problem. The low reactivity was overcome at Oak Ridge, Tennessee, by adding 3000 ppm of sulfur to the feed solution. A similar kind of fluid bed denitrator was in operation at the Springfield Works in the United Kingdom.

Fluidized bed denitration of enriched uranyl nitrate solution was carried out successfully at the Idaho Chemical Processing Plant, where 800 kg of uranium from an enriched solution of 365 g/ml was processed in 14 days. Bjorklund and Offutt[22] described the complete details of the reactor and the operating parameters.

The fluidized bed reactor was 7 ft in height and consisted of four sections. The bottom section was a 6-in.-diameter plenum chamber, and above it was a 6-in.-diameter × 33-in.-long denitration section fitted with a feed spray nozzle, a product overflow thermo-well, and pressure traps. The third section was 7 in. in diameter and 42 in. long and had sintered metal filters. The fourth section, of the same diameter but 5 in. long, was for blowback arrangement of the nozzles. The operating temperature used was 300–400°C. Uranyl nitrate solution, containing 350 g/l of uranium, was sprayed at 10 l/hr through an air-atomizing nozzle over a bed of UO_3 particles expanded to a height of 18 in. at a fluidizing airflow rate of 220–350 cft/hr. The bed, at its static bed condition, weighed 25 kg, and the bed pressure drop was 50 in. of H_2O. The particle size was found to increase at temperatures above 400°C and decrease below 300°C. The average particle size during the run was 0.24 μm. No operating problem was encountered except for some plugging at the product overflow port, and this was overcome by enlarging the size of the conduit, thus minimizing plugging overflow. Outside, the equipment contamination by alpha radiation was minimal. The off-gas from the denitrator was let to the atmosphere through a stack, often condensing it and filtering through a HEPA (high-efficiency particulate air) final filter. There was no major operating problem in running the denitration except for a few minor choking problems in pneumatic instruments and product overflow line. Damerval et al.[25] described fluid bed denitration carried out at Mallinckrodt Chemical Works (St. Charles, Missouri). The fluid bed denitrator finally tested did not have a disengaging zone and cyclone, but had a stainless steel stream-jacketed filter containing 2.75 × 36-in. porous metal tubes (15 in number) located in an enlarged section 27.25 in. in diameter by 5 ft. 10.5 in. long and housed above a 9.56-in.-ID fluidized column 8 ft 10 in. high. The bed was heated by a molten salt heater tube. Steam as well as air were used as the fluidizing media. It was reported that the chemical purity of the UO_3 particles obtained by either stream or air fluidization was quite satisfactory. Particles of uniform size could be produced, and such uniform particle size is desirable for the hydrogen reduction and hydrofluorination reactions that are also usually carried out in fluid bed reactors.

Philoon et al.[19] described a low-cost process for the production of uniform-sized particles of UO_3 on a pilot scale using a fluid bed reactor. A schematic sketch of this pilot plant is shown in Figure 3.4. The product obtained was 99% UO_3 with a residual water and nitrate content of less than 1%. The product did not have the intermediate oxide U_3O_8. A high bulk density of UO_3, ranging from 3.5 to 4.3 g/cc, could be obtained. A high volume ratio of the atomizing air to the liquid feed through the spray nozzle decreased the particle size. The feed rate of UNH and its concentration affected the fluidization near the spray zone and also influenced the particle size of the product. From the foregoing account of UNH denitration in the fluid bed reactor, it is apparent that this nonaqueous route for producing UO_3 from aqueous UNH was well thought out and conceived to operate a pilot-scale plant even three decades ago.

The potentialities of fluid bed calcination to recover uranium values from spent fuel after its aqueous processing were recognized long ago. Lambert and Dotson[26] mentioned the use of calcination of UNH to UN_3 in a fluid bed in General Electric's Midwest Fuel Recovery Plant.

C. Reduction of Oxides of Uranium

1. Oxides of Uranium

There are four[27] established forms of oxides of uranium: monoxide (UO) with a cubic crystal structure, dioxide (UO_2) with a fluorite-type lattice, mixed oxide

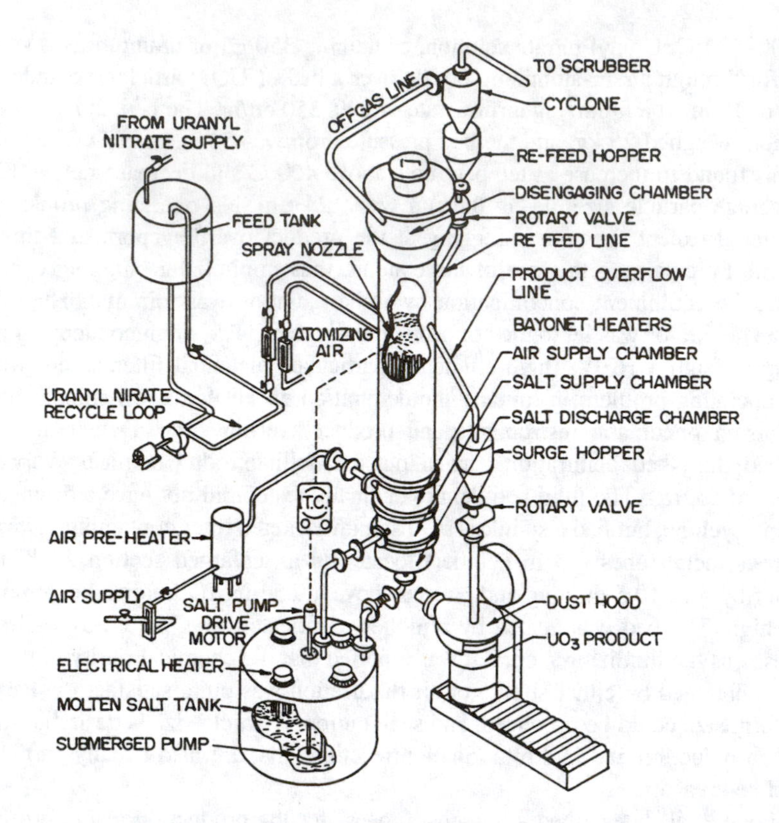

Figure 3.4 Fluidized bed reactor for uranium trioxide plant. (From Philoon et al.[19] With permission.)

(U_3O_8) with an orthorhombic structure, and trioxide (UO_3) with an orthorhombic structure. Oxides of uranium, obtained by thermal denitration of UNH or calcination of ADU, are usually U_3O_8 and UO_3 mixtures. The dioxide of uranium is used as nuclear reactor fuel; its density is close to 11 g/cc and its melting point is 2176°C. It is basic in nature and brown, dark brown, or black in color, depending on the method of preparation. Reduction carried out after calcining U_3O_8 ends up with dark brown or black oxides of UO_2. Chemically reactive UO_2 is desirable for the production of uranium metal. Reactive-grade UO_2 is obtained by the reduction of UO_3 and a ceramic oxide is obtained by reducing U_3O_8. The reduction of U_3O_8 is thermodynamically favored by reductants[23] like ammonia, hydrogen, and CO. Hydrogen is proven to be an efficient reducing agent. The rate of reduction of U_3O_8 by hydrogen is not affected significantly in the temperature range 650–700°C. This is the temperature range chosen for the reduction of U_3O_8 by hydrogen. The transformation of U_3O_8 to the dioxide at 500°C takes place as:

$$U_3O_8 \rightarrow UO_{2.6\pm x} \rightarrow U_4O_9 \rightarrow UO_{2+x} \rightarrow UO_2 \tag{3.14}$$

where x_{max} = 0.16–0.4.

2. Preparation

a. Oxide Criteria

The rate of reduction of U_3O_8 and UO_3 by hydrogen is a function of the temperature, the partial pressure of hydrogen, and the physical properties of the solid. The rate-determining step is a surface reaction between the hydrogen adsorbed on the surface and the oxygen in the oxide.

The particle size of uranium oxides is affected[23] during the calcination of ADU, the reduction of U_3O_8 by H_2, and the oxidation of UO_3. Data on the parameters that affect particle size are especially important in the selection and design of fluidized bed reactors. It can be seen from Figure 3.5 that UO_2 of various particle sizes can be prepared by different methods. It should be noted that the particle size of UO_2 obtained by UO_3 reduction by hydrogen is smaller than its initial size, whereas the particle size of U_3O_8 obtained by air oxidation is increased. The reduction of UO_3 by H_2 is greatly accelerated by 0.15% of sulfate ions. Impurities such as sodium, potassium, and calcium (up to 0.05%) and iron (up to 0.3%) do not affect the reactivity of UO_3.

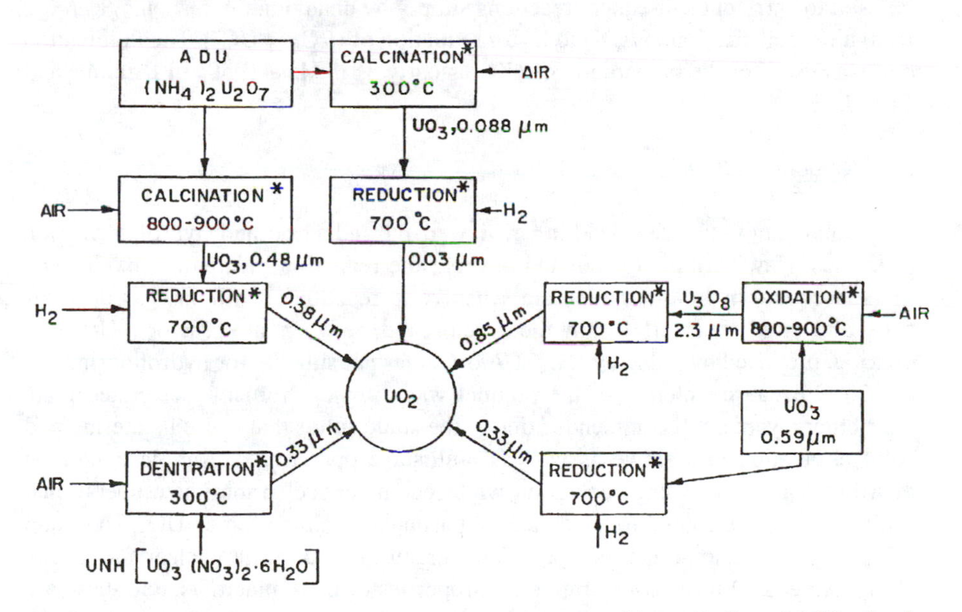

Figure 3.5 Preparation of UO_2 of various particle sizes. Possible application of fluidization is indicated by *.

The temperature of UO_3 reduction depends on the end use of the resultant UO_2. Thus, if UF_4 is required, UO_3 should be reduced at 625°C, which would yield UO_2 particles 40 μm in size with a crystal size of 1 μm or less. Reduction at 800°C would yield UO_2 particles 40–60 μm in size, useful for ceramic fuel elements.

b. Reactor Criteria

It is now fairly clear from the foregoing that UO_2 can be prepared by the reduction of its precursor materials (UO_3 and U_3O_8) in a suitable reactor where the temperature of gas and solid can be precisely controlled. Conventionally, the reductions have been accomplished using multitubular or tray-type furnaces, electrically heated rotating furnaces, rotating kilns, and fluidized bed reactors. The gas–solid contact in all such reactors is not the same, and hence the reaction time and the product quality can ultimately be different. In tray or multitube furnaces, control of temperature at the reaction site is difficult. In a rotary kiln, gas–solid contact is improved by the kiln rotation and the process is continuous. This type of kiln requires a large floor area for its operation and heating by an externally mounted furnace. Energy consumption is increased due to mechanical movement. The refractory lining of the reaction tube must periodically be replaced because of fatigue failure and thermal shock. Operation of a rotary kiln at near isothermal condition requires a good control system, thus adding up to the capital cost of the reactor. A fluidized bed reactor is best suited for carrying out calcination of ADU as well as reduction of U_3O_8 or UO_3 to arrive at a UO_2 product of the required quality. Furthermore, a single reactor can be used to carry out subsequent reactions simply by changing the reactant gas (e.g., from air for calcination of ADU to H_2 for reduction of U_3O_8 or UO_3). The fluidization bed was used for the preparation of UO_2 as early as the late 1950s in Britain[28] and Canada.[29]

c. Reduction Criteria

Yemel'yanov and Yevstyukhin[30] reviewed the reduction and hydrofluorination processes. Hawthorn et al.[15] pointed out that the reduction of uranium oxides and the subsequent hydrofluorination are sensitive to reaction conditions, gas distribution, and solid transport within the reactor and between the reactors. Uranium dioxide, produced by reducing UO_3 at 700°C, was not suitable for hydrofluorination at 450°C because sintering of the product was a major problem. Perforated plate distributors were not recommended due to the static solids that caused sintering and solid circulation in a single as well as multistage operation. A simple cone-type distributor and single-stage operation were recommended to minimize operational difficulties. It is essential to have data on particulate solids such as UO_3, UO_2, and UF_4, as these are important compounds of uranium used in the nuclear fuel cycle. Hawthorn et al.[15] listed some important properties of these materials, and these are presented in Table 3.1.

d. Fluid Bed Reduction

Reduction of UO_3 by hydrogen gas in a fluid bed as well as in a rotary kiln was reported[31] at Eldorado Resources Ltd., Canada. Fluidized bed reduction of UO_3 in a 75-mm-diameter fluid bed reactor at 450–650°C was reported[32] to have been carried out in India. The UO_3 particles used were 19 μm in size with loose and packing densities of 2.2 and 2.6 g/cc, respectively. The UO_3 was obtained by ADU calcination,

Table 3.1 Some Important Properties of Uranium Compounds Processed in a Fluidized Bed

	Compounds		
	UO_3[a]	UO_2	UF_4
Surface area (m^2/f)	0.1–0.5	3.5–5.1	0.1–0.5
Structure	Monoclinic (β = 89.63° ± 0.1)	FCC	Monoclinic
Density (g/cc)			
Tap	4.0–4.3	4.73–4.8	3.5–3.9
Pour	3.6–4.1	4.36–4.42	3.2–3.5
He	6.91	10.2–10.6	6.42
Hg	6.06	5.4–6.4	5.26
Crystal	7.2	10.96 3.5 (fluidized density)	6.70
Remarks	Particle size distribution (as obtained by fluidized bed denitrator): #44 (95%), #60 (4.55%), #72 (5.13%), #100 (12.7%), #120 (22.6%), #170 (17.6%), #200 (5.7%), #240 (6.26%), #300 (5.3%), #360 (10.6%) Reduction reaction: $UO_3 + H_2 \rightarrow UO_2 + H_2O$ ΔH = –24 kcal/mol ΔE = 25–30 kcal/mol	Average pore radius: 200–800 Å Melting point: 2700°C Hydrofluorination reaction: $UO_2 + 4HF \rightarrow UF_4 + 2H_2O$ ΔH = 43 kcal/mol	

[a] # = BSS mesh. FCC = Face centered cubic.
Data from Hawthorn et al.[15]

and the reducing and fluidizing gas stream was H_2 mixed with N_2 in the ratio 3:1. A sintered porous metal disc, with a mean pore size of 20 μm, was used as the gas distributor. UO_2 product analyzing 99.5% was obtained for the bed temperature range 550–600°C, when excess hydrogen used was 61.5–106%. The gas velocity was 5.1–6.3 cm/s, which corresponded to a total flow rate range of 25.5 to 32.5 l/min. at a molar ratio of H_2/N_2 = 3. Thus, the results of preliminary trial runs were encouraging, and further work for large-scale production was recommended.

After calcination, ammonium uranyl carbonate (AUC) gives UO_2 particles that have excellent flowability and are suitable for fluidized bed operation. Further, the UO_2 produced through the AUC route has a better tap density and requires single-stage pelletization. AUC is obtained as a precipitate after the reaction of uranyl nitrate (containing 100–200 g/l U) with ammonia and CO_2 at a temperature of about 60–70°C and a pH of 8–8.5. The reaction is

$$UO_2(NO_3)_2 + 6NH_3 + 3CO_2 + 3H_2O \rightarrow (NH_4)_4\ UO_2(CO_3)_3 + 2NH_4NO_3 \quad (3.15)$$

The calcination of ACU in a fluidized bed yields UO_2 as follows:

$$(NH_4)_4\ UO_2(CO_3)_3 + H_2 \rightarrow UO_2 + 4NH_3 + 3CO_2 + 3H_2O \quad (3.16)$$

Reduction studies on ACU were carried out in Korea[33] using a 17-cm-diameter and 220-cm-high fluidized bed at 520°C by a gas mixture containing 13% H_2 and

87% steam. It was reported that the UO_2 powder produced by AUC reduction was dependent on the characteristic shape of the AUC powder itself.

D. Hydrofluorination of Uranium Dioxide

1. Reaction

The hydrofluorination of UO_2 is carried out to produce UF_4, which is the starting material for the production of uranium metal by metallothermy (i.e., using calcium or magnesium as metal reductant) and also for the preparation of UF_6. The reaction of UO_2 with HF is

$$UO_{2(s)} + 4HF_{(g)} \leftrightarrow UF_{4(s)} + 2H_2O_{(g)}, \Delta H = -52{,}834 \text{ kcal/mol} \tag{3.17}$$

The above reaction is highly exothermic and reversible. Thermodynamic calculations[34] predict the reversibility of the reaction at high temperatures. If the water vapor is not removed, it will react with UF_4 and decompose it. The probability of reversibility of the reaction is increased with increase in temperature and with the presence of large amounts of water vapor. Hence, low temperatures and high concentrations of HF are required to counteract these adverse effects. However, a low temperature slows down the rate of the reaction. In such a situation, a reactor that can provide better gas–solid contact for efficient heat and mass transfer is the best choice. A fluidized bed reactor is the right selection for this application. The heat evolution in the reactor has some adverse effect in that the UO_2 and UF_4 mixture, being a poor conductor of heat with a tendency to form a eutectic at 12 and 18% UO_2 with UF_4, would have a chance to fuse and sinter. Once UF_4 is formed on the UO_2 and also sintered, subsequent reaction of UO_2 with HF is hindered. HF gas diffusion through the UF_4 layer is the main barrier and controls the reaction. Depending on the granulometry, active UO_2 particles can sinter at a low temperature with UF_4. Hence, it is imperative to carry out the reaction isothermally under controlled conditions.

The kinetics of a noncatalytic-type hydrofluorination reaction of UF_4 under nonisothermal conditions had not been studied in much detail until Costa and Smith[36] carried out such studies on hydrofluorination of UO_2. Hydrofluorination was determined to be a first-order reaction, and the activation energy was found to be 6070 cal/g · mol. The HF was assumed to be chemisorbed and the hydrofluorination was presumed to occur on the surface. The rate equation is

$$\left(-r, \frac{\text{g} \cdot \text{mol UF}_4}{\text{cm}^2 \cdot \text{s}}\right) = K_o(\text{cm/s}) \cdot \exp\left(-\frac{E}{RT}\right) \tag{3.18}$$

The temperature and the reactant concentration in the particle (i.e., pellet) can be used to predict the intrinsic reaction rate, and subsequently a conversion versus time plot can be generated. At low temperatures, the apparent rate obtained would be erroneously high due to the physical adsorption of HF superimposed on the chemical reaction. A change in particle (i.e., pellet) porosity would cause a shift in the frequency factor K_o but not in the activation energy (E).

2. *Fluid Bed Hydrofluorination*

A fluidized bed reactor is well suited for carrying out hydrofluorination. It is a general practice in industry to carry out the hydrofluorination reaction in the temperature range 400–600°C. The activation energy for this reaction, depending on the reaction environment, varies from 6 to 15 cal/mol. One of the important factors for fluidization is selection of the particle size of UO_2. At low temperatures (e.g., 450°C), the reaction rate is slow, but it can be increased by selecting fine particles that provide large specific surface area. It may be recalled that fine particles, due to their strong electrostatic force, have a tendency to agglomerate, and this would cause channeling in the bed. A coarse particle, when fluorinated at higher temperatures, can overcome the problems associated with fluidization. If the uranium dioxide is contaminated with higher oxides such as UO_3 and U_3O_8, the product UF_4 is also contaminated with UO_2F_2. Hence, hydrofluorination of UO_2 without any of the higher oxides of uranium is essential in order to obtain the good quality UF_4 that is required for uranium metal production or fluorination of UF_4 to produce UF_6.

The use of a fluidized bed reactor in place of other conventional reactors that have moving mechanical parts is very desirable because there is no need for gas sealing, as in rotating parts. This is actually a great advantage in corrosive media at elevated temperatures. Galkin and Sudarikov[35] reported the use of a fluidized bed reactor for hydrofluorination and pointed out the advantages of a multistage operation in achieving better conversion of UO_2 to UF_4. A three-stage fluidized bed reactor with temperature gradient kept at 400–500–600°C could hydrofluorinate UO_2 in 4 hr, leaving only 0.9% UO_2 in the final product. A single-stage fluid bed reactor could hydrofluorinate UO_2 at 450°C, producing UF_4 and leaving behind 1.4% unreacted UO_2 over a long period of reaction (i.e., 16 hr). A 760-mm diameter, 3-m-high fluidized bed reactor was used for hydrofluorination in two stages. The first stage was at 315°C and the second at 540°C. The reactor was installed at Paducah, Kentucky. Nickel and nickel-based alloys (i.e., Monel, Ni + 30% Cu, and Inconel, Ni + 15% Cr + 7% Fe + 2.5% Ti + 1% Nb + 0.7% Al + 0.7% Mn + 0.4% Si with less than 0.05% C) are the recommended construction materials for hydrofluorination.

Hawthorn et al.[15] described the use of a fluidized bed in the hydrofluorination of UO_2 at Springfields. They pointed out the need for a fluidized bed with a large holdup because of the slow reaction kinetics and the requirement of removal of large quantities of heat. Hence, the fluidized bed reactor used for hydrofluorination was necessarily taller than the hydrogen reduction reactor. The reactor was made of Inconel.

The unreacted hydrogen fluoride from hydrofluorinators is of great environmental concern. Hence, separate development work for hydrogen fluoride recovery was carried out by Hawthorn et al.[15] The unreacted HF would usually be recycled and the dilute off-gas condensed in brine-cooled condensers before being vented. The plant for treating waste hydrogen off-gas was constructed of Monel and was designed for 250 psig irrespective of the actual working pressure. Hydrofluorination of UO_2 in a 125-mm-diameter (5-in.) fluidized bed reactor was reported by Levey et al.[37] The fluidized bed reactor was used for the reduction of UO_3 and also for hydrofluorination, each operated on a 1-hr cycle. A 6-in. fluidized bed reactor was required if hydrofluorination was carried out with 20–35 mesh UO_2. The conversion rate in a fluidized bed is reported to be limited by reagent addition under the condition of consistently good fluidization. A tapered fluidized bed reactor was tested in hydrofluorination in place of a conventional cylindrical fluidized bed. In a cylindrical bed, complete fluidization of materials with an aspect ratio (H/D) of 10 was achieved at a very high superficial gas velocity with bubbling and coalescence of the fluidizing gas. However, complete fluidization in a tapered fluidized bed even at an aspect ratio of 14 could be achieved at just two times the minimum fluidization velocity. Bubbling without coalescence was noted in the tapered fluidized bed. In view of noncaking characteristics and stable fluidization even with very deep beds, a tapered fluidized bed is recommended for slow reactions such as hydrofluorination of UO_2. Instead of a multistage operation with a conventional fluidized bed, a single tapered fluidized bed is recommended when the reaction time is less than 30% of the turnover time.

Levey et al.[37] carried out a numerical analysis to predict the dimensions of a tapered fluidized bed for hydrofluorination and arrived at the following dimensions for a reactor capable of producing 1000 lb/hr UF_4 at 99% conversion utilizing 95% of the reagent HF: length = 16 ft, diameter = 1.13 ft at the bottom and 1.26 ft at the top. The design data reported were for countercurrent operation with UO_2 (20–40#) entering at 205°C and leaving at 565°C. Other studies on hydrofluorination using fluidized beds have also been reported in the literature.[38,39] A two-stage moving bed reactor was reported to have been adopted in South Africa[40] for the reduction of UO_2 by NH_3 in the first (top) stage, followed by hydrofluorination in the second (bottom) stage. In order to operate this reactor successfully, solid movement within the reactor without blocking or channeling was thought to be essential. This was possible with a specified sieve analysis of the feed UO_3 particles. The feed particles should be essentially porous and reactive without a tendency to sinter under a wide range of operating conditions. The advantage of a moving bed reactor relative to a fluidized bed has not been spelled out. No work on the fluidized bed reactor seems to have been carried out to justify the selection and prove the superiority of a moving bed reactor vis-à-vis a fluid bed.

E. Manufacture of Uranium Hexafluoride

1. *Fluorination*

a. *Basic Reaction*

Uranium hexafluoride can be produced by the reaction of fluorine gas with almost all compounds of uranium, such as UF_4, UO_2F_2, UO_2, UO_3, U_3O_8, K_2UF_6 etc. The reaction temperature and the fluorine consumption vary, depending on the type of uranium compound selected for fluorination. The fluorination of uranium oxides is carried out when treatment of uranium wastes that contain enriched U^{235} is warranted. The least amount of fluorine is required when UF_4 is fluorinated with F_2. The reaction

$$UF_4 + F_2 \rightarrow UF_6 \quad \Delta H = -62 \text{ kcal/mol} \tag{3.19}$$

is highly exothermic and also complex in nature. Intermediate compounds like U_4F_{17}, U_2F_9, and UF_5 are formed during fluorination. The fluorination of UF_4 is usually compared with the combustion of carbon, and hence this heterogeneous reaction is assumed to follow diffusion kinetics. The overall rate of fluorination is theoretically dependent on the rate of feed and removal of the reacting gas as well as the product. The mechanism of this gas–solid reaction, which involves fluorine reaction with the external and internal micro- and macropores of the solid, was dealt with in detail by Galkin and Sudarikov.[41]

b. Flame Reactor

Uranium hexafluoride in the early stages of development was prepared at 300°C by the reaction of fluorine with UF_4 contained in a boat and housed in a horizontal tubular furnace. This was a batch-type process and required elaborate arrangements to control the reaction. Later, in industrial practice, flame-type reactors and fluidized bed reactors came to be used for the production of UF_6. A flame-type reactor is used to fluorinate pure UF_4. Finely dispersed UF_4 in fluorine gas is allowed to burn at 1100°C in a flame-type reactor. A typical flame reactor 0.203 m in diameter and 3 m long could process 7.5 tons of uranium per day. The fluorination usually would take place in the top 600- to 900-mm-long zone. This reactor requires an excess amount of fluorine to prevent choking of the exit line due to unreacted UF_4. Fine dispersion of UF_4 in fluorine gas and close control of the feed are essential for smooth performance of the flame reactor.

2. Fluid Bed

a. Uranium Tetrafluoride Fluorination

If the feed uranium tetrafluoride is contaminated with impurities, a flame-type reactor is not recommended. In such cases, the fluidized bed is a wise choice. A fluidized bed reactor is kept at a constant holdup of a bed diluent such as CaF_2 (–40 +200#) and is fluidized by N_2 gas until the reaction temperature is attained. The bed is maintained in the temperature range 343–482°C by heating the wall of the fluidization column externally, using an electrical resistance furnace. During fluorination, the N_2 gas stream is mixed with F_2 (~30%), and UF_4 is continuously fed into the bed of CaF_2. The reaction is allowed to proceed isothermally. Any impurity, such as sodium, calcium, and magnesium, is fixed in the bed as fluoride, and metals such as vanadium and molybdenum escape along with UF_6 as volatile fluorides. The UF_6, contaminated with volatile impurity fluorides, N_2 gas, and unreacted F_2, is cooled to 205°C, filtered by passing through porous Monel metal filters, and the gases are then condensed in primary condensers, where UF_6 is first separated. Unreacted F_2 is trapped by reacting it with UF_4. A secondary condenser finally removes all UF_6. The spent gases leaving the system are scrubbed with potassium hydroxide and then vented to the atmosphere. Vogel et al.[42] investigated the fluorination of UF_4 using the fluidization technique. Their investigations consisted of both laboratory and pilot-plant-scale fluorination of UF_4, either with a bed diluent like CaF_2 or without a diluent. A detailed diagram of a pilot-plant fluid bed fluorinator can be referred[42] to in order to gain an understanding of various systems, such as gas metering, the fluidized bed unit, and condensers for the UF_6. The pilot-plant fluidized bed reactor was 62.5 mm in diameter and made up of Monel. A distributor, which had 24 inverted cones, each with a 1.5-mm (1/16-in.) hole, was used to distribute the fluorine. Uranium tetrafluoride was fed by a 25-mm screw feeder. Nitrogen gas was used for fluidization during heating until the reaction temperature was attained. Addition of uranium tetrafluoride and addition of fluorine to the nitrogen gas stream were carried out simultaneously and the run was limited to 2 hr. The fluorine utilization efficiency was found to increase with increasing temperature (30% at 300°C and 95% at 450°C for an inlet fluorine concentration of 25%). The bed diluent decreased the fluorine utilization efficiency, and the fluorine concentration in the nitrogen stream had little effect. The bed diluent decreased the fluorine utilization efficiency by 50% when the bed diluent level was increased by 50–85%. When the uranium content of the bed was increased, the fluorine utilization efficiency also increased. The lowest portion of the bed heated the incoming gas, and hence little reaction was noticed here. At a temperature of 400°C, the gas velocity had no effect on the fluorine utilization efficiency. In general, it was found that a fluorine utilization efficiency of up to 100%, if not achievable, could be compensated for by an off-gas fluorine reaction with a bed of UF_4.

b. Direct Fluorination

Fluorination of UO_2 directly to UF_4 is an important step in the fluoride volatility process that is used to separate fissionable fuel materials from fission products in the spent fuel obtained from a nuclear reactor. In a developmental study[43] conducted at Argonne National Laboratory, fluorination of unirradiated UO_2 pellets was carried out using a novel two-zone fluidized bed. The fluidized bed reactor (75 mm ID and 145 mm long) was made of nickel and was provided with a 280-mm-ID × 660-mm-long disengaging zone. The reactor at the bottom, that is, the first zone, was filled with pellets of UO_2, and alumina particles were fluidized in the voids of pellet packing. This arrangement at the first zone, known as the fluidized-packed section, was used to oxidize UO_2 to U_3O_8 using O_2 diluted with nitrogen. To study the basics of the fluidized-packed bed, including heat transfer, solid–gas mixing, and elutriation, refer to the work of Gabor and Mecham,[44,45] Ziegler and Brazelton,[46] and Gabor et al.[47] The $O_2 + N_2$ gas mixture also served as a fluidizing medium. The oxidation of UO_2 is mildly exothermic ($\Delta H_{298} = -25$ kcal/mol of U), and heat transfer in the fluidized-packed system is conducive for this purpose. The U_3O_8 which formed on the UO_2 surface eroded and was transported to the upper zone (i.e., second zone) of the fluidized bed of alumina and finally fluorinated by elemental fluorine to form

UF_6. The fluorination of U_3O_8 is highly exothermic ($\Delta H_{298} = -232$ kcal/mol of U). The second zone of unhindered fluidization of alumina helps to control the reaction with good heat transfer.

The oxidation of UO_2 in the first zone and the fluorination of U_3O_8 in the second zone should be controlled in such a way that the U_3O_8 in the first zone is properly transported to the second zone. The formation of intermediate fluoride fines such as UO_2F_2 and UF_4 and their buildup over 20% could result in channeling, caking, and elutriation and render the fluorination reaction inefficient. Various methods, such as vibrating fluidization, pulsed fluidization, and high-velocity fluidization, were tested to transport the U_3O_8 from the first oxidation zone to the second fluorination zone. The pulsed fluidization operation was reported to be successful. The fluorination reaction was accomplished at 500–525°C using 7.17% fluorine in the recycled off-gas. For more detailed research investigations on this two-zone oxidation–fluorination reaction in a fluidized bed, refer to the work of Anastasia and Mecham.[43] The results of these investigations showed the feasibility of producing UF_6 from UO_2 and thus encouraged adoption of the fluidized bed volatility process for recovery of fissionable materials such as uranium and plutonium from spent nuclear reactor oxide fuels.

The literature[48,49] shows the possibility of fluorinating uranium oxides by bromine pentafluoride (BrF_5) instead of elemental fluorine. The fluorination of U_3O_8/PuO_2 with elemental fluorine results in the simultaneous generation of the hexafluorides of uranium and plutonium, thereby necessitating the separation of these two compounds at a later stage. If, on the other hand, BrF_5 is used in place of elemental fluorine for fluorination, uranium can be completely separated by selective fluorination to UF_6, leaving behind nonvolatile PuF_4. Studies by Holmes et al.[50] on an engineering-scale fluidized bed reactor for fluorination of uranium oxide showed encouraging results. At a fluid bed temperature of 300°C and a superficial velocity of 0.89 ft/s for a BrF_5 (29%) + N_2 mixture, up to 164 lb/hr. ft^2 UF_6 at 60% BrF_5 utilization efficiency could be obtained with a batch of 4.4 kg of UO_2 mixed with (–100#) alumina. Holmes et al.[50] adopted the fractional experiment technique and demonstrated the efficient fluorination of UO_2 using BrF_5 in a fluidized bed.

c. *Engineering Studies*

Fluidization as applicable to fluorination and the production of uranium hexafluoride was studied in detail, and much useful basic and engineering data was made available by Corella[51] and Otero and Corella.[52,53] The above work dealing with fluidization and its use in UF_6 generation was carried out at the Materials Division of the Atomic Energy Commission in Madrid. As the fluorination of UF_4 in a fluidized bed is usually accomplished with UF_4 mixed with diluent inert solids, studies were carried out by Otero and Corella[52] to determine the fluidization characteristics of various bed diluents such as corundum (alumina), calcium fluoride, and calcium fluoride–magnesium fluoride eutectic. Mathematical expressions to predict the minimum and complete fluidization velocities were proposed by these investigators.[52] One interesting and important aspect of the fluorination reaction of UF_4 with F_2 in the production of UF_6 is the formation of intermediate compounds such as U_4F_{17} (black), U_2F_9 (black), α-UF_5 (dark grey), and β-UF_6 (dark brown); these compounds are formed even at low temperatures in the absence of fluorine by the recombination of UF_6 with UF_4. The reaction of UF_4 with UF_6 was studied by Broadly and Longton,[54] Labaton,[55] and Labaton and Johnson.[56] The formation of U_4F_{17}, U_2F_9, and β-UF_5 was found to be stepwise, with each stage proceeding at a well-defined reaction initiation temperature.

The fluidized bed fluorinator used by Corella and Otero[53] was made of Monel and consisted of four zones: (1) a gas preheating zone (51 mm ID [59 mm OD] × 550 mm long), (2) a reaction zone (64 mm ID [70 mm OD] × 370 mm long), (3) a filter zone (190 mm ID [194 mm OD] × 630 mm long, and (4) a gas-discharge chamber (180 mm ID [186 mm OD] × 100 mm long). The preheated zone was filled with Monel chips. The reaction zone was provided with a bell-type distributor for issuing the reacting gases, and Monel filters were used in the filter zone. The reactor was provided with an externally heated electrical resistance furnace with a 3.5-kW capacity. The bed of UF_4 was mixed with α-Al_2O_3 to control the reaction temperature. The study was conducted to investigate the effects of various parameters such as linear gas velocity, percentage of diluent mixture, particle size, fluorine concentration, bed temperature, and particle size variation during reaction.

d. Parameters Influence

The particle size does not have a pronounced effect on the conversion because the reaction proceeds not only on the surface but also on the pore surface. During the course of the reaction, the particles disintegrate into fines. Hence, the shrinking core model for the reaction of the solid may not strictly be applicable. The gas velocity, depending on the selected particle size, plays an important role. Particle sizes greater than 185 μm are not recommended unless the fluorine cell does not have the capacity to meet the requirement of fluorine corresponding to the minimum fluidization condition. Particle sizes less than 80 μm also are not advisable due to the problem of elutriation and poor quality of fluidization. The fluorination of UF_4 to UF_6 is seen to occur up to 400°C, and reactions with UO_2F_2 intermediates are found beyond 500°C. Hence, a temperature of 350–550°C is found to be suitable for fluorination. Fluorination kinetics is not affected by various concentrations of the diluent bed materials, but the diluent bed materials are essential for good dissipation of heat in the reactor. If the inlet fluorine concentration is low, then heat dissipation during the reaction is controllable. A reaction temperature of 450°C, a fluidization velocity of 18 cm/s, and 22% inlet fluorine concentration are recommended in order to have total consumption of fluorine. The reaction is of the first-order type with respect to fluorine, and the reaction kinetics is dependent on the amount of active bed materials present in the bed.

e. Reaction Rate

The fluorination of UF_4 in a fluidized bed can be represented by the equation

$$-r\left(\frac{\text{mol } F_2}{\text{hr} \cdot \text{g } UF_4}\right) = -\frac{1}{W}\frac{dn_{F_2}}{dt} = KC_{F_2,0}(1-x) \tag{3.20}$$

If F is the moles of fluorine passed per hour and W is the weight of UF_4 (g) present, then

$$\frac{W}{F} = -\int_0^{x_e} \frac{dx}{r} \tag{3.21}$$

In other words,

$$\ln(1 - X_e) = K\, C_{F_2,0}\, W/F \tag{3.22}$$

and

$$K = K_0\left(\frac{\text{l fluorine at 273 K}}{\text{hr} \cdot \text{g } UF_4}\right)\exp\left[\frac{-E}{RT}\right] \tag{3.23}$$

where X_e is the conversion of fluorine at the exit of the reactor, $C_{F2,0}$ is the inlet fluorine concentration, and K is the reaction velocity constant which varies from 7.9 at 360°C to 54.0 at 550°C. The Arrhenius constant (K_o) = 1.55×10^6. The rate equation at 450°C is given as:

$$-r = 37.3\, C_{F_2}\left(\frac{\text{mol fluorine}}{\text{l at reactor temp}}\right) \tag{3.24}$$

The activation energy is in the range 16–17 kcal/mol in the temperature range 360–550°C when the bed does not have any UO_2F_2 under these conditions.

3. *Fluorization Reactor*

The conversion of uranium tetrafluoride into hexafluoride using a vertical fluorization reactor was discussed by Hua[57] and Hua and Bao.[58] This reactor is used in China as an alternative to the flame or fluidized reactor. In such a vertical fluorization reactor, solid UF_4 is fed by means of a screw feeder down to the bottom, countercurrent to the fluorine gas, which is distributed by a distributor assembly fitted at the bottom of the reactor. The productivity of the reactor has been claimed to be very high. A 200-mm-diameter reactor is capable of producing 4000–5000 kg/m²/hr of uranium hexafluoride. Any unreacted residue is collected above the distributor and subsequently reacted with the incoming fluorine. The temperature of the reactor is controlled by means of air passed through the cooling jacket. The design details of the reactor with regard to solid distribution and gas–solid contact in the dilute phase, the requirement of excess fluorine, and the exact means of temperature control have not been disclosed in the literature. Nevertheless, the reactor is claimed to be handy to operate and easy to maintain. It has a simple structural design. This type of reactor has been adopted in commercial practice. It appears that reactors of this class also belong to a two-zone type of fluidized bed, the bottom zone being fluidized with the unreacted residue and the top zone kept in dilute phase with continuous feeding of solid there.

4. Reactor Control

It is clear from the foregoing discussion of the production of uranium hexafluoride that careful control of temperature is essential for successful operation of a reactor system. Control of the reaction can be accomplished by the following means:

1. By increasing the diluent gas velocity and correspondingly decreasing the fluorine concentration. The diluent gas used is normally nitrogen. An increase in gas flow would result in unstable fluidization and can elutriate the particles, thereby loading the filters and increasing the pressure drops. Condensation of UF_6 in the presence of a large amount of noncondensables is not advisable in order to get the best out of the condenser.
2. The reactor can be cooled by means of a heat exchanger either immersed inside the reactor or kept outside on the reactor. Neither of these heat transfer methods is recommended. The heat transfer equipment kept inside the reactor is prone to corrosion, and the heat-exchanging equipment on the wall of the reactor can hinder the initial heating.
3. The right choice of bed diluent materials, such as CaF_2 and alumina, is important to control the reaction temperature. However, co-fluidization of the diluent materials in the presence of active UF_4 is essential for smooth fluidization.

Fluidized bed reactors are used in the extraction of uranium, from denitration through fluorination, and these processes are nonaqueous in nature. Removal of impurities, which is usually accomplished by conventional solvent extraction methods, can also be carried out by the distillation[59] of UF_6. The use of fluidization technology in various steps of uranium extraction is depicted schematically in Figure 3.6.

III. FUEL MATERIAL PREPARATION

A. Pyrohydrolysis of Uranium Hexafluoride

1. Reaction

The cost of nuclear fuel elements can be significantly reduced by developing new methods for preparing compacted powder fuels.[60] Compacted powder fuels are prepared using vibration compaction or mechanical working. Uranium hexafluoride is usually the starting material for the preparation of enriched uranium oxide

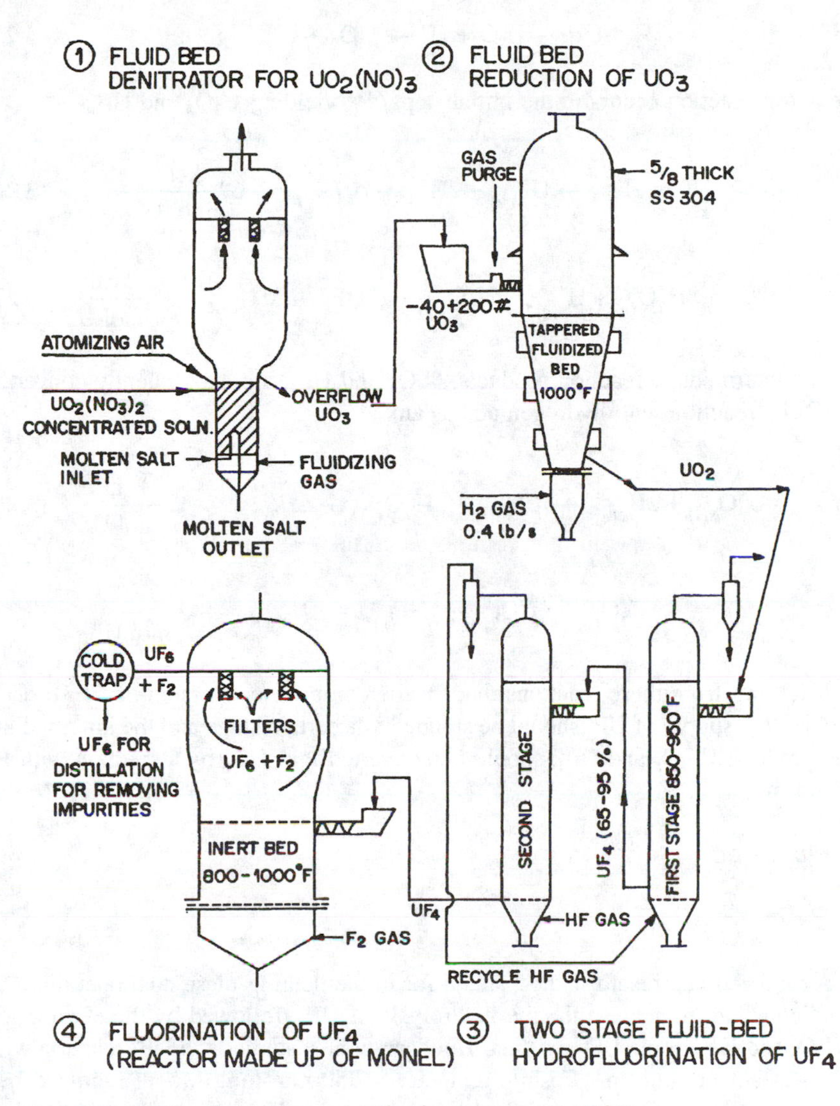

Figure 3.6 Fluidization in uranium extraction by nonaqueous routes.

particles. High-density uranium oxide can be prepared by the oxidation of UF_6 by suitably selecting the oxidizing agent and carrying out the oxidation in an efficient reactor. Application of fluidization for the production of UO_2 has been well demonstrated[61,62] and established. The oxidation of UF_6 by the gaseous oxidizing agent results in a solid product which, when kept in a fluidized state, helps the subsequent gas-phase reaction to utilize the advantageous properties of a fluidized bed. The preparation of dense uranium dioxide particles directly from UF_6 using a fluidized bed reactor was detailed by Knudsen et al.[63] Uranium hexafluoride, when reacted with a steam and hydrogen gas mixture in the temperature range 650–700°C, yields UO_2 particles in accordance with the reaction

$$UF_6 + 2H_2O + H_2 \rightarrow UO_2 + 6HF \tag{3.25}$$

The above reaction occurs in the initial steps,[64,65] yielding U_3O_8 and UF_4:

$$UF_6 + H_{2(g)} \rightarrow UF_{4(s)} + 2HF, \; -\Delta H_{298\,K} = -69\frac{\text{kcal}}{\text{mol U}} \tag{3.26}$$

$$3UF_6 + 8H_2O_{(g)} + H_{2(g)} \rightarrow U_3O_{8(s)} + 18HF, \; -\Delta H_{298\,K} = -8\frac{\text{kcal}}{\text{mol U}} \tag{3.27}$$

The intermediate reaction products, U_3O_8 and UF_4, are subsequently converted to UO_2 by reaction with hydrogen and steam:

$$U_3O_{8(s)} + 2H_{2(g)} \rightarrow 3UO_{2(s)} + 2H_2O_{(g)}, \; -\Delta H_{298\,K} = -13\frac{\text{kcal}}{\text{mol U}} \tag{3.28}$$

$$UF_{4(s)} + 2H_2O_{(g)} \rightarrow UO_{2(s)} + 4HF_{(g)}, \; -\Delta H_{298\,K} = -47\frac{\text{kcal}}{\text{mol U}} \tag{3.29}$$

It follows from these reactions that for the complete conversion of UF_6 to UO_2 particles, the supply of UF_6 should be stopped at a certain stage and the intermediate solid products, U_3O_8 and UF_4, should be converted into UO_2 by a reaction with H_2 and steam.

2. *Fluid Bed*

a. *Reactor*

A fluidized bed reactor is best suited for implementing these dual operations in a cyclic manner to achieve the pyrohydrolysis of UF_6, followed by the elimination of U_3O_8 and UF_4 in the UO_2 product. Knudsen et al.[63] studied the UF_6 reaction with a H_2 + steam mixture in a 75-mm diameter × 750-mm-long Monel fluidized bed with a 60° conical bottom into which an 8-mm centrally placed tube was used to introduce UF_6. The fluidizing gases H_2 and steam entered the annular space surrounding the central UF_6 feed tube. The initial material for the reaction was a bed of UO_2 particles prepared by oxidative fracturing of UO_2 particles. The feed port for the bed materials was kept at a height of 500 mm above the bottom of the fluidized bed reactor. The seed bed materials were first kept in the fluidized bed by passing nitrogen purge along with the fluidizing gas streams (i.e., H_2 + steam) until the reaction temperature was attained. When UF_6 was fed, nitrogen purge was stopped. The reaction proceeded in the temperature range 650–700°C. The reaction continued in two steps: first the production of UO_2 along with the intermediates, U_3O_8 and UF_4, and then cleanup of these intermediates with H_2 and steam in the absence of UF_6. The total reaction period was limited until the product buildup amounted to the initial bed weight. In other words, a second experiment continued with the bed of the preceding product bed material.

b. Product

Uranium oxide particles ranging in size from –18 to +140 mesh with a density of up to 9750 kg/m^3 (equivalent to 89% theoretical density) could be prepared by this technique. The feed rate of uranium hexafluoride in the bed was kept in the range 17–50 g/m. The density of UO_2 particles was lowered (9000 kg/m^3) if the UF_6 flow rates were high. Uranium oxide particles produced by this technique contained residue fluoride analyzed to be 250 ppm. With a steam-to-uranium-hexafluoride ratio of 2:1, UO_2 particles of the highest density could be obtained. UO_2 particles obtained by this technique could be further densified to 10,500 kg/m^3 (96% theoretical density) by sintering in H_2 gas at 1700°C. The fluoride content of the final product was found to be 5 ppm.

The successful pyrohydrolysis of UF_6 directly using steam and a hydrogen gaseous stream in a fluidized bed encouraged further research into the production of a mixed oxide fuel, UO_2–PuO_2, by the pyrohydrolysis of a mixture of the hexafluorides of uranium and plutonium.

c. Bed Depth Effect

Bed depth plays an important role in the density of UO_2 but also has a combined role with the UF_6 feed rate. In general, the higher the bed height, the higher the density of the UO_2 obtained. Although it may be attributed to the higher surface area available at this bed height, the density of UO_2 obtained was not remarkably high at low UF_6 concentrations for similar ratios of UF_6 feed rate to bed depth. An important variable that affected the density of UO_2 particles was positioning of the UF_6 feed nozzle inside the reactor. When the nozzle position was extended more (i.e., ~89 mm), the density of the particles produced was higher. For a short nozzle (20 mm), UO_2 particle density decreased from 8.4 to 8.0 g/cc on increasing the steam/UF_6 mole ratio from 0.2 to 3.1. With a long nozzle, the density of the UO_2 particles varied from 8.7 to 9.2 g/cc under similar conditions.

d. Fine Oxide Preparation

A fluidized bed reactor was reported[63] to have been used for the production of fine UO_2 particles by oxidation and reduction reactions. The fine particles thus produced were used as the seed material in the fluidized bed during the pyrohydrolysis of UF_6. In the oxidation–reduction of UO_2, coarse UO_2 particles were fluidized using an air–nitrogen (10–20%) mixture at 400°C (or 650°C). In a fluidized bed of 76 mm (3 in.) diameter, the superficial velocity used was 0.3 m/s to fluidize a bed of UO_2 material composed of particles of –18 +80 mesh size. The oxidation period ranged from 29 to 58 min and the air used was 0.56–1.20 times the stoichiometric quantity. The oxidation of the UO_2 particles proceeded first with their conversion to the higher oxide, U_3O_8, and the product thus produced was finer than its precursor material. On reducing these finer particles, UO_2 particles of fine size could be obtained. Once the U_3O_8 particles were formed by oxidation in a fluidized bed reactor, subsequent reduction could be accomplished in the same reactor. The bed of oxidized UO_2 particles (i.e., U_3O_8) was kept in a fluidized state by nitrogen purge until a reaction temperature of 650°C was attained. Knudsen et al.[63] reported that the reduction reaction was carried out at 650°C using 100% hydrogen for a period of 30 min. The extent of fines produced was measured relative to the –80 mesh size, and the quantity of –80 mesh particles in the final product accounted for as much as 86.6%. Thus the oxidation–reduction reaction, using a single fluidized bed for the dual reactions, proved to be a viable method for producing fine UO_2 particles starting from a coarse size feed.

B. Stoichiometric Uranium Monocarbide

1. *Uranium Carbide as Nuclear Fuel*

Uranium carbide is an advanced ceramic nuclear fuel useful for high-temperature nuclear reactors, in particular in fast breeder reactors. Uranium carbide (UC) has a higher thermal conductivity and a larger uranium content in 1 mol than UO_2. Furthermore, it has a high melting point and is capable to have high burnup. The properties of the carbide fuel are affected by its carbon stoichiometry and the method of preparation. Hypostoichiometric uranium monocarbide is compatible with stainless-steel cladding, while hyperstoichiometric uranium monocarbide causes embrittlement and failure of the cladding wall due to carbon migration from the fuel. The grain boundaries of hypostoichiometric monocarbide particles contain isolated islands of free uranium metal; the occurrence of these is decreased as the stoichiometric composition for UC is approached. Free uranium metal, if any, would melt and interact with the cladding, causing failure. The preferred carbon content of uranium monocarbide is 4.4–4.8 wt%.

Crane et al.[66] reported some details on the development of UC for use as a nuclear fuel. They reported the preparation of substoichiometric and stoichiometric UC using propane as the carburizing agent. The carbide fuels were prepared by powder metallurgy and skull arc melting with a view to establishing the appropriate conditions to achieve a product of high purity and density and also evaluating fuel behavior, such as irradiation stability, thermal expansion, thermal conductivity, and hot hardness.

Carbide fuels are conventionally prepared by the powder–pellet route.[67,68] This method involves grinding and mixing UO_2 and graphite powders in exact quantities, followed by cold pressing at 50,000 psi and heating the resultant pellets in vacuum at 1700–1800°C for about 2.5 hr. The sintered reaction product is ground to 10–15 μm and compacted by powder metallurgy techniques. The UC produced by this method generally does not conform to the exact stoichiometry, as the final carbon content of the fuel is dependent on the oxygen content of UO_2 and the carbon content before firing in vacuum. Hence, it would be necessary to adjust the final content of carbon in UC by the addition of uranium metal.

2. Reaction

Uranium monocarbide in exact stoichiometry can be prepared by the reaction of the metal with hydrocarbon gases such as propane or methane. The overall gas–solid reaction with uranium metal and propane is

$$3U_{(s)} + C_3H_{8(g)} \rightarrow 3UC_{(s)} + 4H_{2(g)} \tag{3.30}$$

The propane gas can decompose into methane and free carbon. The formation of free carbon by this reaction can be prevented in the presence of hydrogen gas. The reaction of methane with uranium metal would yield UC as well as uranium dicarbide, with 2 mol of hydrogen in each case. The methane also can decompose into free carbon and hydrogen gas. The possible reactions of uranium metal with methane are

$$U_{(s)} + CH_{4(g)} \rightarrow UC_{(s)} + 2H_{2(g)} \tag{3.31}$$

$$UC_{(s)} + CH_{4(g)} \rightarrow UC_{2(s)} + 2H_{2(g)} \tag{3.32}$$

$$CH_{4(g)} \rightarrow C_{(s)} + 2H_{2(g)} \tag{3.33}$$

For all the above reactions, the equilibrium equation is

$$\Delta G = -RT \ln K_{equ} = -RT \ln \frac{P_{H_2}^2}{P_{CH_4}} \tag{3.34}$$

If the total pressure is P_T and equal to 1 atm, then $P_{CH_4} = 1 - P_{H_2}$.

3. Need for Fluid Bed

In the preparation of UC by the reaction of U with CH_4, the decomposition of CH_4 into carbon and H_2 should be suppressed. This can be achieved by careful control of the partial pressure of methane. For example, at 723°C (1000 K), the decomposition of methane can be prevented by maintaining the partial pressure of methane at 0.083 atm when the total pressure is 1 atm. At a methane partial pressure of 0.049 atm and below, the formation of uranium dicarbide can be prevented. If the formation of UC is to be favored, then the partial pressure of methane at 1000 K (723°C) should be in the range 0.91×10^{-6} to 0.0049 atm when the total pressure is 1 atm. In other words, it is implied that the reaction to accomplish the yield of the desired product must be carried out in a reactor where the temperature, the rate of heating, and the time can be controlled to arrive at the desired stoichiometry of the product. It is also important to note that the stoichiometry of UC is controlled by chemical equilibrium and is not dependent on the total amount of carbon and uranium fed into the system. In view of these considerations of thermochemistry, it is imperative to choose a reactor that can bring about efficient gas–solid contact and operate at close temperature control. A fluidized bed reactor is the right choice for this job. Attempts to prepare[69,70] UC by uranium metal–hydrocarbon reaction did not yield encouraging results. Although stoichiomteric UC could be prepared on a small scale (i.e., 100 g per batch), it was not possible to synthesize it on a large-scale (i.e., 100 kg per batch) unit. In a static bed, the absence of intimate contact between the hydrocarbon gas and uranium metal powder, the use of undiluted hydrocarbon gas, and the difficulty of gas diffusion would cause the formation of higher carbides and free carbon. All these problems would be severe when a deep static bed is used. In view of this, a fluidized bed reactor was chosen by Petkus et al.[71] for the synthesis of stoichiometric UC, and it proved to be a successful choice.

4. Fluid Bed Reactor

a. Feed and Gas

The preparation of uranium monocarbide by using a fluidized bed reactor involved three steps: (1) preparation of fine particles of uranium hydride by the reaction of massive uranium metal with hydrogen in a static bed reactor at 250–300°C for about 5 hr under a pressure of 10 psig, (2) decomposition of fine uranium hydride particles in a fluidized bed using propane and hydrogen as the fluidizing gas stream at 500–750°C, and (3) continuation of step (2) for 3–20 hr to convert the uranium metal powder to uranium monocarbide.

Propane gas instead of methane was used, as it was cheaper and had fewer impurities. Hydrogen was mixed with propane to prevent the formation of free carbon consequent to the decomposition of propane.

b. Reactor

The fluidized bed reactor was made of a 25-mm-diameter type 304 stainless-steel tube 152 mm long with a 76-mm-diameter, 225-mm-long disengaging zone. A porous stainless-steel filter with a pore size of 5 mm was used inside the disengaging column to arrest entrainment of the particles. The reactor was not provided with any special distributor. A conical bottom with a ball-check valve severed the gas distribution system and also prevented draining of hot fluidized reactant solid particles. The fluidized bed was heated by a resistance wire wound on the outer wall in the 152-mm zone of the reactor.

The minimum fluidization velocity required to keep the uranium metal powder in a fluidized state using the propane gas mixture was 7.6 cm/s, and this velocity was essential to prevent sintering of the uranium metal particles that formed during dehydriding. An operating fluidization velocity of less than 30 cm/s was recommended to prevent entrainment of fine particles.

c. Product

Stoichiometric monocarbide of uranium was successfully prepared at 600, 700, and 750°C with propane concentrations in H_2 gas of 3, 5, and 2.3 vol%, respectively.

The carbon content of the uranium monocarbide obtained was 4.6 wt% after a reaction time of 10 hr in a fluidized bed.

The uranium monocarbide particles produced by the fluidization technique were fine and pyrophoric in nature. They would react with oxygen and with any gas that contained moisture or oxygen. Hence, it would be essential to house the reactor in a glove box under a helium atmosphere containing 10 ppm water, 40 ppm oxygen, and 160 ppm nitrogen. It would also be essential to remove any trace quantity of oxygen present in the hydrocarbon gas. Hydrogen reaction with trace quantities of oxygen over a palladium catalyst and absorption of the resultant water product over a bed of molecular sieve with a 4-Å pore size is the usual method. If the pore size of the molecular sieve is larger than 4Å, then propane would also be absorbed.

d. Other Carbides

The fluidized bed has been viewed as promising equipment for the synthesis of uranium–plutonium mixed carbide solid fuel. The fact that UC produced by the fluidized bed process is fine and free flowing enables fuel fabrication operations to easily achieve a final density of the compact of over 90%. The excellent control of gas composition in a fluidized bed reactor has further encouraged research into synthesis of mixed ceramic fuels such as uranium–carbide sulfide, uranium–carbide nitride, and uranium–nitride sulfide.

C. Carbides and Nitrides Directly from Uranium Dioxide

1. Reaction Criteria

In the earlier section on the preparation of UC, we saw that UC is prepared finally by the reaction of fine uranium metal particles with graphite particles. The initial preparation of pure elemental uranium metal significantly adds to the cost of fuel production. Although uranium oxide is the cheapest ceramic fuel uranium nitride UN and UC are preferred when a high thermal conductivity fuel with a high uranium density is needed. These ceramic fuels are ultimately required to increase the efficiency of energy conversion by achieving high temperatures in the heat cycle. The preparation of fuels such as UC and UN right from uranium dioxide would be more economical than methods involving elemental reactions.

The reaction of UO_2 with carbon in a flowing stream of nitrogen can be represented as:

$$UO_2 + (2+x)\,C \xrightarrow[1450^\circ C]{N_2} UC_xN_{1-x} + 2CO \tag{3.35}$$

In the above reaction, if carbon and nitrogen partial pressures are carefully controlled, UC solid solution in UN with varying degree of x values can be achieved. In order to drive the carbonitriding of UO_2 in the forward direction, low partial pressures of CO (<2 mm at 1477°C) and modest partial pressures of N_2 (>2 mm at

1477°C) are essential. If the reacted product containing UN at high temperatures is allowed to cool, uranium sesquinitride (U_2N_3) is formed. In practice, the ultimate nitride fuel is $UN_{1.7}$, which is an intermediate between U_2N_3 and UN_2. If the carbon content for the reaction with UO_2 is less, then the product would contain UO_2, and if the carbon content is more, then unreacted free carbon would be present. The nitrogen content can be adjusted by controlling the N_2 partial pressure. Hence, preparation of UC or UN has to be carried out under controlled partial pressures of the reactant as well as the product gases.

2. Fluid Bed

a. Reactor

The fluidized bed reactor was tested and proven to be successful by Hyde et al.[72] for the preparation of UN and carbonitride fuels from UO_2. An inductively heated graphite body reactor 1 in. in diameter with a conical bottom that had a single hole instead of a distributor plate was used to distribute the fluidizing gas. Fluidization was achieved by spouting action. The reactor was fitted with hollow graphite tube with one end closed; the tube was immersed inside a heater with the closed end inside the heated zone. An optical pyrometer was used to measure the temperature. Three types of fluidizing gas streams were used: (1) only nitrogen gas, (2) N_2 gas + 30% H_2, and (3) N_2 gas + 20% Ar. The carburizing agent was either a hydrocarbon like benzene or carbon particles. The fluidized bed was constituted of oxides of uranium, such as UO_3 and UO_2, obtained from normal production, and these oxides were used in green as well as sintered conditions. Thus, four types of uranium oxides were tested in the fluidized bed, and the mean average particle size was 250 μm.

b. Operation

Experiments were conducted in the fluidized bed in three stages. First, the bed of uranium oxides amounting to 40 g (10 cc bulk volume) was brought up to the reaction temperature by induction heating; the heating time was about 30 min. During the second stage, the reactant (benzene) vapor was introduced into the reactor for the reaction, and in the third stage, the product was cooled down in the fluidized state itself without any benzene vapor. The different parameters investigated included the bed temperature, reaction time, fluidizing flow rate, composition of the charge in the bed, and benzene concentration in the feed.

At temperatures of 1100–1200°C, the oxides UO_2 and UO_3 agglomerated during fluidization, and this could be overcome by cooling and reheating or by using hydrogen in the preparatory stage. Hence, a gaseous mixture of nitrogen and hydrogen was recommended for trouble-free fluidization. The optimum reaction temperature range was found to be 1450–1500°C. Below this temperature range, pyrocarbon started coating, and above it the product sintered. A benzene flow rate (mol feed rate of benzene per unit mol of uranium) of 0.08 per hour and a reaction time of 8 hr were found to be sufficient to convert all the oxide into the carbonitride. It was not possible to produce pure UN. During the initial stage, UN formation was faster than carbide formation, whereas the trend was reversed in the final stage. The carbide formed at the terminal stage was at the expense of the nitride. Hence, neither pure UN nor pure UC could be produced. The product formed corresponded to the general formula UC_xN_{1-x} with x ranging between 0.2 and 0.3. The production-grade oxides of uranium (i.e., UO_2 and UO_3) were found to be better than sintered or green oxides of uranium, and an N_2 and H_2 gas mixture was recommended for this hydrocarbon route to produce the uranium carbonitride fuel. The product obtained by the hydrocarbon route had a tendency to stick to the walls, and this adverse feature, in a large-scale operation, may interfere with the fluidization behavior.

c. *Carburizing Agent*

When the carburizing agent consisted of carbon particles instead of a hydrocarbon gas the mixture of UO_2 and carbon particles had to be blended properly. UO_2 (<40 μm) and carbon black (<0.5 μm) were usually mixed, ground in a water medium, dried, and pelletized. A particle size in the range 300–700 μm was selected after crushing and sieving the UO_2–carbon pellets. The fluidized bed reactor was the same as the one used for the hydrocarbon route, but it had a sintered carbon distributor fixed at the bottom of the bed. The amount of carbon mixed with UO_2 had to be below the stoichiometric value (<15%) if the wall of the reactor was of graphite. This was to take care of the carbon that was picked up due to attrition in the course of fluidization. The fluidizing gas would have to be pure and free of moisture and oxygen. Magnesium perchlorate was used to remove moisture. Trace amounts of oxygen were converted to CO on contact with the hot carbon sinter upon entry into the reactor. The reactor had to be heated gradually, as sintering of the charge was inevitable. The fluidizing gas had to be increased sufficiently in order to take care of fluidization of the sintered particles. It should be noted that there was an increase in the density of the charge due to sintering. When the fluidizing gas was argon, UC was formed, while the product was UN with UC as solid solution when nitrogen gas was used for fluidization. At fluidizing gas flow rates above the minimum, the reaction rate was independent of the gas flow rate. This was due to flushing of the CO gas from the core to the surface and the availability of fresh reactant gas for the reaction.

d. *Product Yield*

In order to obtain the best product (i.e., UN), the C-to-UO_2 ratio in the charge would have to be in the range 1.65–1.75. The product normally obtained was 96.4% UN, and the balance was in the form of UO_2 or UC. When the C-to-UO_2 ratio was less than 1.65, the product contained two phases, UO_2 and UN. For ratios between 1.75 and 2.3, the product was a single-phase compound corresponding to a solid solution composition of UC_xN_{1-x} with x ranging from 0.05 to 0.38. When the C-to-UO_2 ratio was 2.4–3.4, a single-phase constant composition product, $U(C_{0.56}N_{0.44})$, resulted along with uncombined carbon amounting to up to 3% by weight. In order to prepare uranium carbonitride fuel rich in carbide content, it would be essential to control both the C-to-UO_2 ratio and the partial pressure of N_2. The initial oxygen content in UO_2 and the amount of carbon required for the desired carbide content in the product should be known. The uncombined carbon can be reduced by using H_2 at 900°C to evolve CH_4. In order to eliminate the higher nitrides, after the completion of the reaction (as indicated by cessation of CO evolution) the reaction product should be heated at 1450°C in an argon gas flow and then cooled in an $N_2 + H_2$ atmosphere. By using either the carbon or the hydrocarbon route, it was not possible to prepare pure UN, but a solid solution of UC in UN with the general composition UC_xN_{1-x} could be prepared. The carbon route appeared to be preferable because x ranged from 0.04 to 1 in the corresponding product compared to nonsystematic changes in x between 0.2 and 0.3 associated with the hydrocarbon route.

e. Recommendations

For commercial-scale production of uranium carbonitride, the carbon route is recommended as it is possible to control the carbon content and to achieve a shorter reaction time. The main advantage of the hydrocarbon route is elimination of the steps involving premixing and pelletization of the UO_2 + C mixture. The success of the fluidized bed in fuel preparation was further tested using a 40-mm cylindrical reactor. Encouraging results with the carbothermic reduction of oxides to carbides as well as nitrides paved the way for technological advancement in the preparation of the carbides of vanadium, zirconium, and molybdenum from their oxides. Hyde et al.[72] reported the preparation of zirconium boride by fluidizing a $ZrO_2 + B_2O_3$ + C mixture using argon gas and recommended this technique for the preparation of sulfides and phosphides.

Advanced nuclear fuels such as UN are in demand for thermoelectric nuclear power systems to generate power in the range of 10 kW(e) to 1 MW(e) for space applications. A liquid lithium-cooled nuclear reactor delivers about 2.5 MW(e) at >1300 K reactor outlet temperature, whereas a silicon–germanium–phosphate thermoelectric converter produces 100 MW(e) with overall system efficiency of 4%. Uranium nitride fuel that has a high metal atom ratio, high density, high melting point, high thermal conductivity, low swelling, low fission gas release, and compatibility with coolant is one of the high-temperature ceramic fuels and is recommended for a nuclear reactor operating at high temperature and fuel burnup (>6 at %). The swelling characteristics of uranium nitride fuels for space applications have been studied by Ross et al.[73] In the preceding few paragraphs, we have outlined the technology of ceramic fuel preparation using fluidized bed technology. Research and development on ceramic fuels are continuously advancing as there is a need for superior quality fuel for high-temperature nuclear reactors. More details on the chemistry and technology of carbides and nitrides of uranium and plutonium as well as the mixed compounds (U,Pu)C, (U,Pu)N, and (U,Pu)(CN) are provided in a monograph by Matzke.[74]

D. Uranium Aluminide

1. Appraisal

Uranium aluminide is an intermetallic fuel. It can be prepared in a single fluidized bed reactor starting from uranyl nitrate solution. The decomposition of uranyl nitrate solution at 400°C can be accomplished over a bed of aluminum powder 200–250# in size fluidized by air. Four steps are involved: (1) denitration, (2) reduction, (3) chlorination, and (4) conversion to the final product by heating. This process permits the production of uranium aluminide without having to produce uranium metal as an intermediate. Grimmett et al.[75] carried out experiments to test the viability of this process. Later, Hogg and Grimmett[76] experimented with a fluidized bed process for the production of uranium aluminide and also studied the basic aspects of the reactions involved in the four steps. All four steps were carried out in series in a single fluidized bed reactor. These steps are described concisely in the following paragraphs to show how the fluidized bed reactor played a prominent role in this novel process.

2. Fluid Bed

a. Reactor

A fluidized bed reactor 50 mm in diameter was provided with a 150-mm-long disengaging section. A slit-type distributor for distributing the fluidizing gas, thermocouples at 6 and 50 mm above the grid plate and in the disengaging zone, a 50-mm spray nozzle above the grid, and external heaters were provided. Provisions for preheating the inlet fluidizing gas were also made for the high-temperature conversion step during preparation of the fuel. The entire system, which consisted of the fluidized bed, the condenser for off-gases, the spray nozzle, the liquid chemical injection system, and the fluidizing gas system, was housed inside a glove box.

b. Reaction Steps

i. Dentitration Step — Uranyl nitrate solution was sprayed over a bed of hot aluminum particles fluidized by air at 400°C. The aluminum particles were coated with UO_3 particles formed by the decomposition of the uranyl nitrate solution. The decomposition reaction was

$$UO_3(NO_3)_{2(l)} \xrightarrow[\text{Air}]{400°\text{C}} UO_{3(s)} + 2NO_{2(g)} + O_{2(g)} \tag{3.36}$$

The UO_3 particles needed to have high reactivity to participate in the subsequent reduction reaction. High reactivity could be achieved by obtaining porous UO_3 particles. Addition of aluminum nitrate solution and excess nitric acid generally improved the reactivity of the UO_3 particles. The denitration process was dealt with in detail in the context of uranium extraction earlier in this chapter.

ii. Reduction — This was the second step and it immediately followed completion of the denitration reaction. The reductant was methyl alcohol vapor injected in the fluidizing gas stream of argon at 360°C. The reduction reaction was

$$UO_{3(s)} + CH_3OH_{(g)} \xrightarrow[\text{Argon}]{360°C} UO_{2(s)} + COH_{2(g)} + H_2O_{(g)} \tag{3.37}$$

The extent of conversion of UO_3 to UO_2 was found to be 85% in 90 min. Total conversion of UO_3 to UO_2 was not essential because UO_3 could also be chlorinated equally well in the chlorination stage. The reduction reaction at the initial stage was assumed to be reaction controlled with a reaction rate constant (K_s) of 0.0190 m/s. In the final stage of reduction, diffusion was the rate-controlling step. The diffusivity of CH_3OH in porous UO_2 particles was estimated to be 0.00108 m^2/s.

iii. Chlorination — The UO_2 coating on aluminum particles was converted to a UCl_4 coating by chlorination in a fluidized bed at 330°C by a mixture of CCl_4 (75 mol%) and Cl_2 (25 mol%). The reaction is

$$UO_{2(s)} + CCl_{4(g)} \xrightarrow[330°C]{Cl_{2(g)}} UCl_{4(s)} + CO_{2(g)} \tag{3.38}$$

It was found essential to use the right proportion of CCl_4 and Cl_2. If CCl_4 was in excess, free carbon would form, thereby leading to the formation of uranium carbide in the high-temperature conversion step. When insufficient CCl_4 was used, the chlorination product of UO_2 was UO_2Cl_2 because the carbon available was inadequate to remove the O_2. If chlorine gas was not used along with CCl_4, UO_2 would be converted mainly to UO_2Cl_2 instead of the desired UCl_4. The optimum reaction temperature was 330°C. Above this temperature, CCl_4 and Cl_2 would react with aluminum particles, and below this temperature, the reaction rate for the conversion of UO_2 to UCl_4 would be slow. It was found necessary to use argon gas as the main fluidizing gas instead of nitrogen gas. Nitrogen, if used instead of argon, can participate in the chemical reaction to yield uranium nitride compounds. The chemical reaction rate constant (K_s) and the diffusivity for CCl_4 in porous UCl_4 were determined to be 0.000196 m/s and 1.38×10^{-3} m^2/s, respectively. Chemical reaction controlled the initial stage of chlorination, followed by diffusion control near the terminal stage of chlorination.

iv. High-Temperature Conversion — This was the fourth step in sequence carried out again in a fluidized bed. After chlorination, the UCl_4-coated aluminum particles were subjected to heating, keeping the bed in the fluidized state using argon as the fluidizing gas. As the bed temperature reached 500°C, UCl_4 reacted with Al, forming $AlCl_3$ and UCl_3. The solid–solid reaction between UCl_4 and Al took place, and the product obtained was UAl_x at the melting point of Al (i.e., 660°C onward). The exact mechanism of this solid–solid reaction has not been established. Hence, the overall reaction at 680°C between UCl_4 and Al can be represented as:

$$3UCl_{4(s)} + (4+3x)\, Al_{(l)} \xrightarrow[680^\circ C]{Argon} 3UAl_x + 4AlCl_3 \tag{3.39}$$

Hogg and Grimmett[76] proposed an equation to predict the time required for the complete conversion of UCl_3 to UAl_x using an unreacted shrinking core model. During the final high-temperature conversion, there would be some possibility of arriving at an impurity like Al_2O_3 due to the presence of unchlorinated UO_2 particles. This alumina, although it could dilute the fuel value, can serve as a fission-gas-retaining medium. Using fluidized bed processes, it was claimed that products such as UAl_2, UAl_3, and UAl_4 were produced. It should be noted that this was a simple process used in a single batch fluidized bed. This process eliminates the production of uranium metal through conventional methods starting with uranyl nitrate and hence reduces the cost of fuel production substantially. From the foregoing, it can be inferred that the fluidized bed process plays a key role in fuel material preparation. The flow diagram in Figure 3.7 provides an overview of these processes utilizing the fluidized bed.

IV. PYROCARBON COATING

A. Reactor Choice

1. Suitability

Fluidized bed reactors, in particular gas–solid ones, have been used extensively in particle growth and agglomeration, coating of heated objects with plastic powders, and coating of numerous particulate solids by protective coating by means of the thermal decomposition of the gaseous or vapor feeds that are injected along with the fluidizing gas. The subject of particle growth and coating in gas fluidized beds was dealt with in detail by Nienow and Rowe.[77] The main application of this technique is in the granulation of fine powder materials using appropriate binders that are sprayed from the top on the surface of the fluidized bed. Nienow and Rowe[77] presented a particle growth model for such a granulator. Another area of application of gas fluidized beds is the coating of hot objects immersed in a bed of fusible polymeric resin. This area of application covers the coating in appliances, power distribution, and pipeline feeds. Studies on fluidized bed coating with plastic materials were reported by Lee,[78] Richart,[79] and Pettigrew.[80] Fluidized bed coatings of metals have become so popular that more than 100 major companies are reported[79] to have adopted this technology. The main advantages of this technology pertain to saving coating material and environmentally safe operation (because this is a physical process that does not involve any chemicals, as conventional coating processes do).

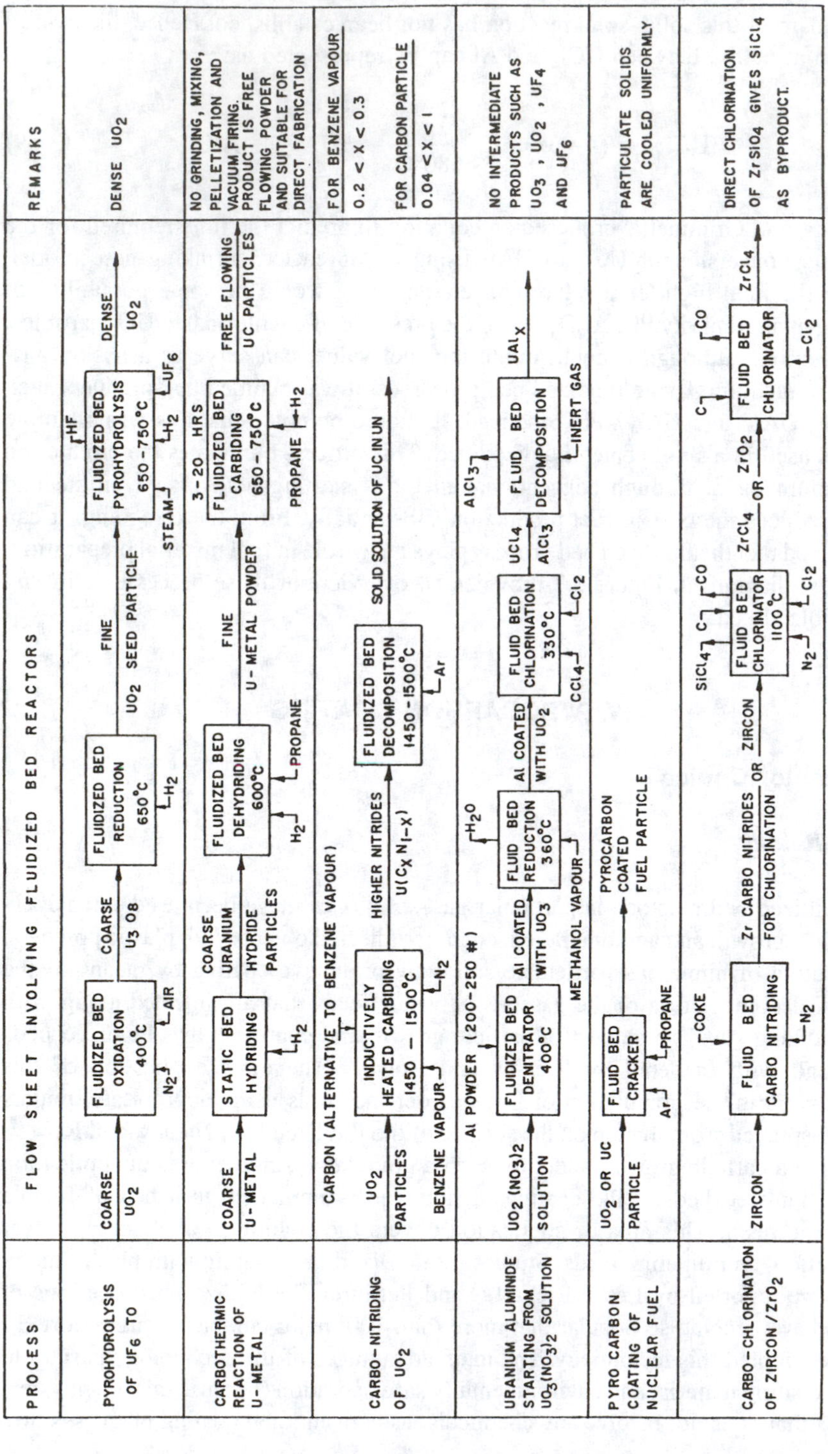

PROCESS	FLOW SHEET INVOLVING FLUIDIZED BED REACTORS	REMARKS
PYROHYDROLYSIS OF UF_6 TO	COARSE UO_2 → FLUIDIZED BED OXIDATION 400°C (N_2, AIR) → COARSE U_3O_8 → FLUIDIZED BED REDUCTION 650°C (H_2) → FINE UO_2 SEED PARTICLE → FLUIDIZED BED PYROHYDROLYSIS 650-750°C (STEAM, H_2, UF_6; HF out) → DENSE UO_2	DENSE UO_2
CARBOTHERMIC REACTION OF U-METAL	COARSE U-METAL → STATIC BED HYDRIDING (H_2) → COARSE URANIUM HYDRIDE PARTICLES → FLUIDIZED BED DEHYDRIDING 600°C (H_2, PROPANE) → FINE U-METAL POWDER → FLUIDIZED BED CARBIDING 650-750°C, 3-20 HRS (PROPANE, H_2) → FREE FLOWING UC PARTICLES	NO GRINDING, MIXING, PELLETIZATION AND VACUUM FIRING. PRODUCT IS FREE FLOWING POWDER AND SUITABLE FOR DIRECT FABRICATION
CARBO-NITRIDING OF UO_2	UO_2 PARTICLES → INDUCTIVELY HEATED CARBIDING 1450 – 1500°C (BENZENE VAPOUR, N_2; CARBON (ALTERNATIVE TO BENZENE VAPOUR)) → HIGHER NITRIDES $U(C_X N_{1-X})$ → FLUIDIZED BED DECOMPOSITION 1450-1500°C (Ar) → SOLID SOLUTION OF UC IN UN	FOR BENZENE VAPOUR $0.2 < X < 0.3$ FOR CARBON PARTICLE $0.04 < X < 1$
URANIUM ALUMINIDE STARTING FROM $UO_3(NO_3)_2$ SOLUTION	$UO_2(NO_3)_2$ SOLUTION → FLUIDIZED BED DENITRATOR 400°C (Al POWDER (200-250 #)) → Al COATED WITH UO_3 → FLUID BED REDUCTION 360°C (METHANOL VAPOUR; H_2O out) → Al COATED WITH UO_2 → FLUID BED CHLORINATION 330°C (CCl_4, Cl_2) → UCl_4 $AlCl_3$ → FLUID BED DECOMPOSITION (INERT GAS; $AlCl_3$ out) → UAl_X	NO INTERMEDIATE PRODUCTS SUCH AS UO_3, UO_2, UF_4 AND UF_6
PYRO CARBON COATING OF NUCLEAR FUEL	UO_2 OR UC PARTICLE → FLUID BED CRACKER (Ar, PROPANE) → PYROCARBON COATED FUEL PARTICLE	PARTICULATE SOLIDS ARE COOLED UNIFORMLY
CARBO-CHLORINATION OF ZIRCON / ZrO_2	ZIRCON → FLUID BED CARBO NITRIDING (COKE, N_2) → Zr CARBO NITRIDES FOR CHLORINATION OR ZIRCON → FLUID BED CHLORINATOR 1100°C (C, N_2, Cl_2; $SiCl_4$, CO out) → $ZrCl_4$ OR ZrO_2 → FLUID BED CHLORINATOR (C, Cl_2; CO out) → $ZrCl_4$	DIRECT CHLORINATION OF $ZrSiO_4$ GIVES $SiCl_4$ AS BYPRODUCT.

Figure 3.7 Fluidization in nuclear material preparation.

2. *Immersed Objects*

The coating of hot immersed objects with plastic materials depends on several parameters, such as the temperature of the immersed object; the fluidizing gas velocity; the fluidized bed temperature; the particle size, shape, and distribution; and the physical properties of the object, the powder, and the carrier gas. There is not much in the literature pertaining to the basics of heat transfer as applied to a fluidized bed coating. Gutfinger and Chen[81,82] developed a mathematical model based on heat conduction in a coating film and presented a correlation to predict the time required to achieve a specified coated film thickness. At the final coating thickness, the Biot number based on the final coating thickness, X_f, i.e., $\mathrm{Bi} = h/kX_f$, is found to be equal to a dimensionless temperature:

$$\theta = \frac{\left(T_w - T_f\right)}{\left(T_f - T_b\right)} \tag{3.40}$$

where h is heat transfer coefficient, T_w is the wall temperature of the heated object, T_f is the melting temperature of the resin, and T_b is the bed temperature. Thus, it can be seen that the heat transfer coefficient plays an important role in determining the ultimate coated film thickness.

3. *Coating Classification*

The type of fluidized coating discussed above is related to physical deposition; the other common type is chemical vapor deposition. In a physical process, vapors of the source material are condensed over the object, often termed the substrate. The physical process of vapor deposition and formation of a coating of the source material on a substrate is popularly known as physical vapor deposition. In chemical vapor deposition, the substrate is coated with the product of the chemical reaction, formed either by a gas–solid reaction as in nitriding or caburizing the metal surface or by oxidation/reduction or thermal decomposition. In the present discussion, the chemical vapor deposition that involves thermal cracking or decomposition of the source material has direct relevance and application in nuclear fuel coatings, and this aspect will be discussed here in some detail. General classification of vapor deposition and applications of the fluidized bed in appropriate types of vapor deposition are presented in Figure 3.8.

B. Classification of Pyrolytic Deposition

1. *Nonmetallic Coating*

In a pyrolytic deposition process, a gas or vapor containing the element to be coated is heated to decompose or allowed to react with the substrate. The decomposed or reacted product nucleates either in the gas phase or on the substrate. A

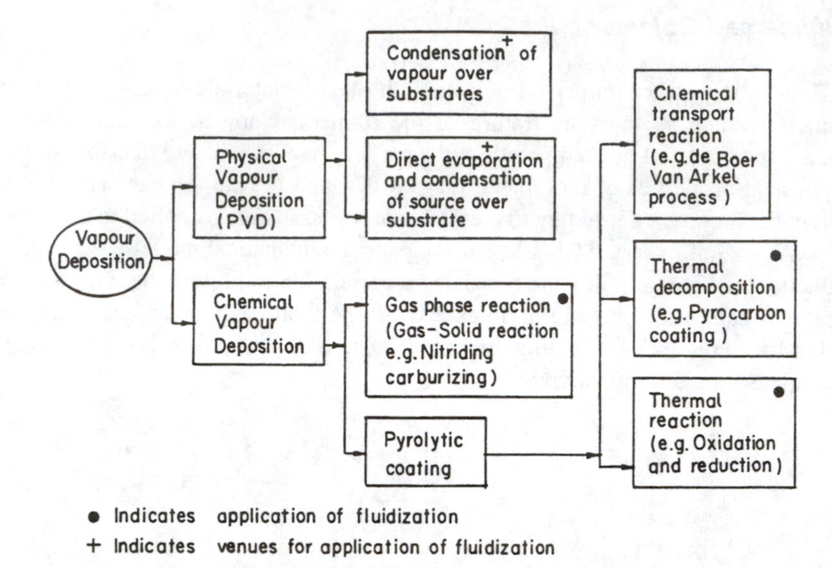

Figure 3.8 Fluidization in coating.

typical product obtained by gas-phase nucleation is in suspended state loosely adhering to the substrate. Pigments and reinforcing agents are prepared in this manner. If nucleation is allowed on the surface of a substrate, it is possible, with proper control of the process conditions, to prepare thin, adherent, strong, and impermeable layers. Many metals and carbides as well as borides of tungsten, vanadium, and zirconium can be coated in this way. Two important nonmetallic coatings of interest, carbon and silicon carbide, will be discussed here.

a. *Carbon Coating*

Carbon in principle can be deposited by heating any hydrocarbon gas, but improper gas-phase nucleation will result in unwanted soot. If a hydrocarbon gas under controlled partial pressure and flow is allowed to decompose on a heated substrate and nucleate, crystallites with a tendency to align themselves can be formed. A coating formed with parallel laminar array will be highly conductive and chemically inactive. A coating of this type will also hinder the passage of heat to the substrate. A coating formed by carbon is termed pyrographite or pyrocarbon, and this type of coating is generally used as a heat barrier in a corrosive environment at high temperatures, such as those prevailing in a rocket nozzle. High purity, absence of pores, and favorable crystal orientation will minimize the chemical reactivity of the pyrolitic-coated carbon. However, a problem exists due to anisotropic thermal expansion. This can be avoided by ensuring structural homogeneity during coating by controlling the gas flow and the partial pressure and also by selecting the right carrier gas. Thus, it is possible to obtain an isotropic pyrocarbon with equal thermal expansion coefficients in all directions. However, such an isotropic pyrolytic carbon, due its higher thermal conductivity, increases the surface chemical reactivity. Hence, isotropic pyrocarbon is preferred for specific applications, in particular in nuclear fuels under intense radiation. Isotropic pyrocarbon is dimensionally stable under neutron bombardment. The suggested reason for this is as follows. Expansion in the *c* direction of the crystallites, caused by displacement of the carbon coating in an isotropic pyrocarbon deposit, is compensated for by a corresponding contraction in the *a* and *b* directions, and this causes the layer to remain in a dimensionally stable condition. The isotropic pyrocarbon coating can, in principle, be prepared by decomposition of higher hydrocarbons as source gases or by reinforcing the anisotropic structure of pyrographite with graphite fibers.

b. Silicon Carbide

Silicon carbide (SiC) is one of the important nonmetallic coatings that imparts refractiveness. As SiC has an isodesmic diamond-like crystal structure and short strong bonds, the coating formed by SiC is usually strong, hard, and impermeable. A SiC coating that conforms to stoichiometric conditions is not difficult to achieve because SiC occurs as a single compound with an equal number of silicon and carbon atoms. Hence, a single component source that has equal atomic proportions of silicon and carbon is best suited for coating applications. Methyl trichlorosilane (CH_3SiCl_3) diluted in a carrier gas like hydrogen is the best choice for this purpose. In the pyrolytic decomposition of methyl trichlorosilane breaking of Si–C bonds can occur, resulting in the formation of carbon and silicon hydride. This would alter the thermal stability and the gaseous diffusion rate of the coating. A slight excess of one of the components, usually silicon, is formed. Under controlled conditions, free silicon formation can be eliminated. The addition of nitrogen as a carrier gas in the required proportion can remove free silicon and also increase the tensile strength, as it modifies crystal grain growth during pyrolytic deposition. Trace quantities of elements such as arsenic and phosphorous, when added during pyrolytic deposition, can also remove free silicon. They inhibit crystal growth and reduce the grain size.

2. Fluid Bed

a. Coating in Fluid Bed

So far, we have discussed pyrolytic carbon and SiC coatings. In the following discussion, we will consider the conditions that must be satisfied in order to prepare a coating with the appropriate properties for a specific application, such as nuclear fuel coating. In this respect, it is important to use the proper chemical reactor. For the purpose of coating carbon or SiC over particulate solids such as nuclear fuel particles, there can be no better choice than a fluidized bed reactor. Carbon and SiC, as a combined coating material, impart extreme refractoriness and chemical resistance. A nuclear fuel is necessarily coated to improve its mechanical, thermal, and irradiation properties. SiC, when coated over the fuel particles, increases strength and abrasion resistance, improves isotropy, and imparts better stability under irradiation. A pyrocarbon layer above the SiC coating protects it from dissociation at high temperatures. Nuclear fuel particles are usually coated with SiC sandwiched between two layers of adherent pyrocarbon.

A fluidized bed reactor for coating particulate materials was developed and the coating technique was discussed by Akins and Bokros.[83] The coating of pyrocarbon and SiC is a main area of research in the program to develop high-temperature gas-cooled reactors. Voice[84] provided an account of the coating methods and the properties of coatings such as pyrocarbon and SiC. He also developed the fluidization technique for coating microspheres of uranium. Several kilograms of such microspheres were reported to have been coated conveniently using a fluidized bed reactor. This technique is superior to any other technique in that the particles are coated uniformly under isothermal conditions. An added advantage of this technique is the capability of mixing several types of hydrocarbon gases along with the fluidizing gas stream, and this helps in forming a homogeneous mixture of a composite coating. Coating fuel particles with a composite of pyrocarbon and SiC is superior to coating individual layers. Pyrocarbon coating is usually accomplished in the temperature range 1400–2000°C. For coating with silicon carbide using methyl trichlorosilane, the recommended temperature is 1600°C.

b. Coating Properties

The structure and properties of a coating depend critically on temperature, reactant concentration, and flow conditions. Anisotropic pyrolytic carbon, in the course of its coating, can be codeposited with SiC by mixing a gaseous reactant source that contains silicon. Carbon and SiC are mutually insoluble, and under certain conditions SiC deposits as needles or shirkers perpendicular to the substrate. This results in an increase in the ultimate tensile strength (UTS). An increase in UTS from 6 to 60 MN m^{-2} can be accomplished by 20% loading of SiC in pyrocarbon. SiC, when encapsulated with graphite, is protected from dissociation and chemical reaction, thereby maintaining its mechanical properties even at temperatures above 2000°C. In view of the importance of isotropic pyrolytic carbon/silicon carbide codeposited in a fluidized bed, studies on the microstructure of such coatings were conducted by Kaae.[85,86] The sample coated was a disc immersed in a bed of fluidized particles, and the fluidizing gas stream was a mixture of propane methyl trichlorosilane and helium. The deposition temperature was kept in the range 1235–1390°C. The microstructure of the coating was found to be dependent on the silicon content. The microstructure of a carbon coating with 10% silicon content was found to be similar to that of pure carbon deposited under the same conditions. When the silicon content was between 16 and 34 wt%, larger silicon carbide particles (~1000 Å) were observed to be distributed nonuniformly.

With specific control of the coating furnace, it would be possible to codeposit SiC with a morphology comprised of platelets or crystal aggregates rather than whiskers or needles with orientation perpendicular to the substrate. This kind of morphology, which is useful in abrasion- as well as oxidation-resistant coatings, is obtained at the cost of strength and thermal expansion. An important application of such a coating is in artificial heart valves. Pyrocarbon as well as codeposited SiC are compatible with living tissue. Hence, prosthetic devices with a pyrolytic carbon coating can be used in the human body indefinitely without reaction.[87] Another important property of pyrocarbon is its lubrication, which is similar to that of graphite. Hence, a coating of pyrolytic carbon over the surface of a loaded bearing would enable it to slide smoothly, and a coating with codeposited SiC would almost eliminate the wear of pyrocarbon. As a result, this kind of durable lubricated surface is far superior to metal-to-metal or plastic-to-metal joints.

c. Recommendations

So far, we have described the importance of carbon and SiC coatings and the use of the fluidized bed as unique choice for such applications. Coating of nuclear fuel particles for high-temperature gas-cooled reactors is an important step in fuel preparation technology. In this context, there are several classified as well as unclassified reports. Although we have not discussed proprietary literature in any detail, this subject continues to be of interest to researchers. Thermodynamic analyses of this type of coating were reported by Minato and Fukuta.[88] More recently, Wood et al.[89] discussed the use of fluidized bed reactors in the context of chemical vapor deposition. In summary, the gas fluidized bed reactor is the right or probably the only choice for the coating of particulate solids by the pyrolysis of one or more gaseous hydrocarbon sources as the reactant.

V. FLUIDIZATION IN ZIRCONIUM EXTRACTION

A. Breaking of Zircon

1. Chlorination

Fluidization technology can be used in various stages in the extraction of zirconium. The principal ores of zirconium are zircon and baddelyte. Zircon is the orthosilicate $ZrSiO_4$, and it is associated with 2–3% of hafnium. Zircon is more abundant than baddelyte. Hence, the starting ore for most zirconium industries is zircon. It is often difficult to open stable zircon and extract the metal values. The various methods used to break down this ore are (1) alkali fusion, (2) carbiding or carbonitriding, (3) thermal decomposition in a plasma furnace, and (4) chlorination. Depending on the desired end product, any of these methods can be adopted. For example, zirconium metal or its oxide, without removal of the hafnium contained, can be sold for nonnuclear applications. In such cases, methods other than chlorination can be used to break down the ore. However, chlorination is an unavoidable step in the production of high-purity zirconium metal. The neutron-absorbing hafnium in zirconium must be removed until its concentration is brought down to 100 ppm or less for the application of zirconium as a fuel-cladding material in thermal nuclear reactors. Zirconium metal must be obtained by the reduction of its tetrachloride by magnesium metal, and this reduction is often called the Kroll process, named for its inventor. Conventional reductants such as H_2, C, CO, Al, and

Ca cannot be used to reduce the oxide of zirconium as the affinity of oxygen for zirconium is very high.

2. Fluid Bed

The various methods commercially adopted are (1) solvent extraction, (2) fractional crystallization, and (3) pyrochemical process. Of these, the environmentally safe and economically competitive route is the pyrochemical route. The process flowsheet starts with direct chlorination of zircon, followed by purification and separation of chlorides of zirconium and hafnium in molten salt media. It is of particular interest to note here that chlorination is an inevitable step in zirconium metallurgy and that it can be carried out in a static or a fluidized bed. Chlorination of purified ZrO_2, obtained by solvent extraction, or direct chlorination of zircon, is preferably carried out in a fluidized bed reactor. ZrO_2 is chlorinated[90] in the temperature range 800–900°C, whereas zircon chlorination[91,92] takes place in the temperature range 1100–1200°C. A static bed reactor for the chlorination of ZrO_2 requires extensive preparation work for the feed pellets. Pellets prepared by blending with carbon and binding with starch require a coking step before chlorination. This adds to the cost of production of zirconium metal, in addition to the lower rate of production with a static bed reactor. A fluidized bed reactor eliminates all these problems and chlorination can be carried out directly by mixing carbon/coke particles of the proper size. Chlorination was covered in greater detail in Chapter 2 as part of the discussion on halogenation. An electrothermal fluidized bed reactor, due to its inherent capability of high-temperature operation (up to 3500°C), can chlorinate zircon directly. This reactor can even be used for carbiding and carbonitriding. An entire section in Chapter 4 is devoted to the function of an electrothermal fluidized bed and its various applications.

B. Miscellaneous Uses

In the process of purification as well as separation of hafnium by the aqueous route, the steps often encountered are drying and calcination. For example, in solvent extraction, the final stage is neutralization of the zirconium-loaded extracts, followed by drying and calcination. Drying is usually carried out using trays stacked in a furnace and calcination is carried out in a rotary kiln. Both operations can be carried out conveniently in a single fluidized bed reactor. The drying operation is an important step after leaching the alkali-fused frit of zircon. Fluidized bed drying is well known and can be conveniently adopted. Fluidized bed dryers were discussed in detail in Chapter 2. In India, alkali fusion of zircon was tested successfully in a fluidized bed. In this process, alkali is sprayed over the hot fluidized bed of zircon. Fusion takes place on individual particles, and the resultant product is a fine particulate solid that can be subjected to leaching conveniently in the next step. An important application of the fluidized bed is in the condensation of sublimable $ZrCl_4$ vapor. Zirconium tetrachloride exiting from the chlorinator is condensed conventionally in space condensers. They occupy a large floor area and condensation efficiency is low. The condensed chloride of zirconium is fluffy. The Kroll reduction furnace for such low-density chloride is deliberately made voluminous. An additional problem with a space condenser is its poor condensation efficiency when the chlorides from the fluidized bed chlorinator, in particular from the electrothermal fluidized bed reactor, emerge along with a noncondensable gas such as nitrogen that has been used as a diluent with chlorine during fluidization. A likely solution to this problem is the use of a fluidized bed condenser. The hot $ZrCl_4$ vapors are cooled down by direct contact with a chilled nitrogen gas stream, which also serves to fluidize the condensed particulate $ZrCl_4$. The product obtained is free flowing, globular in shape, and dense. Thus, the size of the Kroll reactor can be made compact for the same duty and it can be operated continuously. Research in this area has not been given much attention to date even though its importance has long been recognized, as evidenced by an early publication on this topic.[93]

VI. FLUID BEDS IN NUCLEAR FUEL REPROCESSING

A. Fuel Reprocessing

1. *Methods*

Once used in a nuclear reactor, uranium or its compound contains unburned fuel along with the fission products (~3% by weight) and artificially formed fuel such as plutonium. Reprocessing of the highly radioactive spent fuel must be carried out remotely. The purpose of reprocessing spent fuel is essentially to separate valuable plutonium which can be further used in high flux power reactors as well as in breeder reactors. Thus, reprocessing of spent fuel has commercial importance in terms of its fuel value and its vital role in the nuclear fuel cycle. There are, in principle, three ways of processing the fuel: (1) aqueous, (2) nonaqueous, and (3) a combination of both aqueous and nonaqueous.

a. *Aqueous Methods*

In the aqueous method, reprocessing starts with decladding the fuel elements either by chemical or mechanical methods, followed by dissolution of the spent fuel in nitric acid and solvent extraction of nitrate solutions of uranium, the fission products, and the valuable plutonium component. The aqueous route that accomplishes the separation of plutonium and unburned uranium from the fission products in the spent fuel from a nuclear reactor is popularly known as the Purex process[94] and it involves solvent extraction and ion-exchange methods. Details of the Purex process can be found elsewhere in the literature.[94] Currently, most of the fuel reprocessing to recover plutonium from the spent fuel of commercial power reactors is carried out by this popular method. One of the serious disadvantages of the aqueous route of fuel reprocessing is the problem of decontaminating and disposing of the voluminous quantities of highly radioactive liquid effluent. In today's atmosphere of environmental consciousness and strict international safety standards in the disposal of radioactive waste, the aqueous route has many imposed constraints with regard to storage and disposal of fuel reprocessing wastes. Therefore a new concept for fuel reprocessing entirely through a nonaqueous technique was developed as an alternative to the aqueous route.

b. Nonaqueous Methods

The nonaqueous route, widely known as the fluoride volatility process, uses the fluidized bed reactor in several stages. A schematic of a fluidized bed fluoride volatility process for use in fuel reprocessing and in recycling of the recovered fuel is shown in Figure 3.9. The fluoride volatility process involves the following steps:

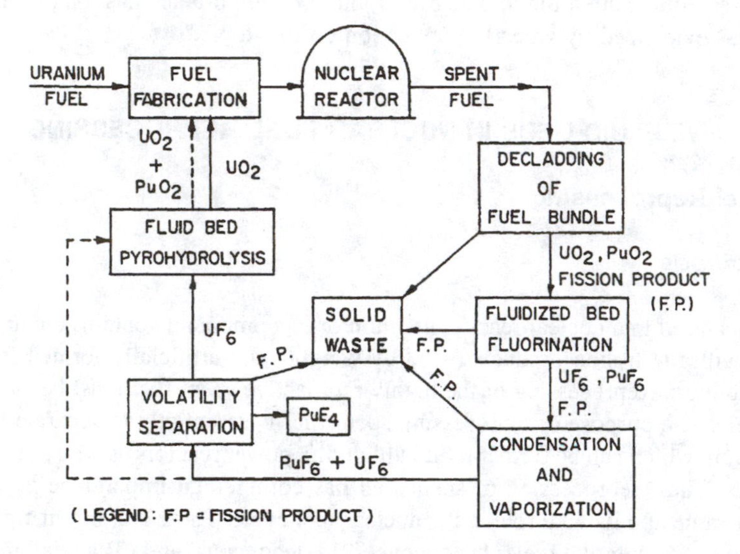

Figure 3.9 Schematic of fluidized bed fluoride volatility process in fuel reprocessing and recycling.

(1) decladding of fuel elements, (2) fluorination of the spent fuel, (3) separation of the hexafluorides of uranium and plutonium, and (4) hydrolysis of the separated hexafluorides to produce the oxide fuel for recycling to the reactor. Fluorination and pyrohydrolysis are conveniently carried out in a fluidized bed reactor to combat the exothermicity of the reaction. The details of these processes were discussed in the preceding section on uranium extraction. The important difference here is the handling of highly radioactive feed materials and the separation of fission products simultaneously. A description of the fully nonaqueous route for processing spent nuclear fuel, in particular through the fluoride volatility process, can be found elsewhere in the literature.[95-97] It appears that there is no commercial-scale fluidized bed reactor in operation using the fluoride volatility process, although a good number of studies have been reported on laboratory or pilot-scale units. Despite the fact that fluoride processes are widely used for uranium enrichment and enriched uranium oxide fuel preparation, there seems to be some degree of apprehension in commercializing the fluoride volatility process. Because the fuel processing method is a

closely guarded secret, many technical details pertaining to this process are not disclosed in the open literature.

c. Combined Methods

Lambert and Dotson[26] discussed a novel combination of the aqueous and fluoride volatility process called Aquafluor, a service trademark of General Electric. This process was conceptually developed as early as February 1963 in light of the knowledge gained from the U.S. Atomic Energy Commission's development work on reprocessing and experience at Argonne National Laboratory on the fluoride volatility process. The Aquafluor process was conceived with the idea of setting up a process and equipment characterized not only by technological excellence but also by commercial competetiveness, guarding against early obsolescence.

The Aquafluor process at the Midwest Fuel Recovery Plant in the United States adopted the following five main steps: (1) leaching of the decladded fuel, (2) aqueous extraction and decontamination of uranium, plutonium, and neptunium (3) selective extraction of plutonium and neptunium by ion exchange, (4) calcination of uranyl nitrate hexahydrate (UNH), and (5) fluorination of UO_3 to UF_6. Added to the main steps were processing and storing the radioactive wastes. Fluidized bed operations were incorporated in four steps: (1) calcination of UNH, (2) fluorination of UNH, (3) calcination of high-level radioactive waste, and (4) fluoride disposal. Application of the fluidized bed for calcination of UNH and fluorination is a reasonably well-accepted concept in the extraction of uranium. However, application of fluid bed processes for waste incineration and fluorine disposal constitutes an upcoming technological advancement, although it was recommended two decades ago. The Midwest Fuel Recovery Plant was set up using fluid bed systems in order to reap the advantages and gain a better understanding of such systems in a radioactive environment. However, information on commercial-scale fluid bed operation remains unpublished and is classified.

2. Fluoride Volatility Process

a. Process Description

The fluidized bed fluoride volatility process was developed for the purification of impure plutonium[98] obtained from impure oxide, casting skull, incinerator ash, and miscellaneous residues. Dow Chemical Company, a contract operator for the U.S. Atomic Energy Commission's Rock Flat Plant near Denver, Colorado, was involved in the development of this purification process. The details of the process were described by Standifer.[98] From the data on fluoride volatility and the fluid bed operating experience gained at Argonne and Oak Ridge National Laboratories, purification of impure plutonium as well as reclamation of plutonium from its scrap was adopted as an alternative to the aqueous route. The reasons for adopting the fluoride volatility process using the fluidized bed were (1) fewer steps and less land needed, plus low levels of exposure to neutron radiation; (2) almost complete recovery of plutonium, which is difficult to achieve in the aqueous route; and (3) minimal aqueous and solid waste.

The process involves direct fluorination of oxides. Hence, scrap metal values are oxidized and impure plutonium compounds are obtained from incinerator ash. Subsequently, hydrofluorination is used to remove silicon contamination. The resultant mixture of PuO_2 and the hydrofluorinated material is then fluorinated in a fluidized bed to PuF_6 and subsequently hydrogen reduced to obtain purified PuF_4, which is finally subjected to bomb reduction using calcium metal for the production of pure plutonium metal. Because plutonium is a highly toxic metal, handling is carried out in glove boxes kept under negative pressure. Furthermore, because plutonium-239 is a fissile material, equipment should be of the proper size to ensure that the mass of the plutonium metal remains below the critical mass.

b. Fluid Bed

The fluidized bed reactor chosen for the fluorination was 50 mm in diameter and 1220 mm long with a 125-mm-diameter disengaging section fitted with porous metal filters. A conical-shaped fluidized bed reactor with a 4° apex angle was used. The reactor outlet was connected with cold traps kept at –70°C by liquid nitrogen and the PuF_6 was condensed. Sodium fluoride traps for collecting uncondensed PuF_6 and KOH scrubbers to remove hydrogen fluoride were used. Unreacted fluorine was removed by reaction with charcoal. The fluid bed reactor was heated externally by a three-zone muffle furnace. The incoming reactant gas was preheated. The initial reaction of fluorine with PuO_2 was carried out by gradually increasing the fluorine concentration. This reaction is highly exothermic. Hence, a bed diluent like alumina was used. In a conical fluidized bed, the diluent bed material was not found necessary due to vigorous good mixing. The reaction of PuO_2 to PuF_4 was rapid, and its exothermicity caused the bed temperature to rise from 450 to 800°C within 1 min. The bed temperature later came down to 550°C in 5 min. Fluorination in the first stage during the conversion of PuO_2 to PuF_4 was carried out with a mixture of argon and fluorine. When the bed was completely converted to PuF_4, only fluorine gas was used. Unreacted fluorine was recycled by a reciprocating diaphragm compressor. The fluid bed, the cold trap, and the reductor were made of nickel 200. All tubings were 10-mm Monel, compression fitted. All valves were also made of Monel. Globe valves of a bellow seal type were operated either manually or pneumatically. Standifer[98] gave a detailed description of the experiments and the results pertaining to oxidation–fluorination/hydrofluorination in a fluidized bed. He concluded that this fluidized bed, in combination with the fluoride volatility process, is economically favorable compared to the aqueous route. However, a cost comparison between a pilot-plant and an industrial-scale unit could not be made because it was difficult to predict labor costs. Nevertheless, this process has been recommended to replace the aqueous process.

B. Novel Fluid Bed Fuel Reprocessing

1. *Uranium–Aluminum/Uranium–Zirconium Fuel*

Processing of metallic fuels such as uranium–aluminum/uranium–zirconium using a fluidized bed reactor was studied by Chilenskas et al.[99] In this process, a fluidized bed with alumina as the diluent was used to control heat dissipation during the hydrochlorination of the fuel immersed inside the bed. Aluminum/zirconium in the fuel was converted into their volatile chlorides, and these could be separated easily from the nonvolatile UCl_3 that also formed during hydrochlorination. Subsequent to the chlorination step, the residual bed containing nonvolatile UCl_3 was converted to gaseous UF_6 by fluorination, thereby emptying the bed of all the solid fuel originally charged into it. Thus, a single fluidized bed could not only handle the dual reactions, namely, hydrochlorination and fluorination, but also served to separate the fuel-alloying elements.

2. *Oxide Fuel*

Bourgeois and Nollet[100] described a fluidized bed dry route for separating uranium and zirconium. Zircaloy-clad uranium oxide fuel is decladded[101] by hydrofluorination in a fluidized bed of alumina. Uranium oxide pellets left after decladding are then subjected to oxidation in the fluidized state using air at 450°C. The volatile $ZrCl_4$ obtained by hydrofluorination can be pyrohydrolyzed to obtain ZrO_2. This is a single-step process to separate uranium and plutonium from a mixed fuel by fluorination using BrF_5 in a fluidized bed. Uranium values are converted into volatile UF_6 and plutonium values remain as solid PuF_4. Studies on fluorination of uranium oxide fuels using interhalogen BrF_5 in a fluidized bed were reported by Holmes et al.[50]

3. *Carbide Fuel*

The potential of a fluidized bed has been established for the reprocessing of carbide fuels[102-104] as well. The carbide fuel, immersed in a fluidized bed of hot alumina, is processed by burning the carbon matrix, leaving behind uranium as U_3O_8. The uranium oxides, upon further treatment either by conventional aqueous acid leaching and solvent extraction or by the nonaqueous route, can be reextracted and purified for fuel recycle. We have already mentioned the use of remotely operated fluidized beds for UNH calcination and fluorination at the Midwest Fuel Recovery Plant[26] in the United States. From the foregoing brief outlines, it may be inferred that the fluidized bed is a favorite choice in fuel reprocessing. As a quick reference on the general application of fluid bed technology in fuel reprocessing, the application areas of this unit are compiled in Figure 3.10.

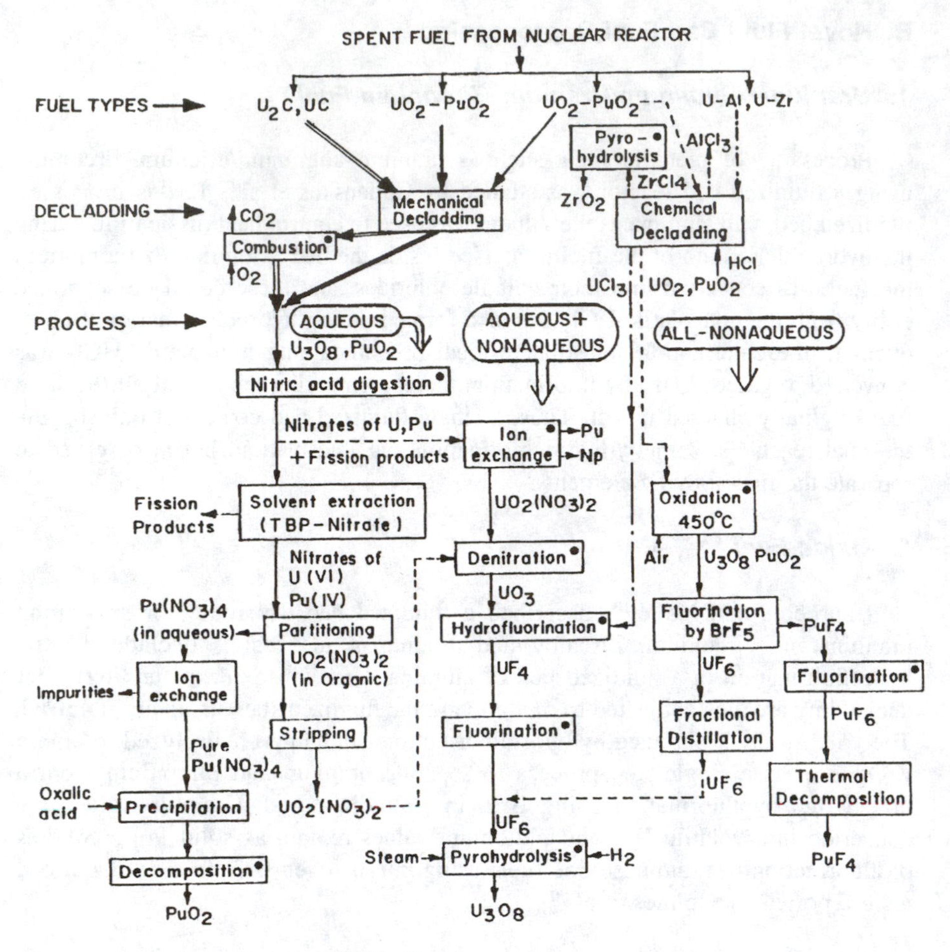

Figure 3.10. Fluidization in nuclear fuel processing.

VII. FLUIDIZATION IN WASTE PROCESSING AND POLLUTION ABATEMENT

A. Fluid Bed

1. Selection

Nuclear wastes of different forms and types are generated by nuclear fuel fabrication, nuclear power reactor operation, and fuel reprocessing plants. The nuclear wastes are radioactive as well as toxic and must be treated safely, without adverse effects to the environment or the biological system. The goal of waste disposal is normally volume reduction of the collected waste and disposal of the treated form of safe and stable waste at minimal quantities in the environment. One important method of treating the waste is to incinerate the solid or liquid form and reduce the volume substantially. Fluidized bed incineration is superior to other conventional waste disposal systems such as rising pot glass, continuous glass, spray calcination, and glassification. The major advantages of fluid bed incineration are efficient combustion, flexibility of operation, the ability to handle different forms and types of feeds, easy control, and low capital and operational cost.

2. Incineration

a. Hazardous Waste

In hazardous waste treatment, fluidized bed incineration helps to reduce the emission of noxious gases and metals, in addition to saving fuel. Application of the fluidized bed for destruction of industrial hazardous wastes started in 1980. A detailed review and the design aspects of the fluid bed incinerator for hazardous waste destruction was presented by Mullen.[105] A critical review of hazardous waste disposal by incineration was provided by Oppelt.[106] Fluidized bed waste incinerators are used for organic wastes such as petroleum tank dregs, waste plastics, waste oil and solvents, and inorganic wastes such as spent activated carbon, precoat filter cake, and carbon black. Chemical process wastes and sludges from the paper and pulp and pharmaceutical industries are also treated by fluidized bed waste incineration. In 1992, around 400 fluidized bed incinerators were in operation worldwide for treatment of industrial and municipal waste, and the size of the reactors varied from 1.2 to 15 m in diameter.

b. Organic Wastes

The fluidized bed incinerator, in view of its excellent heat transfer properties and isothermal operation, has an edge over other conventional incinerators specifically for application in nuclear waste processing.[107] The heating of incinerators which destroy the materials that have less calorific value is accomplished by burning hydrocarbon gases or by using liquid sodium–potassium eutectic metal as the heat transfer waste medium. Organic wastes such as spent tributyl phosphate (TBP) and ion-exchange resins from nuclear fuel reprocessing plants can be successfully incinerated using a fluidized bed. The authors of this book have designed and operated a 100-mm-diameter continuous fluidized bed incinerator for the disposal of spent ion-exchange resin. The reactor was made of Inconel 600 and designed with provisions to handle liquid feed. The liquid feed was sprayed over the hot inert fluidized bed. The inert bed material had to be changed to a reactive bed of dolomite solid when the ion-exchange resin was sulfonated. This enabled the *in situ* capture of sulfur by the dolomite during the incineration of the resin. It was reported that the waste calcination facility located at the National Reactor Testing Facility achieved[26] a volume reduction of six to ten. The off-gas from the incinerator was cleaned up before letting into the atmosphere. The cleaning operation was carried out using a series of cyclones, venturi scrubbers, and filters. Plutonium from the incinerator ash was recovered[108] by oxidizing the ash, followed by dissolution in a nitric acid plus fluoride ion (i.e., HF) solution.

3. *Off-Gas Treatment*

The fluidized bed reactor was tested[26] for removing fluorine and HF from off-gases. The reaction of fluorine in a fluidized bed of charcoal helps in removing 99% of F_2. This method is superior to scrubbing by KOH or capturing F_2 by a bed of alumina. KOH scrubbing leads to aqueous waste disposal problems and absorption by alumina leads to caking of the bed. Lambert and Dotson[26] advocated the use of a fluidized bed as a unit operation for waste disposal.

B. Criteria for Incinerator

1. *Constraints*

The general criterion for an incinerator is that it should destroy and efficiently remove the primary organic hazardous constituents in the waste. The destruction and removal efficiency should be of the order of 99.99%. The off-gas from the incinerator should not contain harmful HCl gas and particulate metals. An incinerator, in principle, cannot be operated at high temperatures. The operating temperature should be below the melting point of the ash and the other constituents of the bed. Hence, the operating temperature of an incinerator is restricted by the nature of the feed and the product, that is, either ash or eutectic formed in the reactor. It may be possible to increase the operating temperature by adding a third component in the reactor that imparts refractoriness to the bed ash or increases the melting point. For example, the formation of a sticky mass with a low melting point by the reaction of alkali metal values with silica can be eliminated by the addition of calcium hydroxide along with the feed. This would result in modification of the incinerator bed components and increase the melting point.

2. *Fluid Bed*

A fluid bed incinerator operates at either a subautogenous or an overautogenous condition, depending on the calorific value of the feed material. In a subautogenous condition, heat has to be supplied to the reactor either by preheating the inlet air or by fuel injection into the bed. This is essential to sustain the incineration. In an overautogenous incinerator, on the other hand, heat must be removed either by direct water injection or by an immersed coil heat exchanger. If the heat is to be recovered efficiently, water injection should be avoided. A fluid bed incinerator has the advantage of accomplishing the desired destruction and removal efficiently by operating at a few hundred degrees below the temperature normally required in other incinerators. The problem of vaporization of metal components, in particular lead and cadmium, is eliminated by close control of the temperature to avoid melting the incinerator bed material and operation at relatively low temperatures. A vaporized metal, after escaping from the off-gas treatment systems, is a cancer-causing agent. Hence, a fluid bed, by not allowing the vaporization of metal (which in a fluid bed finally exists as particulate solid) eliminates serious cancer hazards. As the destruction and removal efficiency of organic wastes is high, no afterburner to further improve the efficiency is required. Thus, a saving in fuel cost together with an additional saving in operating cost makes the fluid bed incinerator a favorable choice compared to any other incinerator. A fluid bed incinerator has to be provided with an arch-type refractory material distributor instead of a metal grid to avoid melting and maldistrubution of the grid plate. Mullen[105] reported that a number of fluid bed incinerators have been in operation for over 20 years without refractory replacement or with only minor repairs.

C. Gas and Liquid Waste

1. Off-Gas

Fluidized beds have proven to be useful for the absorption of sulfur dioxide by alkalized alumina. The reaction of hydrogen gas with sulfur dioxide absorbed by alkalized alumina helps in desorbing the SO_2. The reactions for sorption and regeneration in a fluidized bed are

$$\text{Sorption: } 2NaAlO_2 + SO_2 + 1/2O_2 \rightarrow Na_2SO_4 + Al_2O_3 \quad (3.41)$$

$$\text{Regeneration: } Na_2SO_4 + Al_2O_3 + 4H_2 \rightarrow 2NaAlO_2 + 3H_2O + H_2S \quad (3.42)$$

Town et al.[109] conducted sorption and regeneration studies by removing sulfur from simulated power plant flue gas. The sulfur dioxide removed ranged from 90 to 95% for a flue gas containing 0.2–0.5% SO_2. All the sorbed SO_2 from the alkalized alumina pellets that constituted the fluidized bed could be efficiently removed by reaction with hydrogen at 730°C for 30 min. Fluidized beds are reported[110,111] to have been successfully used to remove submicron particulate solids from gas streams. In a conventional filter system, the filter media have to be removed after some period of service and replaced to avoid excess pressure drop and poor quality filtration. This kind of replacement and repair operation necessitates shutdown of the system. A fluidized bed, on the other hand, uses a granular bed of solid material to filter the submicron particles continuously without any excessive pressure drop due to the dust-laden fluidizing gas stream. Black and Boubel[110] reported 90% effectiveness of a fluidized bed to remove submicron aerosols from the gas stream. In order to increase the efficiency of a fluidized bed filter, Jackson[111] recommended replacement of the single once-through fluidized bed by a multistage type, mainly to increase the gas–solid contact and decrease bypass of the dust–laden gas as bubbles through the bed. Several applications of the fluidized bed for pollution control with the purpose of arresting the degeneration of the environment by various wastes let out by polluting industries were discussed by Hanway.[112] Successful applications of the fluidized bed in the disposal of municipal sewage, petroleum refinery waste, and pulp mill spent effluents have also been reported.

2. *Liquid Waste*

Industrial wastewater treatment produces large amounts of biological sludge which contains mainly dead microbial cells, inorganic salts, inert ash, and a high percentage of water. The sludge has to be dewatered and disposed of. The disposal of biological sludge is a formidable problem because of large space requirements. The disposal is usually effected by onsite or offsite landfill, ocean dumping, and compositing. Due to environmental problems, such disposal methods must be curtailed and new methods developed. One method is incineration, using rotary kilns, single or multihearth furnaces, and moving grate, cyclonic, or fluidized bed incinerators. For wastes such as biological and municipal sludges, which contain high amounts of water, fluidized bed incinerators are usually recommended. Ho et al.[113] studied the kinetics of biological sludge incineration in a fluidized bed and strongly recommended its use for such applications. Baeyens and Geldart[114] developed a model for a fluidized bed incinerator for the complete combustion of hydrocarbon wastes. Design equations were proposed relating the freeboard velocity with the freeboard height required for complete burnout of combustible sludge. The predictions of this model showed reasonable agreement with data obtained from several commercial incinerators.

D. Reclamation from Waste

As the primary resources required for nuclear materials production are rapidly being depleted, it is imperative to look into alternative secondary resources. It is also highly advantageous to recover useful components from the scrap generated at various stages of nuclear fuel fabrication. One of the best examples is reclamation of zirconium metal from its scrap. Fluidized bed chlorination of the scrap to recover zirconium as $ZrCl_4$ is an important and accepted method. Similarly zirconium and uranium can be obtained from uranium–zirconium scraps that contain 6% uranium. Hydrochlorination of the scrap in a fluidized bed followed by fluorination can recover zirconium and uranium as the volatile halides $ZrCl_4$ and UF_6, respectively. The volatile halides can be easily separated by condensation at the exit of the fluidized bed reactor. There is no published work on this. In a recent work,[115] the use of a fluidized bed reactor for burning zirconium scrap into its oxide has been demonstrated and recommended for the recovery and recycle of zirconium in the nuclear industry.

NOMENCLATURE

A	cross-sectional area of the column (m^2)
A_{min}	minimum cross-section ratio (m^2)
$\underline{A}_{min}$	ratio of A_{min}/A_t (–)
A_t	column cross-sectional area at U_t (m^2)
Bi	Biot number, $h/(kx_f)$

$C_{F_2,0}$	initial concentration of fluorine (mol/lit at reaction temperature)
D_p	particle size (m, mm in Equation 3.13)
E	activation energy (J/mol)
F	feed rate of fluorine (mol/hr)
h	heat transfer coefficient ($W/m^2 \cdot K$)
K	reaction rate constant (Equation 3.20)
K_{eq}	equilibrium constant (–)
k	thermal conductivity of resin ($W/m \cdot K$)
K_0	Arrhenius constant [Equation 3.20 l F_2 at 273 K/(hr · g UF_4)]
L	mass flow rate of fluid (kg/m^3)
N	ratio $(L/S)(\rho_s/\rho_f)$
n	Richardson index for voidage factor (Equation 3.5)
n_{F_2}	number of moles fluorine (–)
R	gas law constant, $Pa \cdot m^3/(mol \cdot K)$
r	reaction rate for UF_4 (Equation 3.18) ($mol/cm^2 \cdot s$)
S	solid flow rate (kg/m^3)
T	temperature (K)
T_b	bed temperature (K)
T_f	melting temperature of resin (K)
T_w	wall temperature (K)
U	liquid velocity (m/s)
U_f	fluid velocity (m/s)
U_{mf}	minimum fluidization velocity (m/s, cm/s in Equation 3.13)
U_{rf}	relative fluid velocity as defined by Equation 3.4
U_s	solids velocity (m/s)
$\bar{U}_s$	velocity ratio, U_s/U_t at $\bar{A}_{min}$
U_t	particle terminal velocity (m/s)
V_t	reactor volume (Equation 3.8)
W	weight of UF_4 (kg, g in Equation 3.20)
X_e	conversion of fluorine at the exit of the reactor (–)
x	fractional conversion of fluorine (–)
x_f	final coating thickness (m)
Z	column height (m)
Z_t	column height required to process solid at its mass flow rate corresponding to U_t

Greek Symbols

ΔH	heat of reaction (kcal/mol)
ϵ	bed voidage (–)
θ	dimensionless temperature (–)
θ_p	particle residence time(s)
θ_t	solids residence time(s)
ρ_l	liquid density (kg/m^3)
ρ_s	density of solid (kg/m^3)

REFERENCES

1. Slater, M.J., A review of continuous counter-current contactors for liquids and particulate solids, *Br. Chem. Eng.*, 14(1), 41, 1969.
2. Rickles, R.N., Liquid–solid extraction, *Chem. Eng.*, 72(6), 157, 1965.
3. Kwauk, M., Particulate fluidization: an overview, in *Advances in Chemical Engineering*, Vol. 17, Wei, J., Ed., Academic Press, New York, 1992, chap. 2.
4. Yakubovoch, I.A., Tyuffin, E.P., and Tolkachev, V.A., Separation and washing of sediments in a column containing a fluidized bed, *Khim. Neft. Mashinostr.*, 9, 14, 1970.
5. Korsunski, V.I., D'yachko, A.G., and Svetozarova, G.I., Calculation of the fluidized bed leaching of zinc sinters with considerations of mixing of the solid phase, *Tsvetn. Metal.*, 43(5), 21, 1970.
6. Burovoi, I.A., Dracheva, T.V., and Ibraev, A.Kh., Semi industrial testing of the fluidized bed acid leaching of zinc calcines, *Tsvetn. Metal.*, 44(10), 16, 1971.
7. Kwauk, M., Particulate fluidization in chemical metallurgy, *Sci. Sin.*, 16, 407, 1973.
8. Kwauk, M., *Fluidized Leaching and Washing* (in Chinese), Science Press, Beijing, 1979, 19 (cited in Ref. 3).
9. Kwauk, M. and Wang, Y., Fluidized Leaching and Washing, I, Chem. Eng. Symp. Ser. No. 63, 1981, D4/BB/1-21.
10. Zhang, F. and Zheng, X., Polydisperse behaviour of leached uranium ore in fluidized classifying and washing column, in Proc. China–Japan Fluidization Symp., Hangzhou, China, April 4–9, 1982.
11. Kwauk, M. and Wang, Y., Fluidized Leaching and Washing, I, Chem. Eng. Symp. Ser. No. 3, 1981, D4/BB/1.
12. Legler, B.M., Fluidized bed processing in the nuclear fuel cycle, *Chem. Eng. Prog. Symp. Ser.*, 66(105), 167, 1970.
13. Page, H., Shortis, L.P., and Duke, J.A., The processing of uranium ore concentrate and recycle residues to purified uranyl nitrate solution at Springfields, *Trans. Inst. Chem. Eng.*, 38, 184, 1960.
14. Alexander, R.C., Shortis, L. P., and Turner, C.J., The second uranium plant at Springfields, *Trans. Inst. Chem. Eng.*, 38, 177, 1960.
15. Hawthorn, E., Shortis, L.P., and Lloyd, J.E., The fluidized solids dryway process for the production of uranium tetrafluoride at Springfields, *Trans. Inst. Chem. Eng.*, 38, 197, 1960.
16. Robinson, S.N. and Todd, J.E., Plant Scale Fluid Bed Denitrator, Report MCW 1509, U.S. Atomic Energy Commission, Washington, D.C., 1966.
17. Jonke, A.A., Report ANL-5363, U.S. Atomic Energy Commission, Washington, D.C., 1954.
18. Grimmett, E.S., Calcination of Al-Type Reactor Fuel Waste, Report IDO-14416, U.S. Atomic Energy Commission, Washington, D.C., 1957.
19. Philoon, W.C., Samders, E.F., and Trask, W.T., Uranium trioxide in a fluidized bed reactor, *Chem. Eng. Prog.*, 56(4), 106, 1960.
20. Otero, A.R., Rof, J.S., and Lago, E.C., Fluidized bed calcination of uranyl nitrate solutions, in *Proc. Int. Symp. on Fluidization*, Drinkenburg, A.A.H., Ed., Netherlands University Press, Amsterdam, 1967, 769.
21. Otero, A.R. and Garcia, V.G., Cake formation in a fluidized bed calciner, *Chem. Eng. Prog. Symp. Ser.*, 66(105), 267, 1970.
22. Bjorklund, W.J. and Offutt, G.F., Fluidized bed denitration of uranyl nitrate, *AIChE Symp. Ser.*, 69(128), 123, 1973.

23. Galkin, N.P., Sudarikov, B.N., Veryatin, U.D., Shiskov, Yu., and Maiorov, A.A., in *Technology of Uranium*, Galkin, N.P. and Sudarikov, B.N., Eds., Atomizdat, Moskva, 1964, chap. 10 (translated from Russian–Israel Program for Scientific Translation, Jerusalem, 1966).
24. Gonzalez, V. and Otero, A.R., Formation of UO_3 particles in a fluidized bed, *Powder Technol.,* 7, 137, 1973.
25. Damerval, F.B., Trask, W.T., and Sanders, E.F., Fluid Bed Denitration Process Development, Quarterly Report MCW-1431, 1959.
26. Lambert, R.W. and Dotson, J.M., Fluidized bed applications in the Midwest fuel recovery plant, *Chem. Eng. Prog. Symp. Ser.,* 66(105), 175, 1970.
27. Galkin, N.P., Maiorov, A.A., and Veryatin, U.D., *The Technology of the Treatment of Uranium Concentrates*, Pergamon Press, Oxford, 1963, chap. 8.
28. Smiley, S.H. and Brater, D.C., The development of a high capacity continuous process for the preparation of uranium hexafluoride from uranium oxides and ore concentrates, in Proc. 2nd Int. Cong. on Peaceful Uses of Atomic Energy, Paper 525, United Nations, New York, 1958, 153.
29. Melvanin, F.W., Canadian development work with moving bed reactor for reduction of uranium trioxide and hydro-fluorination to uranium tetrafluoride for subsequent production of metal, in Proc. 1st Int. Conf. on Peaceful Uses of Atomic Energy, Paper 229, United Nations, New York, 1958, 133.
30. Yemel'yanov, V.S. and Yevstyukhin, A.I., *Metallurgy of Nuclear Fuel*, Pergamon Press, London, 1969, chap. 14.
31. Ashbrook, A.W., The refining and conversion of uranium yellow cake to uranium dioxide and uranium hexafluoride fuels in Canada: current processes in advances in uranium refining and conversion, in Proc. of a Technical Committee Meeting on Advances in Uranium refining and Conversion, IAEA TECDOC-420, International Atomic Energy Agency, 1987.
32. Swaminathan, N., Rao, S.M., Sridharan, A.K., Sampath, M., Suryanarayanan, V., and Kansal, V.K., Recent advances and present status of uranium refining in India, in Proc. of a Technical Committee Meeting on Advances in Uranium Refining and Conversion, IAEA TECDOC-420, International Atomic Energy Agency, Vienna, 1987, 21.
33. Chang, I.S., Hwang, S.T., and Park, J.H., Status of uranium refining and conversion process technology in Korea, in Proc. of a Technical Committee Meeting on Advances in Uranium Refining and Conversion, IAEA TECDOC-420, International Atomic Energy Agency, 1987, 63.
34. Galkin, N.P., Mairov, A.A., and Veryatin, U.D., *The Technology of the Treatment of Uranium Concentrates*, Pergamon Press, Oxford, 1963, chap. 8.
35. Galkin, N.P. and Sudarikov, B.N., *Technology of Uranium*, Atomizdat, Moscow, 1964, chap. 11.
36. Costa, E.C. and Smith, J.M., Kinetics of noncatalytic non isothermal gas–solid reactions: hydrofluorination of uranium dioxide, *AIChE J.,* 17(4), 947, 1971.
37. Levey, R.P., Jr., Garza, De La, A., Jacobs, S.C., Heidt, H.M., and Trent, P.E., Fluid bed conversion of UO_3 to UF_4, *Chem. Eng. Prog.,* 56(3), 43, 1960.
38. Anon., AEC releases: details of UF_4 process, *Chem. Eng. News,* 50, 1962.
39. Crohan, C.S., Process and technology, *Chem. Eng.,* July 11, 1960.
40. Ponelis, A.A., Slabber, M.N., and Zimmer, C.H.E., Conversion of non-nuclear grade feed stock to UF_4, in Proc. of a Technical Committee Meeting on Advances in Uranium Refining and Conversion, IAEA TECDOC-420, International Atomic Energy Agency, Vienna, 1987, 111.

41. Galkin, N.P. and Sudarikov, B.N., *Technology of Uranium,* Israel Program for Scientific Translations, Ltd., Jerusalem, 1966, chap. 12.
42. Vogel, G.J., Sandus, O., Steunenberg, R.K., and Mecham, W.J., Fluidized-bed techniques in producing uranium hexafluoride from ore concentrates, *Ind. Eng. Chem.,* 50(12), 1744, 1958.
43. Anastasia, L.J. and Mecham, W.J., Oxidation–fluorination of uranium dioxide pellets in a fluidized bed, *Ind. Eng. Chem. Process Des. Dev.,* 4(3), 338, 1965.
44. Gabor, J.D. and Mecham, W.J., Radial gas mixing in fluidized packed beds, *Ind. Eng. Chem. Fundam.,* 3, 60, 1964.
45. Gabor, J.P. and Mecham, W.J., Fluidized-Packed Beds: Studies of Heat Transfer, Solids–Gas Mixing and Elutriation, Report 6859, U.S. Atomic Energy Commission, Washington, D.C., 1965.
46. Ziegler, E.N. and Brazelton, W.T., Mechanism of heat transfer to a fixed surface in a fluidized bed, *Ind. Eng. Chem. Process Des. Dev.,* 2, 276, 1963.
47. Gabor, J.D., Mecham, W.J., and Jonke, A.A., Heat transfer in fluidized-packed beds as applied to the fluorination of uranium dioxide pellets, *Chem. Eng. Prog. Symp. Ser.,* 60(47), 96, 1964.
48. Gabor, J.D. and Ramaswami, D., Exploratory Tests on Fluorination of Uranium Oxide with BrF_5 in a Bench Scale Fluid Bed Reactor System, Report ANL-7362, U.S. Atomic Energy Commission, 1967.
49. Holmes, J.T., Koppel, L.B., Saratchandran, N., Strand, J., and Ramaswami, D., Pilot Studies on Fluorinating Uranium Oxide with Bromine Pentafluoride, Report ANL-7370, U.S. Atomic Energy Commission, 1968.
50. Holmes, J.T., Koppel, L.B., Gabor, J.D., Ramaswami, D., and Jonke, A.A., Engineering scale studies of the fluorination of uranium oxide with bromine penta fluoride, *Ind. Eng. Chem. Process Des. Dev.,* 8(1) , 43, 1969.
51. Corella, J., Kinetics of Fluorination of Uranium Tetrafluoride in Fluidized Bed (in Spanish), Doctoral thesis, University of Madrid, 1969.
52. Otero, A.R. and Corella, J., Fluidization of mixtures of solids of different characteristics, I. Velocities of fluidization, *Quimica,* 67(12), 1207, 1971.
53. Corella, J. and Otero, A.R., Kinetics of fluorination of UF_4 in a fluidized bed (in Spanish), *Quimica,* 68(1), 63, 1972.
54. Broadley, J.S. and Longton, P.B., The Reaction of UF_4 with UF_6, Report TN-60, United Kingdom Atomic Energy Authority R&DB(c), 1954.
55. Labaton, V.Y., A Kinetic Study of the Fluorination of UF_4 by F_2, Report 8120, United Kingdom Atomic Energy Authority R&DB(c), 1955.
56. Labaton, V.Y. and Johnson, K.D., Fluorides of uranium. 1. Kinetic studies of the fluorination of uranium tetrachloride by fluorine, *J. Inorg. Nucl. Chem.,* 10, 74, 1959.
57. Hua, Z.Z., The conversion from uranium tetra fluoride in a vertical fluorination reactor, *Chin. J. Nucl. Sci. Eng.,* 2, 1, 1982.
58. Hua, Z.Z. and Bao, C.L., The conversion from uranium tetrafluoride into hexafluoride in a vertical fluorization reactor, in Proc. of a Technical Committee Meeting on Advances in Uranium Refining and Conversion, IAEA TECDOC-420, International Atomic Energy Agency, Vienna, 1987, 141.
59. Hauth, J.J., Vibration compacted ceramic fuels, *Nucleonics,* 20(9), 50, 1962.
60. Mears, W.H., Townsend, R.V., Bradley, R.D., Turissini, A.D., and Stahl, R.F., Removal of some volatile impurities from uranium hexa fluoride, *Ind. Eng. Chem.,* 50(12), 1771, 1958.

61. Knudsen, I.E., Levitz, N.M., and Lawroski, S., Preliminary Report on Conversion of Uranium Dioxide in a One-Step Fluid-Bed Process, Report ANBL-6023, Argonne, National Laboratory, Argonne, IL, 1959.
62. Knudsen, I.E., Hootman, H.E., and Levitz, N.M., A Fluid-Bed Process for the Direct Conversion of Uranium Hexafluoride to Uranium Dioxide, Report ANL-6606, Argonne, February 1963.
63. Knudsen, I.E., Levitz, N.M., and Jonke, A.A., Preparation of Dense Uranium Dioxide Particle from Uranium Hexafluoride in a Fluidized Bed, Report ANL-6902, Argonne, National Laboratory, Argonne, IL, 1964.
64. Rand, M.H. and Kubaschewski, O., *The Thermochemical Properties of Uranium Compounds,* Interscience, New York, 1963.
65. Dergazarian, T.E. et al., JANAF Thermochemical Tables, Dow Chemical Company, Midland, MI, 1960.
66. Crane, J., Kalish, H.S., and Litton, F.B., The Development of Uranium Carbide as a Nuclear Fuel, Report UNC-5058, United Nuclear Corporation (Development Division), Washington, D.C., 1963, 47.
67. Roy, P.R., Plutonium metallurgy, *Trans. Indian Inst. Met.*, 39, 88, 1986.
68. Ganguly, C., Sol-gel microsphere pelletization: a powder free advanced process for fabrication of ceramic nuclear fuel pellets, *Bull. Mater. Sci.*, 16(6), 509, 1993.
69. Jakesova, L. and Jakes, D., Carbides, nitrides and silicides of uranium as promising nuclear fuel, *At. Energy Rev.,* 1(3), 3, 1963.
70. Accar, A., *AERE Trans.,* p. 958, 1953.
71. Petkus, E.J., Tevebaugh, A.D., Payne, C., and Bartos, J.P., The synthesis of stoichiomteric uranium mono carbide in a fluidized bed by the uranium metal–hydrocarbon gas reaction, *Chem. Eng. Prog. Symp. Ser.*, 62(67), 76, 1966.
72. Hyde, K.R., Landsman, D.A., Morris, J.B., Seddon, W.E., and Tulloch, H.J.C., The Preparation of Uranium Nitride and Carbonitrides from Uranium Oxide in a Fluidized Bed, U.S. Atomic Energy Research and Development Report ANL-6902, Argonne, National Laboratory, Argonne, IL, 1964.
73. Ross, S.B., El-Genek, M.S., and Mathews, R.B., Uranium nitride fuel swelling correlation, *J. Nucl. Mater.*, 170, 169, 1990.
74. Matzke, H.J., *Science of Advanced LMFBR Fuels*, North-Holland, Amsterdam, 1986, 740.
75. Grimmett, E.S., Ballard, R.K., and Buckham, J.A., *Chem. Eng. Prog. Symp. Ser.*, 63(80), 11, 1967.
76. Hogg, G.W. and Grimmett, E.S., A study of the chemical reactions associated with the fluidized bed production of uranium aluminide, *AIChE Symp. Ser.*, 67(116), 190, 1971.
77. Nienow, A.W. and Rowe, P.N., Particle growth and coating in gas-fluidized beds, in *Fluidization*, 2nd ed., Davidson, J.F., Clift, R., and Harrison, D., Eds., Academic Press, London, 1985, chap. 17.
78. Lee, M.M., Application of electrical insulation by the fluidized bed process, *Electro-Technol.*, 66, 149, 1960.
79. Richart, D.S., A report on the fluidized bed coating system. 2. Plastics for coating and their selection, *Plast. Des. Technol.*, 2, 26, 1962.
80. Pettigrew, C.K., Fluidized bed coating, *Mod. Plast.*, 44, 150, 1966.
81. Gutfinger, C. and Chen, W.H., Heat transfer with a moving boundary application to fluidized coating, *Int. J. Heat Mass Transfer*, 12, 1097, 1969.

82. Gutfinger, C. and Chen, W.H., An approximate theory of fluidized bed coating, *Chem. Eng. Prog. Symp. Ser.*, 66(101), 91, 1970.
83. Akins, R.J. and Bokros, J., The deposition of pure and alloyed isotropic carbons in steady state fluidized beds, *Carbon*, 12, 439, 1974.
84. Voice, E.H., Coating of pyrocarbon and silicon carbide by chemical vapour deposition, *Chem. Engr.*, 785, December 1974.
85. Kaae, J.L., Microstructures of pyrolytic carbon/silicon carbide mixtures co-deposited in a bed of fluidized particles, *Carbon*, 13, 51, 1975.
86. Kaae, J.L., Microstructures of isotropic pyrolytic carbons, *Carbon*, 13, 55, 1975.
87. Bokros, J.C., Lagrange, L.D., and Schoen, F.J., Presented at the Conference on Engineering Medicine, New Hampshire, 1970 (Gulf General Atomic Report GA 10100).
88. Minato, K. and Fukuta, K., Chemical vapour deposition of silicon carbide for coated fuel particles, *J. Nucl. Mater.*, 149(2), 233, 1987.
89. Wood, B.J., Sawurjo, A., Tong, G.R., and Swider, S.E., Coating by chemical vapour deposition in fluidized bed reactors, *Surf. Coat. Technol.*, 49(1–3), 228, 1991.
90. Sehra, J.C., Fluidized bed chlorination of nuclear grade ZrO_2 with CO and Cl_2, *Trans. Indian Inst. Met.*, 27(2), 93, 1974.
91. Manieh, A.A. and Spink, D.R., Chlorination of zircon sand, *Can. Metall. Qu.*, 12(3), 331, 1973.
92. Manieh, A.A., Scott, D.S., and Spink, D.R., Electrothermal fluidized bed chlorination of zircon, *Can. J. Chem. Eng.*, 52, 207, 1974.
93. Cibrowski, J. and Wronski, K., Condensation of sublimable materials in a fluidized bed, *Chem. Eng. Sci.*, 17, 481, 1962.
94. Purex Technical Manual, HW-31000, Hanford Atomic Product Operation, Part II, Richland, WA, March 25, 1955, chap. 6.
95. Vogel, G.J., Carls, E.L., and Mecham, W.J., Engineering Development of a Pilot Scale Facility for Uranium Dioxide and Plutonium Dioxide Processing Studies, Report ANL-6901, Argonne National Lab, 1964.
96. Anastasia, L.J., Gabor, J.D., and Mecham, W.J., Engineering Development of Fluid Bed Fluoride Volatility Processes. Part 3. Fluid Bed Fluorination of UO_3 Pellets, Report ANL-6898, Argonne National Lab, August 1965.
97. Jonke, A.A., Levenson, M., Levits, N.M., Steindler, M.J., and Vogel, R.C., Candidates for second generation fuel reprocessing plants, *Nucleonics*, 25(5), 58, 1967.
98. Standifer, R.L., A fluidized-bed fluoride volatility pilot plant for plutonium purification, *Chem. Eng. Prog. Symp. Ser.*, 66(105), 198, 1970.
99. Chilenskas, A.A., Turner, K.S., Kincinas, J.E., and Potts, G.L., Engineering Development of Fluid Bed Fluoride Volatility Processes. Part 10. Bench Scale Studies on Irradiated Highly Enriched Uranium-Alloy Fuel, USAEC Report ANL-6994, 1966.
100. Bougeois, M. and Nollet, P., Study of the dry chemical treatment of uranium–Zr fuels (in French), *Nucleaire*, 8(3), 165, 1966.
101. Monk, T.H., Pashley, J.H., and Schappel, R.B., Semiworks fluid bed studies of volatility process for zircaloy-clad uranium dioxide power reactor fuels, *Chem. Eng. Prog. Symp. Ser.*, 66(105), 180, 1970.
102. Witte, H.O., Fluidized bed combustion of graphite-base nuclear reactor fuels, *Ind. Eng. Chem. Process Des. Dev.*, 8(2), 145, 1969.
103. Roch, A.P., Kilian, D.C., and Dickey, B.R., Separation of uranium from graphite based fuels by fluidized bed combustion, *Chem. Eng. Prog. Symp. Ser.*, 66(105), 190, 1970.
104. Borduin, L.C., Kilian, D.C., and Rich, A.P., Uranium recovery from graphite based fuels by fluidized bed combustion, *AIChE Symp. Ser.*, 69(128), 130, 1970.

105. Mullen, J.F., Consider fluid bed incineration for hazardous waste destruction, *Chem. Eng. Prog.*, 88(6), 50, June 1992.
106. Oppelt, E.T., Incineration of hazardous waste — a critical review, *J. Air Pollut. Control Assoc.*, 37(5), 558, 1987.
107. Commander, R.E., Lohse, G.E., Black, D.E., and Cooper, E.D., USAEC Report IDO 14662, 1966.
108. Butlar, F.E., Moselay, J.D., and Molen, G.F., Recovery of Pu from Incinerator Ash, Report RFP-396, Dow Chemical Company, 1964.
109. Town, J.W., Paige, J.I., and Russell, J.H., Sorption of sulphur dioxide by alkalized alumina in a fluidized bed reactor, *Chem. Eng. Prog. Symp. Ser.*, 66(105, 260, 1970.
110. Black, C.H. and Boubel, R.W., Effectiveness of a fluidized bed in removing submicron particulate from an air stream, *Ind. Eng. Chem. Process Des. Dev.*, 8(4), 573, 1969.
111. Jackson, M., Fluidized beds for submicron particle collection, *AIChE Symp. Ser.*, 70(141), 82, 1974.
112. Hanway, J.E., Jr., The use of fluidized bed technology in pollution control, *AIChE Symp. Ser.*, 67(116), 236, 1971.
113. Ho, T.C., Paul, K., and Hopper, J.R., Kinetic study of biological sludge incineration in a fluidized bed, *AIChE Symp. Ser.*, 84(262), 126, 1988.
114. Baeyens, J. and Geldart, D., Fluidized bed incineration — a design approach for complete combustion of hydrocarbons, in *Fluidization*, Davidson, J.F. and Keairns, D.L., Eds., Cambridge University Press, Cambridge, 1978, 264.
115. Chakraborty, S.P., Sharma, I.G., and Bose, D.K., A study on thermal oxidation of burnt zirconium fines in a fluidized bed reactor, *Metall. Trans. B*, 26B, 647, 1995.

CHAPTER 4

High-Temperature Fluidized Bed Reactor

I. PLASMA, PLASMA FURNACES, AND PLASMA FLUIDIZED BED

A. Plasma

1. Plasma State

The subject of electrical discharge and plasma has received special attention over the past three decades. Plasma wind tunnels were originally used by NASA for testing materials supposed to meet similar conditions when a space vehicle reenters the earth's atmosphere. Subsequent applications included thermal treatment of powders, particle spheroidization, coating by plasma spraying, plasma vacuum melting, extractive metallurgy, and chemical synthesis. Thermal plasmas have been used in the chemical synthesis of nitrogen oxides from nitrogen and oxygen and for the synthesis of acetylene. Plasma reactors have now found application on a commercial scale in extractive metallurgy as well as in the chemical synthesis of many organic and inorganic compounds.

A gas in its fourth state of matter that has a neutral charge as a whole on a microscopic scale but which contains atoms and molecules in their fundamental or excited states, ions, and electrons is referred to by the term plasma. The occurrence of plasmas is quite common in the universe; they are found in many stars and the sun. On the earth, the plasma state is found to be associated with certain natural phenomena such as lightning. In practice, it is achieved by plasma devices such as fluorescent or neon lamps, electric arcs or discharges, magneto-hydrodynamic power converters, thermonuclear and explosive devices, and so on.

a. Charged Particle

A plasma from a gas is distinguished at ambient temperature and pressure mainly by the presence of charged particles which are responsible for the electrical conductivity of the gas. The quantity of charged particles in a plasma is dependent on the rate of formation of positive ions and free electrons from the parent atoms or molecules and the counterbalancing rate of the elimination of charged particles by recombination to form the initial atoms and molecules. Plasmas or plasma devices are distinguished as low-pressure or atmospheric plasmas. When a gas is heated to a few thousand degrees Kelvin, the molecules dissociate and ionization occurs. At temperatures in the range 5000–35,000 K, the plasma contains heavy particles (molecules, atoms, and ions) and light particles (electrons). The gas will be composed of base nuclei and free electrons at about 10^8 K, corresponding to the commencement of thermonuclear reactions. A low-pressure plasma, also called cold or out-of-equilibrium plasma, contains electrons or light particles at temperatures higher than those of heavy particles (or gas). In thermal plasmas (achieved at atmospheric pressure), the light particle (i.e., electron) temperature is close to the heavy particle (i.e., gas) temperature. For our purposes, it is the thermal plasma which has a wider range of application in combination with the fluidized bed reactor.

b. Collision

The properties of a plasma are influenced by the probability of particle collisions which, in a thermal plasma at atmospheric pressure, are dependent on particle density. A collision or interparticle interaction can take place when the distance between the particles is within a few molecular radii. In a thermal plasma with a particle density range of 10^{14}–10^{16} per cubic centimeter, depending on the temperature, the distance between the particles is large and three orders of magnitude greater than the particle size. Among the particles in a plasma, electrons have a high probability of collision because the heavy particles are almost stationary compared to electrons. The energy transfer between the particles can take place by elastic collision, wherein the total kinetic energy is conserved, and by inelastic collision, during which the kinetic energy is changed, resulting in a corresponding change in the internal energy of the particles. The possible internal energy state of a particle is governed by wave mechanics and is given by the Schrödinger equation.[1] Each particle in a dense plasma may be assumed to follow a Maxwellian energy distribution[2] from which the temperature can be defined. Laboratory or industrial plasmas are optically thin, and Planck's law is not valid for the whole wavelength range. At high temperatures, it is difficult to obtain calorimetric data. Hence, measurements are usually made by spectrometric methods to calculate thermodynamic properties. Beam experiments are used to describe transport properties. The chemical composition of a plasma at a given temperature and pressure is required to predict thermodynamic properties such as specific heat and enthalpy and transport properties such as viscosity and thermal and electrical conductivity.

c. Plasma Systems

In many plasma generator devices, gases such as Ar, He, H_2, and N_2 are commonly used. Such cases are also normally used in fluidized bed reactors, depending on the reactor environment and the reaction conditions or requirements. There are two regimes of plasmas labeled in plasma chemistry: (1) glow discharge, where the electron temperature is 10–100 times the gas temperature, and (2) arcs, jets, or reactors, where the electron temperature is nearly the same as the gas temperature. The latter, usually known as thermal plasma at atmospheric pressure, is commonly used in the laboratory or in industry for physical as well as chemical processes. In general, electrical heating at temperatures above 1370°C (2500°F) is more economical because a huge amount of heat is wasted in gas or hydrocarbon fuel heating. A comparison of the efficiency[3] of heating at high temperatures using natural gas and arc-heated air, as shown in Figure 4.1, reveals that the arc heater can make available

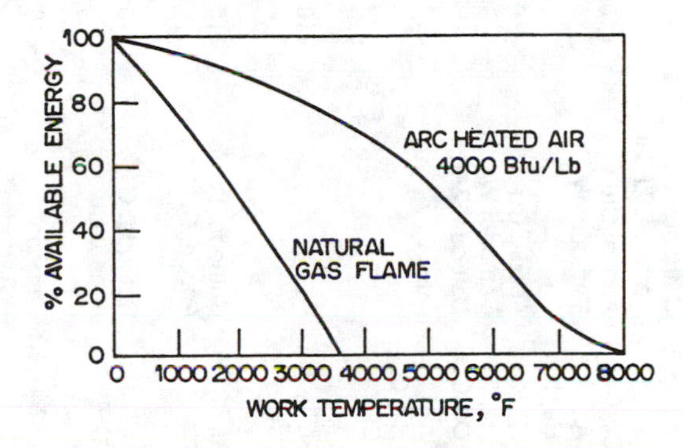

Figure 4.1 Efficiency for high-temperature rating. (From Fey, M.G., *Ind. Heat*, 2, June 1976. With permission.)

more energy than a natural gas flame over a wide temperature range. The electrical energy in a plasma is converted essentially into heat which is transferred to other objects by conduction, convection, and radiation. Thermal plasmas have the advantage of carrying out rapid chemical reaction in a small apparatus in a continuous manner with a good automation. Plasma-generating devices are classified into two main groups: (1) nontransferred arc and (2) transferred arc. In a nontransferred arc generator, a cathode and an anode with an orifice to allow the arc plasma to pass through it are employed, whereas the cathode is spaced[4] away from the anode and the arc is constricted between the electrodes in a transferred arc generator. In plasma generators, the electric coupling is done by either resistive- (arc) or capacitive- (RF plasmas) type methods. Hence, plasma generators can be grouped as generators with electrodes or electrodeless generators. In many chemical reactions, a direct current gas-stabilized plasma is used and is capable of achieving ultra-high temperatures for longer time periods. A list of plasma-generating devices and their characteristics is given in Table 4.1.

2. DC and AC Discharges

Electric coupling by DC arc discharges using a cathode and an anode consists of regions with a steep gradient in voltage close to each of the electrodes. Between the two regions is the main arc, where the voltage falls gradually. The cathodes region acts as an electron emitter, while the anode region acts as a collector. The

Table 4.1 Main Characteristics of Plasma-Generating Devices

Type of Operator	Power Supply	Maximum Power (kW)	Plasma Diameter (mm)	Plasma Length (mm)	Plasma Gas Velocity (m/s)	Gas Nature	Nature of the Electrodes	Residence Time (s)	Application in Industry
Radio frequency	1–20 MHz	1,000	20–200	40–50	5–50	All	Outside the plasma container	$<10^{-2}$	Elemental analysis, optical fiber industry
Gage configuration	DC	100	5–12	20–70	200–1000	Ar, N_2, He, H_2	Anode: Cw Cathode:W Cathode: Zr, Hf	$<10^{-3}$	Plasma spraying
Gage configuration with a two-diameter nozzle	DC	1,000	10–50	30–100	50–400	O_2, CO_2, Ar, N_2, He, H_2	Anode: Cu Cathode: W	$<10^{-2}$	Gas heating
Gage configuration with multiple gas injections	DC	5,000	10–80	30–200	50–400	Ar_4, N_2, He, H_2, O_2, CO_2	Anode: Cu Cathode: W Cathode: Zr, Hf	$<10^{-2}$	Gas heating, production of the reducing gases
Tabular electrodes	DC and AC	1,500	20–120	30–200	20–200	Ar, N_2, H_2, He, O_2, CO_2	Cu–Cu	$<10^{-2}$	Gas heating, production of the reducing gases for blast furnace of sponge iron furnaces
Plasma furnace, three-phase alternating current	AC	150	30–150	100–200	1–10	Ar, N_2, H_2, He	W, C	$<10^{-2}$	Extractive metallurgy
Plasma furnace precessive cathode	DC	1,500		100–400		Ar, N_2, H_2	Cathode: W Anode: Cu	$<10^{-1}$	Extractive metallurgy
Plasma furnace magnetic field expansion	DC	10,000	40–500	100–500	1–10	Ar, N_2, H_2, CH_4	Cathode: C Anode: Cu	$<10^{-1}$	Acetylene production from coal
Plasma furnace Maecker effect	DC	150	40–200	100–500	1–10	Ar, N_2, H_2	Cathode: W\Anode: C, Cu	$<10^{-2}$	Spheroidization magnetite for photocopy zirconia
Plasma furnace rotating wall	DC	150	30–60	100–400		Ar, N_2, H_2		>1	Refractory materials melting
Plasma furnace anode with falling film	DC	1,000	30–100	100–500	5–50	Ar, N_2, H_2, OH_4	Cathode: W Anode: C, Cu	>1	Extractive metallurgy plasma remelting

From Boulos, M. and Fauchais, P., in *Advances in Transport Processes*, Vol. IV, Wiley Eastern Ltd., New Delhi, 1986, chap. 5. With permission.

current density and the heat flux at these regions are very high, and hence they require intense cooling. The arc column is the main body of the plasma where quasi-neutrality prevails. The field strength in the arc column is so low that charged particles in this column are obtained not by field ionization but by thermal ionization. The arcs are usually stabilized by aerodynamically cooled walls or by magnetic fields. The voltage-current characteristics of the arc depend on the arc intensity, the nature of the gas, and its flow rate. These parameters usually influence the arc length and its diameter. The nozzle shape and the stabilization mode also play a role in determining the arc column length and its diameter. In an AC arc, the polarities of the electrodes change in each cycle. If a low-frequency (<100-Hz) field is used in every half cycle, a DC-type discharge is established. If the applied voltage is above the maintenance potential, the discharge is sustained; otherwise, the discharge is extinguished and the charges are swept away by the gas flow. The choice of electrode materials for AC arcs is very limited mainly due to frequent alteration of polarity.

3. *RF Discharges*

With AC discharges at high frequencies, electrons will not reach the electrodes. They will oscillate in the gap between the electrodes and collide with the gas particles. Under such conditions, the electrodes can be placed outside the arc discharge tube. The oscillating field is applied by either an induction coil or two external electrodes. Hence, the plasma can be either inductively coupled (if an induction coil is used) or capacitively coupled (if external electrodes are used). Inductive-type coupling is more common. In a plasma of this type, an external device (i.e., and auxiliary electrode) is required to start the discharge because the electrodes responsible for generating and collecting the charged particles are mounted externally. The frequencies used for RF discharges are in the range 1–50 MHz. For each frequency there is a minimum power that can sustain the discharge.[6] This minimum power in turn is also dependent on the pressure and the type of gas. For example, the minimum power for argon gas is 0.7 kW at 4 MHz and 200 kW at 1000 Hz, whereas it is 250 kW at 4 MHz for hydrogen. The oscillatory field in induction coupling results in the generation of an eddy current on the cylindrical load. This current over the load and the coil radius are interrelated in arriving at an optimum coupling efficiency.[7] Because inductively coupled plasmas are not in contact with electrodes, they are not contaminated by the electrode materials. Hence, many active and corrosive gases such as oxygen, chlorine,[8] and uranium hexafluoride[9] can be employed.

B. Plasma Device System

1. *Gases for Plasma Generation*

a. *Selection Criteria*

The gases commonly used in a plasma generator are hydrogen, nitrogen, argon, and helium. Heat and momentum transfer are influenced by the type and properties of the gas selected. The gas selected can be an inert carrier gas or can take part in the chemical reaction. Noble gases which have low ionization energies and good arc voltages can be employed as carrier gases. The various properties[10] of commonly used plasma gases such as Ar, H_2, O_2, N_2, and air are presented in Table 4.2.

Table 4.2 Properties of Gases Employed in Plasma

Gases	Dissociation Energy (kcal/g · mol)	Dissociated Particle	Ionization Voltage (V)	Open Arc Voltage (V) (at Constant Arc Length and Power)	At 10% Ionization: Arc Temperature (°C)	At 10% Ionization: Heat Constant (W/ft³), NTP
Ar	0	Ar	15.68	18	18,000	75
He	0	He	24.46	26	27,000	110
H_2	104	H	13.53	70	15,000	260
O_2	110	O	13.55	40	16,000	425
N_2	225	N	14.48	40	16,000	425
Air	—	—	—	60	16,000	425

From Bhat, G.K., *J. Vac. Sci. Technol.*, 9(6), 1344, 1972. With permission.

In many industrial applications, electrodes should have long life and the gases used should not attack and damage the electrodes. Furthermore, the gases used for plasma applications should be inexpensive and have good heating value. Figure 4.2 shows the plot[11] of the specific enthalpy of various gases used in plasma as a function of temperature. Ionization and dissociation of the gas have a direct influence on the plasma temperature. Nitrogen gas has proven to have a good heating value even at 10,000 K, with 95% of the energy available for heating.

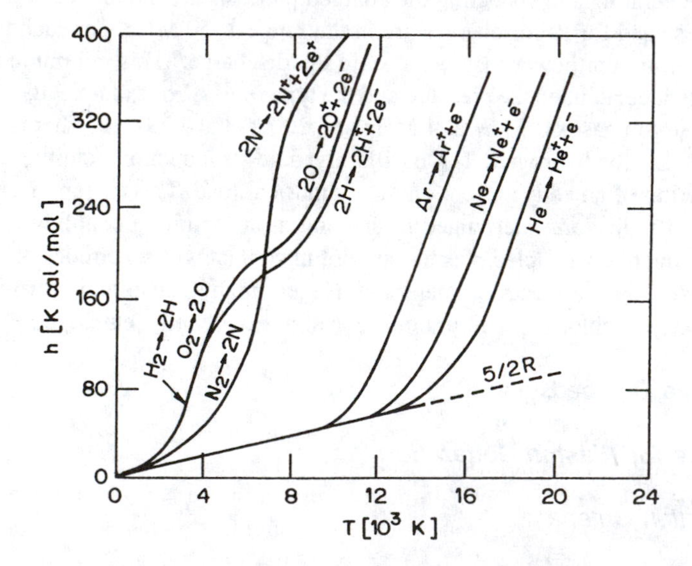

Figure 4.2 Specific enthalapies (*h*) of various gases at atmospheric pressure as a function of temperature (*T*). (From Bhat, G.K., *J. Vac. Sci. Technol.*, 9(6), 1344, 1972. With permission.)

b. Properties

Among the thermodynamic properties, the specific heat and enthalpy of gases are often required in many heat transfer calculations; viscosity, electrical conductivity, and thermal conductivity are often required as useful parameters to assess the transport properties of plasma gases. A large volume of data on thermodynamics and transport properties is available in the literature and was reviewed in a progress report[12] of the IUPAC Subcommission on Plasma Chemistry. The specific heat of nitrogen at constant (atmospheric) pressure is found to have three peaks[11] at temperatures of 7600, 14,500, and 30,000 K, respectively, corresponding to dissociation, first ionization, and second ionization of this gas. The peaks are pronounced for helium and hydrogen, with the highest value corresponding to hydrogen. Thus, the highest heat capacity is obtained with hydrogen gas compared to Ar, N_2, and He gases. The viscosity of gases such as Ar, N_2, H_2, and He in the plasma temperature range varies widely from their respective values at room temperature. Data on the viscosity of these gases for atmospheric plasma can be found in the literature.[12-16] The variation in viscosity with temperature is shown in Figure 4.3.[5] The viscosity

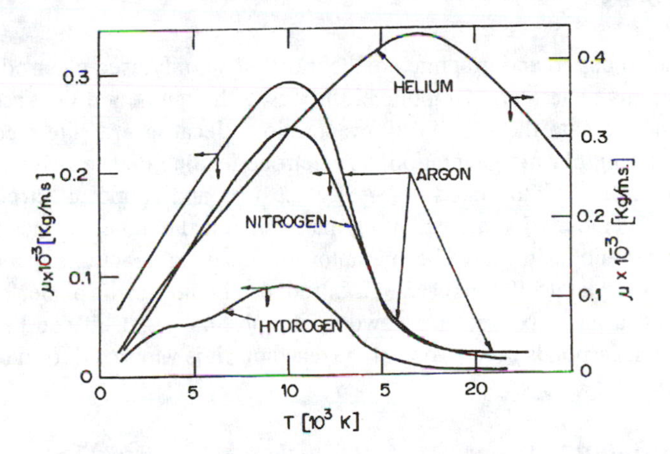

Figure 4.3 Viscosity (μ) of various gases at atmospheric pressure as a function of temperature (*T*). (From Boulos, M. and Fauchais, P., in *Advances in Transport Processes*, Vol. IV, Wiley Eastern Ltd., New Delhi, 1986, chap. 5. With permission.)

of these gases at a temperature of 10^4 K is ten times higher than at room temperature. This has some adverse effect on the mixing of various gases. The viscosity of Ar, H_2, and N_2 increases steeply with temperature and reaches a maximum at about 17,000 K. A further increase in temperature brings down the viscosity of the gases due to ionization. The ionization energy of N_2, H_2, and Ar is of the order of 14 eV and for He is 24.6 eV. The maximum viscosity in the case of helium is obtained at 17,000 K. Ionization is also responsible for changing the electrical conductivity of the plasma gases. Ar, N_2, and H_2 have almost the same electrical conductivity (20 Ω cm^{-1} at 10^4 K), whereas the conductivity of helium at the same temperature has a lower value (i.e., 5 W cm^{-1}). In general, electrical conductivity increases sharply with temperature, retaining a limiting value of 100 Ω cm^{-1} until the point of second ionization is reached. It is interesting to note that the electrical conductivity of a mixture of gases does not depend on the proportion[13] in which they are mixed. The thermal conductivity of a gas can be considered to be the sum of three components: K_{tr} (translational), K_r (reactional), and K_E (internal energy contribution). The translational component in turn is the sum of two components due to electrons (K_{tr}^e) and heavy particles (K_{tr}^h). The internal energy component is due to vibration, rotation, and electronic excitation and is determined[16] from the theory of Euken. The reactional component arises from dislocation and ionization at high temperatures. At low temperatures, the translational contribution is the significant one. The reactional component becomes more active at 10^4 K. At very high temperatures, the translational contribution of electrons is very important. The overall thermal conductivity of a gaseous mixture such as Ar–H_2 is much higher than that of Ar gas even when H_2 is present in small concentrations. This is attributed to the low dissociation energy (4.48 eV) which results in a high reactional conductivity.

2. Electrodes

Tungsten, copper, and graphite are the most commonly used electrode materials in many plasma generators. Graphite electrodes, although they do not require cooling, can contaminate the plasma by evaporation. Because graphite electrodes are consumed during plasma generation, an electrode feeding device is required in this case. Tungsten or 2% thoriated tungsten electrodes and copper electrodes require cooling arrangements. In implementing many chemical synthesis reactions, water-cooled copper anodes[17,18] have been employed. In several reactions that are carried out in the plasma jet itself, the reactant is introduced in the plasma plane. The reactant can be a gas, a gas mixture, or a powder. The plasma gas itself can be a reactant species. When carbon is used as one of the reactants, it is vaporized from the graphite cathode itself.

3. Power Source

In DC plasma generators, thyristors with arc current stabilization are used. Silicon diode power supply sources rated 350 V/600 A are used for small power generators. Power sources of 1000 V/1000 A up to 7 MVA are common in large plasma generators. In the case of high-frequency plasma generators, 10- to 12-kV generators and transformers with constant anode voltage assembled on thyristors or semiconductor diodes are a good choice. Electrovacuum parts are generally used in high-frequency generators.[19] For better economics, AC power can replace DC at large voltages. Electrodeless plasma generators can replace DC or AC electrode plasma generators in many instances. Continuous operation and the introduction of generators that can employ an oxidizing stream such as air or steam are some of the recent developments in plasma technology in the recent past.

4. Arc Plasma Generator

Arc plasma generators can be grouped based on their applications and interactions. Plasma torches or cutters are used for interaction with compact solid matter, and typical applications include welding, cutting, spraying, rock drilling, and preheating of charges. Plasma furnaces are used for melting, refining, reduction smelting, crystal production, and alloying and are thus concerned with plasma interaction with liquid-phase matter. Plasma jet reactors are used for interaction of the plasma with disperse condensed matter, and they find application in ore beneficiation, metal reduction, powder processing, spraying, and refractory material processing. The first design of a plasma torch was created in 1957 by Gage,[20] who used a direct current to generate a plasma between a cathode rod and an anode nozzle. The plasma was extended by forced gas flow. There are several configurations of the cathode and the anode and various modes of gas injection. The cathode can be a rod or a tube, and the anode can be a nozzle or a segmented-type tube. The gas can be introduced axially through the cathode tube or tangentially. The arc generated can be stabilized by a water-cooled wall or a magnetic field produced by either self-induction or an external magnetic coil. Various torch configurations can be found in the literature.[21–23] The plasma flow region can be divided into three zones:[24] (1) an entrance zone where the hydrodynamic boundary layer is formed, (2) a middle or intermediate zone where the interactions between the hydrodynamic and the thermal boundary layers occur, and (3) a turbulent zone. The first zone does not have radial pulsation of the arc, and the third zone consists of arcs divided into many secondary arcs, thereby increasing the thermal anode losses. The arc length in a plasma arc generator is limited to fitting within the first two zones. The anode losses can range from 5 to 80% and are proportional to the current intensity and the arc voltage[25] raised to the power 0.2–0.4. Anode erosion usually occurs at the arc root, and the life[25] of an anode (for copper anodes) is 300–400 hr. Cathode erosion is dependent on the material used. The cathode material, such as tungsten and carbon, can be attacked in the presence of some species such as O_2 in the arc. The loss in thermionic cathode materials is mainly due to their evaporation and sublimation. Zirconium and hafnium cathodes can be used even under oxidizing conditions with air, CO_2, and O_2 for up to 30 hr.[26] The cathode shape can be stick, hollow, or button type. A hollow cathode is used for extremely high current applications (>5000 A). The life of an electrode with a DC arc is twice that of an AC arc. Various type of DC and AC plasma torches are available.

II. PLASMA FURNACES

A. Categories

Plasma furnaces can, in principle, be grouped into two main categories based on the plasma generation mode: (1) DC or (2) RF (capacitive-typing) coupling. These groups are further classified into three categories depending on the residence time for the reacting species. Thus, plasma furnaces can be categorized[27] as (1) short residence time, (2) medium short residence time, and (3) long residence time.

1. DC Plasma Furnace

a. Short Residence Time

In a short residence time plasma reactor, the residence time is less than 0.01 s. Hence, the reaction species, especially if they are comprised of solid particles, should be distributed properly. They can be introduced at the cathode tip itself, where the magneto-hydrodynamic pumping effect[28] would help to ensure proper distribution of the particles. The Ionarc furnace,[29] which is used to break down zircon particulate solids into ZrO_2 and SiO_2, belongs to the short residence time furnace category. This furnace consists of a shielding tungsten cathode and three consumable carbon anodes. The zircon, ground to 20–200 μm, is fed through the ports around the vertical anode. The zircon particles are dissociated in one step to ZrO_2 and SiO_2 at a temperature as high as 13,000 K. Feeding particulate solids into the cathode is difficult due to the pressure fluctuation of the plasma, and about 30–50% of the heat is carried by the solid product as sensible heat. A typical schematic of the Ionarc plasma furnace[29] is presented in Figure 4.4. CSIRO, in Australia, developed[30] a 100-kW four-torch (one cathode and three anodes) plasma reactor for in-flight processing of minerals. The reactor was used to dissociate zircon and had the capacity to process 20–50 kg/hr. The initial Ionarc project was set at 230–450 tons/year (tpy) but was later scaled up to 4500 tpy at a power consumption of 3 kWh/kg. Thus, these furnaces have proven their potential and entered the commercial market, satisfying the

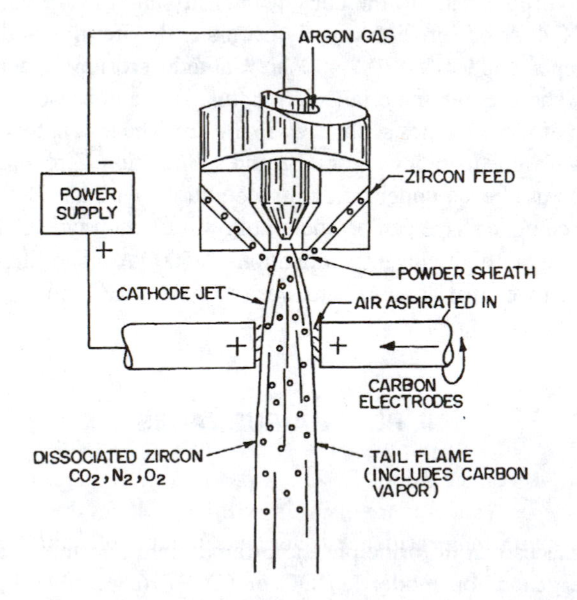

Figure 4.4 Ionarc furnace assembly. (From Wilks, P.H., Ravinder, P., Grant, C.L., Pelton, P.A., Downer, R.J., and Talbot, M.L., *Chem. Eng. Prog.*, 68, 82, 1972. With permission.)

required quality of the product. Furnaces with three transpiration anodes[31] or with three DC plasma jets acting as anodes[32] have also been reported.

b. *Medium Residence Time*

This class of furnaces is grouped into two categories, one based on the solid reactant feed direction (either cocurrent or countercurrent to the plasma) and the second based on the plasma expansion method. A fluidized bed plasma reactor,[33] depicted[5] in Figure 4.5, belongs to this category, and the mean residence time of the solid particles ranges from 0.1 to 1 s in such reactors. In a typical plasma fluidized bed reactor shown in Figure 4.5, aluminum silicate was spheroidized. Extended arc flash reactors[34] used for chromite also belong to this category. The second category of furnaces uses an expanded plasma stream to improve plasma–particle interactions. The usual methods used to improve plasma stream expansion are (1) rotating the water-cooled walls,[35] (2) rotating the cathode[36] in front of a toroidal anode, and (3) rotating the arc[37] by a magnetic field and tangential gas injection.

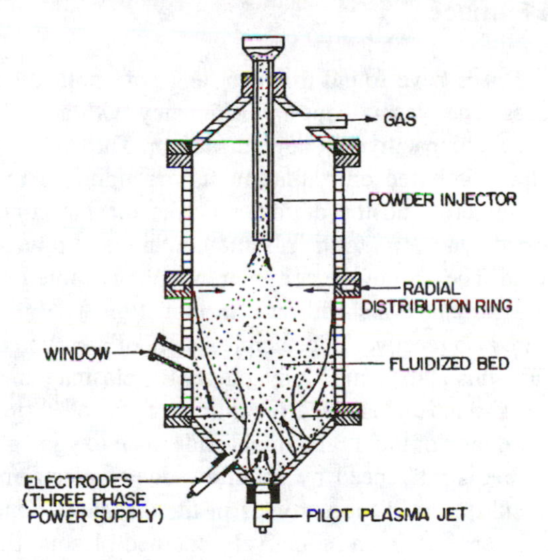

Figure 4.5 Fluidized bed plasma furnace. (From Benet, E., Valbana, G., Foex, M., Daguenet, M., and Dumargue, P., *Rev. Int. Hautes. Temp. Refract.*, 11, 11, 1974. With permission.)

c. *Long Residence Time*

The residence time for the reactant in this type of reactor is greater than 1 s. The furnace can use either a nontransferred or a transferred arc. The plasma stream is usually directed over the charge, which will be molten after prolonged heating. Hence, furnaces with long residence time have found application in the melting of refractory oxides.[38-41] The main advantage of a furnace of this type is the high rate of heat transfer due to containment of the plasma. However, its noncontinuous style of operation is a disadvantage. The falling film[42] type of plasma furnace also has a long residence time. In this type of furnace, the powdered material is introduced tangentially into the plasma stream. The molten material is centrifuged and thrown on the wall of the reaction chamber; subsequently, the liquid film starts falling slowly and is collected into a crucible. This type of furnace, developed by Chase and Skriven,[42] Gold et al.,[43] and Kassabji et al.,[44] has found application in the extractive metallurgy of iron, ferrovanadium, and ferrochromium. These furnaces have certain drawbacks. It is necessary to strictly control feeding of the material, which could extinguish the plasma at high feed rates and could evaporate if fed slowly. In a typical application, the plasma arc[45] is struck between a shielded cathode and an anode that acts as a charge (such as molten MoS_2). Plasma furnaces of this type have been found to have high thermal efficiency.[46-48] Plasma furnaces with a DC plasma torch have been developed, replacing the conventional graphite electrode. These furnaces have wider application in remelting various alloys and scrap materials. Commercial furnaces of this type with a power level of 19.8 MW have been developed.[49]

B. RF Plasma Furnace

DC plasma furnaces have found a wider range of application in industry than RF plasma furnaces. The plasma coupling efficiency with an inductive type of RF plasma is better than a capacitively coupled plasma. The design of the inductively coupled plasma torch is based on requirements pertaining to the type of gas, the power level, the pressure, and the diameter of the plasma confinement tube. An induction coil placed concentric to the plasma containment tube generates an oscillating magnetic field. The wall of the confinement tube is cooled externally by water and internally by a gas called a sheath gas. In induction-type plasma, it is possible to use gases that are also reactive. In principle, gases of any type can be used in this type of furnace, and this is their main advantage. RF plasma generators with power ranges of 200–500 kW have been used for industrial-scale[50,51] production of titania pigments by the oxidation of titanium tetrachloride in an oxygen plasma. The plasma at the discharge zone is influenced by backflow due to electromagnetic pumping, and thus it is difficult to inject the particles in the discharge zone where the recirculation eddy is present.[52-54] In an inductively coupled plasma, there is a minimum power level that is required to sustain the induction charge, and this level is fixed on the basis of the gas type and the pressure and frequency of the electromagnetic field.[50,55] As the frequency is lowered (i.e., from megahertz to kilohertz), the power level required can change severalfold (from 10 kW to hundreds of kilowatts). In order to reduce the minimum power level, it would be necessary to increase the electrical conductivity by either reducing the pressure or adding ionizing impurities such as potassium and cesium. In a capacitively coupled plasma, the minimum power required to achieve a self-sustained plasma discharge is less than that for an inductive-type RF generator. Research on capacitively coupled plasma is ongoing. For low-power operations, such as those used in the laboratory for spectrochemical analysis,[56] inductive-type plasma generators with a power range of 2–3 kW are used. The temperature obtained in such a plasma depends on the type of gas used and its flow rate. The radial temperature changes[57] based on the type of gas used. Among gases such as oxygen, air, and nitrogen, the higher magnitude of temperature in the radial position is obtained with oxygen gas.

III. PLASMA FLUIDIZED BED

A. Plasma–Solid Interactions

1. Heat Transfer to Solid

The efficiency of the interaction between a plasma and solid particles can be determined from the rates at which heat, mass, and momentum are transferred. The plasma jet velocity is usually very high (of the order of 100 m/s), and hence a particle introduced into the stream has a very short residence time. As a result, the net fraction of energy transferred to the solid particle is very small[58] (even less than 5%). The net rate (Q_t) of heat transferred to a solid particle with a surface area S_p, a residence time t_s, and a surface temperature T_s, when it is injected into a plasma stream of temperature T_g, which is confined inside the wall at a temperature T_w, is given by:

$$Q_t = mC_p \int_T^{T_p} dT_p = S_p \int_0^{t_s} \left[h\left(T_g - T_s\right) - \epsilon_p \sigma_s \left(T_s^4 - T_w^4\right) \right] dt \qquad (4.1)$$

where ϵ_p is the particle emissivity and σ_s is the Stefan constant. From Equation 4.1, it can be inferred that the net heat transfer rate is increased if the heat transfer coefficient (h), the residence time of the solid (t_s), the wall temperature (T_w) of the plasma tube, and surface area (S_p) of the particle are increased. An increase in the plasma velocity can increase the heat transfer coefficient but will reduce the particle residence time. If the particle size is reduced, the cost of milling will be correspondingly high. The wall temperature, if increased, could bring down radiative losses. Hence, to improve the efficiency of plasma–particle heat or energy transfer, the following methods can be adopted: (1) reduce the plasma velocity, (2) increase the plasma volume, (3) increase the wall temperature to bring down radiation losses, (4) improved particle injection techniques, and (5) solid flow countercurrent to the plasma jet.

Among the various methods outlined above, the plasma volume can be increased by using several techniques, including (1) introduction of gas into the arc, which results in the plasma jet[59] and also fluid convection;[60] (2) rotation of the wall, which can allow the arc to expand and fill the tube;[61] and (3) mechanical rotation of an orbiting cathode arc onto a ring anode[62] or electromagnetic rotation of the arc between a tubular anode and cathode.[63] An increase in the effective volume of the plasma can increase the capacity as well as the residence time of the particles fed into the plasma.

2. Solid Quenching

Quenching of the desired product obtained from a plasma reactor can be accomplished by using water-cooled copper tubes or a conventional heat exchanger used for high-temperature gases. The maximum heat flux for a fixed surface heat exchanger is limited to 575 W/cm^2. In many plasmas, the heat flux is ten times greater than this value. Hence, it is essential to find an improved heat-exchanging system. Fluidized bed reactors are especially advantageous in quenching the solid products. Nonreacting gases and inert solids can be allowed to directly mix with the solid products. The quenching rate reported in a fluidized bed reactor is of the order of 28×10^6 K/s. An account of the various developments in plasma fluidized bed reactors and their growing application in high-temperature reactions pertinent to extraction and process metallurgy is provided in the following section. No attempt is made in this chapter to describe the transport processes in a plasma reactor or the modeling of a plasma reactor. For a comprehensive review of thermal plasmas and the associated transport processes, refer to the article by Boulos and Fauchais.[64]

B. Plasma and Fluid Bed

1. DC Plasma Fluid Bed

a. Description

The contacting of a nonequilibrium plasma with a granular solid using a fluidized bed reactor has the advantage of both high-temperature heating of the plasma and the conventional good mixing and efficient heat transfer of a fluidized bed reactor. In a plasma fluidized bed reactor, fluidized solids are contacted with the active and current-carrying portion of the plasma. Fluidized bed reactors were initially brought into plasma application mainly for the quenching of the plasma by contacting the fluidized solids with the tail end of the DC or RF plasma jet[65] or just to form a spouted bed by the plasma gas jets.[66] A plasma fluidized bed reactor can be used to carry out gas–solid reactions rapidly. The products are quenched by the fluidized bed of fresh incoming solids. If the reacting solids are in a fused state, they are solidified and agglomerated by the fluidized bed. Because the products are quenched by the solid itself, there is no dilution of the product due to the addition of any external heat diluent material, and thus heat recovery is efficient in the bed. The fluidized bed reactor in conjunction with plasma may not be suitable for efficient solid–solid reactions.

Although a plasma fluidized bed reactor can be smaller in size for carrying out reactions rapidly, it has been a challenge for researchers in this field to generate a stable plasma arc for direct contact between the plasma and the bed of fluidized solids. A plasma-generating device in combination with the fluidized bed is still in the process of evolution. Because development of a reactor of this type needs to take into consideration the engineering aspects of both the plasma and the fluidized bed reactor at high temperatures, this area has recently drawn the attention of many high-temperature scientists and engineers. There are essentially two major parts in a plasma fluidized bed reactor. The first is the plasma-generating part, where the plasma arc struck between the DC electrodes is carried by inert gases like argon or helium or by reactive gases like O_2 or N_2 into the second part of the reactor, which is the fluidized bed. The bed can be fluidized solely by the plasma gas or by an

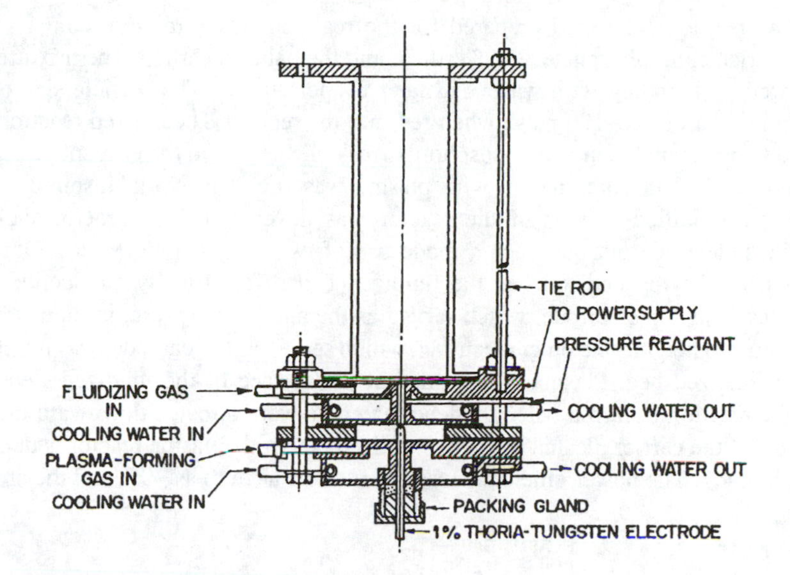

Figure 4.6 Experimental plasma bed apparatus. (From Goldberger, W.M., *Chem. Eng. Prog. Symp. Ser.*, 62(62), 42, 1966. With permission.)

auxiliary gas. A detailed schematic of an experimental plasma bed reactor is shown in Figure 4.6.

b. Testing

Research was carried out at the Battelle Memorial Institute in Columbus Ohio using a plasma fluidized bed reactor.[67] Experiments were conducted to prepare elemental phosphorus by reacting tricalcium phosphate mixed with silica (which acted as a flux) in an argon arc plasma jet. A bed of graphite particles of size –48 +100 mesh was located just above the plasma jet. The mean particle size of tricalcium phosphate and silica (mixed in a 1.8:1 ratio) was 60 μm, and the mixture was carried by an argon gas carrier. The power transferred to the arc was found to be very poor and was at 20% of the generator output. The loss was attributed to the flow of the carrier gas and heat loss to the wall of the water-cooled cathode. The desired product was not found to form. The initial power input was 1.5 kW, and the power efficiency decreased to less than 15% as the fused solids accumulated on the anode. No improvement was observed even when the solids were directly fed into the arc-forming region without using the carrier gas and the power input was increased to 3 kW. Experiments carried out without the graphite bed showed that the reaction occurred mainly in the arc-forming region. Graphite anodes were found to be better than tungsten anodes in this case. A reactor of the type depicted in Figure 4.6 was not found to have satisfactory and hence was replaced by an open-arc-type reactor.

In the open-arc-type fluidized bed plasma reactor, a stable arc was struck between two tungsten electrodes $^1/_4$ in. in diameter and the arc was present even above a fluidized bed of –200 +325 mesh alumina. The electrodes were mounted above the fluidized bed, and the solid feed was directed downward by the plasma carrier gas on the open arc. The feed prepared for the reaction was a mixture comprised of 42.7% tricalcium phosphate, 24.9% silica, and the balance carbon. The mixture was prepared by blending with water and later evaporating it. The particle size of the ground mixture was –100 mesh when fed into the reactor. The desired reaction was found to occur, and elemental phosphorus was collected from the effluent gas. Thus, a fluidized bed in combination with plasma was found to work. Inspired by the results, a modified version of the reactor was developed. This reactor, called a fluidized electrode plasma reactor, made use of two tungsten electrodes. One was hung from the top and touched the fluidized bed of electrically conductive solid particles. This top electrode, which served as the anode, was $^1/_4$ in. in diameter and adjustable in height. The other electrode, which served as the cathode, was immersed in the fluidized bed. Plasma was formed at the surface of the fluidized electrode, and the reaction occurred when the feed materials were directed downward into the plasma by the carrier plasma gas. The fluidized electrode plasma reactor is depicted in Figure 4.7. The power efficiency was found to be as high as 70% and the arc was

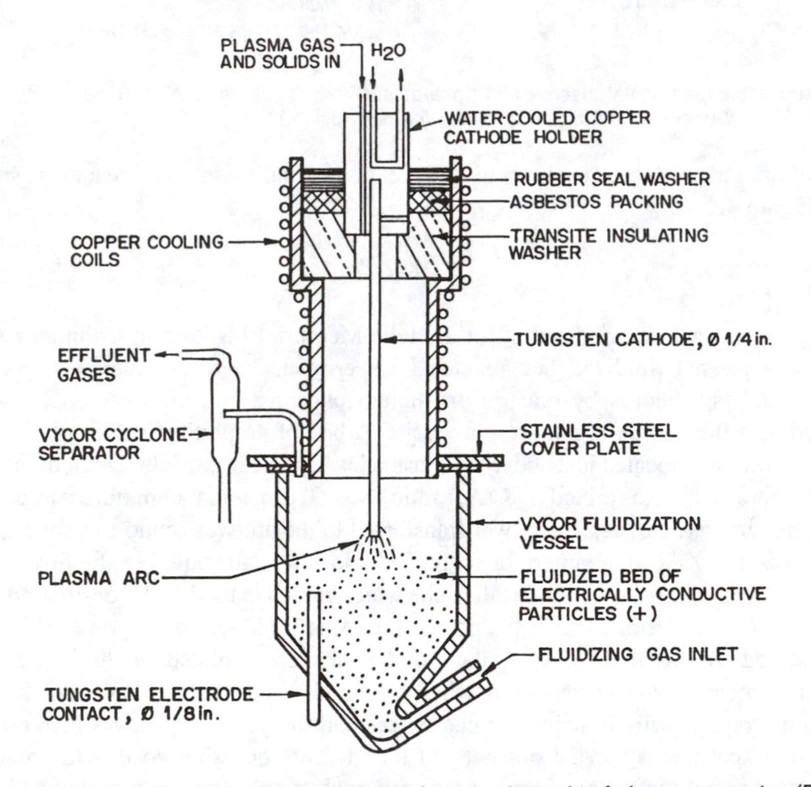

Figure 4.7 Fluidized bed (electrically conductive) as an electrode of plasma generator. (From Goldberger, W.M., *Chem. Eng. Prog. Symp. Ser.*, 62(62), 42, 1966. With permission.)

found to be very stable. The resistance across the arc varied depending on the electrode gap, and the bed temperature was between 1030 and 1200°C. Because the fluidized solid was graphite (–20 +65 mesh), agglomeration and bridging of the bed at the surface was noticed, and this was attributed to the formation of calcium carbide. This reactor required initial heating up to 800°C by a resistive-type heater, and then the arc was struck to generate plasma. In view of the fact that the fluidized bed particles accumulated with a calcium carbide coating, experiments were attempted to study the feasibility of producing calcium carbide by reacting lime with graphite particles. The reactor was found to perform well in reducing calcium carbide. The feed rate of the particles influenced the conversion level, which decreased with increasing feed rate. All the above studies carried out at Battelle Memorial Institute demonstrated the technical feasibility of achieving ultra-high temperatures to carry out chemical reactions in a fluidized bed in combination with a plasma generator. This work later prompted many research workers to further explore the characterization of such ultra-high-temperature reactors. It should be noted here that the fluidized bed electrode plasma reactor in a way resembles the electrothermal reactor (see next section); however, in the former, the top electrode, which is not immersed, is kept just above the bed to strike the arc for generating the plasma region above the bed.

2. *Inductively Coupled Plasma Fluid Bed*

An inductively coupled plasma fluidized bed was examined in a laboratory-size reactor by Manohar and Gleit.[68] The tests were carried out to increase the reaction rate and the yield in chemical synthesis. The reaction of alkali halides in an oxygen plasma in the pressure range 0.1–2 torr was investigated later. The reaction of sodium bromide in an oxygen plasma was studied because the reaction is relatively fast and the compound is available at high purity. An additional advantage is that the compound is not hygroscopic. A 50-W radio-frequency generator was used in a low-pressure plasma reaction. For a horizontal system, the reactor chosen was 3.8 mm OD and 1000 mm long. A solenoid coil consisting of 220 turns of $^1/_4$-in. copper tube served to bring about induction coupling. A 100-mm-long open-ended cylindrical boat was used to hold the sample. The reaction rate in the oxygen plasma was found to depend on the pressure. The position of the boat inside the tube was found to affect the yield. Thus, the yield was high in the tube region where the coil was wound, and the highest yield was obtained at a distance of approximately 50 mm from the center of the coil on the downstream side. In order to increase the reaction rate in the horizontal system, a finer size starting material, periodic mixing by rotating the reactor, and spreading the sample over a larger surface were required. Because a fluidized bed reactor offers the benefit of intimate mixing of plasma gas, experiments were carried out using a reactor of the same size as that used for the horizontal system but with a fluidized bed.

Fluidized bed reactors of different configurations were employed. A cylindrical column 22 mm in diameter fitted with a frit at the center of a 48-cm-long tube could not provide satisfactory performance as the frit in the coil region was overheated. An alternative design with a tapered (venturi-shaped) 260-mm-long bottom portion fitted with a fritted disc and a 520-mm-long upper cylindrical (22 mm in diameter) column failed due to slugging and channeling. A 650-mm-long, 38-mm-diameter column fitted with a 3-cm^2 fritted disc at the bottom proved to be a better choice.

The portion below the fritted disc was slightly tapered. As the fluidization characteristics of particles in the presence of electric fields and charged species are not known, experimental data on the variation in pressure drop with gas flow rate for different bed loadings and particle sizes were investigated. The results showed that particle sizes ranging from 44 to 177 μm and bed loadings ranging from 2.15 to 6.4 g could be fluidized well. The fluidized bed reactor could achieve a yield five times higher than that of the conventional horizontal-type reactor under similar experimental parameters. Thus, on a laboratory scale, the use of a fluidized bed for reaction of solids in a low-temperature plasma proved to be viable, and recommendations[68] were made to employ units of this type to achieve high reaction rates and yield.

C. Plasma Fluidized Bed Characteristics

1. Interparticle Forces and Minimum Fluidization Velocity

In a fluidized bed, the interparticle forces and the fluid-dynamic forces are considered, based on present knowledge, to be responsible for bubble formation and bed stabilization.[69] The presence of a high-frequency electromagnetic field[70] has been proven to be effective in suppressing bubble formation and bed stabilization, but the effect of an ionizing gas under a strong electromagnetic field in the fluidized bed reactor has not been investigated. Wierenga and Morin[71] attempted to ionize the gas in the interstitial interslug and the interbubble regions of a subatmospheric fluidized bed reactor by inductively coupling the fluidizing gas with a high-frequency plasma generator. Fluidized bed reactors made up of glass columns 10 and 20 mm in diameter were used to fluidize glass beads ranging in diameter from 0.17 to 3 mm with argon as well as hydrogen gas at pressures ranging from 100 to 6600 Pa. The Reynolds number at U_{mf} was in the range 0.004–0.2. The pressure drop across the fluidized bed was measured by using pressure probes just below the porous distributor plate and also above the bed. The fluidizing gas in the interstitial, interbubble, and interslug regions was ionized by microwave. The power circuit consisted of a 0 to 500-W microwave generator operating at 2.45 GHz, transmission lines, and a resonant cavity. The details of coupling the microwave signal to the plasma by the resonant cavity and the operation of the resonant cavity in several modes can be found in the literature.[72,73] The fluidized bed reactor was mounted coaxially with the resonant cavity. Four resonant modes of the cavity were employed for either hydrogen or argon gas. The fluidized bed of glass beads was melted at the central regions and ilmenite was fused, indicating that the temperature attained was above 2000 K. Glowing bubbles in a fluidized bed was observed. The fluidization characteristics under the plasma-free condition showed the validity of the Wen and Yu[74] correlation for predicting the minimum fluidization velocity for Reynolds numbers ranging from 1×10^{-4} to 4×10^{3}. Under the plasma condition, however, there was a practical problem in measuring the temperature profile in the reactor. The average temperature of the bed predicted by the Ergun[75] equation using an iterative method for a given mass velocity and pressure drop showed that the bed temperature would increase with both pressure and power level. For the same pressure and power level, the average bed temperature predicted for a bed fluidized by hydrogen gas was higher. The variation in pressure drop across a 4-cm-high sphere of glass beads 0.45 mm in diameter fluidized by argon at various pressures but at a constant power density of 29.9 W/cm³ is shown in Figure 4.8. The temperature dependence of gas viscosity was responsible for increasing the pressure drop at elevated temperature.

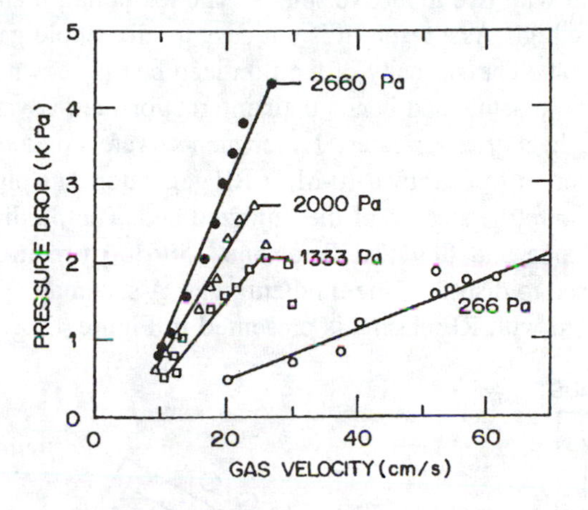

Figure 4.8 Pressure drop versus gas velocity in the presence of argon plasma (0.45–mm glass sphere, bed height 4 cm). (From Wierenga, C.R. and Marin, T.J., *AIChE J.*, 35(9), 1555, 1989. With permission.

The minimum fluidization velocity (U_{mf}) was found to be influenced conspicuously by particle size. For example, U_{mf} for argon gas varied as 1.75 d_p, while it varied as 2.24 d_p with the ionized plasma gas. This result indicated that interparticle forces were affected by particle size in the presence of an ionized gas. It was found that U_{mf} decreased with increasing bed temperature.[76] The same effect would be expected in a plasma fluidized bed, but the data of Wierenga and Morin[71] showed an opposite trend. This unexpected result was attributed to the effect of interparticle forces, which play a more significant role in plasma fluidization than in conventional fluidization. The interparticle forces can be enhanced by softening the bed material followed by adhesion and by an electromagnetic field. For the above studies, it was apparent that the plasma fluidized bed had different characteristics than those associated with conventional fluidization. The hydrodynamics of such systems needs to be evaluated precisely using the exact bed temperature, which should be measured instead of predicted using the Ergun equation.[75]

2. *Plasma Interaction with Fluid Bed*

a. *Tests*

Plasma interaction with a fluidized bed was investigated by Arnould et al.[77] with reference to its application for limestone decomposition. Cavadias and Amouroux[78] studied the preparation of all oxides in low- and high-pressure plasma reactors. Because the use of a catalyst like WO_3 enhanced the N_2 fixation yield from 8 to 19%, an atmospheric fluidized bed plasma reactor using a WO_3/Al_2O_3 catalyst was developed for nitrogen oxide synthesis. Lateral injection of an N_2–O_2 gas mixture into an inductive-type plasma was adopted. The fluidized bed was heated by a high-frequency plasma[79] which entered horizontally above the distributor plate. The plasma generator had a power level of 5 kW, and the torch was made of a 30-mm quartz tube wound with five inductive spirals. The temperature and the viscosity of the plasma gas are high. As a result, it is not easy to mix a cold gas or solid particle in the plasma stream. The viscosity of the fluidized bed is shown to be of the same order as that of the plasma, and hence their interactions are advantageous for both mixing and high heat transfer rates. Experiments were conducted with thermal plasma of Ar, N_2, or air formed by a 40-MHz RF generator. The plasma was directed at various angles onto the surface of the fluidized bed. The fluidized bed consisted of a preheating zone and fluidized limestone 250–300 μm in size. A ball-type distributor was used to distribute the fluidizing gas. A schematic of such a fluidized bed reactor coupled with RF plasma is presented in Figure 4.9.

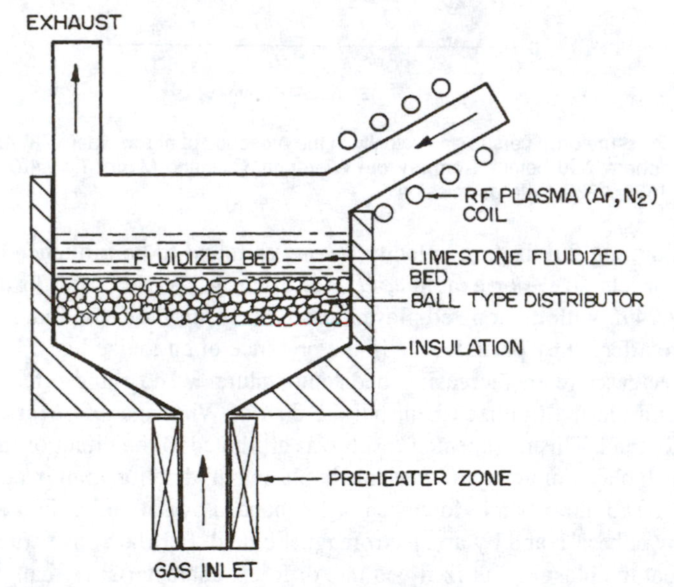

Figure 4.9 Plasma fluidized bed with ball-type distributor. (From Arnould, Ph., Cavadias, S., and Amouroux, J., in *Proc. 7th Int. Symp. on Plasma Chemistry,* Timmermans, C.J., Ed., Eindhoven Univ. Tech., Eindhoven, 1985, 195. With permission.)

b. Results

The temperature of the fluidized bed, as measured by Ni–Cr thermocouples, ranged from 500 to 1100°C even through the bed showed high temperature due to the fusion of limestone. The minimum fluidization velocity (U_{mf}) measured indicated the validity of the Wen and Yu[74] correlation. The minimum fluidization velocity can be predicted by a more refined method from the work of Lippens and Mulder.[80] The viscosity of the fluidized bed, which would depend on U_{mf}, was found to be close to that of the plasma. The heat transfer coefficient varied with the power, the gas flow rate, the plasma temperature, the fluidized bed temperature, and the gas flow in the plasma as well as in the fluidized bed. The heat transfer coefficient was compared with the correlation of Baerus and Fetting[81] and a wide difference was observed. Typical values of the experimental data and the predicted heat transfer coefficient are given in Table 4.3. The heat transfer rate data shown in Table 4.3 indicate the possibility of achieving a high heat transfer coefficient. The good mixing of the fluidized bed and the plasma may be attributed to this. Arnould and Amouroux[82] later developed a mathematical model for limestone decomposition in a plasma fluidized bed and showed that external mass transfer would be the rate-determining step. The mathematical model can establish correlations to predict both the particle temperature and the mass transfer coefficient. In a normal gas fluidized bed, the distributor plays an important role, and this aspect is presented in Chapter 5. In a plasma fluidized bed, the choice of distributor plate material is often a problem. As the quality of fluidization is a very important factor for efficient reactor performance, proper grid design and grid selection for high temperatures are essential. Unfortunately, scant information is available on distributors for high-temperature operation.

Table 4.3 Heat Transfer Data for Plasma Fluidized Bed

	Plasma				Fluidized Bed		Heat Transfer Coefficient (W/m² K), $h = Q/A(T_p - T_b)$	
Sr. No.	Power (W)	Gas	Gas Flow Rate (l/min)	Temperature (T_p, K)	Gas Flow Rate (l/min)	Temperature (T_b, K)	Exp	Predicted[81]
1	760	Ar	20	1730	10	1073	1508	5313
2	1345	Ar	20	4654	6	1123	423	6270
3	642	N_2	22	1780	7	773	837	2326
4	675	N_2	21	1650	6.75	1040	1189	1295
5	700	Air	20	1700	10	925	651	3221
6	675	Ar	12	2900	8	1040	353	4004

From Arnould, Ph., Cavadias, S., and Amouroux, J., in *Proc. 7th Int. Symp. Plasma Chemistry*, Eindhoven Univ. Tech., Eindhoven, 1985, 195. With permission.

3. Plasma Spouted Bed

a. Tests

A plasma spouted bed reactor is a type of reactor in which a distributor plate is not required. This type of reactor is similar to a conical vessel fluidized bed reactor. In a plasma spouted bed reactor, the spouting at the conical bottom of the reactor is generated by the plasma jet itself. Hence, a plasma spout reactor can be fitted at the conical bottom directly by plasma-generating torches. A plasma spouted bed reactor was studied by Jurewiez et al.[83] using alumina powder as the spouted bed and an argon or nitrogen DC plasma jet. Detailed information on the spouted bed can be found in the literature.[84] The main advantage of a plasma spouted bed is that the particles come into contact with the tail end of the plasma for a short time, attain high temperatures, and are rapidly quenched in the bed and reenter the hot plasma. However, the reactor is less energy efficient because the radiation heat and the convective heat are not fully utilized. Hence, Jurewiez et al.[83] attempted to develop an energy-efficient plasma spouted bed reactor for applications in thermal processing of powders, gasification of coal, and mineral treatments such as chlorination and roasting. The performance of the plasma spouted bed reactor must be improved with respect to its particle elutriation, attrition, and recirculation characteristics.

Studies were carried out on a reactor made of Pyrex except for the brass spout nozzle section wherein a DC plasma torch was housed. The anode nozzle of the plasma was 7.33 mm in diameter. A schematic diagram of the plasma spouted bed reactor is shown in Figure 4.10. Alumina was used as the spouted bed. Three size fractions of alumina in the overall range 53–1000 μm were used. Argon and an argon/nitrogen gas mixture at flow rates of 47 and 37.3 l/min, respectively, were used, and the power level varied from 9 to 28 kW. Experiments could be carried out only for short periods (5–10 min) for each run because the reactor was made of perspex. Temperatures were measured using a radiation pyrometer as well as chromel–alumel thermocouples.

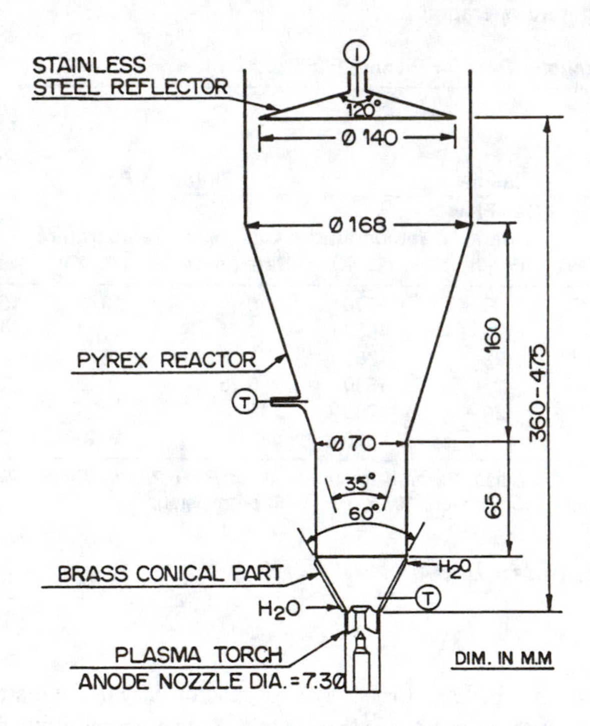

Figure 4.10 Plasma spouted bed reactor. (From Jurewiez, J., Proulx, P., and Boulos, M.L., in *Proc. 7th Int. Symp. on Plasma Chemistry,* Timmermans, C.J., Ed., Eindhoven Tech. Univ., Eindhoven, 1985, 243. With permission.)

b. Results

Reactor stability was tested visually for different loadings of the alumina powder when the plasma was ignited. A luminous plasma core was seen when the bed was not sufficiently loaded. When the plasma core was just submerged, the appropriate condition for a high rate of solid circulation and fluidization was achieved. When the bed height was beyond a certain level, the bed was unstable and bubbles and slugs were prominent. Measurement of the bed temperature profile showed the uniformity of temperature in the bed except in the water-cooled brass cone through which the plasma was emerging into the bed. Particle attrition, due to particle–particle or particle–jet interactions or due to the thermal shock of the particles during their short flight into the plasma jet region, was quite significant even for 10 min of operation. Thus the reactor, which had excellent particle recirculation and mixing characteristics, did not have desirable characteristics with respect to particle attrition and elutriation. Concentrated research must be carried out on the development of such a novel reactor. In order to characterize the fluid bed performance, in particular the bubbles, glow discharge probes were developed.[85] A needle placed in front of a conducting surface and connected to a high voltage source could detect the clear passage of a bubble when the signal ratio of bubble passage past the needle to particle flow past the same point was large. The signal ratio thus obtained was affected by the applied voltage, the gap between the needle and the metal surface, and the degree of bed expansion. The microdischarge corresponded to a plasma condition, and voltages up to 6 kV were reported to have been used.

4. Mechanism of Plasma Jet Quenching

a. Plasma Gas Temperature

The problem of quenching a plasma in a fluidized bed was investigated and the pertinent energy balance presented by Kolinowski and Borysowski.[86] The quenching of a plasma using particulate heat transfer agents such as Fe_2O_3, Fe_3O_4, and Al_2O_3, which have a wide particle size range (0.5–3.3 mm), was studied by Lobanov and Tsyganov[87] for various gas flow rates and temperatures.

The mechanism of quenching a gas plasma jet in a fluidized bed is based on the heat transfer from the gas entering the plasma jet to the gas–solid particle boundary. It was shown[88] that radiative heat transfer accounted for only 0.3% for quartz sands smaller than 200 μm and 0.1% for larger particles. Thus, determination of the heat transfer coefficient from knowledge of convective and radiative heat transfer has significance in plasma fluidized bed quenching. The temperature (T_g) of the plasma gas along a nozzle of diameter d_o varies with distance (X) from its initial temperature (T_o) according to the equation

$$T_g = T_o d_o/(X + d_o) \tag{4.2}$$

The gas velocity (V_g) along the jet distance (X) changes according to the expression

$$V_g = 2.73\ V_o r_o/C\, X \tag{4.3}$$

where V_o, r_o, and C are, respectively, the initial isothermal gas velocity, the jet radius, and a coefficient that takes into account jet expansion and gas leakage through the gas–solid boundary layer of the jet. The viscosity coefficient for nitrogen gas has been evaluated over the temperature range 800–6000 K using the experimentally tested equation

$$\mu_g = \mu_o \, (T_g/T_o)^{0.83} \tag{4.4}$$

where μ_o corresponds to 6000 K.

b. Effect of Variables

The geometric characteristics of the torch flame immersed in a fluidized bed under quasi-stationary conditions of discharge show that the initial part of the flame is not large but is commensurate with the diameter of the nozzle. Theoretical and experimental results on the variation of the axial velocity along the plasma stream[89] showed a maximum deviation of 40% at the flame and a minimum of 19.8%. An important finding was that the velocity profile (expressed as a dimensionless quantity) at the initial portion of the submerged stream and the stream passing through the fluidized bed did not seem to coincide. An electro-spark reactor[90] similar to the electrothermal fluidized bed was characterized with respect to the power parameters. The current-versus-voltage relationship showed the possibility of establishing a stable and efficient plasma. The total resistance of the static bed of the electro-spark reactor was found to decrease with increase in bed height, particle diameter, and inner diameter of the electrode.

5. Plasma Jetting Fluid Bed

a. Reactor

A plasma jetting fluidized bed reactor was developed by Kojima et al.[91] with the objectives of delineating the basic principles of plasma-enhanced chemical vapor deposition on particulate solids and evaluating the efficiency of such a reactor. Fluidized bed reactors, in general, are favored for coating particulate solids by chemical vapor deposition. The possible coating processes in nuclear industries that use fluidized bed reactors were discussed in Chapter 3. A plasma jetting reactor in combination with a fluidized bed has the advantage of exciting the reacting species in the plasma and coating the solid particles which are periodically exposed to the reaction zone due to the high rate of solid circulation between the jet and the annulus region that forms between the jet and the reactor wall. A 49.5-mm-ID stainless-steel fluidized bed was combined with a nontransferred-type DC plasma generator constructed using a 10-mm diameter tungsten cathode, a water-cooled copper anode, and a nozzle 4 mm in diameter and 10 mm long. A schematic diagram of such a plasma jetting fluidized bed reactor is shown in Figure 4.11.

b. Methane Decomposition Studies (DC Plasma)

Studies were carried out on the coating of alumina particles with carbon by the decomposition of methane. Hydrogen gas was also used to effect the decomposition

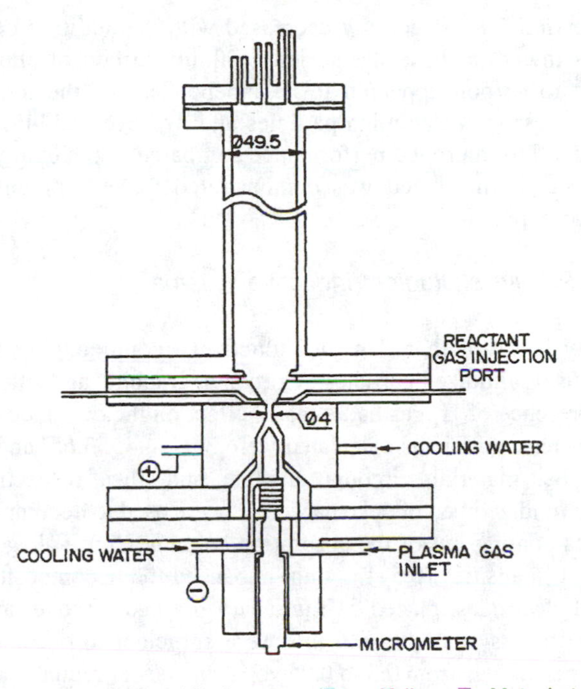

Figure 4.11 Plasma jetting fluidized bed reactor. (From Kojima, T., Matsukata, M., Arao, M., Nakamura, M., and Mitsuyoshi, Y., *J. Phys. II (Paris),* 1, C2-429, 1991. With permission.)

of methane. Alumina particles ranging in size from 250 to 500 μm in 200-g batches were used to evaluate reactor performance. The reactor was initially tested for current-voltage behavior by injection of methane gas either through the argon plasma or through the side port, forming a vortex around the plasma jet. Voltages up to 70 V and currents up to 50 A were used. The I-V characteristics were not distributed when methane was introduced through the side port, and hence the plasma jet was stable. However, the I-V characteristics for the same current corresponded to higher voltages when methane was passed through the plasma-forming argon gas. This indicated that methane, when passed around the plasma jet, could not reach the electrode zone. A detailed study on the conversion of methane in an empty bed and in a fluidized bed of alumina revealed that the conversion level for methane decreased when fluidized alumina particles were present as compared to an empty bed. In an empty bed reactor, the highest level of methane conversion occurred when this gas was passed through the argon plasma. When hydrogen was mixed, the conversion decreased. On the other hand, addition of hydrogen to methane, injected through the side port and forming a vortex around the plasma jet, enhanced methane conversion. This effect was reversed in a bed of fluidized alumina particles. In a fluidized bed which was coupled with the plasma jet, the mean temperature was around 570 K, and methane decomposition could hardly take place at this temperature. Hence, it was surmised that the main reaction zone was the plasma jetting zone, where the excited methane species reacted and decomposed on the alumina particle surface. In an empty plasma jetting reactor, the selectivity to carbon was not affected by the level of conversion. The selectivity decreased with the addition of hydrogen and with side port injection. In a plasma jetting fluidized bed of alumina, however, the selectivity to carbon appeared to be independent of the mode of methane injection. The presence of alumina particles and hydrogen addition increased the selectivity and led to improved performance compared to the empty reactor. Thus, a plasma jetting fluidized bed was demonstrated to be a potential reactor for chemical vapor deposition.

c. Methane Pyrolysis Studies (Inductive Plasma)

The pyrolysis of methane in an inductively coupled plasma reactor was investigated[92] using fluidized particles of zirconia, graphite, and silica to quench the plasma. The presence of H_2 gas had a major effect on the decomposition of CH_4 to C_2H_2. The transfer coefficients calculated were 3.97–4.45, 3.65, and 4.32 kcal/(m^2. hr. K) for the bed materials zirconia, graphite, and silica, respectively. Electrical discharges in a fluidized bed of coke particles[93] were used to decompose the residual nitrogen oxides coming out of the absorption column. High voltages (150–220 V) and low current intensities (0.5–1.0 amp) enhanced the decomposition rate. Brush and perforated electrodes, placed 21 mm apart in a bed of coke (apparent density 1.5 kg/m^3), and a gas velocity >0.6 m/s were sufficient to decompose and bring down the nitrogen oxides from 0.3 to 0.05 vol%; the power required to process every 1000 Std. m^3 of the flue gas was 20 kW.

d. Local Characteristics

Masahiko et al.[94] studied the local gas mixing behavior at the bottom of a jetting plasma fluidized bed with the main objective of monitoring the vertical progress of methane conversion in a DC plasma. This study was an extension of investigations on the same type of apparatus used by Kojima et al.[91] The purpose of the investigation was to determine experimentally the reaction near the plasma jet zone where a high-temperature field and a steep temperature gradient were present. Thus, the bottom spouting zone could not be regarded as the isothermal part of the reactor. Reacting species at the plasma spout would be excited, transported along with the plasma gas, and the reaction product would be quenched. The progress of the reaction was monitored by a specially developed gas-sampling probe. Helium was used as the tracer gas in the argon gas plasma to study gas mixing in the plasma and in the annulus region formed between the spout and the reactor wall. The fluidized bed was at atmospheric pressure and always at a temperature below 480 K for a bed of alumina at 200 g loading with an argon plasma powered by a generator with a capacity of 24 V and 40 A.

Gas mixing in the jetting zone of the plasma (i.e., at the nozzle), as determined by helium concentration measurements, showed that the gas exchange under non-plasmatic conditions occurred between 1.5 and 5 cm above the jet, while such exchange occurred intensively even at 0.5–2 cm under the plasmatic condition. In the presence of alumina solid particles during fluidization, the concentration profile

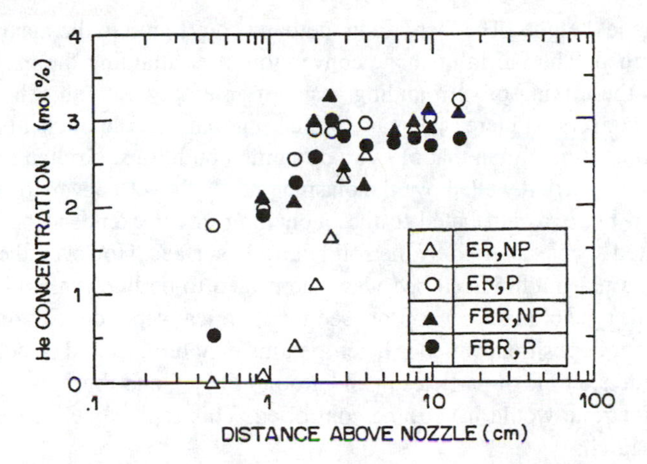

Figure 4.12 Variation in helium concentration with distance above nozzle. ER = empty reactor, FBR = fluidized bed reactor, NP = nonplasmatic condition, P = plasmatic condition. (From Masahiko, M., Hikaru, O.H., Toshinori, K., Yusuke, M., and Korekazu, U., *Chem. Eng. Sci.*, 47(9–11), 2963, 1992. With permission.)

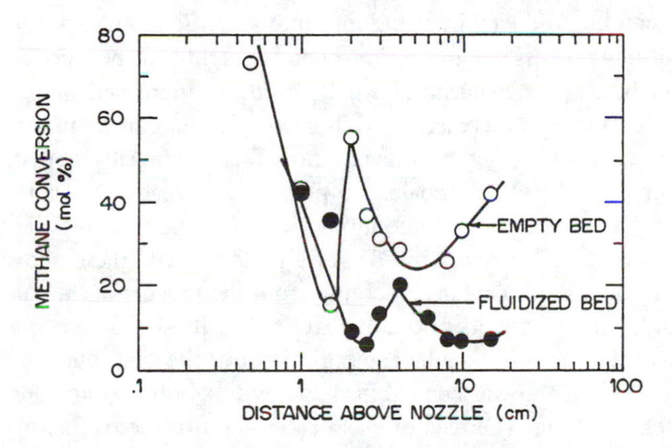

Figure 4.13 Progress of methane conversion versus distance above nozzle. (From Masahiko, M., Hikaru, O.H., Toshinori, K., Yusuke, M., and Korekazu, U., *Chem. Eng. Sci.*, 47(9–11), 2963, 1992. With permission.)

along the nozzle overlapped for both plasmatic and nonplasmatic conditions. Gas mixing in the jetting region was believed to occur due to the circulating solids which drew the gas into the jet. Gas mixing due to the plasma phenomenon was assumed to be negligible. Typical experimental results in the variation of the helium gas concentration along the nozzle are shown in Figure 4.12. Methane conversion along the jet resulted in an interesting plot, as shown in Figure 4.13 for the case of an empty reactor and a fluidized bed reactor. The conversion decreased up to a height of 1.5 cm in both cases by the same order of magnitude. However, above a height of 1.5 cm, the methane conversion increased and reached a maximum at 2 cm in the empty reactor and 4 cm in the fluidized bed reactor. These heights corresponded to the plasma jet heights. The increase in methane conversion to the maximum could not be explained. The fall in methane conversion after attaining the maximum was attributed to the mixing of surrounding gases in an empty column. Above a height of 8 cm, the conversion increased again, indicating that the role of chemical kinetics was more predominant than that of hydrodynamic conditions. Excited species were thus found to be carried well above the plasma jet. The lower conversion of methane in a fluidized bed was attributed to the quenching effect exercised on the plasma and the excited species carried by the solid particle surface. However, the selectivity of carbon in a plasma fluidized bed was determined to be higher than in an empty bed. Thus, this type of reactor was proposed for chemical vapor deposition. However, chemical vapor deposition, which is heterogenous in nature, should be achieved with a longer residence time of particles in and around the jet, and thus gas–solid contact around the jet zone would have to be controlled. This topic should be emphasized in future research.

6. Radially Coalesced Plasma

The arc which generally constricts due to thermal convection to surrounding gas has a tendency to coalesce due to electromagnetic attraction. Thus, large volumes of discharge can be produced by using multiple arcs. Such an expanded volume of plasma is useful in process metallurgy and chemical synthesis because the probability of interaction between a material and the plasma is increased mainly due to the longer residence time. The coalesced volume of plasma can be made stable if the arcs are electrically isolated so as to avoid the effect of current fluctuation of an arc over the other arcs. A radially coalesced plasma was made by using six pairs of electrodes, and a coalesced discharge up to 20 cm in diameter was incorporated in a fluidized bed reactor.[95] A boron nitride sieve plate placed 10 cm above the plasma chamber distributed the argon gas emerging from the plasma at the rate of 20 l/s to fluidize 7-mm spherical pellets of titania (TiO_2). The fluidized bed consisted of two regions measuring 1.8 m in length. The region above the distributor behaved like a spouted/fluidized bed and was conical in shape, with a bottom diameter of 7 cm and a top diameter of 44 cm. The reactor was made of saffil fiber refractory and coated with a refractory wash to withstand abrasion. The main fluidizing zone was provided with an enlarged settling zone to arrest the carryover of fines from the reactor. Experiments were conducted on 1-kg batches. Pellets of titania and ilmenite were successfully fluidized. In an argon–nitrogen plasma, titanium nitride could be produced successfully. Due to the wall effect, the plasma fluidized bed was found to behave somewhere between a spouted and a fluidized bed. The reactor was tested to fluidize molybdenite (MoS_2) pellets, but the test was not successful because the pellets disintegrated. The wear and tear on the boron nitride sieve plate was minimal. The reactor was operated for 30 min, as the gas available was sufficient to run the experiments only for that period. The maximum temperature of 1673 K occurred in the reactor at a height of 6 cm above the sieve plate. This could be enhanced up to 2500 K by recirculating the hot gas. Reactions with high enthalpy could not normally be carried out successfully during in-flight reaction in the plasma jet. This was due to the alert residence time of the particles in the high-velocity plasma jet. To this effect, a coalesced arc plasma fluidized bed could pave the way for improved performance of the plasma jet reactor.

Multiple arc discharges for application in metallurgical reduction and melting were developed by Harry and Knight.[96] The multiple arc discharges can be generated from a single power supply, and this allows a reduced load of current on the electrodes. The total current thus can be increased. The electrodes, when separated by a distance less than the arc column length, tend to coalesce and form a large volume of ionized gas. This coalesced plasma can be used for melting, liquid-phase smelting, and also for reaction with a particulate phase. In arc furnaces, graphite electrodes may be replaced by nonconsumable multiple electrodes. Multiple electric discharges of this type can be used for gas heating and also for in-flight chemical reactions. Multiple electrode discharges can replace graphite electrodes in the manufacture of steel from scrap, and savings of up to 10% of the overall cost of steel production can be achieved. Multiple plasma torches from a single power supply can be used in plasma furnaces to provide a larger reaction zone and a uniform temperature at very high power levels. Arc lengths of 0.5 m at a current level of 200 A per torch can be used to achieve plasma furnace temperatures of up to 1200°C to process materials at feed rates of up to 50 kg/hr.

D. Feeding Methods of Particulate Solids in Plasma

1. *Gas–Solid Feeding*

Most in-flight reactions of particulate solid materials in a plasma gas require an efficient feeding device because of the mixing problem and the short residence time in the high-velocity plasma jet. The solid reactant can usually be in the form of consumable electrodes which have to be replaced periodically. Additionally, a precision mechanism is required to move the electrode continuously at a rate corresponding to the electrode material consumption. The electrode burnoff rate would impose severe operating limitations on the plasma jet reactor. Giacobbe[97] gave an account of the literature dealing with powder feeding devices and the companies that have developed various gas–solid blending devices for feeding the reactant directly into the plasma. For fine powders, such as carbon black, blending is difficult due to the tendency of such powders to agglomerate, and hence it is often essential to look for a reliable gas–solid feeding system. Such a feeding system is critical especially when the gas-to-solid ratio in the feed is to be maintained and controlled to conform to the stoichiometric requirement. Giacobbe[97] described the development of gas–solid feeders and evaluated their performance.

2. *Feeder Types*

Three types of feeders have been described as advanced powder feeders: (1) impeller-type feeder, (2) slotted cylinder powder feeder, and (3) slotted disc powder feeder. Schematic diagrams of these feeders are shown in Figure 4.14. In an impeller-type feeder, powder material from a reservoir was fed into a fluidizing chamber through a rotary valve. A set of rotating impellers in the fluidizing chamber delivered

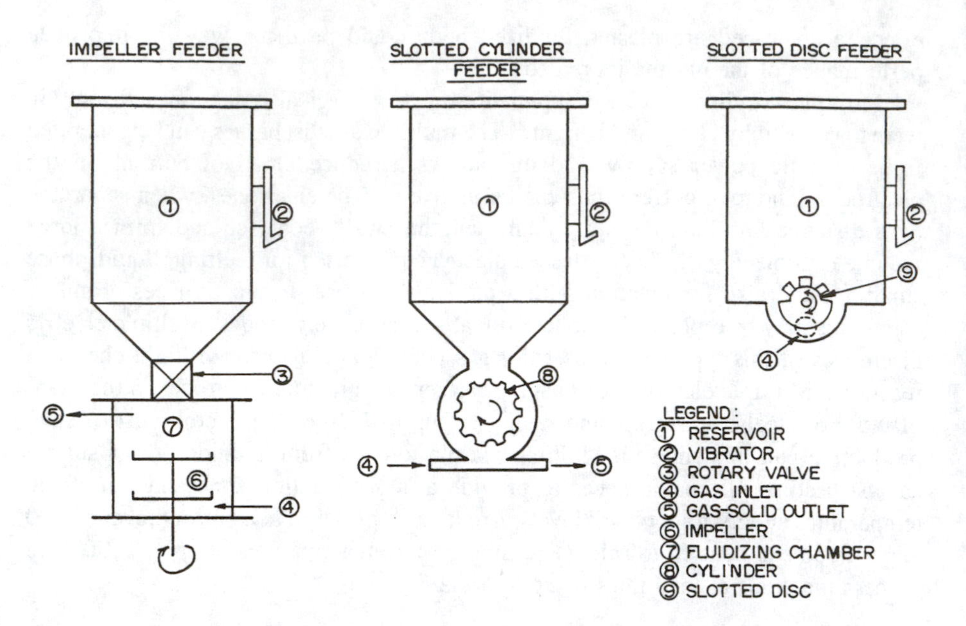

Figure 4.14 Various types of feeders for gas–solid mixture into plasma jet.

the gas–solid mixture into the reactor. The feeder performance was not satisfactory as it could not feed the optimum ratio of gas–solid at the high flow rates required by large plasma reactors. The average feed rate, as determined by emptying the powder from the reservoir for a set of fixed operating conditions, often led to erroneous data mainly due to the powder holdup in the rotary valves and the impellers in the fluidizing chamber. Hence, the feed rate could not be evaluated precisely. Even a slotted cylinder powder feeder, which could be considered to be superior to the impeller type, failed to perform satisfactorily. In this type of later generation feeder, a cylinder with slots made 15° to the axis of the cylinder served to discharge the powder rotation of the cylinder into a tube where the gas carried the particles into the plasma reactor. This feeder was considered to perform better than a rotary valve feeder because the feed rate of solids could be predicted precisely from the slot volume, the powder density, and the rpm of the cylinder. However, this type of feeder failed to provide reliable data when powders such as carbon black, which stick to and close the slots in the cylinder, were used. Hence, a third version of the feeder, known as a slotted disc feeder, was developed. This type of feeder has slots on the disc which during rotation drags the powder from the reservoir and lets it out in the gas stream, which dislodges all the solids in the slots of the disc. This type of feeder can perform satisfactorily and allows the feed to enter the plasma jet without pulses when the rpm of the disc is kept at 20. In order to overcome this critical speed barrier, either the diameter of the slotted disc or the size of the slots must be altered. Alternatively, multiple slotted disc feeders operated at a synchronized speed were suggested by Giacobbe.[97] A fluidized bed device was described as a satisfactory for fine particulate solids in an inductively coupled RF plasma. The performance characteristics of such a feeding device and its calibration are described in detail[99] for application in spectrochemical analysis of solid elements such as silicon, zinc, and copper.

IV. APPLICATION OF PLASMA FLUIDIZED BED IN MATERIALS PROCESSING

A. Scope of Plasma Application in Extractive Metallurgy

Application of plasma for high-temperature chemical synthesis of compounds such as nitrogen oxides and acetylene[100] and pyrolysis of methane[101] has been primary interest ever since plasma was first explored for high-temperature gaseous reactions. Subsequently, its manifold applications in powder processing and metallurgical extraction have aroused the interest of many research scientists. A compilation of the work carried out since early 1967 on the processing of powders in high-temperature plasma was presented by Waldie,[102] highlighting the processing methods and economics. This comprehensive review covered plasma spraying, particle spheroidizing, ultrafine powders, refractory carbides and nitrides, and metallurgical extraction. A brief review by Waldie,[103] provided an account of the particle dynamics and heat and mass transfer during plasma interactions. There are numerous applications of plasma in extractive metallurgy. Plasma applications in iron ore reduction, ferrovanadium and ferrochromium production, dissociation of zinc, magnetite spheroidization, molybdenum production, and molybdenum disulfide decomposition are covered in detail in a chapter devoted to plasma applications in extractive metallurgy by Ettlinger et al.[104] For successful application of plasma in extractive metallurgical processes, technoeconomic studies on the process and a commercially viable plasma reactor should be conducted.

Metallurgical processes usually require a large and high-power reactors. Hence, those ventures into the development of plasma fluidized beds are advised to consider raw material availability and marketability of the processed material. It has been established that a plasma can be used on a commercial scale with a high level of efficiency in production. The plasma fluidized bed reactor, which is emerging as an improved multipurpose high-temperature reactor, could obviously replace the conventional reactor. However, developmental work must be undertaken before it is used in huge plants such as those in the steel industry; 100-MW plasma reactor would be required for a modest steel production rate of 250,000 tpy. The conventional fluidized bed is reported[105] to have been combined with a plasma smelting furnaces (Hi-Plas) to achieve integrated production of iron and ferroalloys. This process has been found suitable for metal fines, ore fines, and dust. The uses of plasma in extractive metallurgy were reviewed by Gauvin and Choi,[106] with special emphasis on its application in the production of refractory metals such a molybdenum, zirconium, titanium, tungsten, chromium, and vanadium. Various types of plasma reactors for the above applications have been discussed. Special mention should be made of the plasma production of zirconium, which eliminates the Kroll process. The Kroll process, which is the final step in zirconium production, is cost and labor intensive in addition to consuming magnesium metal. This step alone accounts for about 35%

of the total cost of zirconium production. In plasma production, $ZrCl_4$ is directly decomposed in a continuous manner to produce dense zirconium ingots. The technical feasibility of this process has been well demonstrated.[107,108] This process has the potential advantage of producing zirconium metal without using the cumbersome Kroll process. Titanium tetrachloride can also be decomposed in a similar manner to produce titanium metal directly. A fluidized bed apparatus for producing titanium metal was described by Parker.[109] In this apparatus, titanium metal powder and glass beads were fluidized by $TiCl_4$ and H_2 which were preheated to 60°C. A glow discharge was produced between the electrodes immersed in the bed of electrically conducting titanium particles. The flow of current was not direct between the electrodes, but the potential applied between the electrodes generated sparks, and glow discharge was established across the conducting titanium particles, which grew in size due to the deposition of the titanium metal formed when $TiCl_4$ was reduced by H_2 gas. Any unreacted $TiCl_4$ or H_2 was purified and recycled.

B. Plasma Fluid Bed Processes

1. *Particulate Processes*

a. Spheroidizing

When spheroidized, that is, made into a spherical shape, particles have several attractive features, such as maximum apparent density, free-flowing characteristics, lowest surface area per unit volume, close size range, and controlled porosity. All these features impart desirable geometrical and mechanical characteristics to materials used in applications such as fluidization and making nuclear fuel, porous sintered bodies, lubricant powders, high-temperature filters, and maximum flow sintered filters. The methods most commonly used to produce spherical particles are (1) the consumable wire method and (2) atomization. The consumable wire method can produce particles ranging in size from 5 to 150 μm but not in a close size range. Furthermore, current fluctuation can shift the size range adversely. The atomization technique is limited to materials with a low melting point and has not been used for high melting point or refractory metals. Potter[110] gave an account of spheroidizing particles using a DC plasma. By this process, particles of metals and nonmetals which include alumina, zirconia, columbium, zirconium, uranium dioxide, uranium monocarbide, tantalum, and zirconium diboride can be spheroidized efficiently. Particles can be processed at a rate of 3 to 20 lb/hr, with sizes ranging from 6 to 200 mesh.

Particles of tungsten (19.3 g/cc) as small as 0.5 μm and beryllium (1.85 g/cc) particles 35 μm in size were reported to have been successfully produced by the DC plasma spheroidizing technique. In this technique, the material to be spheroidized is allowed to pass through the plasma arc struck between the cathode and the anode. The particle is carried into the plasma gas by a carrier gas which is usually the same as the plasma-forming gas, and the carrier gas flow accounts for 5–10% of the plasma flow. The solid particle, during its flight in the plasma gas, melts, and when it emerges as a liquid droplet from the flame, it is collected into the cooling chamber. The atmosphere can be either air or an inert gas, depending on the active nature of the spheroidized particles. An RF plasma, like a DC plasma is capable of spheroidizing particles. However, there are limitations due to the low enthalpy of the gas and the problem of extinguishing the plasma flame by the carrier gas. Plasma spheroidization can be carried out by coupling the plasma with a fluidized bed, which acts as an efficient quenching medium and collects the particle, maintaining its spherical shape without sticking or agglomeration. However, a problem is always anticipated in fluidizing particles in the submicron range mainly due to their cohesive force.

b. Coating

Fluidized bed reactors are reported[111] to have been used for purifying metal powders by glow discharge. Metal powders such as Sm–Co, Fe, and W were continuously purified and chemically activated by fluidizing the powders in a monoxidizing atmosphere in the glow discharge obtained between two electrodes under some specified pressure where the selected range for the product of the electrode distance and the pressure is 0.01–10 torr. The impurities on the metal were released by ion impact and pure metal was continuously removed. Spherical grains (~500 μm) of SiO_2 and ZrO_2 were made successfully made[112] in a plasma fluidized bed reactor using a mixture of oxygen and nitrogen gas. Pure ZrO_2 as well as yttria- or calcia-stabilized ZrO_2 were used. The refractory oxides SiO_2 and ZrO_2 evaporated in a plasma and condensed in a finely divided form. A plasma spouted bed was tested by Horio et al.[113] for the granulation of Al–SiO_2 and Al–Fe mixtures. The reactor was a conical–cylindrical type. The cylindrical portion was 50 mm in diameter and the orifice provided at the conical bottom was 4 mm in diameter. An argon plasma, powdered at 100 V and 40 A, was tested successfully with a spouted bed of SiO_2 particles ranging in size from 420 to 500 μm.

Plasma techniques in conjunction with a fluidized bed were also examined for particle coating. Surface treatment of particles/powders with isostatic polypropylene in a low-temperature plasma of oxygen and nitrogen, produced by the action of a high-frequency discharge in a fluidized bed, was investigated by Derco et al.[114] The reaction product was characterized by infrared nuclear magnetic resonance, and electron spin resonance spectra and the influence of exposure time on coating was investigated. Investigations of a similar type were reported by Conte[115] for coating with thermoplastic, thermosetting, and polyphenylene sulfide polymers using an electrostatic fluidized bed and plasma spray techniques.

2. Advanced Materials Processing

a. Fine Powders

Ultrafine powder materials are conventionally prepared by chemical vapor deposition using empty space or aerosol reactors. Fluidized bed reactors have now been used for chemical vapor deposition processes. The fluidized medium is agglomerated fines that will have improved surface properties in a fluidized state. Morooka et al.[116]

summarized the use of the fluidized bed for the processing of fine particles as advanced materials. A fluidized bed reactor can be conveniently used for Geldart Group A and B powders (see Chapter 1 and Table 1.2 on types of fluidization). However, powders that are cohesive in nature are usually difficult to fluidize. Gibilaro et al.[117] handled powders larger than 10–30 μm. For advanced ceramic applications, powders of a size below 1 μm are required. Interestingly, fine powders that have a high degree of cohesive force have been reported[118-120] to have good fluidization behavior due to the tendency of these powders to form agglomerates which can subsequently fluidize well. However, Van der Waals force[121] – dominant agglomeration and shearing-force-dependent disintegration ultimately determine the fluidization characteristics. The dynamic equilibrium between particle agglomeration and disintegration depends on many parameters, such as particle size, surface condition, nature of the material, temperature, and humidity and velocity of the gas. When particles are fine and not fluidizable, the vibration technique[122] and the addition of coarse particles[123] to the fine to promote turbulence may be helpful in solving fluidization problems. Thus, it seems feasible to use fluidization for the processing of ultrafine powders. Submicron-size aluminum nitride particles were prepared[124,125] using a fluidized bed by the reaction of 15-μm aluminum particles with a mixture of N_2 and NH_3. The reaction occurred mostly at the freeboard zone of the reactor. Pure aluminum particles fluidized by N_2 gas at ambient temperature were carried to the freeboard zone, which was kept at a temperature in the range of 1423–1823 K, and the reaction of the molten aluminum metal at this temperature occurred with the N_2/NH_3 gas mixture injected in the freeboard zone. AlN particles with 100% conversion of aluminum particles were obtained at 1523 K with 5 mol% NH_3 in N_2 gas. The AlN particles were in the form of fine cracked spherical shells, which were ground gently to obtain 0.2-μm AlN particles. The AlN spherical shell appeared to form when the molten aluminum inside the spherical shell escaped through the cracks on the AlN surface. These cracks were presumed to have formed due to the differential thermal expansion of AlN and molten aluminum. Horio et al.[126] proposed the use of the fluidized bed technique to produce nitrides of silicon, titanium, or aluminum in the form of submicron particles. Kato et al.[123] and Chiba et al.[127] used the fluidized bed reactor to decarburize SiC particles to remove residual carbon without burning the particles. It is essential to decarburize commercial SiC which contains 20–30% by weight of excess carbon. All the above processes require high-temperature operation, and the combination of plasma with a fluidized bed can deliver products of good quality at a reasonably high rate.

Design data for a semicommercial fluidized bed reactor for polycrystalline silicon were reported[128] for a production capacity of 16 tpy. A plasma fluidized bed 10 ft in diameter operating with CO plasma at 13,000 kW was reported[129] to produce 150 tons/day (tpd) of CaC_2 by the reaction of powdered carbon (fed at 57 tpd) with lime (containing 98.7% calcium oxide, fed at 130 tpd). The average fluidized bed temperature was 800–1900°C. The flow of CO for the plasma generator was kept at 1600 ft^3. Once formed, the pellets of CaC_2 had a thin film of liquid on the surface, and this facilitated the surface reaction of lime and carbon. The hot product was cooled by preheating the feed plasma gas, and the cooled product was shipped as such without any further processing.

b. *Particle Nitriding*

Okubo et al.[130] successfully nitrided titanium particles using a nitrogen plasma at reduced pressures in a fluidized bed reactor. Titanium powder (ranging in size from 46 to 74 μm) was fluidized by Ar and N_2 gases in a 27.5-mm-ID, 750-mm-long quartz glass reactor. Quartz particles 1 mm in size were used as the packing for the distribution of the fluidizing gas. The bed of titanium charge was 50 g, which was equivalent to 40 mm of static bed height in the reactor. The bed was heated by an electric heater and maintained at 923 K. Heating due to plasma was negligible. The plasma was coupled by radio-frequency power (13.56 MHz) with the help of 4-mm copper tube discharge coils with six turns placed 50 mm above the static surface of the bed. The reactor was evacuated by a rotary pump to 1 Pa pressure and the plasma generated was at the freeboard region. The minimum fluidization velocity of titanium particles corresponded to 0.22 m/s, and this agreed well with the predictions of Kusakabe et al.,[131] who studied the fluidization characteristics of fine particles at reduced pressures. The superficial velocity was kept at four to five times the minimum fluidization velocity for the nitriding studies. Titanium particles were fluidized by argon gas initially and then the gas was changed to N_2 during the nitriding reaction.

A schematic of the experimental apparatus used for low-pressure plasma nitriding is shown in Figure 4.15. The extent of nitriding of titanium particles using plasma

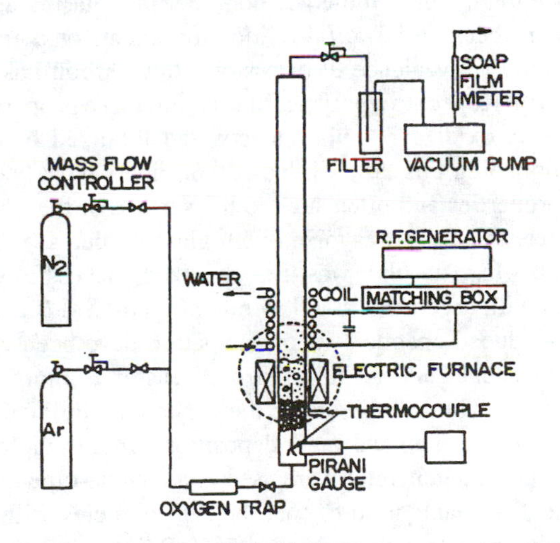

Figure 4.15 Experimental setup for low-pressure plasma nitriding. (From Okubo, T., Kawamura, H., Kasakabe, K., and Morooka, S., *J. Am. Ceram. Soc.*, 73(5), 1150, 1990. With permission.)

as well as thermal process (i.e., without plasma) was determined by the nitrogen concentration in TiN. Figure 4.16 depicts the variation of nitrogen concentration with reaction time. The results established that plasma nitriding increases the nitrogen concentration tremendously. The mechanism of nitriding was presumed to have three steps during the reaction: (1) nitrogen absorption or transport to the surface,

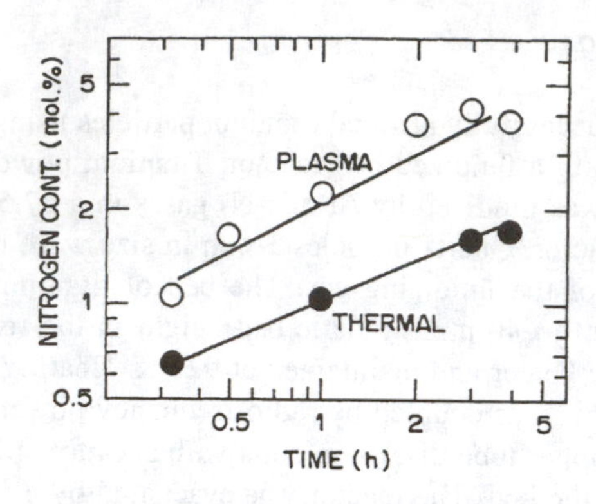

Figure 4.16 Nitrogen concentration versus time during nitriding of titanium particle. (From Okubo, T., Kawamura, H., Kasakabe, K., and Morooka, S., *J. Am. Ceram. Soc.*, 73(5), 1150, 1990. With permission.)

(2) diffusion of the gas through the nitride layer, and (3) reaction with titanium metal. The diffusion step was found to control the reaction rate. The above studies established that the nitriding rate was enhanced in the presence of a plasma in a fluidized bed even if the bed was not heated by the plasma. They further established the viability of fluidization of fine particles at low pressures. An experimental apparatus similar to the one described in Figure 4.15 was used by Kawamura et al.[132] for the surface treatment of milled carbon fiber in a plasma-activated fluidized bed reactor. Carbon fiber is a powerful reinforcing agent for composite materials. Milled fibers have the convenience of extrusion. Raw carbon has poor wettability and dispersibility with respect to a polymer matrix, but these properties are enhanced if the fiber surface is oxidized. A plasma-activated fluidized bed reactor is most suitable for accomplishing this task. Milled carbon fibers have been found to have poor fluidization properties and often stick to the surface of the column wall of the fluidized bed. Hence, this fiber was mixed with glass particles (volumetric fraction of the fiber kept at 0.1). The fiber was then oxidized successfully by oxygen gas diluted to 9.3 vol% with argon gas, at a flow rate of 2 cm^3 Std./s at 100 Pa pressure. Plasma and photo-induced chemical vapor deposition have been employed to coat the surface of fluorescent ZnS. A fluidized bed reactor is more suitable for this purpose. Nakamura and Hirate[133] described the conceptual fluidized bed reactor for plasma and photo-induced chemical vapor deposition. An ultraviolet lamp for photo reaction and an RF plasma generator were used in a conical-type quartz fluidization vessel. The chemical reactants used to coat the ZnS particles with SiN were SiH_4, SF_6, and NH_3. Silicon nitride, when coated over ZnS, increases the life span of the fluorescent particle.

c. Diamond Synthesis

Artificial diamonds are synthesized by high pressure and high temperature, thermal or plasma chemical vapor deposition, ion plating, and ion beam evaporation. Preparation of abrasive-grade diamond particles can be attempted by microwave as well as thermal filament chemical vapor deposition. Fluidized bed reactors have been successfully tested for imparting a coating of a thin film of diamond or for the growth of diamond over particulate solids such as α-Al_2O_3, tantalum, molybdenum, silicon, or diamond particles themselves. The continuous removal of by-products during diamond synthesis is essential,[134,135] and fluidized bed reactors are therefore advantageous; additional advantages accrue because of the high rate of particle-to-particle interaction and efficient heat and mass transfer. Hydrocarbon gases such as methane, ethane, acetylene, or butene, along with hydrogen gas, can be allowed to react in a fluidized bed coupled with a thermal plasma to accomplish this type of high-temperature reaction. Matsumoto et al.[135] claim to have grown diamond particles 30 μm in size in about 3 hr using a 600-W microwave plasma at 0.06 atm. Diamond particles of various shapes have been produced[136] by reacting a mixture of 0.5% methane in hydrogen over a bed of 500 to 1000-μm silicon particles in a 10-mm-ID reactor; 200-W microwaves were employed at a gas pressure of 2–3 kPa and the reaction period was 3–10 hr.

C. Salt Roasting in Spout Fluid Bed

1. *Spout Fluid Bed Reactor*

A systematic study on the application of a plasma spout fluid bed reactor for salt roasting of vanadium concentrate was reported by Munz and Mersereau.[137] Vanadium can be extracted from its concentrate by roasting with a sodium salt such as sodium carbonate in an oxidizing condition at 1170 K and recovering the vanadium content as water-soluble sodium metavanadate. The reaction for salt roasting is

$$Na_2CO_3 + O_2 + V_2O_3 \xrightarrow{1170\text{ K}} 2NaVO_3 + CO_2 \tag{4.5}$$

The roasting can be carried out using either a rotary kiln or a fluidized bed reactor. Fluid bed roasting requires a relatively shorter residence time and does not feed of agglomerated concentrates. Thus, a fluidized bed reactor is more economical than a rotary kiln for this salt roasting. However, heating the reactor to high temperatures often presents technical problems. In this regard, a plasma introduced as a spout into a fluidized bed serves as an excellent heat source for efficient heating of the whole bed. However, a plasma spouted fluid bed is reported[138] to be unstable even though it can achieve high conversion in a shorter residence time. For the salt roasting of vanadium ore, a DC power source at 70 kW was used, and the torch was made of a conical thoriated tungsten cathode and an annular copper anode. The fluidized bed was made of 304 SS, with a test section 127 mm in diameter and 150 mm long with a disengaging height of 254 mm and an enlarged section diameter of 254 mm. The plasma was generated by an argon and nitrogen gas mixture, and air was used as the fluidizing as well as the oxidizing gas. Air was distributed through 0.75-mm-diameter holes. The hole density was 0.25 per square centimeter. The plasma entered through a 24.4-mm opening. The bed material used was a vanadium concentrate with a mean diameter of 250 μm.

2. Performance Stability

The stability of the spout fluid bed was investigated with a charge of 4 kg, which corresponded to an aspect ratio (i.e., *H/D*) of 1. The plasma torch at 65% efficiency could heat the bed and maintain the bed isothermally at 1148 K with a maximum spout-to-bed temperature difference of 30 K. The bed temperature could be controlled by regulating the plasma power and the flow of argon and nitrogen gases. The heat transfer mechanism was attributed to the motion of the charge which came in contact with the hot plasma for a short time, heated up, and subsequently transferred the heat to the main bed. In order to maintain the bed in a stable condition, agglomeration of the particles had to be avoided, and this could be achieved by eliminating the particles melting in the plasma during their short contact time. However, stable operation and high conversion of the salt-coated concentrate had opposing effects because stability of operation was achieved at low temperatures whereas high conversion was favored at high temperatures. At a low concentrate-to-salt ratio, bed instability was brought about by the formation of fused material around the spout, thereby hindering particle heat transfer to the jet. As a result, the center of the bed was overheated while the remaining part of the bed was a cooler state. This made the bed reach a nonisothermal condition. Elutriation of a salt-free charge for a 20-min run was about 1% of the charge, whereas it varied with run time with a salt-coated charge. Hence, the Na:V molar ratio of the charge decreased depending on its initial value and the run time; the highest value of the Na:V ratio was 40. Reactor performance on batch operation was unsatisfactory due to the bed instability caused by particle agglomeration and loss of bed material by elutriation. However, this difficulty can be overcome, and successful operation of a plasma spout fluidized bed can be achieved by simulated continuous operation.

D. Miscellaneous Applications

1. Carbothermy

The use of plasma in a transferred arc argon reactor was demonstrated[139,140] for the carbothermic reduction of niobium pentoxide as well as pyrochlore. Because pyrochlore is an important source mineral for niobium, it is used as the raw material for the production of ferroniobium. Process development studies for the production of ferroniobium via aluminothermy as well as carbothermy have been of considerable interest for several years. The literature on these reduction reactions was summarized by Munz and Chin.[139] Although aluminothermy is currently used in the production of ferroniobium, a technoeconomic study[141] showed that carbothermic reduction is a viable and competitive alternative. Hence, the need for a high-temperature reactor for carbothermic reduction was determined, which obviously resulted in the application of a transferred arc argon plasma reactor. The reactor used consisted of a thoriated tungsten cathode and a water-cooled copper anode above which the reaction samples in the form of pellets were kept after loading in a graphite crucible. Carbothermic reduction carried out at temperature ranges of 1625–2855 and 1520–2440 K, respectively, for Nb_2O_5 and pyrochlore charges established that the reaction products were mainly NbC and CO. The main influencing parameters for the Nb_2O_5 reaction were the contact between the oxide and carbon and the optimum ratio of C to Nb_2O_5, which was found to be ten. These data provided valuable information on the application of a plasma fluidized bed reactor to carry out such reactions without resorting to pellet preparation. However, not much is known to date on this subject. Hence, there is room for further research on the plasma fluidized bed as applied to many carbothermic reduction reactions.

2. *Carburization*

Microdischarge between a metallic electrode such as iron and a fluidized bed of graphite was studied[142] with special reference to modeling the carburizing of the metal (i.e., the metallic electrode). The microdischarge plasma temperature in a fluidized bed was found to go up to 3600–6000 K between the Fe metal and the graphite particles. Fluid bed ore reduction and plasma smelting of Fe and Mn ore on an integrated pilot scale were described by Naden and Kershaw.[143] Gasification of coal[144] in an air plasma fluidized bed at a temperature of 4000 K could produce gases containing 19.53% ethylene, 21.30% H_2, 15.67% CO, 0.52% CH_4, 1.04% O_2, and 41.96% N_2 and HCN by volume. When coal was gasified with water plasma at 3500 K with a contact time of 0.2 s, an exit gas containing 59.8% H_2 and 35.6% CO (by volume) was produced. Gasification in a plasma fluidized bed is economically attractive and has its own advantages, as 90% of the electrical energy supplied is consumed by this process. A moving fluidized bed apparatus for hydrogen pyrolysis of coal in a plasma environment at ~2273 K was reported.[145]

3. *Gasification*

A laboratory type of plasma chemical reactor using a spouted bed was explored[146] for gasification of coal fines in an argon plasma. The specific expenditure of energy ranged from 14.9 to 12.0 kWh per kilogram of the volatile carbonaceous compounds. The entire energy of the plasma was utilized in the reaction zone, and reaction equilibrium was reached due to the insulation effect of the spouted bed. The optimum temperature of the waste gases was 800–850°C. Three types of spouted bed reactors were used[147] for plasma pyrolysis of brown coal by a low-temperature hydrogen plasma. The products of gasification were C_2H_2, CO, C_2H_4, CH_4, CO_2, and C_6H_6. Coal was added countercurrently to the top of the hydrogen plasma, and highest yields of C_2H_2 (25.3% based on converted coal) and CO (19.8%) were obtained.

The plasma fluidized bed, by virtue of its many applications in the emerging frontiers of advanced materials processing, has recently attracted the attention of many researchers and is now a new field of research interest for fluidization engineers. Because plasma in the form of a spout is introduced in a conventional fluidized bed, spout fluidization behavior at elevated temperatures must be investigated in the future. The exiting models of the fluidized bed must to be tested for their validity in the plasma environment, and model concepts may have to be reassessed for plasmatic reactions. Efforts to understand the basics of plasma spout fluidized beds are already underway in many research institutes. Currently, this area remains a topic of academic interest. With increasing input from plasma technology and fluidization engineering, it is time to couple them and investigate the combined effects.

V. ELECTROTHERMAL FLUIDIZED BED

A. Description

1. Principle

When electricity is passed through a fixed bed of electrically conducting particulate solids, the bed offers resistance to the flow of current; this resistance depends on many parameters, including the nature of the solid, the nature of the linkages among the particles within the bed, the bed voidage, the bed height, the electrode geometry, etc. If the same fixed bed is fluidized by passing a gas, the resistance of the bed increases; the resistance offered by the conducting particles generates heat within the bed and can maintain the bed in an isothermal condition. A fluidized bed thus heated by the passage of current through the bed of conductive particulate solids is termed an electrothermal fluidized bed or electrofluid reactor. A typical electrothermal fluidized bed reactor is depicted in Figure 4.17.

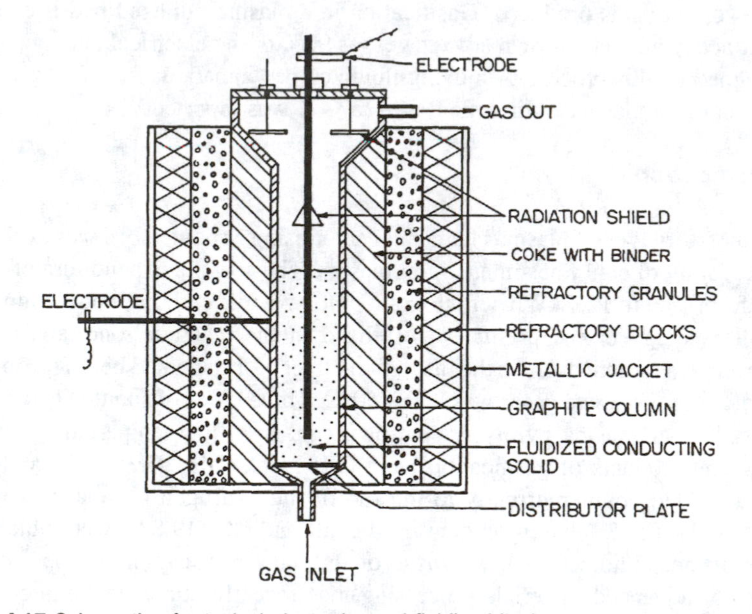

Figure 4.17 Schematic of a typical electrothermal fluidized bed.

2. Advantages

Electrofluid reactors have potential application in many chemical, metallurgical, and metalworking industries. In many high-temperature reactions, electrofluid reactors offer *in situ* heating during the reaction and the heating is stopped automatically when the charge is consumed. In other words, these electrothermal fluidized bed reactors for chemical reactions save energy because no external heating or transfer of heat is required. An externally mounted furnace, especially for high-temperature reactions with gases/halogens, requires careful maintenance and regular replacement of the heating elements. In an electrofluid reactor, such problems are totally eliminated. A mandatory requirement for a reactor of this type is a bed of solid that is electrically conducting and does not volatilize during heating. Thus, it appears that the reactor is of no use if nonconducting solids are to be heated. This problem can be overcome by mixing a nonconducting solid with a conducting solid. This principle of heating and reacting minerals in an electrothermal fluidized bed is used in pyrometallurgical operations, and the relevant details will be presented later as part of the discussion on direct chlorination of refractive materials/silicate ores such as zircon.

An earlier major industrial application of the electrothermal fluidized bed reactor was in the production of hydrogen cyanide.[148-150] By using an electrothermal fluidized bed, a faster reaction at high temperature was achieved without using the costly platinum catalyst which otherwise is essential for this reaction. Application of this reactor was subsequently reported in many other fields. The electrothermal fluidized bed has been used in coal industries mainly for the gasification of coal char[151] in the production of synthesis gas which can be transported through pipelines both conveniently and economically.

3. Configurations

Schematics of various configurations of the electrothermal fluidized bed reactor are shown in Figure 4.18. In Figure 4.18A, the reactor vessel or the fluidizing column itself serves as one of the two electrodes; the other electrode is centrally suspended from the top of the reactor and submerged inside a fluidized bed of conducting solids. As the electrothermal fluidized bed operates at high temperatures (up to 4500°C), it is essential to have good insulation and also to select a proper high-temperature-resistant reactor vessel. For most experimental or large-scale reactors, graphite is chosen as the construction material and the column is generally insulated with lampblack and suitable refractory granules. Additional insulation between the outer metallic jacket and the refractory granules is provided with materials such as sovlit crumb. In a typical[152] experimental 100-mm-diameter graphite column electrothermal fluidized bed with a concentrically suspended graphite electrode, the outer insulation provided consists of 8 mm of lampblack and 100 mm of refractory granules. In order to reduce radiation heat losses, radiation shields in the form of graphite plates above the bed and graphite central electrodes may be used.

There are some merits and drawbacks associated with each of the electrode configurations. In a concentric-type configuration, the radial distance is fixed and fluidization takes place in the annulus. Any variation in the area of the current flow can be affected by either increasing the bed height or changing the submerged height of the electrode inside the fluidized bed. In the case of parallel plates and multiple electrodes (Figure 4.18B–C) for a constant bed height, the bed resistance can be varied by changing both the plate distance and the submerged height of the bed. For

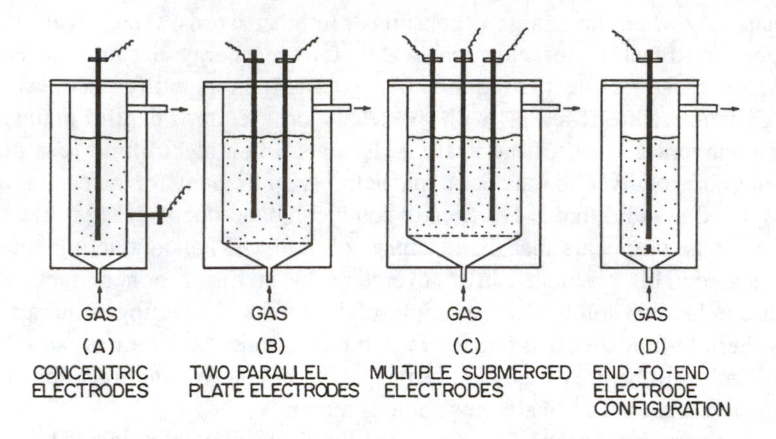

Figure 4.18 Various configurations of electrothermal fluidized bed and electrode arrangements.

an end-to-end electrode (Figure 4.18D), adjustments of the electrode gap and the submergence height are possible. Fluidized beds of a cylindrical column or vessel are common for concentric and end-to-end electrode configurations. For parallel plates or multiple electrodes, rectangular or square cross-sectional beds are generally used.

The use of multiple electrodes in electrothermal fluidized beds for miscellaneous applications such as carbothermic reduction of metallic oxides, formation of carbides and nitrides, and halogenation of various metallic oxides was reported three decades ago by Goldberger et al.[150] Both concentric and end-to-end electrodes were used by Kavlick et al.[153] in their studies on coal char gasification using an electrothermal reactor. They reported that high-voltage–low-current operation in an end-to-end electrode configuration limits the I^2R loss for a given power input, and this appeared to be the only advantage of such operation. Further, they reported that with an end-to-end electrode, the possible increase in the area of cross-section for current flow is limited and thus the resistance is generally high. For example, with an end-to-end electrode with a $1^1/_2$-in. diameter stainless-steel rod and a gap spacing of 1–12 in., the DC resistance obtained ranged from 50 to 300 Ω. Any further increase in resistance required a generator with a higher voltage output for the reactor. In a concentric-type electrode configuration, the overall resistance can be manipulated within 0.5–2 Ω for the same $1^1/_2$-in. stainless-steel electrode immersed 60 cm into the bed of coal char contained in a 150-mm-diameter 316 SS column. This low resistance caused enough I^2R loss to generate sufficient heat to bring the temperature in the range of 1750–1900°F at a test pressure of 1000 psi.

B. Construction

1. Temperature Range

The temperature ranges attainable with various thermal-generating means are depicted in Figure 4.19. It can be seen that the possible operating temperature range for an electrothermal fluidized bed lies between the highest temperature generated in an electric resistance furnace and that generated in an arc furnace. It is well known

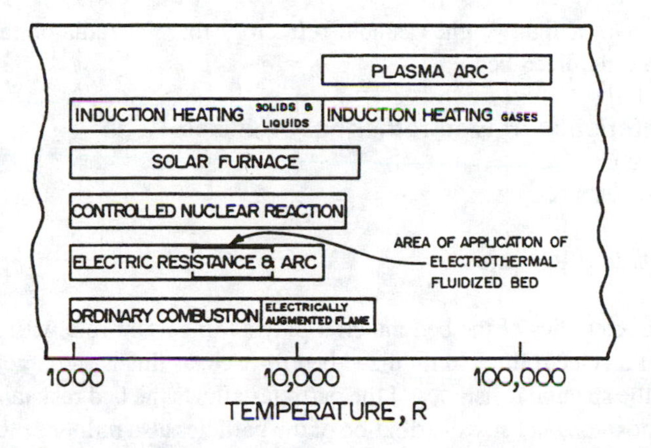

Figure 4.19 Temperature range for various thermal generators. (From Goldberger, W.M., Hanway, J.E., and Langston, B.G., *Chem. Eng. Prog.*, 61(2), 63, 1965. With permission.)

that most of the high-temperature generation for industrial processing is by electrical power and that the range of temperature attainable in an electrothermal fluidized bed reactor is from 1110 to 3900 K. Although this range is probably not the ultimate for high-temperature operation, an electrothermal fluidized bed can be successfully utilized in modern technology because of a limitation in terms of construction materials with other types of high-temperature generating devices. Furthermore, electrical power dissipates into heat and it utilized more efficiently at the reaction site in an electrothermal fluid bed. Hence, conventional heat generation and transfer through the containing wall is made redundant by an electrofluid bed reactor.

2. Furnace Details

Details for construction of a 60-cm-diameter fluidized bed furnace with a multiple electrode were given by Goldberger et al.[150] The design concept is the same as that shown in the schematic diagram in Figure 4.18C. Information on insulation material is given in detail. The wall of the reactor is generally made of graphite as it is versatile and compatible with many high-temperature reactions of industrial interest. The atmosphere for the reaction is restricted to the neutral or the reducing type. The wall itself can act as an electrode for the reactor. The fluidized solids can be graphite, carbon, or any other high-melting-point, electrically conducting particles. The other electrodes, which is usually immersed in the bed, can also be graphite or a high-melting-point metal. SS 316 can be used when the operation temperature is below its melting point. The power source can be either AC or DC. Voltages applied in an electrothermal fluidized bed are not above 100 V in many industrial applications. The insulation materials, starting next to the wall in series, are usually coke with binder, bubble alumina, MgO, and castable refractory. The feeding arrangement to the reactor can be either in-bed or type-feed. In-bed feeding of charge at high temperatures requires a good mechanical system separated by insulators from the electrical motors which drive the solid feeding devices. The top of the bed should be covered with a lightweight castable refractory thermal radiator facing the hot surface of the fluidized bed.

C. Characteristics of Electrothermal Fluidized Beds

1. *Bed Resistance*

a. Manipulating Variables

The characteristics of the bed material play an important role with regard to the resistance of an electrothermal fluidized bed furnace; as this is a charge resistor type of furnace, the specific resistivity of the particles affects the bed resistance. The size, shape, composition, and size distribution of the particles also influence the magnitude of the bed resistance. In addition, when the bed is fluidized, the voidage generated increases the bed resistance. The total resistance of the bed is the sum of the interelectrode resistance (i.e., the resistance between the electrode and the bed) and the bed resistance. The area of contact between the bed material and the electrodes changes, depending on the electrode submergence and the amount of charge in the fluidized bed. Hence, the electrical resistance and the power level can be manipulated by adjusting these variables. It has been suggested that this kind of adjustment can vary the power level without affecting the applied voltage or the current drawn from the power source.

b. Gas–Solid System

The type of gas and its conductivity also play a key role in determining the electrical resistance of the fluidized bed. The particle shape in particular influences the particle–particle and particle–gas transfer of electric charge. Furthermore, the conductivity of the gas decreases at high temperatures; it also depends on the gas pressure. When the applied voltage is high, arcing between the electrode and the bed as well as particle-to-particle arcings ionize the gas, thereby bringing down the bed resistance. Arcing inside the bed, in principle, is not desirable as it would lower the electrical and thermal efficiency.

Several workers have attempted to measure the electrical resistivity of a fluidized bed of conducting solids. In many cases, the bed was of a carbonaceous material such as calcined coke, coal char, graphite, or carbon power. The fluidizing column, operated at room temperature, was made of cylindrical or rectangular Plexiglas; a 12-V dry cell battery was used, and the electrodes were either graphite or a metallic material in the form of slabs or cylindrical rods. The fluidizing gas at room temperature was generally air and was nitrogen gas at higher temperatures. At temperatures above 1000°C, an inert gas such as argon was employed.

c. Mechanism of Electricity Transfer

An earlier study on the conductivity of a fluidized bed of coke by Goldschmidt and LeGoff[154] showed a continuous increase in the resistance of the bed from its settled condition to the fluidized state. The bed resistance reached a maximum close to the minimum fluidization velocity. Three types of mechanisms of conductivity were suggested to be responsible for the current flow: (1) the network of continuous chains of touching particles which were electrically conducting, (2) diffusion-type sharing of the charge between colliding particles, and (3) arcs between the particles. The results of Goldschmidt and LeGoff[154] indicated that the contributions of diffusion-type current flow and arcing between particles was negligible compared to current flow through the chains of touching particles. In order to further explore the mechanism of electrical energy transfer in a fluidized bed, Graham and Harvey[155] carried out studies on the resistance of fluidized beds of coke and graphite. These experiments were confined to a 50-mm-ID fluidizing column provided either with two pairs of 20-mm graphite electrodes at 180°C kept apart flush with the wall or with two rectangular 1.75-in. × 3.75-in. and 8-in. × 4-in. cross-sections with two cylindrical graphite electrodes mounted at the center. Compressed air at 10 psig was used as the fluidizing gas and was distributed by a porous stainless-steel distributor. The power source was a 0- to 12-V DC supply. Bed resistance was determined by measuring the current and applied voltage. In order to make accurate measurements, a known resistance was added in series to the bed resistance.

d. Ohmic Law

Bed resistance can be predicted by ohmic law, and the specific resistance of the bed can be expressed in the case of parallel flat plates as:

$$\rho_b = \frac{R_b A_E}{l} \tag{4.6}$$

and for coaxial electrodes as:

$$\rho_b = \frac{2\pi L R_b}{\ln \dfrac{d_e}{d_i}} \tag{4.7}$$

where R_b is the bed resistance between the electrodes, l is the distance between the electrodes (i.e., length of the current path), A_E is the mean surface area of the electrodes, L is the electrode immersion depth, d_i is the outer diameter of the inner electrode, and d_e is the inner diameter of the reactor (or inner diameter of the outer electrode).

Inaccuracies in the measurement of bed resistance have been reported[155] to lead to errors if the electrode is close to the gas-distributing grid (mainly due to the end effect) and also if the bubbles formed disturb the bed. Fluctuations in bed resistance from 0 to 20 cycles/s were reported by Graham and Harvey.[155]

e. *Model for Bed Resistance*

Because most of the work[155-160] on measurement of the overall resistance of fluidized beds was confined to small beds, it was necessary to systematically study and analyze the bed resistance by adopting a fundamental approach. Thus, a model to predict the electrical resistance of even an irregularly shaped conductor was proposed by Knowlton et al.[161] On the basis of the assumption that a fluidized bed surrounding an electrically conducting particle is of constant resistivity with a steady current flow, the potential field in such a conductor was expressed using the Laplace equation and current flow was predicted from the stream function. Using a numerical method, equipotential points and current flow in a cylindrical electrode bed were mapped. The predictions were later tested by comparison with experimental results obtained by using a 150-mm-diameter bed of calcined coke fluidized by N_2 gas, with U_{mf} at 0.13 ft/s. For regular-shaped electrodes in contact with conducting particles, the well-known resistance equation $R_b = \rho_b l/A_E$ can be used, but it would be necessary to determine the current flow path to exactly predict the bed resistance for irregularly shaped electrodes. For example, the resistance of a cylindrical electrode with a conducting band of height Z_C with an insulated length Z_2 in contact with the bed and submerged at a height Z_1 above the grid inside a bed of height H could be predicted by the expression

$$R_b = \frac{\rho_b}{2\pi\left[Z_C - r_1\dfrac{(Z_1 + Z_2)}{(r_2 - r_1)}\right]}\ln\left(\frac{r_2 Z_C}{r_1 H}\right) \tag{4.8}$$

where r_1 and r_2 are, respectively, the radius of the electrode and the fluidized bed, and the bed height $(H) = Z_1 + Z_2 + Z_C$. It was suggested that the bed resistance determined by using the field theory was a good estimate and that it was accurate for electrodes of irregular shapes. However, application of the field theory is based on the assumption of homogeneous and constant resistivity, which is not always valid.

f. *Measurements*

Measurement of the electrical resistance of a fluidized bed of conducting particles is usually carried out by using Ohm's law and knowledge of the voltage applied and the current flow. In order to eliminate the resistance between the current-supplying electrodes and the bed, the voltage drop across the bed is measured using separate electrodes. Measurements unaffected by any loss due to the voltage drop across the current-supplying electrodes were carried out by inserting voltage-drop-measuring electrodes at different points[158,162] inside the bed.

2. *Effect of Gas Velocity*

The electrical resistance of a bed of conducting solids changes significantly with the fluidizing gas flow rate. There are two schools of thought on this aspect. One school[155,162] observed a sharp increase in resistance from the static bed onward when the gas flow rate was increased; a maximum occurred close to the incipient fluidization velocity, followed by a decrease at higher velocities. At gas flow rates sufficient to initiate slugs, the resistance again increased. The other school[152,158] observed a similar increase in resistance up to a maximum but did not encounter a decrease beyond this point. Borodulya[152] attributed this result to the experimental conditions where the bed height was equal to the column diameter and hence there was no scope for the slugs to form. The results of Graham and Harvey[155] on the electrical resistance of –65 +100# coke for an electrode submerged 1, 2, 3, and 4 in. in a 4-in. × 8-in. rectangular column showed relatively higher values of resistance in static/settled bed conditions for greater submerged heights, and this trend was reversed in a fluidized bed (i.e., for low submerged heights of an electrode, the resistance of the bed was high). A typical plot[152] of the change in resistivity of a graphite bed with increasing superficial gas velocity is shown in Figure 4.20 for current densities of 0.004 and 0.4 amp/cm² for four size fractions of solids where U_{mf} varies from 2.9 to 18.3 cm/s.

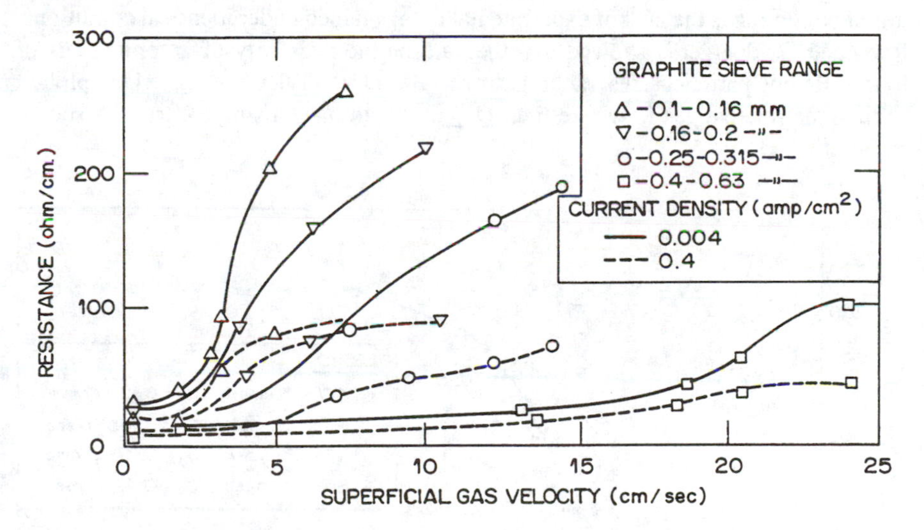

Figure 4.20 Effect of gas velocity on the specific resistance of graphite beds. (From Borodulya, U.A., Zabrodsky, S.S., and Zheltov, A.I., *AIChE Symp. Ser.*, 69(128), 106, 1971. With permission.

3. *Effect of Particle Size*

From the plot in Figure 4.20, it can be seen that electrical resistance increases with superficial gas velocity and that resistance is relatively higher with fine than with coarse particles. An increase in current density brings down the resistance of the bed for all particle size ranges, keeping the trend the same as far as the effect of particle size is concerned.

The electrical resistivity (ρ_b) of a bed of particulate solids is

$$\rho_b = \rho_p \frac{d_p}{d_c} \tag{4.9}$$

For fine particles, the contact area diameter (d_c) is less, and thus ρ_b is more. Although the above expression is valid for equal sized particles in a cubic arrangement with the particles touching each other, it can also be extended to the case of a particulate fluidized bed. Because fine particles are easily separated by the fluidizing gas, the air gap adds to the specific resistance. Hence, a bed of fine particles always exhibits higher specific resistance. There seems to be no model that can be used to develop a single correlation to predict bed resistivity in terms of relevant parameters such as superficial gas velocity, particle size, current density, etc. The main reason for this shortcoming is the lack of experimental data obtained under identical conditions. Reed and Goldberger[158] showed that the data on the resistivity of graphite beds of four different particle sizes ranging from –35 +150 (Tyler mesh), when plotted against the relative gas flow rate (i.e., Q/Q_{mf}), could have a single curve, as shown in Figure 4.21.

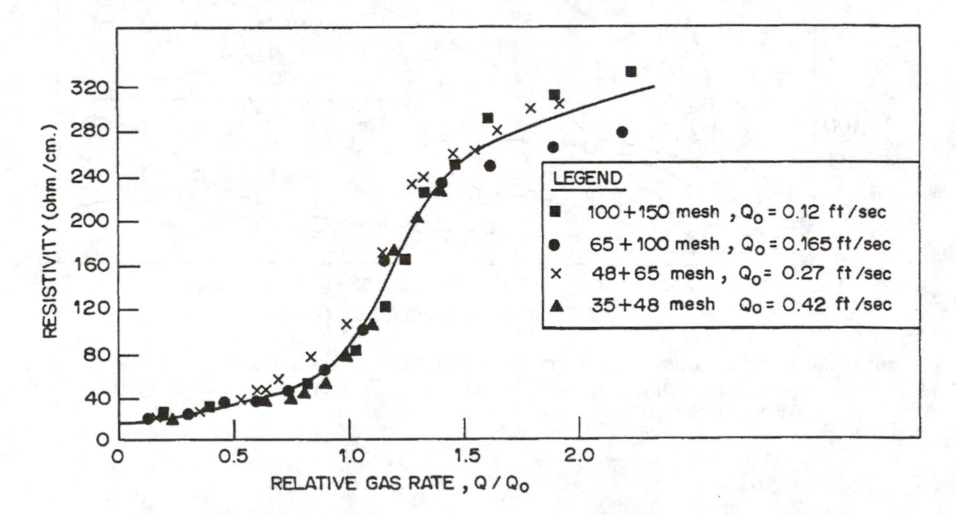

Figure 4.21 Resistivity versus relative gas flow rate for graphite bed. (From Reed, A.K. and Goldberger, W.M., *Chem. Eng. Prog. Symp. Ser.*, 62(67), 71, 1966. With permission.)

4. *Effect of Bed Voidage*

Goldschmidt and LeGoff[154] proposed that electrical conduction is due to a chain of conducting and touching particles in a fluidized bed, and they developed a mathematical expression for prediction of the ratio of the resistance of a fluidized bed to that corresponding to the incipient state of the fluidized bed in terms of bed porosity (ϵ) at velocity U and incipient bed porosity (ϵ_{mf}) at velocity U_{mf}. The theory is quite simple and is outlined here.

For a two-phase fluidized bed, the fraction of bubbles occupying the bed is given by:

$$\delta = (\epsilon - \epsilon_{mf})/(1 - \epsilon_{mf}) \tag{4.10}$$

According to Bruggeman,[163] the resistance ratio (R/R_{mf}) of a particulate bed of solids to a two-phase fluidized bed is

$$\frac{R}{R_{mf}} = (1-\delta)^{3/2} \tag{4.11}$$

Since for a fluidized bed[164]

$$\frac{U}{U_{mf}} = \frac{(1-\epsilon_{mf})}{(1-\epsilon)}\frac{\epsilon^3}{\epsilon_{mf}^3} \tag{4.12}$$

it can be shown by algebraic manipulation that

$$\frac{R}{R_{mf}} = 1 - 3/2\left[\frac{\epsilon_{mf}}{(3-2\epsilon_{mf})}\left(\frac{U}{U_{mf}} - 1\right)\right] \tag{4.13}$$

According to Equation 4.13, R/R_{mf} increases with U/U_{mf}. A typical set of experimental values is shown in Table 4.4, and it can be seen that as the column diameter is increased by a factor of 2, the resistance ratio also increases approximately 2.5–3 times. The reason for this is not explained by Equation 4.13.

Table 4.4 Data on Variation of R/R_{mf} with U/U_{mf} in an Electrothermal Fluidized Bed

	R/R_{mf}	
U/U_{mf}	d_t = 5 cm	d_t = 10 cm
1.25	1.4	3.46
1.5	2.41	6.55
1.75	3.36	10.1
2.00	4.29	13.1

5. Resistivity and Bed Status

Prediction of the resistivity of a conducting bed of fluidized solids as a function of the operating gas velocity is rather difficult. In an attempt to predict the resistivity without considering the gas velocity, but as a function of parameters which would give maximum and also minimum resistivity, Jones and Wheelock[165] conducted experiments on a single grade of calcined coke of average particle size ranging from 81 to 175 μm at ambient temperature and pressure over the current density range 0.2–10 mA/in.^{2}. The fluidizing gas was N_2 and the column diameter was 50 mm. The minimum fluidization velocity for the particle sizes of the various fractions fluidized by nitrogen gas varied from 0.029 to 0.153 ft/s. The experimental runs showed that the resistivity of the bed increased from the settled bed to the minimum fluidization condition and then rose steeply until a maximum was reached. After this maximum with further increases in gas velocity, a falling trend with a minimum followed by a rising trend was observed by Jones and Wheelock.[165] Hence, they felt it would be appropriate to correlate the resistivity of the bed for three conditions as follows.

1. Bed resistivity in settled condition:

$$\log\rho_{bs} = 1.922 - 0.676 \log D_t - 0.300H - 0.307 \log D_p \tag{4.14}$$

where ρ_{bs} is in Ω-in., D_p (in micrometers) = $\Sigma(ndp)_i/\Sigma n_i$, and H and D_t (column diameter) are in inches. The above correlation was obtained by stepwise regression analysis.

2. Peak resistivity of the bed (ρ_{max}):

$$\rho_{max} = 7690 \frac{\sigma^{0.435}\rho_g^{0.050}}{D_t^{0.658} d_p^{0.700}} \tag{4.15}$$

3. Minimum resistivity of the bed (ρ_{min}):

$$\frac{\rho_{min}}{\rho_{bs}} = \frac{18.66\ \sigma^{-0.210}}{D_t^{0.765}(\tan\theta)^{3.47}} \tag{4.16}$$

for 15 μm < σ, standard deviation of d_p < 32 μm, 35.9° < θ, angle of repose < 40.9°.

The increase in the resistivity of a fluidized bed after a trough has been reached is due to the formation of slug. As pointed out earlier, the occurrence of falling resistance after a maximum is entirely due to the experimental conditions and follows the first school of thought. Pulsifer and Wheelock[162] reported the electrical resistance of carbonaceous solids (graphite and calcined coke) at ambient and elevated temperatures. Their conclusion was that bed resistance, after reaching a maximum, falls at higher gas velocities and starts rising again from the point corresponding to the onset of slugging.

6. *Effect of Particle Shape*

The effect of particle shape on bed resistance has not been explored in detail. A qualitative study reported by Graham and Harvey[155] showed that the ratio of the resistance of the bed to the resistance of the settled bed (R/R_o) had higher values for spherical coke particles than for crushed (irregularly shaped) graphite particles. A typical plot for a 1.75-in.-high bed with an electrode immersion depth of 1.75 in. is shown in Figure 4.22. It can be seen from this figure that the largest separation

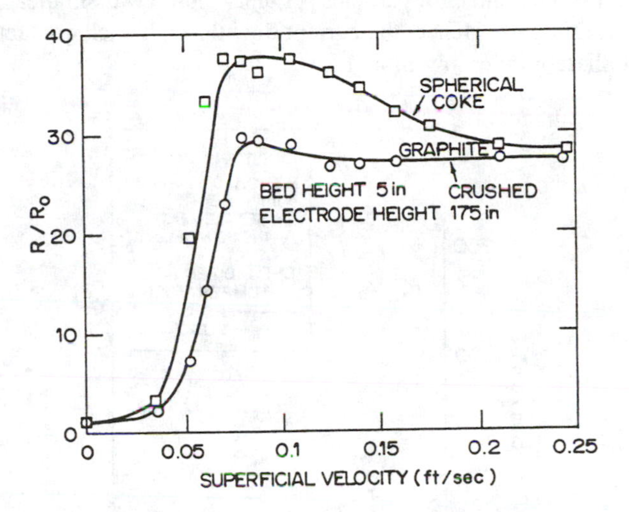

Figure 4.22 Relative resistivity (R/R_o) versus superficial velocity for spherical and crushed graphite particle. (From Graham, W. and Harvey, E.A., *Can. J. Chem. Eng.*, 43(3), 145, 1965. With permission.)

in R/R_o occurred near the maximum value of this ratio. The reason for this difference could not be explained. However, it was recognized that current flowed through a chain of touching particles of conducting type. For irregularly shaped particles, the contact area could increase, thereby decreasing the resistance.

7. *Effect of Current Density*

In general, the resistivity of a bed of solid particles decreases with increasing current density. The results obtained by various workers are interesting and often contradictory. For example, Reed and Goldberger[158] reported that with a rectangular apparatus, the resistivity of the bed remained constant in the current density range 25–100 mA/in.^{2}. However, the resistivity of a bed of –100 +150 graphite particles in a cylindrical vessel that was heated externally and also vibrated during fluidization decreased over the current density range 0.3–2.5 mA/in.^{2}, and the change in resistivity was found to be negligible if fluidizing gases such as Ar, N_2, and CO were used in the temperature range 600–1000°C. At current densities above 2 A/in.^{2}, visual indication of the occurrence of small arcs showed the possibility of gaseous conduction. Boroduyla et al.[152] measured the charge in a bed of synthetic graphite particles ranging in size from 0.10 to 0.16 mm, with U_{mf} = 2.9 cm/s in a 96-mm-diameter cylindrical column fitted with an axial electrode. The area for the current path was taken midway between the electrodes, and the results were obtained at 20 and 2000°C for the current density range 0.004–0.4 A/cm^2 for the static bed, the incipiently fluidized bed, and the fluidized bed. The results are shown in Figure 4.23. It can be seen that resistance decreased 1.5 times for the fixed bed and 2–3 times for the fluidized bed over the current density range 0.004–0.4 A/cm^2. The greater decrease in the case of the fluidized bed could be attributed to a smaller number of contact points (as the number of particles per unit volume was smaller) and a smaller area for the current path. Hence, the current flow through each contact point would be increased, thereby releasing more heat.

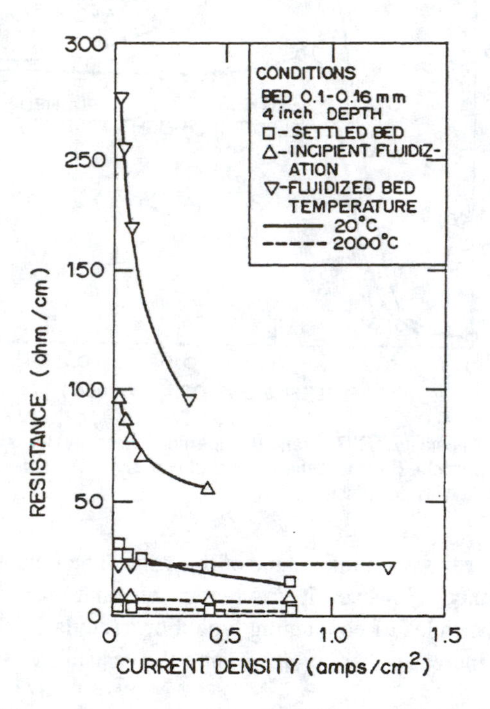

Figure 4.23 Bed resistance versus current density. (From Borodulya, U.A., Zabrodsky, S.S., and Zheltov, A.I., *AIChE Symp. Ser.*, 69(128), 106, 1973. With permission.)

The fall in bed resistivity with an increase in current density is attributed mainly to photoionization, which is caused by vaporization of carbon (at a high temperature >2000°C) due to the heat generated at the contact points and further ionization of this vapor along with the fluidizing gas. The other mode of ionization is the thermal mode, which requires temperatures of the order 10,000 K. Surface ionization demands a large potential difference. Collision ionization is not possible for particles and electrical circuits separated by infinitesimal distances.

Another possible mechanism is due to the effect of the electric field itself. The electric field is weak at low current densities, and the current divides into several irregular paths through chains of particles. As the current density is increased, the field becomes stronger and draws together all the conducting chains of the current path, thereby reducing the area of the current path. This would then result in an increase in resistance, and hence the path would be heated up. As the chains of conducting particles get overheated, the resistance ultimately decreases and more current flows; this again increases the field strength, resulting in a further reduction in the flow area, where a high level of overheating is followed by a drastic drop in resistance.

8. *Effect of Bed Height and Diameter*

The resistivity of a deeper (150-mm-high) fluidized bed at a relatively low current density range (0–0.25 A/cm^2) in a 96-mm-diameter column was reported[152] to be lower than that of a shallow (75-mm-high) bed. At higher current densities when the mechanism of current conduction is due to ionization, the effect of bed height on bed resistivity was found to be negligible. In a settled bed, however, the resistivity of the bed remained constant at all current densities and bed heights. Bed diameters in the range 50–250 mm, at a constant current density (0.004 A/cm^2), did not influence the resistivity of the bed significantly when the bed was not fluidized. However, when the bed was fluidized at the incipient condition, that is, at low bed expansion, the resistance of the bed started to fall as the column diameter was increased. The resistance of a well-developed fluidized bed increased, reached a maximum, and then decreased when the column diameter was increased progressively. The reason for this could be attributed to slugging in narrow-diameter columns and bubbling in wider diameter columns.

9. *Effect of Temperature*

The effect of the bed temperature on the resistivity of a bed of synthetic graphite at high temperatures was reported by Borodulya et al.[152] At a constant superficial gas velocity, the resistivity of the bed decreased monotonically with temperature, as shown in Figure 4.24. The resistivity at a high current density (1.0 A/cm^2) was lower than that obtained at a low current density (0.004 A/cm^2) at temperatures ranging from ambient to 2500°C. Results obtained under similar conditions but with a constant mass flow rate showed a rise in bed resistivity with temperature up to a maximum; thereafter, there was a sharp drop in bed resistivity. Experiments conducted[158] in a vibrating cylindrical column with an externally heated bed up to a temperature of 1000°C showed that the resistivity of a 50-mm-deep fluidized bed of graphite particles with a size range of –100 +150# was unaffected by the type of fluidizing gas; the trend for the change in resistivity of the fluidized bed is shown in Figure 4.25. The decrease in bed resistance could not be attributed to the ionization of the gas in the temperature range 600–1000°C, because the resistivity of the bed remained the same for different fluidizing gases. Although it is known that resistivity of graphite particles decreases with increase in temperature, properties such as hardness are reported to have a major effect on bed resistivity.

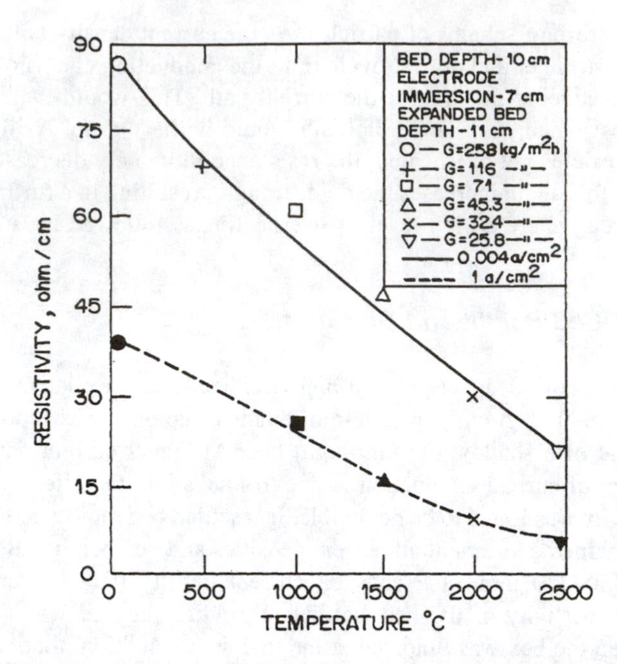

Figure 4.24 Effect of temperature on bed resistivity at constant superficial gas velocity. (From Borodulya, U.A., Zabrodsky, S.S., and Zheltov, A.I., *AIChE Symp. Ser.*, 69(128), 106, 1973. With permission.)

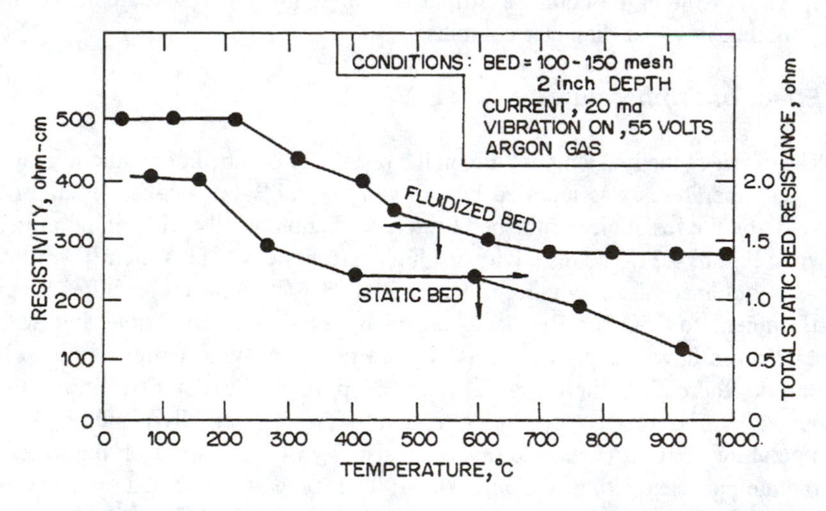

Figure 4.25 Effect of temperature on resistivity of static and fluidized bed. (From Reed, A.K. and Goldberger, W.M., *Chem. Eng. Prog. Symp. Ser.*, 62(67), 71, 1966. With permission.)

Pulsifer and Wheelock,[162] in their studies on the electrical resistance of fluidized carbonaceous materials, noted hysteresis in the plot of bed resistivity versus temperature for partially gasified coal and also for graphite. They observed a generally decreasing trend in bed resistivity with increasing temperature. When measured during cooling, the resistivity increased, but the values obtained were lower than those obtained during heating. This resulted in a lower bed resistivity when the bed reached room temperature (i.e., resistivity lower than that originally measured at room temperature). This type of hysteresis during the heating and cooling cycle could not be explained theoretically. However, they attributed it to gas desportion and adsorption or to a change in the structure of the bed material. Another important observation was that bed resistivity at ambient temperature was restored to its original value when the cooled bed was kept for a sufficient time at this temperature. A similar trend was observed when the bed material was graphite. The resistivity of a fluidized bed of coal char decreased much faster than that of the fixed bed at temperatures in the range 25–525°C.

At high temperatures, arcing was encountered, as reported by Ballain and Pulsifer.[166] In an attempt to increase the bed temperature beyond 965°C (1800°F), the applied voltage was increased from 180 to 210 V and a sudden increase in current flow was noted, indicating a decrease in resistance. When the voltage was reduced from 160 V to 150 V, current flow decreased, showing that resistance had gone up again. The cycle of increasing current with increased applied voltage and decreasing current with decreased voltage at high bed temperatures could be repeated as depicted in Figure 4.26. These observations indicated a change in the current flow mechanism, and this was confirmed to be due to arcing. The electrode temperature was not changed during this operation. In an electrothermal fluidized bed reactor, the axial electrodes such as those made of stainless steel and copper were found to melt even though the bed temperature was far below their melting points.

Investigations carried out by Ballain and Pulsifer[166] in a 4-in.-ID reactor made of 446 SS and equipped with a $^1/_4$-in. axial electrode showed that the difference between the electrode and the bed temperatures increased as the extent of electrode immersion was increased and that this difference was also influenced by power input, electrode diameter, and gas velocity. The heating up of the axial electrode was brought about by the high current density due to the small available area in the vicinity of the surface of the electrode. This electrode overheating would increase further if the electrode diameter were to be reduced. Interestingly, the tip of the axial electrode was heated severely because of the large diverging flow area from a small tip. As a large current flowed from this tip, it was heated up, and it was further heated up when the applied voltage was increased, thereby causing melting of the electrode progressively from the tip. Overheating at the tip could be reduced by completely immersing the electrode inside the bed. In such a case, where the electrode was not partially submerged, the total electrode length was exposed to an equal area of the reactor wall, and the difference in the electrode temperature and the bed temperature was lower and also uniform.

All the above investigations and results were qualitative and were confined to a small reactor. Results on larger reactors and on reactors of different geometries would help in understanding the phenomenon of high temperatures in electrofluid reactors.

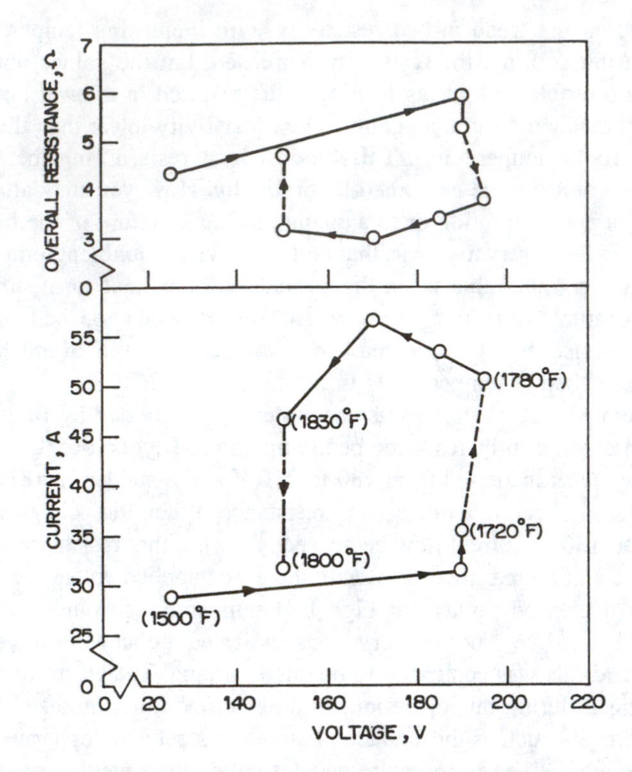

Figure 4.26 Variation of resistance, current, and voltage during bed heatup. (From Ballain, M.D. and Pulsifer, A.H., *Chem. Eng. Prog. Symp. Ser.*, 66(105), 229, 1970. With permission.)

10. Effect of Pressure

The electrical resistivity of a bed of highly volatile bituminous coal was studied[167] using a 4-in.-ID reactor provided with an axial $^1/_2$-in. stainless-steel electrode for pressures ranging from ambient to 1000 psi. The results showed that bed resistivity increased with pressure. However, current density seemed to play an important role. As the current density was increased from 0.02 to 0.5 A/in.^{2}, bed resistivity started falling with increasing pressure; the pressure has no effect on bed resistivity at a current density of 0.5 A/in.^{2}. This can easily be inferred from Figure 4.27. An explanation for the change in the bed resistivity of a fluidized bed with pressure was suggested by Lee et al.[167] According to Paschen's law, the sparking potential (V) is proportional to the pressure (P) and the gap between electrodes (d). In a fluidized bed, the gap (d) is the distance between particles and is proportional to bed expansion (or the bed expansion ratio, $E = H/H_{mf}$). Bed resistivity for an axial electrode is proportional to the product of the electrode immersion height (L) and the resistance (R). The current (I) is proportional to the bed height (L) for a deeply immersed electrode. It can be shown that:

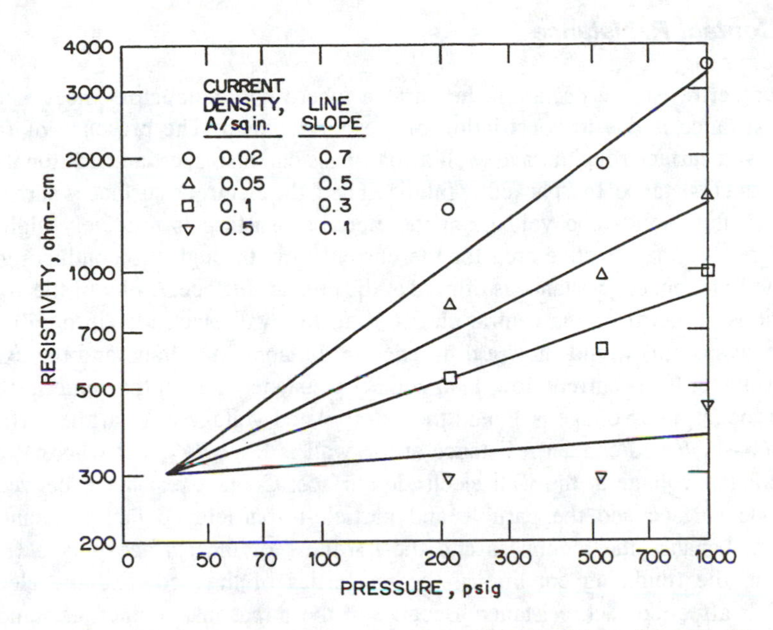

Figure 4.27 Resistivity versus pressure in electrothermal fluidized bed at 0.5 ft/s. (From Lee, B.S., Pyrcioch, E.J., and Schora, F.C., Jr., *Chem. Eng. Prog. Symp. Ser.*, 66(101), 75, 1970. With permission.)

$$\rho \,\alpha\, LR \,\alpha\, L\frac{V}{I} \,\alpha\, L\frac{V}{L} \,\alpha\, V \,\alpha\, Pd \,\alpha\, PE \tag{4.17}$$

The above relationship shows that ρ increases with pressure and expansion. Under constant pressure, however, bed resistivity increases with bed expansion. It has been shown that Paschen's law is valid up to a pressure of 100 psi at low current densities. Hence, a modified Pashcen law was proposed, incorporating bed expansion (E) in place of electrode distance (d) or particle distance. At high pressures, current flow is suggested to occur by sparking. The reason is that when a fluidized bed is expanded, each particle is surrounded by a cover of gas, and the size of this cover increases with gas flow. This would ultimately lead to breakage of the continuous link of conducting particles, thereby changing the mode of current flow from simple conduction to sparking or arcing. The reason for the increase in bed resistivity with pressure at low current densities has not been explained clearly. However, it has been shown that fluidization at 1000 psi is smooth and uniform without slugs, even when the gas velocity is 16 times higher than that corresponding to minimum fluidization. Current flow has been suggested to be analogous to spark discharge between spherical electrodes in a gaseous atmosphere. An increase in pressure suppresses sparking so that the voltage gradient required for ionization increases with pressure. However, beyond a certain pressure, there will not be any increase in the sparkling voltage gradient, and bed resistivity is independent of pressure at high current densities.

11. Contact Resistance

Contact resistance occurs at the surface where two conductors meet or touch; this resistance is due to constriction of the current path. The presence of foreign bodies and surface roughness as well as hardness can also exert an additional effect on contact resistance. In a packed or fluidized bed, the electrode surface is surrounded by particulate solids and voidage at the electrode surface is relatively high. This would reduce the effective area for the current flow through the conducting solid particles, and hence resistance is offered at the contact surface. A plot of the voltages at various points from the central electrode to the wall electrode (in a cylindrical geometry system) would show a difference in the applied voltage and the extrapolated voltage. If I is current flow, then contact resistance (R_{ce}) at the center electrode (when the applied voltage is V_o and the extrapolated voltage is V_1 at the surface) is $R_{ce} = (V_o - V_1)/I$ and contact resistance at the wall is $R_{cw} = (V_2 - 0)/I$ where V_2 is the extrapolated voltage at the wall electrode surface. Contact resistance between the electrode surface and the particle and particle-to-particle contact resistance are different because the toughness and the hardness of the surface play a role. In addition, the fluidizing conditions, the properties of the bed, and the electrode materials affect contact resistance. Because of these reasons, contact resistance is a complex property.

Only a few workers have reported contact resistance in an electrothermal fluidized bed reactor. Reed and Goldberger[158] reported that contact resistance decreased with increasing current density and was sensitive to the electrode surface condition. Glidden and Pulsifer[168] found that contact resistance at the surface of a graphite electrode was 1 or 2% of the total interelectrode resistance. Chen et al.[169] reported some details on the bed and the contact resistance of calcined coke and graphite with different electrode materials such as silicon carbide, stainless steel, brass, and graphite. Nitrogen gas was used as the fluidizing gas in 15.2-cm-diameter Plexiglas column as the test cell. They mainly reported the center electrode-to-bed contact resistance, because the bed-to-wall electrode contact resistance was less than 1% of the total interelectrode resistance. They observed that contact resistance increased with gas velocity and that the highest change occurred near incipient fluidization. The behavior of the contact resistance between U_{mf} and the velocity at which the bed was fully fluidized (U_f) was erratic and the measured voltage changed markedly with a small change in velocity. Further, they noticed that contact resistance decreased in succession for silicon carbide, stainless steel, brass, and graphite. This decrease was attributed to the decreasing hardness of the material. Hence, the highest contact resistance was obtained with silicon carbide electrodes and the lowest with graphite. The change in the magnitude of contact resistance (taken as the ratio to total resistance) relative to the velocity ratio U/U_{mf} is depicted in Figure 4.28. From this plot, it can be seen that contact resistance is linearly proportional to bed resistivity for gas velocities less than U_{br} where the visible bubble is first sighted. Beyond U_{br}, a simple relationship does not hold.

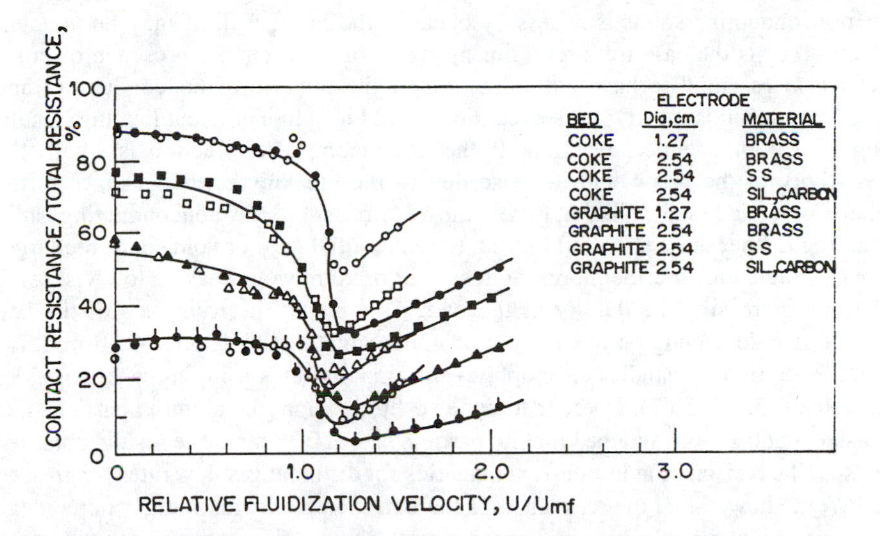

Figure 4.28 Contact resistance versus relative fluidization velocity. (From Chen, T.P., Yuan, E., and Pulsifer, A.H., *AIChE Symp. Ser.*, 73(161), 51, 1977. With permission.)

12. Effect of Distributor

The distributor has a profound effect on the electrical resistance of the bed. Graham and Harvey[155] blocked the active site of a porous gas distributor and measured bed resistance in a 3-in. × 1.6-in. rectangular fluidized bed of –100 +150 graphite particles for two immersed electrode heights of 2.0 and 5 cm inside a bed of initial static height of 10.6 cm. In all cases when the distributor malfunctioned due to blockage by mercury, the resistivity of the bed decreased abruptly. In the case of a porous distributor, the bubbles near it are very small and too many number, and hence particulate fluidization prevails. This would then cause an increase in bed resistance to a very high level. Above the grid, however, the bubbles coalesce, grow in size, reduce in number, and rise faster. The resistance of the bed is thus reduced. This has been experimentally tested by Graham and Harvery,[155] who injected bubbles and saw through an oscillograph a fall in bed resistance and a corresponding increase in current pulse during the period of bubble injection. They attributed this to the flow of gas from the dense phase to the bubbles, thereby increasing the density of the dense phase, which in turn caused the resistance to decrease and the current to increase. This could also explain the fact that in a bed of angular particles wherein bubbles can easily form resistance can be expected to fall as compared to a bed of spherical particles.

13. Effect of Nonconducting Solids

The literature on this subject is scant. Most studies are confined to conducting solids fluidized in the reactor. However, in many chemical reactions, the presence of nonconducting solids is necessary because the reactant itself may be nonconducting, as is the case for zircon during its chlorination in the presence of coke. It is also essential to have cofluidization of the mixture of nonconducting and conducting solids; otherwise, segregation would lead to inefficient heating, resulting in a poor chemical reaction. If the conducting solid fraction is small, the resistivity of the bed would increase due to the breaking of the linkages in the chain of conducting solids between the electrodes. If the nonconducting solid fraction is finer in size, it would fill up the interstitial gaps or voidage of the larger conducting solids. Furthermore, the resultant mixture would have a low viscosity, thereby increasing the fluidity of the bed. As a result, the resistance of the bed would rise depending on the amount of nonconducting solids present. Borodulya et al.[152] carried out studies on graphite–alumina mixtures using three fractions of alumina (25, 33, 50%). In order to achieve fluidization, the alumina particle size was kept at 0.22 mm and the graphite particle size at 0.33 mm. The results obtained on specific resistance at low current densities for different gas flow rates (expressed as Q/Q_o) showed that the resistance of the bed increased with a larger percentage of nonconducting alumina and increased with the gas flow rate. A sharp increase was noted at U_{mf} (i.e., $Q/Q_o = 1$). Typical results are shown in Figure 4.29.

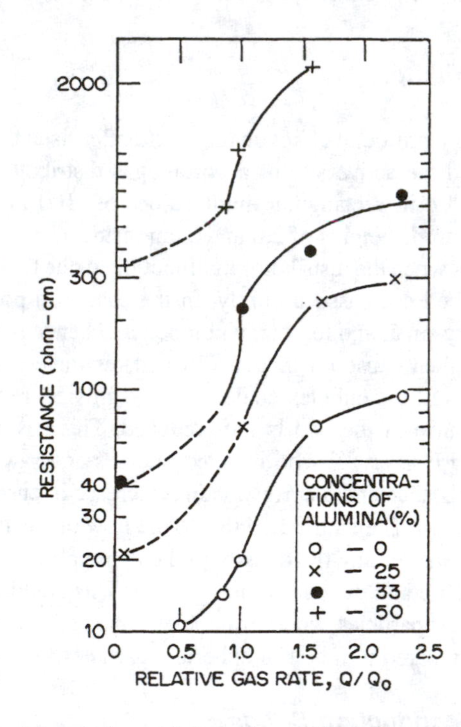

Figure 4.29 Specific resistance of graphite–alumina mixture at room temperature. (From Borodulya, U.A., Zabrodsky, S.S., and Zheltov, A.I., *AIChE Symp. Ser.*, 69(128), 106, 1973. With permission.)

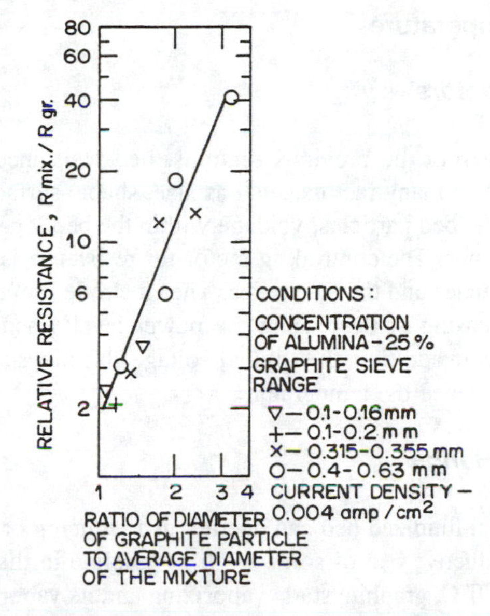

Figure 4.30 Relative resistance versus ratio of diameter of graphite particles to average diameter of the mixture (graphite + alumina) gr = graphite. (From Borodulya, U.A., Zabrodsky, S.S., and Zheltov, A.I., *AIChE Symp. Ser.*, 69(128), 106, 1973. With permission.)

By adding finer nonconducting solids to coarse conducting solids, the chain of conducting bridges is broken. The variation in the relative resistance of graphite (R_{gr}) to that of a graphite–alumina mixture (R_{mix}) with the ratio of graphite particle size to alumina particle size is shown in Figure 4.30.

An empirical correlation has been proposed for the variation of the specific resistance of the mixture (ρ_{mix}) in terms of the volume fraction (V_{nc}) of the nonconducting material and the specific resistance (ρ_c) of the conducting material. The correlation is expressed as:

$$\frac{\rho_{mix}}{\rho_c} = \frac{1+2\,V_{nc}}{1-2\,V_{nc}} \tag{4.18}$$

where

$$V_{nc} = \frac{V}{V+V_c}$$

is the volume fraction of the nonconducting solid, V is the volume of nonconducting material, and V_c is the volume conducting material.

D. Operating Temperature

1. Controlling Factors

As can be seen from the previous sections, bed resistance in an electrofluid reactor is dependent on many factors, such as size, shape, surface nature, hardness, size distribution of the bed particles, voidage within the bed, operating velocity, and interelectrode resistance. The controlling factor for resistance is the area of contact between the bed particles and the electrodes. The electrode immersion height is also responsible for increasing or decreasing the power level when a constant-voltage power source is used. In addition, the applied voltage also plays a key role in altering the power level and hence the temperature.

2. Temperature Range

An electrothermal fluidized bed can operate in a reducing or nonoxidizing environment with a conductive bed of solids such as graphite in the temperature range 980–4425°C. At 3870°C, graphite starts vaporizing and its vapor pressure is 133 Pa. Because graphite vapor is conducting, this vaporization is not a constraint or limiting factor with regard to the operating temperature. However, in many cases the fluidized bed of solids would soften beyond a certain temperature; thermodynamic as well as kinetic considerations for the reaction also impose a temperature limit. When the gases used for fluidization are oxidizing in nature, the attainable temperature of operation is limited and is lower than that which could be obtained with a reducing or neutral operation. However, with the use of certain carbides and oxidation-resistant materials, the temperature of operation can be raised to the range 1100–1500°C. It should be emphasized in this context that concentrated efforts should be made to develop oxidation-resistant high-temperature electrode materials.

3. Limiting Cases

At high temperatures, the resistance of a bed of particulate solids decreases. Furthermore, the conductivity of gases decreases, in addition to which there is an increased incidence of arcing and ionization. As a result, the overall bed resistance is drastically reduced at high temperatures, and this warrants a reduction in the applied voltage to control the power supply. In order to achieve a desired power level, a high current at a low voltage must be used. The lower limit of temperature is dictated mainly by economic considerations. Where fossil fuel is cheap, thermal energy can be used without incurring much expense. However, if a reaction occurs at a temperature below 980°C and if there is a need to obtain clean products without contamination by combustion products, the electrothermal reactor can be used with advantage. It should be noted that the electrothermal bed generates the required heat at the reaction site itself, and this saves energy consumption. For halide conversion, the use of the electrothermal fluidized bed reactor is recommended even for low-temperature operations.

E. Power Loading and Control

The main characteristics of an electrothermal fluidized bed are rapid heating and prompt control. The power loading depends on the temperature required and the size of the reactor. Goldberger et al.[150] reported that the power level required for a typical 150-mm (6-in.) electrothermal fluidized bed reactor was 3 kVA at 980°C and 10 kVA at 2750°C. The voltage used in this power range varied from 40 to 80 V. In the case of a reactor with a 2-in. diameter, a power loading of 350–450 kVA was used. However, voltage requirements in excess of 100 V are seldom encountered.

The electrothermal fluidized bed can be controlled in the following three ways:

1. *Adjusting the gas flow*: Because the conductivity of the bed depends on the extent of voidage or gas packets inside the bed, any variation in the gas flow rate would change the power level; hence the temperature can be controlled by adjusting the fluidizing gas flow rate. The flow rate required for optimum performance corresponds to a velocity which equals or slightly exceeds the minimum fluidization velocity.
2. *Adjusting the electrode submergence*: The power level can also be controlled by varying the electrode immersion level inside the bed, because the conductivity of the bed is dependent on the area of contact between the conducting particles and the electrode. The total resistance of the bed falls as the electrode is submerged deeper inside the bed. In other words, the surface area of the electrode available for current flow increases with electrode submergence, leading to a reduction in resistance. The resistance falls from 1 to 0.2 Ω when the submerged area of the electrode increases from 100 to 500 $in.^2$.
3. *Adjusting the applied voltage*: While the first two methods are practicable even when the applied voltage is constant, the third method is to change the applied voltage itself to obtain the desired power level. It should be noted that changing the power level by using the first two methods is more affordable or economical than the third method.

Knowledge of the power–temperature relationship and the variation in bed resistance from start-up of the reactor to attainment of the desired temperature is essential for a clear understanding of the workings of the electrothermal fluidized bed reactor. Kavlick and Lee[170] reported such data for a 150-mm (6-in.) reactor used for high-pressure gasification of coal char to produce synthesis gas. The hydrogasified coal char was fluidized during the heat-up period with nitrogen gas, and the power to the reactor was supplied by a 300-kW, 1200-rpm shunt-wound DC generator across the wall electrode and the 38-mm ($1^1/_2$-in.) SS axial electrode. A typical power–temperature relationship and the corresponding variations in resistance are shown in Figures 4.31 and 4.32.

F. Electrically Stabilized versus Electrothermal Fluidized Bed

It should be pointed out here that an electrofluid reactor or an electrothermal fluidized bed is not the same as that used with an AC or DC electric field to control the bubbles in a gas fluidized bed. The application of a high-strength electric field to

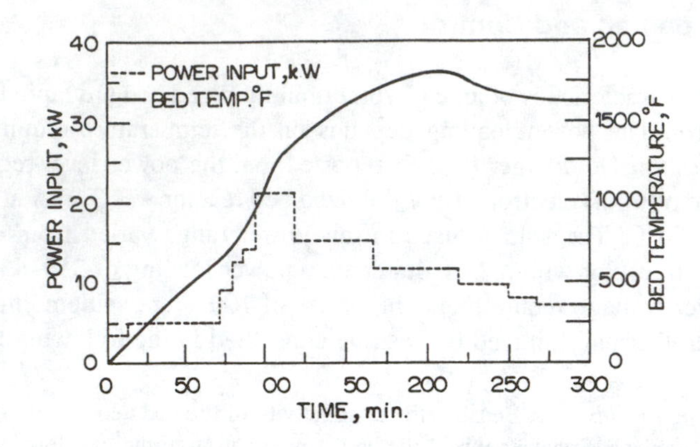

Figure 4.31 Power and temperature relationship of nitrogen as fluidizing gas at bed height of 5 ft and electrode immersion of 24 in. (From Goldberger, W.M., Hanway, J.E., and Langston, B.G., *Chem. Eng. Prog.*, 61(2), 63, 1965. With permission.)

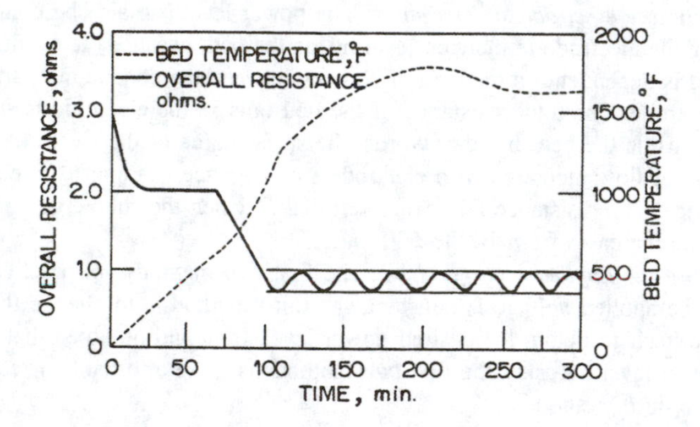

Figure 4.32 Relationship of overall resistance and bed temperature. (From Kavlick, V.J. and Lee, B.S., *Chem. Eng. Prog. Symp. Ser.*, 66(105), 145, 1970. With permission.)

a gas fluidized bed improves the heat and mass transfer characteristics because the bed is uniformly fluidized with good control or bubble size. The electric field applied to a gas fluidized bed becomes an important second independent variable; the first is the gas velocity. Heat and charge transfers in an AC electrofluidized bed were comprehensively studied and reviewed by Colver and Bosshart.[171] The stabilizing influence of an electric field in a gas fluidized bed in dissipating the bubbles can be explained in terms of electromechanics pertaining to the formation of an electric suspension. It is believed that electrically induced cohesion between adjacent particles may be presumed to act as surface tension to reduce the bubble size.

The various mechanisms by which electrical charge can be transported in a fixed bed, a fluidized bed, and an electrical suspension were summarized by Colver.[172] The mechanisms were generalized (for a DC current) to be due to (1) particle transport, (2) particle rotation, (3) particle surface (or bulk) electrical conductivity, (4) particle single or multiple contact, (5) fluid ionic transport, and (6) fluid convective transport. In an electrothermal fluidized bed, these mechanisms of charge transport must be considered to arrive at generalized electrical conductivity. In a uniformly fluidized bed, assuming negligible charge transport due to linear and rotational movement of particles, the net electrical conductivity can be evaluated in terms of the conductivity due to the particle surface (σ_3) and that due to particle-to-particle contact (σ_4) as:

$$\sigma_{mix}(\text{fluidized bed}) = \frac{\sigma_3\sigma_4}{\sigma_3 + \sigma_4} \tag{4.19}$$

In a hypothetical fluidized bed without particle contact, $\sigma_4 = 0$, resulting in zero net conductivity. The method for determining the net electrical conductivity of an electrothermal fluidized bed has not been fully explored to date, and much remains to be studied on this subject.

In conclusion, it is important to emphasize that the electrothermal fluidized bed combines the advantages of fluidization and rapid generation of heat at the site where the reaction takes place. In a nonreacting system, the bed is used to generate heat and maintain high temperatures, which is not possible with conventional resistance-type furnaces. For a high-temperature endothermic reaction in a gas–solid system, the electrothermal reactor is the best choice if the bed is electrically conducting. Such a reactor can also perform the job with good control. Although this reactor was originally developed for use in coal industries, it has been found to have potential in many pyrometallurgical applications.

G. Applications

1. General

The first commercial application of an electrothermal fluidized bed was for the production of hydrogen cyanide, which formed upon contact of ammonia and a hydrocarbon such as propane over an electrically heated fluid bed of carbon particles. The temperature maintained in the electrothermal fluid bed ranged from 1260 to 1600°C for this reaction. The hydrogen cyanide yield ranged from 85 to 90% even when the residence time of the gases was just 0.1–0.5 s. Goldberger et al.[150] reported that an experimental electrothermal reactor could be operated at 3600°C for an extended period. Thus, this type of reactor, because of the combination of convenience and high-temperature capability, can replace the conventional electric furnace in several carbothermic reactions. The use of an electrothermal fluidized bed reactor for the production of elemental phosphorus by the carbothermic reduction of phosphate shales at temperatures in the range 1100–1300°C was reported; the yield of phosphorus by this technique was 75%. Continuous operation and excellent control of the reaction were achieved.

2. *Classification*

The application of an electrothermal fluidized bed reactor can be classified into three major groups: (1) chemical industries, (2) environmental control, and (3) metallurgical industries.

In chemical industries, its applications are numerous, and it is widely used for hydrocracking hydrocarbons and for pyrolysis. Its use for hydrogasification of coal char has been well recognized, and it is recommended as a potential chemical reactor for spark-discharge-activated chemical reactions.

As far as application in environmental engineering is concerned, the reactor's capability to produce high temperatures can be exploited for testing many refractory components for space and other applications, particularly under specific gas environments at intermediate furnace temperatures (see Figure 4.19). It can also be used in the carburizing of metals, the heat treatment of complex metallurgical shapes at elevated temperatures, and even for firing ceramic pieces.

3. *Chlorination*

The potential of an electrothermal fluidized bed for metallurgical use is unlimited because it combines the advantages of a fluidized bed with high temperature capability. Because it can be operated continuously with close control of temperature, it can be used for the carbothermic reduction of many refractory oxides, the production of metallic carbides/nitrides, and high-temperature halogenation. An example of high-temperature halogenation is the chlorination of ZrO_2 or $ZrSiO_4$ in the presence of carbon. The chlorination of zircon is endothermic and the reaction is feasible in the presence of carbon. The product is a volatile tetrachloride of zirconium. The electrothermal fluidized bed is ideal for this type of reaction. The results[173] of the chlorination of zirconium oxide, carried out in an electrothermal fluidized bed, are shown in Table 4.5.

Table 4.5 Results on Chlorination of ZrO_2[150,173] (ZrO_2 [–100#] Mixed with –65 +100# Carbon

	Temperature (°C)				
	500	650	1000	1200	1200
Conversion of ZrO_2 (%)	45	61	73	82	94[a]

Note: Superficial velocity of chlorine = 0.046 m/s.

[a] With 100# pelletized mixture of ZrO_2 + carbon.

a. *Zirconium Dioxide Chlorination*

The chlorinator for the chlorination of ZrO_2 in an electrothermal fluidized bed reactor was made of graphite, with a 125-mm top ID tapered down to 100 mm at the bottom, with an overall length of 600 mm. The power supply to the reactor was through a 220-V AC power source controlled by a 10-kW variable electrode. The bed was initially charged with 1600 g of –6 +100# petroleum coke and fluidized by

N_2 gas. When the desired reaction temperature was reached, chlorine gas was admitted at the rate of 10–11 g/min (3.55–3.65 cm/s, space velocity). The ZrO_2 was fed through a hole in the bottom electrode. As the reaction progressed, petroleum coke was depleted and was replenished by adding 100-g batches intermittently.

b. Zircon Chlorination

The direct chlorination of zircon ore in an electrothermal fluid bed was also studied. By using this method of direct chlorination, preparation of intermediate zirconium carbide by the expensive electric arc method could be eliminated. It should be noted that the chlorination reaction of zirconium carbide or carbonitride is highly exothermic and can be accomplished in the temperature range 500–600°C by using a conventional fluidized bed reactor. The use of a conventional fluidized bed reactor is essential because of the highly exothermic nature of the reaction, which warrants quick removal of heat.

The direct chlorination of zircon is thermodynamically feasible in the presence of carbon at 1000–1200°C, and the reaction is endothermic. Supply of heat of this reaction by either an electric furnace or by burning excess carbon is not economical. The combustion product of carbon contaminates the reaction product if heat is supplied by burning excess carbon. Heating with flue gas at high temperatures is uneconomical because only a fraction of sensible heat can be transferred to the fluid bed. In order to have an economical operation, a heat recovery unit in the post-reactor section must be added.

The limitation of heat transfer efficiency at high temperatures is not a constraint in an electrofluid reactor because the heat energy is directly transferred to the reactant system. Hence, the electrofluid reactor appears to be promising for the direct chlorination of zircon. Furthermore, it is necessary to have either CO or carbon for the reaction. The use of CO as a reducing agent is not economical, as it was reported[174] that the reaction rate slowed down after 5% of zircon was chlorinated. At high temperatures in a corrosive medium, special construction materials are required. Because refractory materials are to be used in most high-temperature operations, the generation of heat by an external furnace and transferring the same to the chlorination reaction is uneconomical.

Manieh et al.[175] studied the chlorination of zircon in an electrothermal fluidized bed reactor made up of a 6.4-cm-ID, 11.4-cm-OD, and 55.9-cm-long graphite column fitted with a disengaging section with a 11.4-cm ID, 16.5-cm OD, and 15.2-cm length. A gas distributor with 1-cm holes were used. The reactor wall was one of the electrodes. An axial electrode 1.9 cm in diameter, made of a dense graphite rod, was used. The reactor was heated up by a 25-kVA power supply and was controlled by temperature signal feedback from the reactor. Chlorination studies were conducted over the temperature range 1000–1200°C and the reaction time varied from 0.5 to 2 hr. A chlorine utilization efficiency of up to 100% was reported. Thus, the electrothermally heated fluidized bed was found useful for direct chlorination of zircon. The chlorination reaction was found to be of zero order with respect to chlorine concentration and of first order with respect to the surface area of zircon particles; it was also weakly dependent on the carbon/zircon molar ratio. The reaction rate was observed to follow the equation

$$r_{Cl} = 4\, r_{Zr} = k_n\, A\, V_R \frac{R_{av}^{0.23}}{R_i} = 94.5\, A\, V_R \left(R_{av}\right)^{0.23} \cdot \exp(-10560/RT) \qquad (4.20)$$

where A is the equivalent surface area of zircon particles (m^2), V_R is the bed volume per initial surface area (l/m^2), k_n is the reaction rate constant = $k_o e^{-E/RT}$, k_o is the frequency factor (g mol/l · hr), R_{av} is the average carbon-to-zircon molar ratio, and R_i is the stoichiometric carbon-to-zircon molar ratio. The activation energy was 10,550 cal/g · mol and the reaction was controlled by chemisorption.

4. *Miscellaneous Applications*

The electrothermal fluidized bed[176] was studied for the production of hydrogen-rich synthesis gas by reacting coal char and steam. A bench-scale reactor 30 cm in diameter was used. The use of this type of reactor for the production of an H_2–CO mixture and methane was been described. The need to understand the electrical characteristics of such systems was emphasized for future research.

The possibility of producing carbon disulfide by the reaction of carbon and sulfur vapor or carbon and hydrogen sulfide was discussed long ago by Johnson.[177] Johnson[177] predicted that this type of reactor could also be exploited for the chlorination of TiO_2 and SiO_2 in the presence of carbon. Pyrolysis of acetic acid and acetone to yield ketones is another application of an electrothermal fluidized bed reactor. Application of the electrothermal fluidized bed, as pointed out earlier, was explored mainly for coal char gasification, and some interesting literature by Pulsifer et al.[178] and Beeson et al.[179] may be referred to for this application.

NOMENCLATURE

A	equivalent surface area of zircon particle (m^2)
A_E	mean surface area of electrodes (m^2)
C	coefficient of jet expansion (–)
C_p	specific heat of particle (J/kg · s)
d_0	plasma nozzle diameter (m)
d_c	contact area diameter (m)
d_i	outer diameter of the inner electrode (m)
D_p	number mean size of particle (m)
d_p	particle size (m)
d_r	inner diameter of the reactor (m)
E	activation energy (cal/g · mol in Equation 4.21)
$\overline{E}$	bed expansion (–)
H	bed height (m)
H_{mf}	bed height at minimum fluidization velocity (m)

I	current density (A/m^2)
k_n	reaction rate constant, (g · mol/l · hr in Equation 4.21)
k_o	frequency factor (g · mol/l · hr in Equation 4.21)
L	electrode immersion length (m)
P	pressure (Pa)
Q	gas flow rate (m^3/s)
Q_{mf}	gas flow rate at the minimum fluidization velocity (m^3/s)
Q_t	net rate of heat transferred to solid particles (J/s)
R	bed resistance (Ω)
r_1	radius of the electrode (m)
r_2	radius of the fluidized bed (m)
R_{av}	average carbon-to-zircon molar ratio (–)
R_b	bed resistance between electrodes (Ω)
r_{cl}	reaction rate with respect to chlorine for the chlorination of zircon (g mol/l · hr)
R_i	stoichiometric carbon-to-zircon molar ratio (–)
R_{mf}	bed resistance at minimum fluidization velocity (Ω)
r_o	plasma jet radius (m)
r_{zr}	reaction rate with respect to zircon in chlorination of zircon (g mol/l · hr)
S_p	surface area of particle (m^2)
T_g	plasma gas temperature (K)
T_o	plasma gas temperature at distance $x = 0$ (m)
t_s	residence time of partice (s)
T_w	wall temperataure (K)
U	superficial gas velocity (m/s)
U_{mf}	minimum fluidization velocity (m/s)
V	volume of nonconducting particle (–)
V_c	volume fraction of conducting particle (–)
V_g	plasma gas velocity along jet distance x
V_{nc}	volume fraction of nonconducting particle (–)
V_o	initial isothermal gas velocity (m/s)
V_R	bed volume per unit surface area (l/m^2)
x	resistance from nozzle entry (m)
Z_1, Z_2, Z_C	submerged height, insulated height, and conductivity band height of cylindrical electrode (m)

Greek Symbols

δ	bubble-volume fraction in a fluidized bed (–)
ϵ	bed voidage (–)
ϵ_{mf}	bed voidage (–)
ϵ_p	particle emissivity (–)
μ_0	viscosity of gas at temperature T_0 (Pa · s)
μ_g	viscosity of gas at temperature T_g (Pa · s)

ρ_b	specific electric bed resistance (mho)
ρ_{bs}	specific electric resistance of bed of solid particles (mho) (in Equation 4.14 Ω-in.)
ρ_{max}	peak bed resistivity (Ω-in. in Equation 4.15)
ρ_{min}	minimum bed resistivity (Ω-in. in Equation 4.15)
ρ_{mix}	specific resistivity of mixture (ohm-m)
ρ_p	specific electric resistance of particle (mho)
σ	standard deviation in Equation 4.15
σ_3	electrical conductivity due to particle surface (Ω-m)
σ_4	electrical resistivity due to particle contact (Ω-m)
σ_s	Stefan constant
ϕ	angle of repose (°)

REFERENCES

1. Schrödinger, E., *Wave Mechanics*, Blackie and Son, London, 1928.
2. Bell, A.T., Fundamentals of plasma chemistry, in *Techniques and Applications of Plasma Chemistry*, Hollahan, J.R. and Bell, A.T., Eds., John Wiley, New York, 1974.
3. Fey, M.G., Electric arc (plasma) heaters for the process industries, *Ind. Heat.*, 2, June 1976.
4. O'Brien, R.L., *Plasma Arc Metal Working Procedures*, American Welding Society, New York, 1967.
5. Boulos, M. and Fauchais, P., Transport processes in thermal plasmas, in *Advances in Transport Processes*, Vol. IV, Mujumdar, A.S. and Mashelkar, R.A., Eds., Wiley Eastern Ltd., New Delhi, 1986, chap. 5.
6. Pool, J.W., Freeman, M.P., Doak, K.W., and Thorpe, M.L., Simulator Tests to Study Hot Flow Problems Related to a Gas Corrector, NASA CR-230g, 1973.
7. Eckert, H.U., Analysis of thermal induction plasma dominated by radiation conduction losses, *J. Appl. Phys.*, 41, 1520, 1970.
8. Mackinnon, I.M. and Reuben, B.G., *J. Electrochem. Soc.*, 122, 806, 1975.
9. Roman, W.C. and Zabielski, M.F., Spectrometric gas composition measurements of UF_6 RF plasmas, presented at 30th Annual Gaseous Electronic Conference, Palo Alto, CA, October 1977.
10. Bhat, G.K., New developments in plasma arc melting, *J. Vac. Sci. Technol.*, 9(6), 1344, 1972.
11. Drellishak, K.S., Ph.D. thesis, Northwestern University, Evanston, IL, 1963.
12. Report of the IUPAC (International Union of Pure and Applied Chemistry) Subcommission on Plasma Chemistry, Thermodynamic and Transport Properties of Pure and Mixed Thermal Plasmas at LTE, *Pure App. Chem.*, 54, 1221, 1982.
13. Gorse, Thésé 3iéme Cycle Universite de Limoges, September 9, 1975.
14. Buttler, J.N. and Borokaw, R.S., *J. Chem. Phys.*, 26, 1636, 1957.
15. Vanderslice, J.T., Weisman, S., Mason, E.A., and Fallon, R.J., Thermal conductivity of gas mixture in chemical equilibrium, *Phys. Fluids*, 5, 155, 1962.
16. Emmons, H.W., *Modern Development in Heat Transfer*, Ibelec, W., Ed., Academic Press, New York, 1963.
17. Stokes, C.S., Chemical reactions with the plasma jet, *Chem. Eng.*, p. 190, April 12, 1965.

18. Stokes, C.S., Chemistry in high temperature, *Plasma Jets Adv. Chem. Ser.*, 80, 390, 1965.
19. Rykalin, N.N., Plasma engineering in metallurgy and inorganic materials technology, *Pure Appl. Chem.*, 4B, 179, 1976.
20. Gage, R.M., Arc Torches and Process, U.S. Patent 2,806,124, 1957.
21. Pfender, E., Electric arcs and arc gas heaters, in *Gaseous Electronics,* Vol. 1, Hirsh, M.N. and Oskam, H.J., Eds., Academic Press, New York, 1978.
22. Pfender, T., Generation and Properties of Equilibrium Plasma, Course of Fundamentals and Application of Plasma Chemistry, University of Minnesota, Minneapolis, 1981.
23. Kassabji, F. and Fauchais, P., Les generateurs a plasma, *Rev. Phys. Appl.*, 16, 549, 1981.
24. Leontiev, A.J. and Voltchkow, E.P., Charactéristiques electriques et thermiques d'un plasmatron de haute enthalpie, in *Investigations Experimentales des Plasmatrons*, Joukov, M.F., Ed., Novossibirsk, Nauka, 1977.
25. Fauchais, P., Les chalumeaux à plasma à arc electrique, in *Les Hautes Temperatures et Leurs Utilisations en Physique et an Chimie*, Vol. 1, ed. Masson et Cie, Paris, 1970.
26. Kimblin, S.U., Erosion des électrodes et processus d'ionisation entre les électrodes de l'arc dans le vide et à la pression atmostheriqqu, in *Investigations Experimentales des Plasmatrons*, Joukov, M.F., Ed., Novossirbirsk, Nauka, 1977, 226.
27. Bonet, C., Thermal plasma technology for processisng of refractory materials, *Pure Appl. Chem.*, 52, 1707, 1980.
28. Maecker, H., Zeitschrift für Physik, Electronendichte und temperatur in der säule des Rochstromkohlebogens, 136, 119, 1953.
29. Wilks, P.H., Ravinder, P., Grant, C.L., Pelton, P.A., Downer, R.J., and Talbot, M.L., Plasma process for zirconium dioxide, *Chem. Eng. Prog.*, 68, 82, 1972.
30. Sinha, H.N., From zircon to high purity zirconia for ceramics, *Miner. Process. Extract. Metall. Rev.*, 9, 313, 1992.
31. Sheer, C., Korman, S., and Kang, S.F., Investigations of Convective Arcs for the Simulation of Re-entry Aerodynamic Heating, Afors-TR-74-1505, Contract FO44-620-69-C-01404, 1974.
32. Bayliss, R.K., Bryant, J.W., and Sayce, I.G., Plasma dissociation of zircon sands, in Proc. (S-5-2) III Int. Symp. on Plasma Chemistry, Limoges, France, July 12–19, 1977.
33. Bonet, E., Valbona, G., Foex, M., Daguenet, M., and Dumargue, P., High temperature processing of super refractories in a plasma fluidized bed reactor, *Rev. Int. Hautes Temp. Refract.*, 11, 11, 1974.
34. Pickles, C.A., Wang, S.S., McLean, A., Alcock, C.B., and Segworth, R.S., A new route to stainless steel by the reduction of chromite ore fines in an extended arc flash reactor, *Trans. Iron Steel Inst. Jpn.*, 18, 369, 1978.
35. Whyman, D., A rotating wall D.C. arc plasma furnace, *J. Sci. Instrum.*, 44, 525, 1967.
36. Tylko, J.K., High Temperature Treatment of Materials, U.S. Patent 3,783,167, February 14, 1972.
37. Arc Coal Process Development, Final Report, April 1972, AVCO Corporation Systems Division, Lowell, MA.
38. Sayce, I.G. and Selton, B., Preparation of ultrafine refractory powders using the liquid-wall furnace, *Spec. Ceram. Res. Assoc.*, 5, 157, 1972.

39. Schnell, C.R., Hamblyn, S.M.L., Hengartner, K., and Wissler, M. (Lonza Ltd.), The industrial application of plasma technology for the production of fumed silica, presented at the Symposium on Commercial Potential for Arc and Plasma Processes, Atlantic City, NJ, September 8–11, 1974.
40. Yerouchalmi, D. et al., Développement du four rotatif à haute température chauffé axialement par un plasma dàres et essais de fusions d'oxydes ultra-réfractaires, *High Temp. High Pressures*, 3, 271, 1971.
41. Howie, F.H. and Sayce, I.G., *Rev. Int. Hautes Temp. Refract.*, 11, 169, 1974.
42. Chase, J.D. and Skriven, J.F., Process for the Beneficiation of Titaniferous Ores Utilising a Hot Wall Continuous Plasma Reactor, U.S. Patent 3,856,918, 1974.
43. Gold, R.G., Sandall, W.R., Cheplick, P.G., and McRac, D.R., Plasma reduction of iron oxide with hydrogen and natural gas at 100 kW and 1 MW in Proc. Int. Round Table on Study and Application of Transport Phenomena in Thermal Plasma, IUPAC Odeillo, France, September 12–16, 1975.
44. Kassabji, F., Pateyron, B., Aubreton, J., Boulos, M., and Fauchais, P., Conception of 0.7 MW plasma furnace for iron oxide reduction, *Rev. Int. Hautes Temp. Refract.*, 18, 123, 1981.
45. Kunabek, G.R., Gauvin, W.H., Irons, G.A., and Choi, H.K., The industrial application of plasmas to metallurgical processes, in Proc. 4th Int. Sympon. Plasma Chemistry, Zurich, August 27 – September 1, 24, 1979.
46. Gauvin, W.H., Kubanck, G.R., and Irons, G.A., The plasma production of ferromolybdenum — process development and economics, *J. Met.*, 42, January 1981.
47. Pateyron, B., Aubreton, J., Kassabji, F., and Fauchais, P., Some new design of reduction plasma furnaces including the hollow cathode system electrical transfer to the bath and falling film, in Proc. 5th Int. Symp. on Plasma Chemistry, Waldec, D., Ed., University of Edinburgh, 1981, 167.
48. Mehmetoglu, M.T., Characteristics of Transferred-Arc Plasma, Ph.D. thesis, McGill University, Montreal, 1980.
49. Borodachyov, A.S., Okorokov, G.N., Pozdief, N.P., Tulin, N.A., Fiedler, H., Kruller, F., and Schorf, G., Plasma steel making furnaces — development and application experience, in Proc. World Electrochemical Congress, Paper N:LA 26 Session IV, Moscow, June 1977.
50. Rykalin, N.N., Plasma engineering in metallurgy and inorganic materials technology, *Pure Appl. Chem.*, 48, 179, 1976.
51. Rykalin, N.N., Thermal plasma in extractive metallurgy, *Pure Appl. Chem.*, 52, 1813, 1980.
52. Boulos, M.I., Flow and temperature fields in the fire ball of an inductively coupled plasma, *IEEE Trans. Plasma Sci.*, PS-4, 28, 1976.
53. Boulos, M.I., Heating of powders in the fire ball if an induction plasma, *IEEE Trans. Plasma Sci.*, PS-6, 93, 1978.
54. Boulos, M.I., Gagne, R., and Barnes, R.M.C., Effect of swirl and confinement on the flow and temperature fields in an inductively coupled plasma, *Can. J. Chem. Eng.*, 58, 367, 1980.
55. Pool, J.W., Freeman, M.P., Doak, K.W., and Thorpe, M.L., Simulator Tests to Study Hot Flow Problems Related to a Gas Corrector, NASA CR-230g, 1973.
56. Barnes, R., Recent advances in emission spectroscopy; inductively coupled plasma discharges for spectrochemical analysis, *CRC Crit. Rev. Anal. Chem.*, p. 203, 1978.
57. Donoski, A.V., Goldfarb, T.M., and Klubnikin, V.S., in *Physics and Technology of Low Temperature Plasmas*, Dresvin, S.V., Ed., 1972, English edition translated by Cheron, T. and edited by Eckert, H.V., Iowa State University Press, Ames, 1977.

58. Rykalin, N.N., Plasma engineering in metallurgy and inorganic materials technology, *Pure Appl. Chem.*, 4B, 229, 1976.
59. Gage, R.M. and Molrtan, H.S., U.S. Patent 3,130, 292, 1960.
60. Shear, C., Korman, S., Angier, D.J., and Cahn, R.P., Arc vaporiztion of refractory powders, in Proc. Fine Particles Symp., William, E.K., Ed., Electrochem. Soc., Boston, 1973, 133.
61. Whyman, D., A rotating wall D.C. plasma Furnace, *J. Sci. Instrum.*, 44, 525, 1967.
62. Tylko, J.K., High Temperature Treatment of Materials, Canadian Patent 957733 (granted to Tetronics Ltd.), 1974.
63. Fey, M.G., Electric arc heaters for the process industries, in Industrial Electric Heating Conf., Cincinnatti, OH, February 1976.
64. Boulos, M. and Fauchais, P., Transport processes in thermal plasmas, in *Advances in Transport Processes*, Mujumdar, A.S. and Mashelkar, R.A., Eds., Wiley Eastern Ltd., New Delhi, 1986, chap. 6.
65. Goldberger, W.M. and Oxley, J.H., Quenching the plasma reaction by means of the fluidized bed, *AIChE J.*, 9, 778, 1963.
66. Jurewicz, J.B., Proulx, P., and Boulos, M., The Plasma spouted bed reactor, in *Proc. 7th Int. Symp. on Plasma Chemistry*, Timmermans, C.J., Ed., Eindhoven Univ. Tech., Eindhoven, 1985, 244.
67. Goldberger, W.M., The plasma bed, performance and capabilities, *Chem. Eng. Prog. Symp. Ser.*, 62(62), 42, 1966.
68. Manohar, H. and Gleit, C.E., Fluidized plasma: solid reactions, *Chem. Eng. Prog. Symp. Ser.*, 67(112), 55, 1971.
69. Yates, J.G., *Fundamentals of Fluidized Bed Chemical Processes*, Butterworth, London, 1983.
70. Moissis, A.A. and Zahn, M., Boundary value problems in electrofluidized and magnetically stabilized beds, *Chem. Eng. Commun.*, 67, 181, 1988.
71. Wierenga, C.R. and Morin, T.J., Characterisation of a fluidized bed plasma reactor, *AIChE J.*, 35(9), 1555, 1989.
72. Morin, T.J., Chapman, R., and Hawley, M.C., Flow calorimetry of electrical discharges, *Chem. Eng. Commun.*, 73, 183, 1989.
73. Harrington, R.F., *Time-Harmonic Electromagnetic Fields*, McGraw-Hill, New York, 1961.
74. Wen, C.Y. and Yu, Y.H., Mechanics of fluidization, *Chem. Eng. Prog. Symp. Ser.*, 62(62), 100, 1966.
75. Ergun, S., Fluid flow through packed columns, *Chem. Eng. Prog.*, 48, 89, 1952.
76. Pattipati, R.R. and Wen, C.Y., Minimum fluidization velocity at high temperatures, *Ind. Eng. Chem. Process Des. Dev.*, 20, 705, 1981.
77. Arnould, Ph., Cavadias, S., and Amouroux, J., The interaction of a fluidized bed with a thermal plasma, in *Proc. 7th Int. Symp. on Plasma Chemistry*, Timmermans, C.J., Ed., Eindhoven Univ. Tech., Eindhoven, 1985, 195.
78. Cavadias, S. and Amouroux, J., Synthesis of nitrogen oxides in plasma reactors, *Bull. Soc. Chim.*, 2, 147, 1986.
79. Cavadias, S., Heating of a fluidized bed by the injection of a thermal plasma; application to the synthesis of nitrogen oxides, in *Proc. 6th Int. Symp. on Plasma Chemistry*, Boulous, M. and Munz, R.J., Eds., McGill University, Montreal, 1983, 169.
80. Lippens, B.C. and Mulder, J., Prediction of the minimum fluidization velocity, *Powder Technol.*, 75, 67, 1993.
81. Baerus, M. and Fetting, F., Abschrecken eines heissen gasstrahis in einer gaswirbelsschicht, *Chem. Eng. Sci.*, 20, 273, 1964.

82. Arnould, P. and Amouroux, J., Operation and study of plasma fluidization process, *Bull. Soc. Chim. Fr.*, 6, 985, 1987.
83. Jurewiez, J., Proulx, P., and Boulos, M.I., The plasma spouted bed reactor, in *Proc. 7th Int. Symp. on Plasma Chemistry,* Timmermans, C.J., Ed., Eindhoven Univ. Tech., Eindhoven, 1985, 243.
84. Mathur, K.B. and Epstein, N., *Spouted Beds,* Academic Press, New York, 1974.
85. Crescitelli, S., Egiziano, L., and Macchiaroli, B., Glow discharge probes for fluid bed applications, *Quad. Ing. Chim. Ital.*, 10(2), 23, 1974.
86. Kolinowski, B. and Borysowski, J., Chemical engineering problems of cooling a plasma in fluidized bed, in *Proc. Chem. Plazmy. Ogolnopol. Symp. Nausk,* Szymanski, A., Ed., Wydawn Univ., Warsaw, 1972, 197.
87. Lobanov, N.F. and Tsyganov, V.I., Development and study of a reactor for quenching of a plasma flow in a fluidized bed, in *Sovers h. Konstr. Mash. Appar. Khim. Proizvod,* Bykov, Yu.M., Ed., Mosk. Inst. Khim., Mashinostr., Moscow, 1982, 96.
88. Zabrodsky, S.S., Andryushkevich, M.B., and Burachonok, I.N, Mechanism of quenching high temperature gas (plasma) jets in a fluidized bed, *Inzh. Fiz. Zh.*, 33(3), 419, 1977.
89. Minaev, G.A., Evaluation of the geometric characteristics of the torch flame under quasistationary conditions of its discharge into a fluidized bed, *Khim. Promst. (Moscow),* 6, 456, 1975.
90. Parkhomenko, V.D., Dovgal, M.A., Gan z, S.N., and Oleinik, E.E., Power parameters of an electrospark reactor, *Khim. Promst. (Moscow),* 47(3), 218, 1971.
91. Kojima, T., Matsukata, M., Arao, M., Nakamura, M., and Mitsuyoshi, Y., Development of a plasma jetting fluidized bed reactor, *J. Physi. II (Paris),* C2-429, 1991.
92. Amouroux, J. and Talbot, J., Conditions for the pyrolysis of methane in an inductive plasma of argon, *Ann. Chim.*, 14(3), 219, 1968.
93. Bykov, E.A., Anokhin, V.N., Efremov, A.N., and Mukhlenov, I.P., Decomposition of nitrogen oxides during an electrical discharge in a fluidized bed of solid particles, *Zh. Prikl. Khim. (Leningrad),* 42(8), 1710, 1969.
94. Masahiko, M., Hikaru, O.H., Toshinori, K., Yusuke, M., and Korekazu, U., Vertical progress of methane conversion in a D.C. Plasma fluidized bed reactor, *Chem. Eng. Sci.*, 47(9–11), 2963, 1992.
95. Patterson, M.C.L., Charles, J.A., and Fray, D.J., Application of a radially coalesced plasma to extraction metallurgy, in Proc. Pydrometallurgy 87, Inst. of Mining and Metallurgy, London, September 21–23, 1987, 862.
96. Harry, J.E. and Knight, R., Multiple arc discharges for metallurgical reduction or metal melting in plasma processing and synthesis of materials, in *Mat. Res. Soc. Symp. Proc.*, Vol. 30, Szekely, J. and Apelian, D., Eds., North-Holland, New York, 1984, 245.
97. Giacobbe, F.W., Advanced powder feeding device for use in gas/solid plasma synthesis and processing applications, in *Mat. Res. Soc. Symp. Proc.*, Vol. 98, Apelian, D. and Szekley, J., Eds., Materials Research Society, Pittsburgh, 1987, 405.
98. Nimalasiri, S.K. and Guevremont, R., A fluidized bed sampling system for the direct introduction of solids into an inductively coupled plasma. I. Performance characteristics, *Spectrochim. Acta B*, 41B(9), 865, 1986.
99. Guevremont, R. and Nimalasisri, S.K., A fluidized bed sampling system for the direct introduction of solids into an inductively coupled plasma. II. Calibration and internal reference methods, *Spectrochim. Acta B*, 41B(9), 875, 1986.
100. John W.B., Fluidized Bed Process and Apparatus, U.S. Patent 3,370,918, February 27, 1968.

101. Amouroux, J. and Talbot, J., Conditions for pyrolysis of methane in an inductive plasma of argon, *Ann. Chim. (Paris)*, 14(3), 219, 1968.
102. Waldie, B., Review of recent work on the processing of powders in high temperature plasmas. I. Processing and economic studies, *Chem. Eng.*, 259, 92, 1972.
103. Waldie, B., Review of recent work on the processing of powders in high temperature plasmas. II. Particle dynamics, heat transfer and mass transfer, *Chem. Eng.*, 261, 187, 1972.
104. Ettlinger, L.A., Nainan, T.D., Ouellette, R.P., and Cheremisinoff, P.N., *Electrotechnology Applications*, Ann Arbor Science, Ann Arbor, MI, 1980, chap. 2.
105. Naden, D., Fluidized bed reduction and plasma smelting technology for metals production from ore fines and dusts, *Steel Times Int.*, 10(2), 22, 1986.
106. Gauvin, W.H. and Choi, H.K., Plasma in extractive metallurgy, in *Material Research Society Symp. Proc.*, Vol. 30, Szekely, J. and Apelian, D., Eds., North-Holland, New York, 1984, 77.
107. Kyriacou, A., Characteristics of a Thermal Plasma Containing Zirconium Tetrachloride, M.Eng. thesis, McGill University, Montreal, 1982.
108. Spiliotopoulos, P., Characteristics of Zirconium Tetrachloride Thermal Plasmas, M.Eng. thesis, McGill University, Montreal, 1983.
109. Parker, I.M., Process and Apparatus for Chemical Reactions in the Presence of Electric Discharges, British Patent 1,462,056 (Cl. C22 B5/12), January 19, 1977.
110. Potter, E.C., Direct current plasma — a new spheroidizing process, *Met. Prog.*, 90(5), 127, 1966.
111. Kiyoshi, I., Apparatus and Method for Activating and Purifying Metal Powder, German Offen. 30 pp B22 F 001-00, 1976 (CA85 (12):82196 K).
112. Bonet, C., Vallbona, G., Foex, M., Daguenet, M., and Dumargue, P., High temperature processing of super refractories in a plasma fluid bed reactor, *Rev. Int. Hautes Temp. Refract.*, 11(1), 11, 1974.
113. Horio, M., Hanaoka, R., and Tsukada, M., A plasma spout/fluidized bed as a new powder processing reactor, *Proc. Jpn. Symp. Plasma Chem.*, 1, 213, 1988.
114. Derco, J., Lodes, A., Lapeik, L., and Blecha, J., Application of plasma chemical processes in treatment of properties of macro molecular materials. II. Surface treatment of polypropylene by oxygen nitrogen plasma in the fluidized bed, *Czech. Chem. Zvesti*, 38(4), 463, 1984.
115. Conte, A.A., Jr., Painting with polymer powders, *Chem. Technol.*, 4(2), 99, 1974.
116. Morooka, S., Okuba, T., and Kusakabe, K., Recent work on fluidized bed processing of fine particles as advanced materials, *Powder Technol.*, 63, 105, 1990.
117. Gibilaro, L.G., Di Felice, R., and Foscolo, P.U., On the minimum bubbling voidage and the Geldart classification for gas-fluidized beds, *Powder Technol.*, 56, 21, 1988.
118. Chaouki, J., Chavarie, C., Klvana, D., and Pajonk, G., Effect of interparticle forces on the hydrodynamic behavior of fluidized aerogels, *Powder Technol.*, 43, 117, 1985.
119. Morooka, S., Kusakabe, K., Kobata, A., and Kato, Y. Fluidization state of ultra fine powders, *J. Chem. Eng. Jpn.*, 21, 41, 1988.
120. Kono, H.O., Huang, C.C., Morimoto, E., Nakayama, T., and Hikosaka, T., Segregation and agglomeration of type C powders from homogeneously aerated type A–C powder mixture during fluidization, *Powder Technol.*, 53, 163, 1987.
121. Visser, J., Van der Waals and cohesive forces affecting powder fluidization, *Powder Technol.*, 58, 1, 1989.
122. Mori, S., Haruta, T., Yamamoto, A., Yamada, I., and Mizutani, E., Vibrofluidization of very fine particles, *Kagaku Kogaku Ronbunshu*, 15, 992, 1989.

123. Kato, K., Takarada, T., Koshinuma, A., Kanazawa, I., and Sugihara, T., in *Fluidization VI*, Grace, G.R., Schemilt, L.W., and Bergougnou, M.A., Eds., AIChE, New York, 1989, 351.
124. Hotta, N., Kimura, I., Tsukuno, A., Saito, N., and Matsuo, S., Synthesis of aluminum nitride by nitridation of floating alumina particles in nitrogen, *Yogyo Kyokai Shi*, 95, 274, 1987.
125. Kimura, I., Ichiya, K., Ishii, M., Hotta, N., and Kitamura, T., Synthesis of fine aluminum nitride powder by a floating nitridation technique using a nitrogen/ammonia gas mixture, *J. Mater, Sci. Lett.*, 8, 303, 1989.
126. Horio, M., Tsukada, M., and Naito, J., in *Fluidization VI*, Grace, G.R., Schemilt, L.W., and Bergougnom, M.A. Eds., AIChE, New York, 1989, 335.
127. Chiba, S., Honma, S., Nishiwaki, N., Hayashi, T., and Endo, K., in Proc. 21st Autumn Meet. Soc. of Chem. Eng. Jpn., 1988, 610 (in Japanese).
128. Takamizawa, M. and Hongu, T., Chem. Eng. Symp. Ser. 20, Soc. Chem. Eng. Japan, 1983, 69 (in Japanese).
129. Ranney, M.W., *Microencapsulation Technology*, Noyes Development Corporation, Park Ridge, New Jersey, 1969, 119.
130. Okubo, T., Kawamura, H., Kasakabe, K., and Morooka, S., Plasma nitriding of titanium particles in a fluidized bed reactor at a reduced pressure, *J. Am. Ceram. Soc.*, 73(5), 1150, 1990.
131. Kusakabe, K., Kuriyama, T., and Morooka, S., Fluidization of fine particles at reduced pressure, *Powder Technol.*, 58(2), 125, 1989.
132. Kawamura, H., Okuba, T., Kusakabe, K., and Morooka, S., Plasma surface treatment of milled carbon fibre in a fluidized bed reactor, *J. Mater. Sci. Lett.*, 9, 1033, 1990.
133. Nakamura, J. and Hirate, T., A Study on the Surface Treatment of ZnS Fluorescent Powder, Report No. CPM 87-31, Inst. of Electronics Inform. Commun. Eng., 1987 (in Japanese).
134. Matsumoto, S., Sato, Y., Kamo, M., and Sedaka, N., Synthesis of Polycrystalline Diamond, Jpn. Kokai Tokkyo Koho (Japanese Patent unexamined application) S 59-137311,69, 1984.
135. Matsumoto, S., Sato, Y., Kamo, M., and Sedaka, N., Synthesis of Polycrystalline Diamond, Jpn. Kokai Tokkyo Koho (Japanese Patent unexamined application) HI 157497, 1989.
136. Takarada, T., Ebihara, F., and Kato, K., in Proc. 54th Meet. Soc. Chem. Eng. Jpn., 1989, 486 (in Japanese).
137. Munz, R.J. and Mersereau, O.S., A plasma-spout fluid bed for the recovery of vanadium from vanadium ore, *Chem. Eng. Sci.*, 45(8), 2489, 1990.
138. Kreibaum, J., Plasma Spouted Bed Calcination of Lac Doré Vanadium Concentrate, M.Eng. thesis, McGill University, Montreal.
139. Munz, R.J. and Chin, E.J., The carbothermic reduction of niobium pentoxide and pyrochlore in a transferred arc argon plasma, *Can. Metall. Q.*, 30(1), 21, 1991.
140. Munz, R.J. and Chin, E.J., The carbothermic reduction of niobium pentoxide and pyrochlore in the presence of iron in a transferred arc argon plasma, *Can. Metall. Q.*, 31(1), 17, 1992.
141. Liang, A. and Munz, R.J., Ferroniobium production by plasma technology, a techno-economic assessment in 2nd World Conf. on Chemical Eng., Vol. III, Montreal, October 4–8, 1981, 252.
142. Jyukhai, G.G. and Shimanovich, V.D., Experimental study of microdischarge between a metallic electrode and a fluidized bed of graphite particles, *Vestsi Akad. Navuk B SSR Ser. Fiz, Energ. Navuk*, 3, 100, 1976 (in Russian).

143. Naden, D. and Kershaw, P., Fluid bed ore reduction and plasma smelting technology for metal production from ore and metal fines and EAF dusts, in Proc. Steelmaking Conf. 69, 5th Iron and Steel Congr., Bk. 1, 1986, 51.
144. Krukovskii, V.K., Kolobova, E.A., Valibekov, Yu V., Shoikhedbrod, S.P., Novikova, G.A., and Kononov, V.V., Coal Gasification in a Plasma of Steam and Air Deposited Doc., VINITI, 2860, USSR (Ru), 1981.
145. Apparatus for Hydrogen Pyrolysis of Coal and Lignite, Institut National des National des Industries Extractives, Belg. 886, 903 (Cl. C10G), December 29, 1980, 9.
146. Bal, S., Musialski, A., and Swierczek, R., Gasification of coal fines in a plasmochemical reactor with a spouted bed, *Koks Smola Gaz*, 16(5), 123, 1971.
147. Brzeski, J., Opalinska, T., and Resztak, A., Plasma pyrolysis of brown coal in a spouted bed, *Koks Smola Gaz*, 29(8–9), 215, 1984.
148. Shine, N.B., Fluochmic process for hydrogen cyanide, *Chem. Eng. Prog.*, 67, 52, 1971.
149. Johnson, H.S., Reactions in a fluidized coke bed with self resistive heating, *Can. J. Chem. Eng.*, 39, 145, 1961.
150. Goldberger, W.M., Hanway, J.E., and Langston, B.G., The electrothermal fluidized bed, *Chem. Eng. Prog.*, 61(2), 63, 1965.
151. Paquet, J.L. and Foulkes, P.B., Calcination of fluid coke in an electrically heated fluidized bed, *Can. J. Chem. Eng.*, 43(4), 94, 1965.
152. Borodulya, U.A., Zabrodsky, S.S., and Zheltov, A.I., Electrical properties of fluidized and settled beds of graphite particles at temperatures up to 2500°C, *AIChE Symp. Ser.*, 69(128), 106, 1973.
153. Kavlick, V.J., Lee, B.S., and Schsora, F.C., Electrothermal coal char gasification, *AIChE Symp. Ser.*, 67(116), 228, 1971.
154. Goldschmidt, D. and LeGoff, P., Electric resistance of fluidized beds — average resistance of conducting particles fluidized by air — preliminary results, *Chem. Eng. Sci.*, 18, 805, 1963.
155. Graham, W. and Harvey, E.A., The electrical resistance of fluidized beds of coke and graphite, *Can. J. Chem. Eng.*, 43(3), 145, 1965.
156. Cibrowski, J. and Wlodarski, A., On electrostatic effects in fluidizedbeds, *Chem. Eng. Sci.*, 17, 23, 1962.
157. Goldschmidt, D. and LeGoff, P., Electrical methods for the study of a fluidized bed of conducting particles, *Trans. Inst. Chem. Eng.*, 45, T196, 1967.
158. Reed, A.K. and Goldberger, W.M., Electrical behaviour in a fluidized bed conducting solids, *Chem. Eng. Prog. Symp. Ser.*, 62(67), 71, 1966.
159. Lee, B.S., Pyrcioch, E.J., and Schora, F.C., The electrical resistivity of a high pressure fluidized bed, *Chem. Eng. Prog. Symp. Ser.*, 66, 75, 1970.
160. Graham, W. and Harvey, E.A., Electrical conductivity of a fluidized bed of coke and graphite, *Can. J. Chem. Eng.*, 43, 146, 1965.
161. Knowlton, T.M., Pulsifer, A.H., and Wheelock, T.D., Prediction of fluidized bed resistance using field theory, *AIChE Symp. Ser.*, 69(128), 94, 1973.
162. Pulsifer, A.H. and Wheelock, T.D., The electrical resistance of gas fluidized beds, in *Fluidization*, Davidson, J.F. and Keairns, D.L., Eds., Cambridge University Press, Cambridge, 1978, 76.
163. Bruggeman, D.A.G., The calculation of various physical constants of heterogeneous substances. I. The dielectric constants and conductivities of mixtures composed of isotropic substances, *Ann. Phys. (Leipzig)*, 24, 636, 1935.
164. Leva, M., *Fluidization*, McGraw-Hill, New York, 1959, 94.

165. Jones, A.L. and Wheelock, T.D., The electrical resistivity of fluidized carbon particles: significant parameters, *Chem. Eng. Prog. Symp. Ser.*, 66(105), 157, 1970.
166. Ballain, M.D. and Pulsifer, A.H., Electrode temperature and resistance of an electrothermal fluidized bed, *Chem. Eng. Prog. Symp. Ser.*, 66(105), 229, 1970.
167. Lee, B.S., Pyrcioch, E.J., and Schora, F.C., Jr., The electrical resistivity of a high pressure fluidized bed, *Chem. Eng. Prog. Symp. Ser.*, 66(101), 75, 1970.
168. Glidden, H.J. and Plusifer, A.H., Electrode contact resistance in a fluidized bed, *Can. J. Chem. Eng.*, 46, 476, 1968.
169. Chen, T.P., Yuan, E., and Pulsifer, A.H., Bed and contact resistance in an electrically conducting fluidized bed, *AIChE Symp. Ser.*, 73(161), 51, 1977.
170. Kavlick, V.J. and Lee, B.S., High pressure electrothermal fluid bed gasification of coal char, *Chem. Eng. Prog. Symp. Ser.*, 66(105), 145, 1970.
171. Colver, G.M. and Bosshart, G.S., Heat and charge transfer in an A.C. electrofluidized bed, in *Multiphase Transport*, Vol. 4, Veziroglu, T.N., Ed., Hemisphere Publishing, Washington, D.C., 1980, 2215.
172. Colver, G.M., Bubble control in gas fluidized beds with applied electric fields, analogies and mechanisms, in 2nd Powder and Bulk Solids Conf., Chicago, May 10–12, 1977.
173. Spink, D.R., Cookstoss, J.W., and Hanway, Jr., J.E., The fluidized bed chlorination of zirconium bearing materials, paper presented at the Annual Meeting of Am. Inst. of Mining, Metall. and Petrol Eng., New York, February 1964.
174. Almond, J.C., Ph.D. thesis, University of Washington Seattle, 1965.
175. Manieh, A.A., Scott, D.S., and Spink, D.R., Electrothermal fluidized bed chlorination of zircon, *Can. J. Chem. Eng.*, 52, 507, 1974.
176. Pulsifer, A.H. and Wheelock, T.D., Production of hydrogen from coal char in an electrofluid reactor, *Ind. Eng. Chem. Process Des. Dev.*, 11(2), 229, 1972.
177. Johnson, H.S., Reactions in a fluidized coke bed with self resistive heating, *Can, J. Chem. Eng.*, 39, 145, 1971.
178. Pulsifer, A.H., Knowlton, T.M., and Wheelock, T.D., Coal char gasification in an electrofluid reactor, *Ind. Eng. Chem. Process Des. Dev.*, 8(4), 539, 1969.
179. Beeson, J.L., Pulsifer, A.H., and Wheelock, T.D., Coal char gasification in a continuous electrofluid reactor, *Ind. Eng. Chem. Process Des. Dev.*, 9(3), 460, 1970.

CHAPTER 5

Fluid Bed Design Aspects

I. OPERATING VELOCITY

A. Classification

In fluidization literature, it is not an easy task to define an operating velocity universally; it cannot have a unique value, as does the minimum fluidization velocity. The operating velocity in the case of gas fluidization is based on the conditions pertaining to (1) hydrodynamics, (2) heat transfer, and (3) reaction. The velocity at which the best hydrodynamic condition is obtained in a cold model system can differ widely from the operating velocity based on heat transfer or reaction conditions. Hence, it is a complex parameter which needs be chosen properly after understanding the whole system adequately. Empirically, it has been a standard practice to use an operating velocity in a fluidized bed that is three to five times the minimum fluidization velocity. This choice has no theoretical basis, but is widely accepted as a rule of thumb. In this section, we will consider some important aspects that govern the choice of an operating velocity.

In Chapter 1, we described flow phenomena and various operating velocities and flow regimes. Nevertheless, operating velocities that pertain to favorable hydrodynamic conditions will be described here to highlight their relative importance.

B. Hydrodynamics Operating Velocity

1. Range

The most important fundamental velocity in fluidization is the minimum fluidization velocity. It is not recognized as a velocity for operating purposes. The very fact that it is the minimum velocity required for fluidization means that a fluidized bed can have an operating velocity above it. The operating velocity (U) is normally expressed as a ratio to U_{mf}. When the operating velocity is increased, the ratio U/U_{mf} can assume any value up to U_t/U_{mf}. That is, the operating velocity increases in sequence through velocities corresponding to minimum fluidization (U_{mf}), minimum bubbling (U_{mb}), turbulent fluidization (U_{tr}), and particle carryover (U_t). The correlations pertinent to predicting these velocities were presented in Chapter 1. The criteria for fixing an operating velocity responsible for maintaining uniform/smooth/stable fluidization are discussed in the following sections.

2. Stability of Operation

a. Complexity

The term stable operation in fluidization literature is somewhat vague and has not been defined like other parameters such as U_{mf}. In principle, the velocity corresponding to stable operation should create a fluidization condition free of gas channeling and bring about good gas–solid contact conforming to optimum energy consumption. In other words, the energy imparted to the fluidization of the bed materials should be efficiently dissipated to maintain a well-agitated condition of the bed. Because fluidized beds are favored due to their inherent properties of self-induced mixing and isothermal operation, a stable fluidization condition should necessarily be able to satisfy these conventional demands. However, the formation of criteria to evaluate the operating velocity, starting from the first principles, is always a challenging task for researchers. The various aspects relevant to prediction of the stable operating velocity on the basis of the hydrodynamics of fluidization are as follows.

b. Pressure Drop Criterion

In a fluidized bed, the fluidizing gas encounters mainly two types of resistance, one across the distributor and the other across the bed. Either or both resistances can control the stability or the uniformity of fluidization. The resistance across the distributor depends on its type, the free open area, and the gas flow through it, whereas bed resistance is dependent on the type of bed material and its quantity or holdup.

c. Grid/Distributor Pressure Drop Criterion

Bed resitance for a smooth fluidized bed remains constant and the distributor pressure drop varies, depending on the operating velocity. Hence, at a given velocity, which may be designated the stable operating velocity, the distributor drop and the bed drop should remain constant. In other words, the total pressure drop, which is the sum of the two pressure drops, should not change with time once the velocity is fixed. However, change can occur due to the unusual behavior of the bed caused by channeling. Hence, the total drop can vary and create a condition of unstable fluidization. Siegel[1] developed a criterion to fix the pressure drop ratio of a fluidized bed that does not channel at the incipient state. This criterion, called Siegel's criterion, is based on the constant total pressure drop concept, and the required pressure drop ratio according to this criterion is

$$\frac{\Delta P_d}{\Delta P_b} = \frac{1}{\bar{n}}\left(\frac{U_{mf}}{U_t}\right)^{1/\bar{n}} \frac{1}{\left(1-\epsilon_{mf}\right)} \tag{5.1}$$

where the exponent $\bar{n}$ is obtained from the Richardson–Zaki[2] equation:

$$\frac{U}{U_t} = \epsilon^{\bar{n}} \tag{5.2}$$

Siegel's criterion implies that the pressure drop ratio ($\Delta P_d/\Delta P_b$) is fixed for a given gas–solid system, because the ratio U_{mf}/U_t is a function of the Archimedes number (Ar). As the bed pressure drop is constant for $U \geq U_{mf}$, the distributor pressure drop is also constant for a given Ar. The operating velocity (U_{opt}) can be predicted from fluid mechanics theory. In other words, $U \alpha \sqrt{\Delta P_d}$ for a perforated distributor and $U \alpha\ \Delta P_d$ for a porous distributor. According to Siegel's equation, $\Delta P_d/\Delta P_b$, the pressure drop ratio, increases with Ar. Hence, it can be inferred that the operating velocity would also increase with Ar to maintain channel-free fluidization and that the magnitude of this increase would depend on the nature of the distributor.

Siegel's theory was further analyzed by Shi and Fan,[3] and they proposed that in order to achieve complete fluidization the operating velocity should satisfy the following condition:

$$\left(\Delta P_t\right)_{U>U_{mf}} = \left(\Delta P_t\right)_{U_{mf}} \tag{5.3}$$

The above analyses to define the operating velocity for either a channel-free condition or complete fluidization are based solely on theoretical considerations and do not take into account available experimental data. Studies on the operating characteristics of multiorifice distributors by Fakhimi and Harrison[4] and Sathiyamoorthy and Rao[5,6] and tuyere distributors by Whitehead et al.[7] on large-scale fluidized beds showed that all the gas-issuing sites in these distributors do not deliver with equal flow rate at U_{mf}. They start operating gradually and become fully operational at a velocity U_M, discharging with equal flow rate through all the gas inlet sites. The velocity U_M confirms the operation of a distributor without malfunction. However, it cannot provide information as to whether the bed is fluidized uniformly. Sathiyamoorthy and Rao[8] analyzed the pressure drop criterion, keeping in mind the Siegel model[1] and the model of multiorifice distributor design. They attempted to arrive at an equation that could predict the operating velocity to create hydrodynamically stable fluidization. Thus, it was shown by Sathiyamoorthy and Rao[6] that:

$$\frac{U_M}{U_{mf}} = 1 + \left[c\left(\Delta P_d/\Delta P_b\right)\right]^{1/c} \tag{5.4}$$

where $c = 2$ for a multiorifice distributor. If the distributor-to-bed pressure drop ratio in Equation 5.4 is expressed as per Siegel's criterion (Equation 5.1), the operating velocity ratio (U_M/U_{mf}) that can initiate and maintain stable fluidization[8] is given by:

$$\frac{U_M}{U_{mf}} = 2.65 + 1.24 \log_{10} \left(U_t/U_{mf}\right) \tag{5.5}$$

From Equation 5.5, it can be inferred that for fine particles (i.e., $U_t/U_{mf} \approx 10$), $U_M/U_{mf} \approx 3.89$.

Criteria for uniform fluidization pertinent to nonaggregative-type fluidization were developed by Mori and Moriyama[9] by considering a nonfluidized bed to consist of two parts, one fixed and the other fluidized. Their theory takes into consideration the operating number of nozzles and the pressure drop ratio ($\Delta P_d/\Delta P_b$) under the conditions of a last nozzle operation. They cite experimental evidence in support of their theory but could not specify a parameter that would indicate the fraction of the stationary bed in a fluidized bed.

Pressure drop studies[10] across perforated-type distributors in a gas fluidized bed showed that the distributor drop in the presence of the bed was higher than that the empty bed drop and that the drops had no difference at a velocity called the real minimum fluidization velocity, which was higher than the conventional U_{mf}. Real fluidization was found to occur when the hydrodynamic condition at all the gas-issuing sites of the perforated plate distributor remained the same. No attempt was made to suggest a correlation for this real minimum fluidization velocity.

d. Critical Grid Resistance Ratio

Uniformity of fluidization in the case of multiorifice distributors was suggested[11] to occur when the gas flow overcomes a certain critical resistance (ΔP_{ds}) at the distributor; this resistance is caused by stagnant zones. If ΔP_{ds} is known, then the following relationship can be used:

$$U_c/U_{mf} = \left(1 + \Delta P_{ds}/\Delta P_d\right)^{0.5} \tag{5.6}$$

where U_c is the critical velocity for complete fluidization and ΔP_{ds} and ΔP_d are given by:

$$\Delta P_{ds} = \frac{P_s(1-\epsilon) g\, d_o}{4\phi}\left[1 - \frac{7}{3}\sqrt{\phi} + \frac{4 \cdot l}{d_o}\right] \tag{5.7}$$

where l is the depth of the orifice up to which particle spilling occurs, and:

$$\Delta P_d = \rho_g\, U_c^2/2\phi^2 \tag{5.8}$$

The above correlation to predict U_c is not convenient because U_c appears on both sides of Equation 5.6. However, this approach is a new way to characterize the multiorifice gas distributor and find the critical velocity for complete fluidization.

C. Backmixing Critical Velocity

An important gas velocity in the fluidized bed is the one responsible for inducing backmixing. In other words, a well-mixed fluidized state occurs in a bubbling fluidized bed above a certain critical velocity ratio (U/U_{mf}). Each bubble, while moving upward, displaces an equal volume of the dense phase in the direction of its motion. This implies that a part of the solid-rich dense phase (i.e., emulsion phase) descends. A critical velocity is reached when the emulsion phase velocity (U_e) is the same as the interstitial gas velocity (U_{mf}/ϵ):

$$U_e = U_{mf}/\epsilon \tag{5.9}$$

If the volume of the displaced emulsion phase per bubble volume is R, then by mass balance:

$$U_e = R(U - U_{mf}) \tag{5.10}$$

Eliminating U_e by using Equations 5.9 and 5.10, one obtains for the critical condition:

$$\left(U/U_{mf}\right)_{cr} = \frac{1+\epsilon R}{\epsilon R} \tag{5.11}$$

For $\epsilon = 0.4$ and $R = 1$, $(U/U_{mf})_{cr} = 3.5$, and for $\epsilon = 0.4$ and $R = 0.5$, $(U/U_{mf})_{cr} = 6.0$.

It can be seen from the above examples given by Stephens et al.[12] that the critical velocity is quite sensitive to the bed voidage (ϵ) and the ratio R. Further, for small U_{mf} values, U/U_{mf} exceeds the critical velocity ratio, thereby assuring a well-mixed fluidized bed; for higher U_{mf} values, the critical velocity ratio is not so easily attained without reaching the slugging bed condition. The critical velocity ratio corresponds to plug flow condition for a gas–solid system that has high U_{mf} values.

D. Operating Velocity Under Condition of Particle Attrition or Agglomeration

Determination of the operating velocity based on channeling or full fluidization or the operation of all the gas-issuing orifices with an equal discharge rate can lead to instability of the fluidized bed operation if the particles in the bed tend to grow in size due to agglomeration or tend to reduce in size due to severe attrition. In such events, the fluidization characteristics change mainly due to changes in the particle size, and this is reflected in a change in the value of the dimensionless parameter, the Archimedes number (Ar). Sathiyamoorthy et al.[13] analyzed the effect of changing

Ar on the distributor-to-bed pressure drop ratio, $(\Delta P_d/\Delta P_b)$, and derived the following relationship for this situation at U_{mf}:

$$(\Delta P_d/\Delta P_b)_{mf} = [\mathrm{Re}_{mf}/(1 - \epsilon_{mf})]\ [\delta\epsilon_{mf}/\delta A_r) \cdot (\delta \mathrm{Re}_{mf}/\delta A_r)] \tag{5.12}$$

Substitution of the above equation in Equation 5.4 for $U = U_M$, and evaluation of the derivatives of ϵ_M and Re_M with respect to Ar led to the value of the Reynolds number (Re) which agreed well with many of those reported in the literature for Re_{opt} corresponding to U_{opt} at which the maximum heat transfer coefficient in a gas fluidized bed is attained. These findings were found to be consistent[13] with experimental data on U_{opt}.

E. Operating Velocity for Minimizing Solid Leakage Through Distributor

1. *Solid Leakage or Weeping Through Grids*

When gas fluidized beds are operated with perforated or nozzle-type distributors, fluidized solid particles pass through the gas inlet points, thereby obstructing gas flow and allowing the solid to collect in the plenum. Over a period of time, the solid collected at the plenum gradually increases and cleanup of the plenum becomes necessary. Furthermore, these solid particles can cause plugging of the active gas-issuing sites in the distributor. As a result of this, the distributor can impede fluidization. It is therefore essential to operate the fluidized bed at least at the minimum gas velocity, which would minimize or eliminate solids leakage through the distributor. Solids leak through the distributor by two modes: dumping and weeping. Dumping of solids occurs during the start-up of fluidization. Kassim[14] reported that it occurred during the initial stages of bubble formation at the office, whereas Brien et al.[15] observed its occurrence all across the cross-section in a haphazard manner. Dumping of solids is attributed to the momentary higher pressure above the distributor compared to that at the plenum, and this pressure reversal was also confirmed[16,17] in large-scale fluidized beds. Weeping of solids through the distributor holes normally occurs after the initiation of fluidization but at a reasonably high velocity. It occurs all along the periphery of the orifices. The transition from dumping to weeping has not been discussed clearly in the literature. Hence, the selection of an operating velocity for this transition is a difficult task.

2. *Operating Velocity at a Desired Solid Weeping Rate*

Kassim[14] reported that weeping can be eliminated at a certain critical velocity. However, Brien et al.[17] and Serviant[18] concluded that this critical velocity cannot be realized. According to Brien et al.,[17] weeping occurred even at a hold velocity of 43 m · s^{-1} or at a $\Delta P_d/\Delta P_b$ ratio of 1.5. They attributed weeping to fluctuations in pressure transmission caused by erupting bubbles at the bed surface. They confirmed this by incorporating vertical internal wave breakers which reduced weeping and also the pressure fluctuations transmitted from the bed surface. They observed that the weeping rate decreased with an increase in the hole velocity and proposed a correlation to predict the weeping rate as:

$$W\,(\text{kg m}^{-2}\,\text{s}^{-1}) = 6.34 \times 10^{9}\, V\,(\text{m s}^{-2})^{-7.82} \tag{5.13}$$

The assumption made in the development of the above correlation was that all holes were fully active. A correction would be needed if any inactive orifices were present. Hence, the selection of an operating velocity corresponding to a specified weeping rate is a practical problem.

Otero and Munoz[19] found that the flow of solids occurred even through the bubble-cap distributor plate. The flowback rate of solids through a distributor plate was found to be influenced by bed height,[17] bed internals,[20] and the diameter of the column.[21] However, correlations to determine the flowback rate of solids through a distributor plate are scant. Sathiyamoorthy[22] and Sathiyamoorthy et al.[23] studied the flowback rate of fluidized solids through various types of multiorifice distributor plates and proposed a correlation:

$$W\left(\text{kg m}^{-2}\text{s}^{-1}\right) = \alpha_d\left[U/U_{mf}^{-0.456}(H/D)^2\right] \tag{5.14}$$

which is valid for $1 \leq (U/U_{mf}) < 9$, $1 < H/D \leq 4$. In this expression, α_d is a distributor parameter and is defined as:

$$\alpha_d = 0.2895\,(\phi\, d_o/d_p)^{0.7} \tag{5.15}$$

over the interval $7 \times 10^{-3} < (\phi\, d_o/d_p) \leq 95 \times 10^{-3}$; ϕ is the fraction of free open area in the distributor.

It can be seen from Equation 5.14 that the operating velocity ratio (U/U_{mf}) can be predicted for a specified minimum flowback of solids through the distributor if the aspect ratio (H/D) and the distributor parameters are fixed. However, the value of U/U_{mf} does not provide any information about the quality of fluidization.

F. Operating Velocity Based on Maximum Bubble Size

There is a limiting bubble size depending upon which the operating velocity can be fixed. For example, a bubble can be stable or unstable. Harrison et al.[24] stated that if the upward rising velocity of the bubble (U_{br}) is greater than the particle terminal velocity (U_t), then particles would be drawn through the bottom of the bubble and the bubble can break. The following criteria for bubble stability were suggested:

$$U_{br} < U_t \text{ for a stable bubble}$$

$$U_{br} = U_t \text{ for a maximum size of a stable bubble}$$

$$U_{br} > U_t \text{ for a unstable bubble}$$

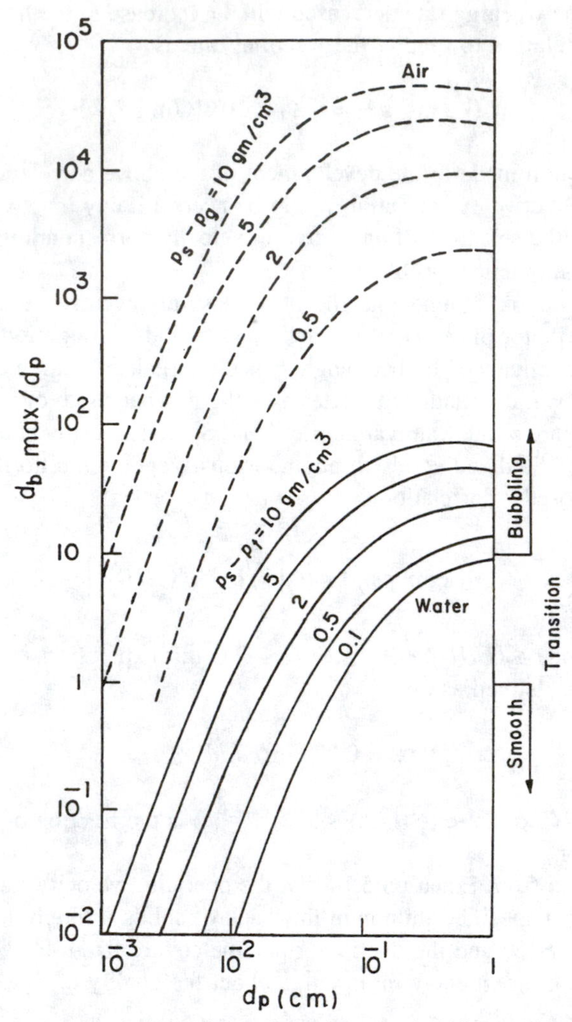

Figure 5.1 Maximum stable bubble size and condition for bubbling and smooth fluidization. (From Kunii, D. and Levenspiel, O., *Fluidization Engineering*, Wiley, New York, 1969. With permission.)

Based on these criteria, an approximate method of determining a stable bubble size was proposed by Davidson and Harrison;[25] this is shown in a graphical form in Figure 5.1.[26] A transition from smooth fluidization to bubbling occurs around $1 < d_{max}/d_p < 10$. Figure 5.1 indicates the condition for smooth fluidization of a gas (air) fluidized bed and the bubbling fluidization condition of a liquid fluidized bed. From Figure 5.1, for a given particle size (d_p) and for smooth fluidization, $d_{b,max}/d_p$ can be determined and $d_{b,max}$ can be predicted. This stable $d_{b,max}$ is a function of the operating velocity, which can be predicted from an appropriate correlation. The maximum bubble size[27] is given by:

$$d_{b,\max}(\text{cm}) = 0.65\left[\frac{\pi}{4}d_t^2\left(U - U_{mf}\right)\right]^{0.4} \tag{5.16}$$

This relationship is valid for $0.5 < U_m < 20$ cm/s, $0.006 < d_p < 0.045$ cm, $U - U_{mf} \leq 48$ cm/s, $d_t < 130$ cm. The operating velocity (U) for a known $d_{b,\max}$ can be obtained by using Equation 5.16, if the column diameter (d_t) and the particle size (d_p) are prefixed.

The operating velocity thus determined would assume a range of values as d_p changes for the same gas–solid system. This is because there is a maximum stable bubble size for each particle size, and this in turn is influenced by the corresponding operating velocity.

The various operating velocities discussed so far are based on hydrodynamic conditions, and these conditions correspond to smooth or stable fluidization. The variety of operating velocities described arise mainly due to the various ways of defining a smooth or stable fluidized bed. Although the operating velocity predicted on the basis of hydrodynamic conditions can, in principle, maintain stable fluidization, it may or may not yield the best results in terms of heat transfer and chemical conversion.

G. Operating Velocity for Optimum Heat Transfer

1. Maximum Heat Transfer

It is an established fact that the heat transfer rate in a gas fluidized bed increases with the fluidizing gas velocity until an optimum velocity is reached; on increasing the velocity further, the heat transfer rate decreases. Because it is difficult to ascertain precisely the gas–solid mixing or the particle movement in the immediate neighborhood of a heat transfer surface, it is not easy to devise an analytical method to evaluate the maximum heat transfer and the corresponding optimum gas velocity. The difficulty is mainly due to the opposing effects of particle mobility and voidage, both of which increase with the fluidizing gas velocity. Numerous correlations are available in the literature for predicting the maximum heat transfer coefficient and the optimum velocity. These correlations were mentioned in Chapter 1 as part of the discussion on heat transfer aspects in a fluidized bed.

2. Optimum Velocity

Although numerous correlations are available for predicting $\mathrm{Re}_{\mathrm{opt}}$, no systematic approach seems to have been adopted to identify the situation in which the maximum heat transfer rate is obtained. Sathiyamoorthy[22] and Sathiyamoorthy et al.[28] investigated the effect of various types of multiorifice distributors on heat transfer from the immersed surface in a gas fluidized bed and found that the maximum heat transfer occurred when all gas-issuing orifices in the distributor had just become fully operational. This was experimentally verified in terms of the number of operating orifices and the wall heat transfer coefficient, and a typical result[28] is depicted in Figure 5.2. The experimentally measured optimum gas velocity for $\mathrm{Ar} \leq 1000$ showed that the $\mathrm{Re}_{\mathrm{opt}}$ value based on U_{opt} coincided with the Re_M value based on U_{M}, which can be predicted by using Equation 5.5. The results[28] are shown in Figure 5.3. Sathiyamoorthy et al.[28] analyzed experimental data reported in the literature along with their own data and correlated the results on the variation of $\mathrm{Re}_{\mathrm{opt}}$ over a wide range of Ar values. The correlations proposed to predict $\mathrm{Re}_{\mathrm{opt}}$ are

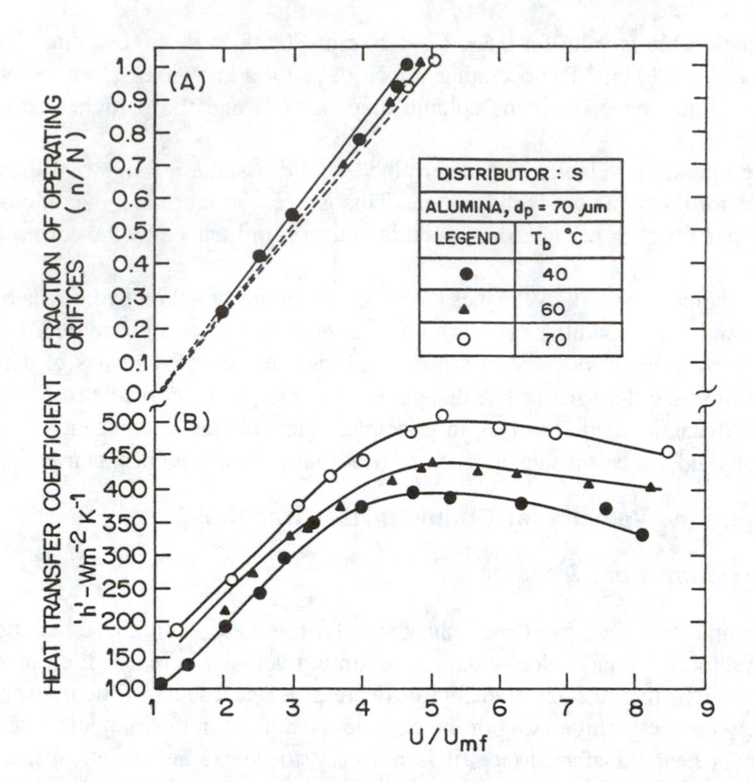

Figure 5.2 Variation of (A) fraction of operation orifices (*n/N*) in a multiorifice plate distributor and (B) wall heat transfer coefficient for horizontally immersed heater with U/U_{mf}. Percent free area of distributor = 0.088, number of orifices = 55, size of orifice = 0.8 mm. (From Sathiyamoorthy, D., Rao, Ch.S., and Raja Rao, M., *Chem. Eng. J.*, 37, 149, 1988. With permission.)

$$\mathrm{Re_{opt}} = 0.008\ \mathrm{Ar}^{0.868} \text{ for } 1 \le \mathrm{Ar} \le 3000 \tag{5.17}$$

$$= 0.13\ \mathrm{Ar}^{0.52} \text{ for } 3000 \le \mathrm{Ar} \le 10^7 \tag{5.18}$$

A fluidized bed can be operated at U_{opt} if it is used for heat exchange purposes. Most correlations for predicting U_{opt} were developed for nonreacting systems which transfer heat either from the wall of the heater immersed inside the bed or from the bed to outside media. The conditions under which the optimum operating velocity is determined are obviously physical in nature, and hence the use of such correlations for a chemically reacting system involves uncertainties and enhanced design risks.

H. Operating Velocity Dependent on Chemical Reaction

1. Stoichiometric Considerations

An optimum operating velocity in a chemical reactor should satisfy the requirements of stable fluidization, optimum heat transfer, and the best conversion of the reactant in question. From the preceding sections, it can be inferred that selection

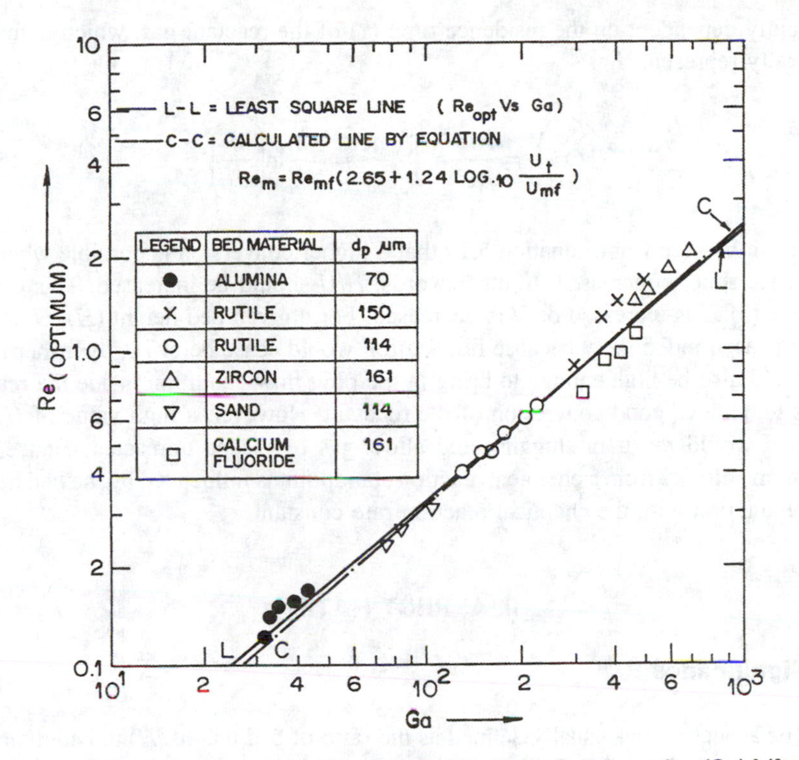

Figure 5.3 Variation of optimum Reynolds number (Re_{opt}) with Galileo number (Ga) [$d_p^3 \rho_g(\rho_s - \rho_g)g/\mu_g^2$]. Re_{opt} corresponds to U_{opt}, at which wall heat transfer coefficient in a gas fluidized bed is maximum. (From Sathiyamoorthy, D., Rao, Ch.S., and Raja Rao, M., *Chem. Eng. J.*, 37, 149, 1988. With permission.)

of an optimum velocity depends entirely on the requirements of the designer, and it is difficult to arrive at a unique operating velocity that satisfies all three conditions. For example, if the operating velocity is fixed at U_{opt} for maximum heat transfer, then despite the fact that complete fluidization can be obtained at this velocity with a multiorifice-type distributor, the flow rate thus fixed may be insufficient or excessive insofar as the stoichiometry of the reaction is concerned. If U_{opt} is insufficient with regard to the requirement of the stoichiometry of the reaction, it would be necessary to increase the operating velocity. In that case, the hydrodynamics of fluidization can change and the optimum condition in turn can be distributed. If U_{opt} corresponds to an excessive supply of the reactant, poor conversion would result because of an insufficient residence time. Thus, it is necessary to compromise between the two conditions to arrive at the best operating velocity for a chemical reactor. The aspect ratio also plays a key role in determining the chemical conversion. This aspect will be considered later in this section.

2. Conversion Consideration

Let us now examine the influence of the operating velocity in the specific case of a first-order irreversible reaction of type A → B. The conversion of the reactant is mainly dependent on the residence time (τ) of the reactant gas, which is mathematically represented as:

$$\tau = \frac{V}{\upsilon} = \frac{AH}{AU} = \int_{c_o}^{c} \frac{dc}{-r} = -\frac{1}{k}\ln\left(\frac{c}{c_o}\right) \tag{5.19}$$

It can be seen from Equation 5.19 that a higher conversion is possible when the residence time is increased. In other words, H/U should be increased. This can be achieved if H is increased or U is decreased. For a given bed height (H), U cannot be decreased indefinitely because fluidization would cease below U_{mf}. Furthermore, U should also be high enough to bring in the plug flow condition inside the reactor so as to achieve good conversion of the reactant. However, a high value of U in a tall bed would result in slugging and allow gas to escape unreacted. Hence, the optimum velocity from a chemical reaction standpoint is influenced by the bed height in conjunction with the chemical reaction rate constant.

II. ASPECT RATIO

A. Significance

The aspect ratio is usually defined as the ratio of bed height (H) to bed diameter (D). Although the effect of the aspect ratio on fluidized bed performance is an important topic, not much research work seems to have been carried out on this aspect. A bed can be termed deep if the aspect ratio is well above unity, and it is shallow when the ratio falls below unity. In a deep or tall bed, bubbles rise faster, form slug, and then escape with much of the reactant gas unreacted. In a shallow bed, the gas can short-circuit, thereby resulting in poor conversion. If the reactant gas is allowed to stay inside the bed for a long time by breaking the bubbles and reducing their rise velocity by means of bed internals, then the conversion can be enhanced. In such a case, the aspect ratio may not play a prominent role in influencing reactor efficiency. However, the bed internals cannot always be incorporated due to some practical reasons. For example, in high-temperature chlorination of zircon, there are no stable internals that can withstand the aggressive reaction environment. In such situations, the aspect ratio should be selected judiciously.

B. Distributor Effect

The distributor design has a major impact on the bed height particularly for establishing complete or uniform fluidization. The criterion of Siegel as given by Equation 5.1 predicts the ratio $\Delta P_d/\Delta P_b$ in terms of the velocity ratio (U_M/U_t), which is a function of the Archimedes number only. Hence, the ratio has a fixed value for a given gas–solid system (i.e., for a given Ar). This implies that to maintain smooth fluidized bed, the distributor pressure drop must be changed if the bed height or the aspect ratio is changed. In other words, if the bed height is increased, the distributor pressure drop should also be increased either by increasing the operating velocity or by reducing the free open area. The operating velocity cannot be increased indefinitely in view of the need to conserve the reactant gas. The alternate choice is to decrease the free open area of the distributor. It follows from this analysis that distributor design plays an important role in fixing the aspect ratio.

C. Bubble Size Effect

Mori and Wen[27] proposed a correlation for estimation of the bubble diameter (D_B) incorporating the aspect ratio as a variable. The correlation is

$$\frac{d_{BM} - d_B}{d_{BM} - d_{Bo}} = \exp\left(-0.3h/D_t\right) \tag{5.20}$$

The aspect ratio (i.e., h/D_t) can be predicted for a bubble diameter (d_B) which can be assumed to be a certain fraction of the maximum bubble diameter (d_{BM}), which need not necessarily be stable. Harrison et al.[24] expressed the maximum stable bubble size by the equation

$$d_{BS} = \frac{1}{g}\left(\frac{U_t}{0.71}\right)^2 \tag{5.21}$$

As the maximum bubble size (d_{BM}) can be less than d_{BS}, Mori and Wen[27] pointed out that d_{BM} is a fictitious maximum bubble size that is attainable due to coalescence.

D. Key Influencing Parameters

The correlation contained in Equation was 5.5 proposed by Sathiyamoorthy and Rao[8] based on criteria for developing smooth fluidization; the corresponding distributor-to-bed pressure drop ratio showed that the operating velocity for complete fluidization was independent of the aspect ratio. However, the same authors pointed out that bed height does not play a vital role in the stable operation of fluidization corresponding to distributor-to-bed pressure drop ratios of 0.12–0.24. The effect of the aspect ratio on bed expansion in particulate fluidization was studied by Mazumdar and Ganguly,[29] who found that the degree of expansion was independent of the aspect ratio for unicomponent system of crushed solid particles fluidized by water. The degree of bed expansion was defined as the ratio of the expanded bed height (H_e) above the static bed height (H_s) to the static bed height, i.e., $(H_e - H_s)/H_s$. They prefer to use this degree of bed expansion instead of bed voidage for the sake of accuracy in developing a correlation in terms of the particle Reynolds number (Re_p). Studies[30] on gas distribution in shallow packed beds showed that maldistribution was reduced if the bed depth was increased. In other words, for larger values of the reduced diameter ($D/d_{packing}$), greater reduced packed bed heights ($H/d_{packing}$) were suggested for uniform gas distribution. This implied that the aspect ratio should be high for a packed bed to obtain better distribution of gas in a large-diameter column. Hence, in a tall fluidized bed, a dead or packed zone of defluidized solid of sufficient height can act as a distributor and facilitate better fluidization.

E. Predictions

On the basis of good data collection on successful and unsuccessful operation of large-scale fluidized bed reactors and the analysis of data on the distributor-to-bed pressure drop ratio as a function of the aspect ratio, Qureshi and Creasy[31] arrived at the following correlation for predicting the critical pressure drop ratio, $(\Delta P_d/\Delta P_b)_c$:

$$(\Delta P_d/\Delta P_b)_c = 0.01 + 0.2\,[1 - \exp(0.5D/Z)] \tag{5.22}$$

A stable or successful fluidized bed operation was reported to be achieved if the pressure drop ratio was above a critical value. This conclusion was reached on the strength of 21 sets of data, all taken from the literature. The correlation was proposed as a guideline, and it was not obtained based on any regression data. Geldart and Baeyens[32] later pointed out that the correlation given by Equation 5.22 does not agree with the data of Geldart and Kelsy[33] for a low aspect ratio ($H_{mf}/D < 0.5$) and proposed a correlation for low values of the aspect ratio in order to have stable fluidization:

$$\left(\frac{\Delta P_d}{\Delta P_b}\right)_c \geq \exp\left(-3.8H_{mf}/D\right) \tag{5.23}$$

Investigations into the dependency of the critical distributor-to-bed pressure drop ratio on the aspect ratio to maintain stable fluidization led many researchers to arrive at the following important conclusions:

1. A high critical pressure drop ratio is desirable for shallow beds.
2. Shallow beds need to be operated at low values of the critical pressure drop ratio.

A distributor pressure drop of 35 cm of the water column was recommended by Agarwal et al.[34] for a fluidized bed with a diameter up to 4 m.

III. DISTRIBUTORS

A. Introduction

It is a well-established fact that an intimate gas–solid contact is achieved in a gas fluidized bed due to the self-induced mixing caused by the bubbles. The characteristics of the bubbles and their population density are influenced by several parameters, such as the physical properties of the gas and the solid, geometry of the bed, operating velocity, pertinent thermal conditions, and the gas distributing grid. The bubbles that form initially at the gas-issuing sites of the distributor plate or the gas jet emerging from the distributor have a remarkable influence on the hydrodynamics of the bed. If a gas fluidized bed is used mainly for heat exchange, it is essential to have prior knowledge of the distributor type, its selection criteria, and its design. This exercise has to be carried out because distributors have been found to influence the heat transfer coefficient, in particular its maximum value. Fluidized beds have been studied and modeled over several years without much attention focused on the type of distributor used and its influence on the performance efficiency of a fluidized bed reactor. The literature on distributors started to accumulate over the past three decades, and this topic has been given special attention at many international meetings. As a result, the missing link between fluidization hydrodynamics and distributor design characteristics has now been reasonably well established. However, much still remains to be done on the theoretical aspects of distributor-related phenomena in order to arrive at the conditions for complete fluidization as discussed in the preceding section on operating velocity. In principle, the distributor plate, in conjunction with the thermal and chemical conditions of the bed, plays a key role in fixing the operating velocity. A few reviews[35-42] on distributors are available in the literature. All of them essentially conclude with recommendations regarding distributor design and emphasize the need for more concentrated research on distributors, in particular those intended for use in large industrial-type reactors. The available information on distributors is summarized in this chapter in an effort to convey the importance of this aspect of the fluidized bed which is often ignored or given insufficient weight in many monographs and texts on fluidization.

B. Functions of Distributor

A distributor is a flow restrictor and a bed-supporting device. In addition, it has other important functions, such as (1) initiation of smooth fluidization, (2) prevention of solids flow into the plenum chamber during either downtime or operation, (3) maintaining stable operation throughout the operating periods, (4) minimizing the attrition of solid particles and the erosion of the bed internals, and (5) preservation of the distributor surface without a dead or defluidized zone of particles.

C. Importance of Distributor

1. Hydrodynamic Factors

The hydrodynamics of a fluidized bed is influenced by the distributor plate. A gas fluidized bed can be considered to be made up of three zones: the grid zone, the constant bed density zone, and the bubble-erupting zone. The hydrodynamics of each zone is different. The grid zone is influenced by the type and design of the distributor plate. Any change in grid zone dynamics subsequently affects the other two zones. A great deal of information on distributors can be obtained from the review articles by Saxena et al.,[36] Sathiyamoorthy and Vogelpohl,[37] and Werther and Schoessler.[42] The importance of the distributor is highlighted in this section.

2. Interfacial Area Factor

In dealing with the design of a fluidized bed reactor, certain important parameters such as transfer units based on mass transfer and reaction are to be evaluated. All these transfer units are functions of the interfacial area, which is influenced by the bubbles. The total surface area available for the transfer operation is

$$a_R = \int_o^H a' d h \tag{5.24}$$

where

$$a' = \frac{nA_b}{A_t \delta h} = \left(\frac{nV_b}{A_t \delta h}\right) \frac{A_b}{V_b} = \epsilon_b \cdot \frac{6\lambda}{d_v} \tag{5.25}$$

The bubble holdup (ϵ_b) is a function of the visible bubble velocity (U_b), which in turn is a function of the distributor design parameters. Werther[43] proposed an equation for U_b:

$$U_b = \Psi (U - U_{mf}) \tag{5.26}$$

where Ψ is a nonequality factor with a value of 0.67 and 0.76, respectively, for porous and perforated (also nozzle) distributors. Thus it is clear that the interfacial area is affected by the distributor type and the transfer units change accordingly. Studies[44] on gas entry effects revealed that gas bubbles formed at the distributor site accounted only for one-third of the gas injected through the orifice and that at a height of 0.25 m above the grid plate, the visible bubble flow accounted for three-fourths of the gas flow through the orifice. The grid zone itself is now considered to be made up of two zones; one is known as the jet stem with stable jets and the other as the bubble-forming region. A bubble which is born near or at the distributor grows in size and travels up; its size and rise in velocity depend on the height as well as its initial size. Except for a porous distributor plate, level above which the height is measured to calculate the bubble size is independent[45] of distributor type, such as perforated or tuyere. Baur[46] suggested that the total interfacial area for an industrial type of distributor is the sum of the areas at the grid zone and at the bubbling zone. From the above discussion, it can be inferred that the distributor plays an important role in determining the number of transfer units in a fluidized bed. Thus the distributor type ultimately affects the overall conversion. Development of a model for predicting the reactant conversion in a fluidized bed should therefore take into consideration the distributor type. This aspect will be discussed later in this chapter when dealing with models and their analysis. In describing the operating velocity in an earlier part of this chapter, it was pointed out that the velocity at which all gas-issuing sites are fully operational is a function of the distributor design parameters, and this velocity has great relevance with respect to the stability of fluidization and the maximum heat transfer rate.

3. Influence on Bed Behavior

a. Two-Phase Fluidization

Because gas flow in a gas fluidized bed is prone to a greater degree of maldistribution compared to liquid flow in a liquid fluidized bed, much attention has been focused in the literature on gas distributors. The effect of distributors in liquid fluidized beds has not been studied much. In recent years, this type of reactor has drawn the attention of biotechnologists in the context of the development of immobilized biocatalyst bioreactors. A recent study[47] on the hydrodynamics of a liquid fluidized bed, taking into consideration the effect of the distributor, showed that particle density plays a prominent role with regard to the distributor region flow behavior. Distributor-induced flow disturbances are significant for low-density ($\approx$1.61 g/cm^3) particles. A distributor that has a low hole density and a high pressure drop also has a similar effect close to the distributor site. Application of a dispersion-type model at the distributor region was cautioned against. However, a continuously stirred tank reactor model has found to be applicable for this region. Hence, for a liquid fluidized bed, application of a two-zone model (namely, a continuously stirred tank reactor) close to the distributor and a dispersion model above it is recommended. This two-zone model can be simplified to a single-zone model in the case of high-density particles (2460 kg/m^3) as distributor-induced disturbances are negligible in this case.

b. Three-Phase Fluidization

The three-phase fluidized bed has drawn much attention recently because of its growing application in biochemical processes. The modeling of such reactors is usually carried out by considering the entire bed as a single zone that has the same hydrodynamic condition. Alvarez-Cuenca[48] and Alvarez-Cuenca et al.[49] recently showed that two well-defined zones, a grid zone and a bulk zone above it, exist in a three-phase fluidized bed. The grid zone is found to have a high mass transfer rate and is observed to occupy about 25% of the bed height. A shallow bed of small particles (1 mm) is influenced by the distributor and the mass transfer rate is enhanced considerably. Because the distributor affects fluidization and leads to the formation of two distinct zones of different hydrodynamics, a two-zone model was proposed[50] for predicting mass transport in a three-phase fluidized bed. The plug flow condition was presumed to prevail in the grid zone and the dispersed condition in the bulk zone. The volumetric mass transfer coefficient ($K_l a$) in a three-phase fluidized bed is an important basic parameter with regard to evaluation of reactor performance. This parameter is strongly influenced by the bubble flow pattern, which in turn is dependent on particle size and the velocity of the gas as well as the liquid. Because the bubble flow pattern or the flow regime is complex, the parameter $K_l a$ also behaves in a complex manner. As a result, a number of correlations are reported in the literature.[51] Studies by Tang and Fan[52] with low-density solid particles showed a discrepancy with those reported in the literature, which was attributed to the type of distributor plate used. Another parameter that can change $K_l a$ pertains to the wettability[53] characteristics of the solid, but this aspect does not appear to have been studied extensively. Hence, the distributor effect must be resolved properly after accounting for all other parameters. The distributor zone is often considered to have a negative feature in that it can generate more fines due to vigorous gas–solid interaction, resulting in particle attrition. However, this view was shown to be incorrect by the study of Kimura and Kojima,[54] who examined the grid zone contribution with regard to silicon production by fluidized bed chemical vapor deposition. Their results indicate that the major contribution to the production of fines stems from the bubbling zone and not from the grid zone.

D. Types of Distributors

Distributors of various types and their relative merits and drawbacks were discussed by Kunii and Levenspiel.[26] Design and construction details for distributors were presented by Basu.[39] Distributors can be classified into various types depending on their design, geometry, and construction. Alternatively, distributors can be classified on the basis of their functional characteristics and the nature of the fluidizing fluid, an this type of classification is discussed in the following section.

1. Conventional Distributors for Gas or Liquid

The most commonly used distributor in the laboratory is the porous-type distributor, which can be made of either a sintered ceramic or a metal. A distributor of this type initiates smooth fluidization at the expense of high pressure drop. Such a distributor is not normally used in an industrial-scale reactor because of its high pressure drop and the problems associated with its preparation for the desired pore size and free-flow area for fluid flow. Furthermore, such distributors, when used at high temperatures, can malfunction due to variation in pore size caused by expansion or sintering. Fabrication of a large-diameter porous plate by joining smaller pieces entails the occurrence of inactive pores in the welded joints. Despite these disadvantages, this type of distributor is preferred by many fluidization engineers. In studies involving laboratory-scale equipment, the use of distributors of this type is always advised because they can develop near ideal fluidization better than any other type of distributor plate. Some of the alternatives to porous distributors are wire mesh screens or filter cloth papers used in multilayers to a desired thickness to achieve smooth fluidization. Perforated to sieve plate distributors can easily be fabricated with controlled free open area. Figure 5.4 shows various types of distributors for gas or liquid.

A perforated-type distributor can be used as a single plate in small-diameter units as well as in shallow beds. In other cases, it has to be given bottom support or it can be used as a double perforated plate with one plate placed above the other. In order to take care of thermal expansion in high-temperature applications, the plate has to be concave or convex, but the holes should be distributed properly to ensure uniform fluid distribution. For example, a smaller number of holes at the periphery than at the center region is required in the case of a concave perforated plate and the converse holds for a convex distributor plate. Perforated plate distributors are

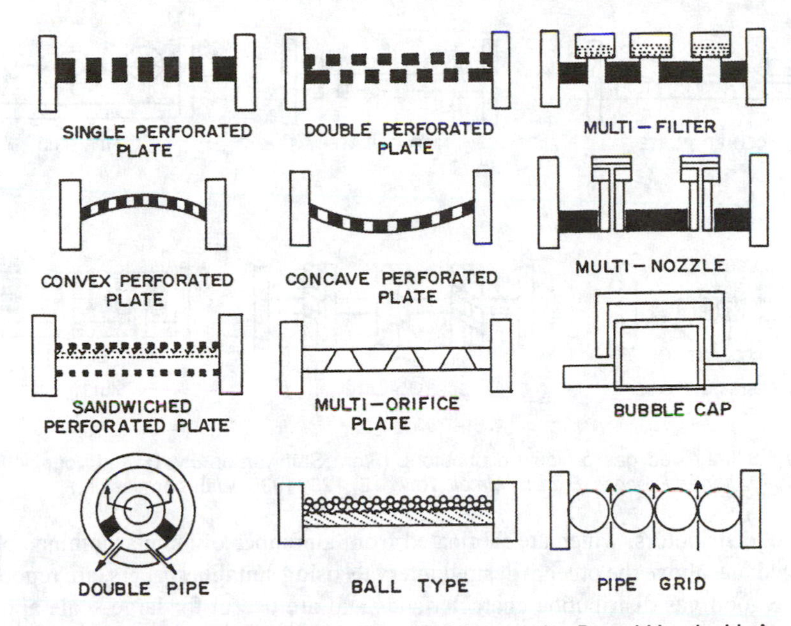

Figure 5.4 Distributors for gas or liquid. (From Sathiyamoorthy, D. and Vogelpohl, A., *Miner. Process. Extract. Metall. Rev.*, 12, 125, 1995. With permission.)

prone to solid leakage, and the fabrication cost to fix the size and number of holes is sometimes prohibitive. Although multihole distributors have such disadvantages, they are preferred because they can be fabricated and used to promote fluidization close to the ideal condition. Multifilter, multinozzle, and bubble cap distributors can avoid solid drains and can be used in large and deep beds. Where the incoming gaseous reactants are to be premixed and preheated, sandwiched-type distributors can be used. If the gases are dust laden, ball-type distributors can be employed. For larger units, pipe grid distributors may be the simplest to use. All the above types of distributors except the double pipe require a plenum chamber.

2. *Improved Gas–Liquid Distributors*

A simple perforated plate can be modified to an inclined plate, which can eliminate particle draining and also vertical gas jets. Figure 5.5 shows a distributor of this type along with other improved versions of gas or liquid distributors relative to those depicted in Figure 5.4. A nipple-type distributor plate can be an alternative choice to a multihole or orifice plate distributor because it can be fabricated economically and the pressure drop can be controlled by simply plugging the nipples at desired locations. Pipe grid distributors, through which the liquid is passed and distributed into a fluidized bed, do not required plenum chambers and have the added advantage of controlling the fluid supply through individual pipe grids. A heat-resistant grid allows the coolant fluid to pass through and protect the distributor plate from thermal shocks caused by explosive reactions. This kind of distributor can eliminate any thermal gradient in the distributor zone. Screw-type distributors are simple to fabricate and can be a good choice in place of bubble cap distributors.

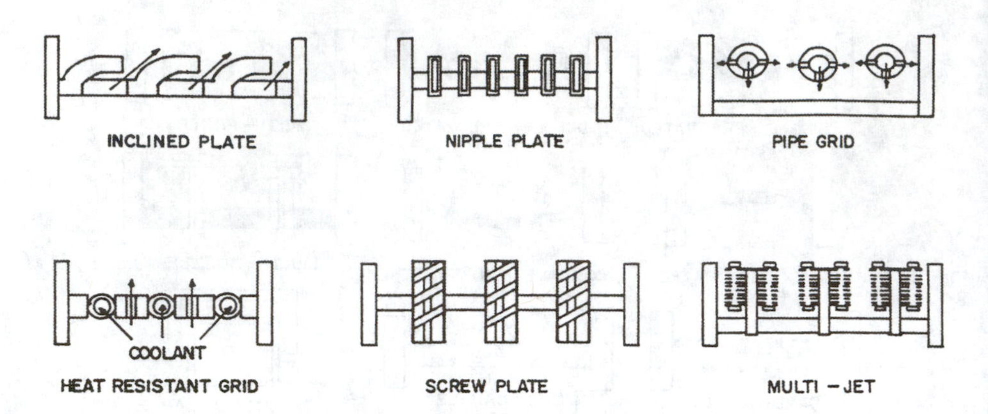

Figure 5.5 Improved gas or liquid distributors. (From Sathiyamoorthy, D. and Vogelpohl, A., *Miner. Process. Extract. Metall. Rev.*, 12, 125, 1995. With permission.)

Multijet distributors, which are fabricated from a number of annular laminae plates arranged one above the other at desired intervals using suitable spacers, are reported[55] to have good gas distribution characteristics and are useful for large-scale applications. However, this class of distributors may not be economical as both the material and laminae fabrication costs are high.

3. Common Distributors for Gas and Liquid

Reactors for gas–liquid and solid contact can be classified according to the flow direction and the continuous phase of the fluid. The classification of such reactor types can be found elsewhere in the literature.[56] There are several types of distributor plates for two-liquid distribution based on their location and fluid-directing principle. Schematic diagrams of a variety of simple laboratory-purpose distributors are given in Figure 5.6. A simple way to distribute the gas–liquid either in a fixed or fluidized bed is to saturate the liquid with the gas and then distribute the dissolved gas-rich liquid. This type of distribution procedure is more suitable for bioxidation processes such as nitrification of wastewater. In an alternative procedure, the gas can be predistributed below the liquid distributor, as shown in the gas–liquid distributor depicted in Figure 5.6A. In order to improve gas distribution inside the plenum for a distributor of this type, the plenum chamber can be filled with packing material. For a laboratory-type reactor, a concentric/annular distributor, as shown in Figure 5.6B, may be simple. However, such a distributor must be fitted with a conical bottom; for a large-scale reactor, such a distributor may have to be used in multiples, with several conical sections joined together to form a larger equivalent diameter for the reactor.

A venturi type[57] of distributor (Figure 5.6C) can be a simple and quick choice for a laboratory-scale reactor. A cocurrent distributor,[58] as depicted in Figure 5.6D, is a conventional choice but may not distribute the fluid as efficiently as the other classes of distributors described earlier. In general, it is a good practice to distribute the gas in the liquid before the liquid is distributed into the bed. Hence, gas predistributed between the sandwich grid[59] (Figure 5.6E) and gas distributed by a ring-type sparger

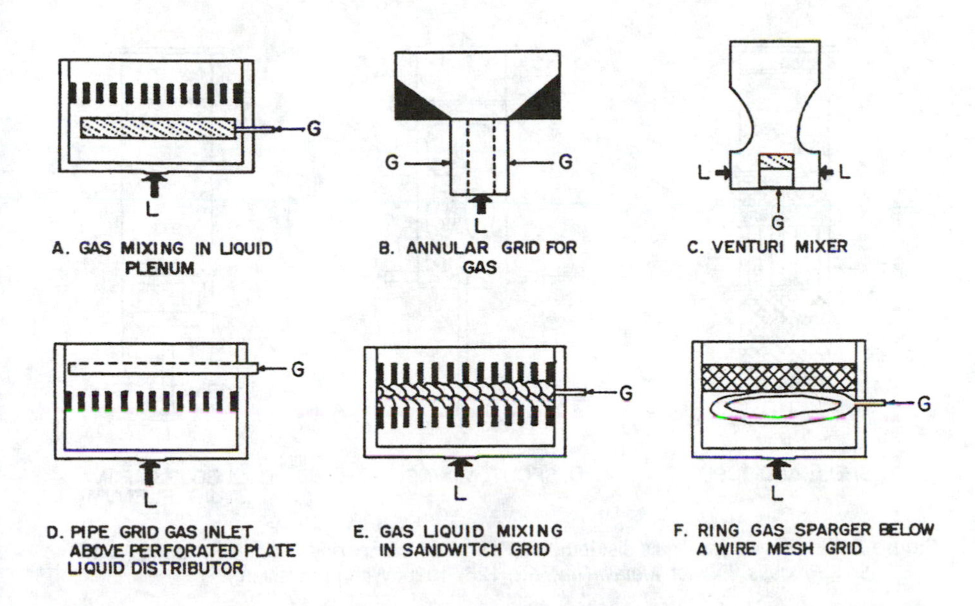

Figure 5.6 Gas–liquid distributors. (From Sathiyamoorthy, D. and Vogelpohl, A., *Miner. Process. Extract. Metall. Rev.*, 12, 125, 1995. With permission.)

below a wire mesh liquid distirbutor[60] (Figure 5.6F) are normally expected to give good results in respect to mass transfer rates. Kim and Kim[61] injected the gas into the orifice or perforation through which liquid also passed, and this caused turbulent mixing of the gas and the liquid at the hole. In an attempt to achieve better gas–liquid distribution, various research workers implemented several different ideas, and this gave rise to the various laboratory-scale distributors. However, it is difficult to assert that these distributors can create conditions compatible with the desired hydrodynamics on the reactor. Hence, the search for improved versions of the gas–liquid distributor has stimulated further research in this area.

4. Advanced Gas–Liquid Distributors

An advanced version of a gas–liquid distributor is a shell-and-tube type similar to the one shown in Figure 5.7A; a distributor of this type was used by Fan et al.[62] for three-phase fluidization. This type of distributor is essentially made up of three sections: (1) a plenum chamber for the liquid, (2) a gas–liquid distributor, and (3) a fixed bed. A more detailed description of the liquid and gas distributor is given in the literature.[63] Although this type of distributor is considered to be an improvement for the distribution of gas–liquid, its fabrication is similar to a mini shell-and-tube heat exchanger. Hence, it is not economical in terms of cost. Another class of an improved type of distributor that exhibits the advantages of both spouting and fluidization is shown in Figure 5.7B. In normal fluidization of coarse particles, only solid movement is achieved and solid circulation is largely absent. In spout fluidization, this disadvantage is overcome and better solid circulation with the advantage of a high flow rate is possible. Kono[64] proposed a spout mixing distributor for a three-phase fluidized bed.

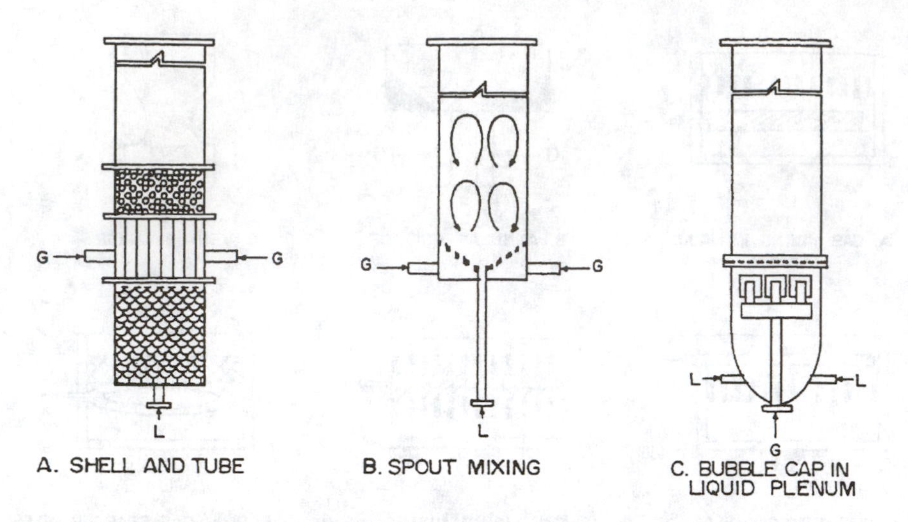

Figure 5.7 Improved gas–liquid distributors. (From Sathiyamoorthy, D. and Vogelpohl, A., *Miner. Process. Extract. Metall. Rev.*, 12, 125, 1995. With permission.)

Not much information is available regarding the design of the spout mixing distributor. A bubble-cap distributor[65] immersed in a liquid plenum, as depicted in Figure 5.7C, is a simple but improved type of gas–liquid distributor. The use of distributors of this type is not recommended for high fluid flow rates or for corrosive fluids because of erosion and corrosion problems.

5. *Pressure-Drop-Dependent Distributors*

a. *Classification Criteria*

Flow resistance of a fluidized bed does not vary and does not increase above the bed weight per unit area, whereas flow resistance of a distributor plate varies continuously with increasing flow rate. This implies that the fluidization characteristics can be manipulated by varying the distributor pressure drop, which is usually expressed in terms of its ratio to the bed pressure drop. In the literature, this pressure drop ratio is taken as a standard to classify distributors into two main categories: low pressure drop and high pressure drop. For a bed to operate in a stable condition, pressure drop ratios ranging from 0.015 to 0.4 were suggested by Hiby[66,67] and Gregory.[68] If the pressure drop ratio is less than 0.2, the distributor is designated as low pressure drop distributor, whereas it belongs to the high pressure drop category if the ratio is 0.4 and higher. Distributors with pressure drop ratios ranging from 0.2 to 0.4 have not been mentioned separately, presumably because they may be taken as moderate pressure drop ratio distributors. However, this classification criterion still is not well established.

b. *Low-Pressure-Drop Distributor*

Rowson[69] pointed out that a large pressure drop cannot be used in some practical operations. Hence, it is necessary to achieve stable fluidization with the minimum possible pressure drop. Hiby[66] and Gregory[68] recommended the low-pressure-drop distributor to achieve smooth fluidization. Hiby claimed that radial migration of gas due to the high fluidizing gas velocity is responsible for smooth fluidization. Deep fluidized beds, when operated with sieve plate or multihole distributors, also fall in the low-pressure-drop category or distributors. The distributor pressure drop in a multiorifice distributor is proportional to the square of the gas velocity through the orifice.[70] The low-pressure-drop ratio is, therefore, achieved with a plate that has a large free open area or a low operating velocity. Gvozdev et al.[71] cautioned against the use of a free open area in the range 2.5–4.0 as there is a significant reduction in pressure drop across the distributor. The gas, before entering the low-pressure-drop distributor, should not be maldistributed inside the plenum chamber. In several cases, the plenum chamber is filled with packing materials to avoid gas maldistribution prior to its entry through the distributor. Richardson[72] pointed out the need for gas predistribution in the case of a multiorifice distributor that has a moderate pressure drop. Zenz[73] showed that a high degree of imbalance in gas flow occurs in a pipe grid distributor which does not require a plenum chamber. The location of the gas entry point below the distributor also has considerable significance in terms of flow distribution below the distributor. This aspect will be considered as part of the discussion on the design aspects later in this section.

c. High-Pressure-Drop Distributor

Higher pressure-drop-ratio distributors are normally porous or sintered plate distributors and are not usually employed in industrial-scale units. Porous plate distributors offer high pressure drops compared to multihole/orifice, multijet, or bubble-cap distributors. Hence, they are not preferred for industrial-scale operation. In the porous plate distributor, the distributor pressure drop is proportional to the gas velocity through the pore. The reason for a high pressure drop is the low free area of the distributor, which causes the gas velocity to increase tremendously. Avery and Tracey[74] suggested that is necessary to operate shallow fluidized beds under a high distributor-to-bed pressure drop ratio. Gregory[68] pointed out that many catcrackers operate successfully at a distributor-to-bed pressure drop ratio as high as 0.4. In order to avoid solid flow through a simple perforated-type distributor, a high gas velocity is required. In such cases, the distributor-to-bed pressure drop ratio is large.

E. Pressure Drop

1. Selection Criteria

a. General Rule

It was mentioned in the preceding section that the distributor pressure drop is usually expressed as the ratio of the bed pressure drop, which remains constant for a well-fluidized bed. As a rule of thumb, this pressure drop ratio was suggested to be 0.1 for a deep bed or 10 to 12 in. of water column for the distributor pressure drop in the case of a shallow fluidized[34] bed. The pressure drop across the distributor in a fluidized bed can also be taken as 100 times its free expansion value[72] to arrive at smooth fluidization. On the other hand, the distributor-to-bed pressure drop ratio is reported[66,68] to range from 0.015 to 0.4 to achieve smooth fluidized bed operation. In general, it is complicated task for the fluidization engineer to choose an appropriate distributor pressure drop as there is a wide range of values. It is also not appropriate to fix an arbitrary pressure drop for any design.

b. *Kinetic Energy*

Kunii and Levenspiel[26] proposed a criterion for the kinetic energy of the gas jet issuing from the orifice of the distributor for stable operation. According to this criterion, if the kinetic energy of the gas jet is greater than the bed weight per unit area, the gas jet can penetrate the bed, thereby causing channels inside the bed. If, on the other hand, the kinetic energy of the jet is smaller, solid particles from the bed can drain through the distributor hole. In order to use a conservative value, the recommended range for the kinetic energy of the gas jet is 50–75% of the fluidized bed pressure drop, that is,

$$\rho_f U_o^2/2g = (0.5 \text{ to } 0.75)\left(W_{\text{bed}}/A\right) \tag{5.27}$$

c. *Stable State*

When the distributor-to-bed pressure drop ratio is chosen, the underlying conditions for which it is deemed to be valid should be stated explicitly. Unfortunately, the conditions vary widely, as there is no unique way to explain the stability of fluidization. In fact, distributors play a definite role in creating a stable atmosphere in a fluidized bed; this was brought out for first time by Hiby[66] in a convincing way. According to Hiby, a fluidized bed fitted with a constant-opening distributor can remain in a stable state if a local decrease in the bed pressure drop is counteracted by a change in the distributor pressure drop. Medlin et al.[75] predicted that the convective instability caused by fluctuating solid circulation can be suppressed by either increasing the pressure drop across the distributor or decreasing the bed diameter. They recommended a bed pressure drop of 10% across the distributor for stability of operation. Zuiderweg[76] and Lago et el.[77] supported a magic figure of 10% of the bed pressure drop for the distributor drop for good gas–solid contact and conservative design. Hiby's, approach[66] to describe the stable operation of a fluidized bed based on variation of the distributor pressure drop is rather qualitative.

d. *Operation of Gas-Issuing Ports*

A more quantitative treatment of the stability of fluidized bed operation, based on the operating characteristics of tuyere-type distributors in large fluidized beds, was developed and described by Whitehead and Dent.[78] According to their theory, which was supported by experimental findings, the stability of a fluidized bed under a constant flow rate is achieved when a decrease in the bed pressure drop caused by inactive gas-issuing orifices is equalized by a corresponding increase in the distributor pressure drop due to excess gas flow through the active holes in the same distributor. Their findings were based on the complete operation of all gas-issuing sites, and the distributor pressure drop ratio was found to range from 0.015 to 0.014, corresponding to an operating velocity ratio ranging from 2 to $8U_{mf}$. The findings of Whitehead and Dent signaled a new way of viewing and analyzing a stable fluidized bed operation. In his analysis to explain channel-free stable fluidization, Siegel[1] concluded that the distributor-to-bed pressure drop ratio would lie in the range 0.12–0.24 for a fluid–solid system where Ar ranges from 1 to 10^4. Even though Siegel's analysis[1] was based on theory without the support of his own experimental evidence, his findings may explain the wide range in distributor-to-bed pressure drop ratio for stable operation reported in the literature.

2. *Significance of Pressure Drop*

As previously discussed, the distributor-to-bed pressure drop ratio plays an important role in fixing velocity commensurate with uniform fluidization. In the context of fixing a constant pressure drop ratio, the operating velocity cannot be changed at will unless proper adjustments are made in the bed height or the distributor design parameters. In other words, once the pressure drop ratio ($\Delta P_d/\Delta P_b$) is fixed from the standpoint of achieving smooth fluidization, an alteration in bed height (e.g., by increasing the quantity of the bed material/inventory) warrants an increase in ΔP_d so as to keep $\Delta P_d/\Delta P_b$ constant. This necessitates an alteration of the distributor design by changing the free-flow area or a change in the flow rate. The latter parameter cannot be easily altered, as stoichiometry and hydrodynamics impose restrictions on its choice. Hence, it is the distributor design that must be manipulated to maintain bed stability. Selection of an appropriate distributor-to-bed pressure drop ratio is vital for stable fluidization. Prediction of this important ratio and the optimal approach for distributor design are discussed in the next section.

3. *Prediction of Pressure Drop Ratio*

Prediction of the distributor-to-bed pressure drop ratio that corresponds to stable or smooth fluidization is not possible by any straightforward method and has been a subject of controversy. This is mainly due to the varied and often disputed conditions for which a correlation is proposed. Furthermore, additional problems are encountered in fixing a criterion for the development of a correlation for a two- or three-phase fluidized bed when the direction of fluid flow also has an influencing role. In view of this complex situation, most of the correlations available in the literature[1,6,7,9,13,14,26,31,76,78-86] for normal gas fluidization are presented in Table 5.1. Each correlation was developed or proposed for that condition under which the respective researcher was satisfied with the quality of fluidization achieved. In other words, the criteria for all the correlations presented in Table 5.1 are not the same, and they vary widely in terms of the underlying principle that governs smooth/uniform fluidization.

Table 5.1 Ratio of Distributor-to-Bed Pressure Drop for Smooth Fluidization

$\Delta P_d/\Delta P_b$ Ratio	Remarks	Reference
$\{\Gamma^2[U_{mf}/U)(C_{d1}/C_{d2})(\Omega_2/\Omega_1]\}^{-1}$, where C_d is the drag coefficient and Ω is the resistance coefficient	Γ = velocity ratio of fluid flow between active and nonactive orifice; suffix 1 for active and 2 for inactive orifices	Zabrodsky[79]
$(H_{min}/H_{mf})/\{1-(U_{mf}/U_m)^2\}$	H_{min} is the height of solid above a hole to avoid channeling and ΔP_b is at U_{mf}	Zabrodsky[80]
$0.012(1-1.4U_{mf}/U)$	Distributor type not mentioned	Zuiderweg[76]
0.3	Perforated plate	Zenz[81]
$0.046(D/H)$	Perforated and porus plates	Kelsey[82]
0.1	Distributor not mentioned	Kunii and Levenspiel[26]
$0.0062\, N^{0.22}[1-\epsilon_{mf})(1-1.4U_{mf}/U)^{-1}]$	N = number of tuyeres, ϵ_{mf} = bed voidage	Modified form of Whitehead and Dent[7] by Baskakov et al.[83]
$\{\epsilon_{mf}\Psi(U/U_{mf})\Psi K\}/\{\alpha(1-\epsilon_{mf})(U/U_{mf})\Psi\}$	$\epsilon = \epsilon_{mf}(U/U_{mf})\Psi$ $\Delta_d^P \sim (U/U_{mf})^\alpha$	Modified form of Hiby[67] by Baskakov et al.[83]
$b(D/H)^{1/2}$	$b = 0.03$ for $U \le U_{mf}$ and 0.016 for $U \ge U_{mf}$ for nozzle/tuyeres	Whitehead[84]
$0.08(D/H)$	Perforated plate	Kassim[14]
$(U_{mf}/U_t)_{\frac{1}{n}}/[n(1-\epsilon_{mf})]$ where $n = f(\text{Ar})$	Porous plate; for perforated plate 50% of the predicted value	Siegel[1]
$c\epsilon_b((1-\epsilon_b)/[1-(U_{mf}/U)^2]$ where c is constant	Perforated and porous plate	Mori and Moriyama[9]
$2[U_{mf}/(U_m-U_{mf})]^2$	Multiorifice plate	Sathiyamoorthy and Rao[6]
$0.01+0.2[1-\exp(-0.5D/H_{mf})]$	All types, U/U_{mf} has no effect	Qureshi and Creasy[31]
$0.185(P/H_{mf})(P/d_p)^{0.33}\sin^{-1}(U_{mf}/U_m)/[1-(U_{mf}/U_m)^2]$	Perforated plate	Fakhimi et al.[85]
$(\Delta P_d/\Delta P_b)_{min}+(\epsilon_b+0.365H_s/H_{mf})$	ϵ_b = bubble voidage, H_s = spout height	Yue and Kolaczkowski[86]
$[\text{Re}_{mf}/(1-\epsilon_{mf})]/[(\delta\epsilon_{mf}/\delta\text{Ar})/\delta\text{Re}_{mf}/\delta\text{Ar})]$	ϵ_{mf} and Re_{mf}, voidage and Reynolds number at U_{mf}	Sathiyamoorthy et al.[13]

From Sathiyamoorthy, D. and Vogelpohl, A., *Miner. Process. Extract. Metall. Rev.*, 12, 125, 1995. With permission.

4. *Minimum Operating Velocity Criteria*

a. *Tuyeres Operation*

Whitehead[84] pointed out long ago that the practice of fixing a constant value for the pressure drop ratio ($\Delta P_d/\Delta P_b$) should be viewed cautiously, as this ratio varies from 0.02 to 0.5 for industrial units. The pressure drop ratio was derived by assuming complete operation of the gas-issuing sites and the occurrence of smooth fluidization. Hence, it is quite reasonable to express the distributor-to-bed pressure drop ratio in terms of the operating velocity ratio that can establish smooth fluidization. There are a few correlations available in the literature, but their applicability for practical design purposes has certain limitations. For example, Whitehead and Dent[7] developed a correlation to predict the velocity (U_m) that would sustain the operation of all tuyeres; this correlation can be expressed as:

$$\frac{U_m}{U_{mf}} = 0.7 + \left[0.49 + 3.23\times 10^{-3} N^{0.22} D_s \rho_s \left(K_D/U_{mf}\right)^2\right]^{1/2} \tag{5.28}$$

For velocities less than U_m, the flow discharge is not equal through all the tuyeres. U_m is determined by decreasing the flow rate from a well-fluidized state. Another velocity (U_M) at which all tuyeres just become fully operational can be determined by increasing the flow rate gradually until the static bed is transformed to a fluidized state. The U_M thus obtained is greater then U_m. Equation 5.28 can be used if N and the distributor flow factor (K_D) are known. In designing a distributor, one of the objectives is to predict the number of tuyeres or orifices (N), but the value of N must be assumed if Equation 5.28 is to be used for predicting the operating velocity. Thus, use of this correlation has limitations.

b. Multiorifice Plate

The operating characteristics of a multiorifice type of distributor were studied by Fakhimi and Harrison,[4] and they determined the velocity (U_M) required to bring all the orifices of the distributor to a fully operational state during increasing flow. A theoretical model was proposed to predict the fraction of functioning orifices, and the velocity (U_M) corresponded to a situation where the value of the fraction reached unity. The correlation proposed was

$$U_M \Big/ U_{mf} = \left[1 + \left(2\phi^2 / U_{mf}^2 \rho_g\right)\left(H_s \rho_s\right)\left(1 - \epsilon_{mf}\right)\left(1 - \frac{2}{\pi}\right)\right]^{1/2} \tag{5.29}$$

In Equation 5.29, the spout height (H_s) above the operating orifice is an unknown parameter; it was later evaluated by Fakhimi et al.[85] as:

$$H_s = 0.51\ (P/d_p)^{0.33}\ \sin^{-1}\ (U_{mf}/U) \tag{5.30}$$

Prediction of the operating velocity ratio (U_M/U_{mf}) by using Equation 5.29 in conjunction with Equation 5.30 still requires knowledge of an important parameter: the fraction of free open area of the distributor (ϕ). In fact, this parameter must be predicted on the basis of the information available with respect to the ratio U_M/U_{mf}. On the other hand, prior information regarding either U_M/U_{mf} or ϕ is required for the use of Equation 5.29 to approach a design solution. The model of Fakhimi and Harrison[4] was supported well by their experiments when the orifices in the distributor plate are operational up to 75%. Kassim[14] later reported some discrepancies when he tested this model with particles that have a low minimum fluidization velocity.

Sathiyamoorthy and Rao[6] carried out a detailed study on the basics and operating principles of multiorifice-type distributors and found that the operating velocity ratio was a function of the distributor-to-bed pressure drop ratio. Certain aspects of this model were mentioned as part of the discussion on operating velocity. An important point here is that the fluidized bed, at an operating velocity U_m, need not necessarily initiate smooth fluidization. This is due to the fact that a distributor plate provided with only a few orifices can be operated at a velocity U_m, satisfying the condition of complete operation of all the orifices in the plate. However, the bed near the distributor can have dead zones and the grid zones will have high-intensity jets. In order to satisfy the condition of complete operation of orifices and channel-free fluidization, to complying with Siegel's criterion, Sathiyamoorthy and Rao[8] proposed the correlation contained in Equation 5.5 to predict the operating velocity. It is interesting to note that the ratio U_M/U_{mf} is a function of the ratio U_t/U_{mf}, which is a function of only the physical properties of the gas–solid mixture or the Archimedes number. As mentioned earlier, the distributor-to-bed pressure drop ratio can be evaluated if the ratio U_M/U_{mf} is known, and this pressure drop ratio has a single or unique value for a given gas–solid system.

Based on the performance of a multiorifice distributor in a two-dimensional fluidized bed for Group B Geldart[87] powders Yue and Kolaczkowski[86] proposed a correlation for predicting the ratio U_M/U_{mf} as:

$$\frac{U_m}{U_{mf}} = \left[1 + \Delta P_{b,mf}\left(\epsilon_B + 0.363\ H_g/H_{mf}\right)\right]\left[2g\ C_d\ \phi^2/\left(\rho_s U_{mf}^2\right)\right] \qquad (5.31)$$

Equation 5.31 requires an iterative procedure to solve for U_m because the bubble voidage (ϵ_b) and the spout height (H_s) are functions of the velocity (U_m). Also, Correlation 5.31 is applicable exclusively to Group B powders. Baskakov et al.[83] examined the theory of orifice operation and postulated that a nonoperating orifice requires an excess pressure drop ($\Delta P_{d,z}$) to overcome the blockage due to solid particles and that this excess pressure drop is zero when the orifices are brought into operation. Their correlation is

$$\frac{U_M}{U_{mf}} = \left[1 + \Delta P_{d,z}/\Delta P_d\right]^{1/2} \qquad (5.32)$$

The excess pressure drop ($\Delta P_{d,z}$) attains its maximum value at a stagnant zone during its destruction and depends on the previous history of the bed, gas–solid properties, and distributor parameters. Because $\Delta P_{d,z}$ is not be reproducible, the ratio U_M/U_{mf} also cannot be reproduced. When all the stagnant zones are destroyed, $\Delta P_{d,z}$ vanishes. Therefore, Equation 5.32 is valid only up to $U = U_m$. In order to use Equation 5.32, the value of $\Delta P_{d,z}$ must be known. Baskakov et al.[83] developed the following expression for this quantity:

$$\Delta P_{d,z} = \rho_s(1 - \epsilon)\ g\ (d_o/4\phi)\ [1 - (7/3)\ \phi^{1/2} + 4z/d_o] \qquad (5.33)$$

The term z is the depth up to which solid particles penetrate an orifice of diameter d_o, and it cannot be easily predicted. Thus, Equation 5.32 can be used in conjunction with Equation 5.33 when all the parameters are known exactly.

c. Recommendations

Prediction of a minimum operating velocity such as U_M or U_m in the case of a multiorifice distributor in a gas–fluidized bed appears to have been analyzed fairly well in the fluidization literature. However, no such velocity has been evaluated for a three-phase system where the fluid flow direction (i.e., for gas and/or liquid) can be either cocurrent or countercurrent. In the absence of such information for such a system, the results pertaining to a two-phase system can be extrapolated with some caution, at least for the cocurrent upflow three-phase fluidized bed when the liquid constitutes a continuous phase. This aspect will be considered when we deal with specific design aspects of distributors.

IV. OPTIMAL DESIGN APPROACH

A. Reaction Kinetics with Hydrodynamics-Satisfied Design

1. *Kinetic Approach*

The design of a fluidized bed reactor is mainly based on its hydrodynamics and the chemical kinetics of the reacting species. These two factors should be mutually adjusted in such a way that a fluidized bed reactor is tuned to give its best or optimum performance. It may be appreciated from the discussions in the earlier sections that a distributor plate plays a key role with regard to the hydrodynamics of a fluidized bed reactor. The very fact that good fluid–solid contact has to be achieved implies the need to bring about an optimum level of bubble-induced mixing in a gas fluidized bed reactor. As the bubble dynamics is largely dependent on the operating velocity, gas–solid properties, bed geometry, distributor, etc., it is necessary to choose these parameters judiciously to achieve the desired performance. For a given gas–solid system and a given reactor geometry, the operating velocity and the distributor are largely responsible for establishing the appropriate hydrodynamic condition that is conducive for the chemical reaction to occur. As pointed out in the proceeding sections, the operating velocity cannot be fixed arbitrarily and is closely related to the distributor type and its design parameter. Let us now consider an example to illustrate the optimal design of a fluid bed reactor. More specifically, the gas fluidized bed reactor has direct relevance to the distributor design factors. In the case of a simple first-order irreversible reaction, the residence time (τ) required to achieve a desired conversion (x) is given by:

$$\tau = -(1/k \cdot e) \ln (1 - x) \tag{5.34}$$

where the efficiency factor (e) for an ideal plug-flow-type reactor is unity. Because the fluidized bed reactor is a nonideal type of reactor, the efficiency factor tends to be less than unity. It is largely influenced by the hydrodynamics and hence by the distributor. The residence time (τ) is also expressed as:

$$\tau = \frac{V}{\nu} = \frac{AH}{AU} = \frac{H}{U} \tag{5.35}$$

If the velocity U corresponds to an optimum operation, τ can be an optimum value for a given H, which has to be fixed on the basis of the quantity of the material to be processed. It is often necessary to process a given quantity for the best possible conversion. In such cases, the bed height for a given diameter of the reactor and the conversion level are fixed beforehand.

2. Hydrodynamics Approach

The ultimate option to manipulate reactor performance is the operating velocity (U), which should enhance the exchange process and bring the efficiency factor (e) close to unity. It amounts to looking for an optimum velocity (U_{opt}) and its relation to a gas distribution system. The next section focuses on distributor design. The importance of the parameter U_{opt}, in relation to distributor design is highlighted. For the sake of simplicity, the discussion is confined to a gas fluidized bed with a multihole/orifice type of distributor, which is the most widely accepted in industrial practice.

3. Distributor Design Model

a. Ratio of Pitch to Orifice Diameter

Distributor design can be established reasonably well if four basic parameters are predicted on the basis of the functional or operating characteristics of the distributor plate in conjunction with its promotional effect in maintaining stable fluidization. These four parameters are the free area fraction (ϕ), the number of fluid inlet points or orifices (N), the orifice size (d_o), and the orifice spacing (P). The free area fraction (ϕ) is expressed as:

$$\phi = N(a/A) = N(d_o/D)^2 \tag{5.36}$$

where the size of the orifice (d_o) has to be fixed based on the economics of fabrication as well as the flowback of solids through the orifices down to the plenum chamber. The diameter of the column depends on the bed inventory and is related to the bed height, which is expressed as the aspect ratio. Selection of the aspect ratio was highlighted earlier in this chapter. In order to estimate the open area fraction, it is necessary to know the optimum or minimum operating velocity (U_m) at which complete fluidization is achieved and also the corresponding distributor pressure drop (ΔP_d). The governing equation for the velocity ($U_{o,M}$) through an orifice in terms of the distributor pressure drop when the corresponding superficial velocity of fluidization is U_m is given by:

$$U_{o,M} = C_d(2\Delta P_d/\rho_f)^{1/2} \tag{5.37}$$

where $U_{o,M}$ is related to U_M by:

$$U_{o,M} = \frac{q_m}{a} = Q_M \Big/ (Na) = \frac{AU_M}{Na} = \frac{U_M}{Na/A} = \frac{U_M}{\phi} \tag{5.38}$$

Equation 5.38 should be corrected if the variation in density of the fluid at the entry and exist of the orifice is appreciable. For simplicity, we will proceed with Equation 5.38 as such for substitution in Equation 5.37. On doing so, the following expression for U_M is obtained:

$$U_M = \phi C_d (2\Delta P_d/\rho_f)^{1/2} = (N/A)a\ C_d (2\Delta P_d/\rho_f)^{1/2} \quad (5.39)$$

where (N/A) is the number of orifices per unit area and is related to the orifice spacing (P) as:

$$\frac{N}{A} = C_p / P^2 \quad (5.40)$$

The factor C_p is dependent on the orifice arrangement. It is unity for a square array and equals $2/\sqrt{3}$ for a triangular array of orifices. The ratio of orifice spacing (P) to orifice size (d_o) can be obtained by substituting Equation 5.40 in Equation 5.39 and rearranging the terms:

$$\left(\frac{P}{d_o}\right)^2 = \frac{\pi}{4} \cdot C_p \cdot C_d \left[\left(2 \cdot \Delta P_d / \rho_f\right)\right] / U_M \quad (5.41)$$

The distributor pressure drop (ΔP_d) is expressed as its ratio to the bed pressure drop (ΔP_b); it can also be expressed in terms of the velocities U_M and U_{mf} for the condition of stable fluidization.[69] The pressure drop ratio is

$$(\Delta P_d/\Delta P_b) = 2[U_{mf}/U_M - U_{mf})]^2 \quad (5.42)$$

where the bed pressure drop for a fluidized bed is

$$\Delta P_b = H(\rho_s - \rho_f)(1 - \epsilon_{mf})g \quad (5.43)$$

Substituting Equations 5.42 and 5.43 in Equation 5.41 and eliminating ΔP_d, we get:

$$(P/d_o)^2 = \beta_d (H/D)^{1/2} \quad (5.44)$$

where

$$\beta_d = K_d\ \{[(\rho_s/\rho_f) - 1](1 - \epsilon_M)]\}^{1/2} / \{(\mathrm{Fr}_D)^{1/2} (U_M/U_{mf})[(U_M/U_{mf}) - 1]\} \quad (5.45)$$

The distributor constant (K_d) in Equation 5.45 is given by:

$$K_d = (\pi\ C_p \cdot C_{d_o}/2) \quad (5.46)$$

The parameter β_d is a function of the fluid–solid properties and the ratio U_M/U_{mf} is also a function of the gas–solid properties. Thus, for a specified fluid–solid system, the parameter β_d is a constant. If the orifice (d_o) is fixed based on the economics of fabrication and the column diameter (D) is also fixed, then according to Equation 5.44 the orifice spacing (P) is proportional to $H^{1/4}$ for a given fluid–solid system. This implies that for deep or tall beds, the spacing between the orifices should be large. Because the number or orifices per unit area or the orifice number density is inversely proportional to the orifice spacing, the distributor plate for deep beds should be provided with a relatively smaller number of orifices. This implies that the distributor plate in a deep fluidized bed plays only a limited role in establishing stable fluidization. The converse is applicable for a shallow bed; in other words, the distributor in a shallow fluidized bed has a strong role in establishing stable fluidization. The constant K_d is a function of the orifice discharge coefficient (C_{d_o}), which is determined by the distributor plate thickness (t) and the orifice diameter (d_o) according to Qureshi and Creasy.[31] The discharge coefficient (C_{d_o}) is expressed as:

$$C_{d_o} = 0.82\,(t/d_o)^{0.13} \tag{5.47}$$

b. Guidelines for Fixing Ratio of Pitch to Orifice Diameter

The ratio of orifice spacing to orifice size defined by Equation 5.44 varies with gas–solid properties which can be conveniently expressed in terms of the Archimedes number. In other words, P/d_o can vary with the aspect ratio for a given Archimedes number. A typical plot depicting the trend in variation of P/d_o with H/D for the Ar range 10^2–10^5 (e.g., water–balltoni system, with ρ_s = 2623 kg/m^3 and d_p in the range 186–1860 μm) is shown in Figure 5.8. It is important to select the appropriate orifice size (d_o), which as a rule of thumb is usually taken as three to four times the particle size. In consideration of the economics of fabrication, the lower size limit was also suggested[39] as 1500 μm. From the plot of P/d_o versus H/D (Figure 5.8), it is clear that P/d_o decreases for higher Ar for a given value of the aspect ratio. This result indicates that a distributor with a high orifice number density is required for coarse or heavy particles. In other words, coarse particles should be fluidized with a distributor of a large free area which offers low resistance to flow. Conversely, fine or light particles which correspond to low Ar values should be fluidized using a high resistance distributor plate. These points were covered earlier when discussing the significance of the ratio $\Delta P_d/\Delta P_b$ but in an implicit manner. So far, the distributor design aspects have been confined to particulate beds of solids that are well fluidizable. In other words, the design concepts are pertinent to Group B particles of the Geldart classification. For particles of Group C or for Ar < 100, the cohesive nature of the fine particles prevents channel-free fluidization. Hence, proper care should be exercised when a distributor design is proposed for this class of particles.

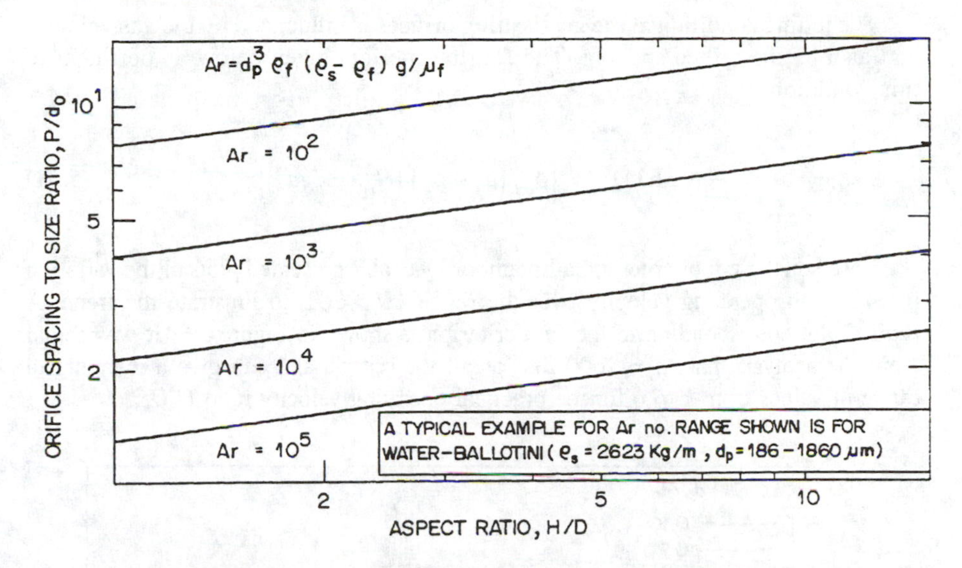

Figure 5.8 Orifice spacing to size ratio (P/d_o) at the minimum required fluid flow rate for varying aspect ratio and Archimedes number (Ar). (From Sathiyamoorthy, D. and Vogelpohl, A., *Miner. Process. Extract. Metall. Rev.*, 12, 125, 1995. With permission.)

c. *Multiple Choices for Operation and Selection of Ratio of Pitch to Orifice Diameter*

An important consideration in the distributor design approach concerns the condition of smooth fluidization achieved at a specific velocity, which for a perforated or multiorifice distributor is the minimum required to sustain all fluid inlet points in operative mode. What would happen if the operating velocity is increased beyond the minimum value? How would the increase in flow affect fluid bed performance? What new sets of conditions would influence the distributor design procedure? The answers to these questions are complicated. These aspects were considered and analyzed by Sathiyamoorthy et al.[88] in the context of four major criteria. The first criterion considers the elimination of bubble coalescence at the distributor site, and for this condition:

$$P \geq 0.0061\ U_{mf}(\alpha - 1)^2 \tag{5.48}$$

where $\alpha = U/U_{mf}$

The second criterion was formulated to eliminate the dead zone of solid particles on the distributor. According to this criterion:

$$P \leq 0.0061\ U_{mf}(\alpha - 1)^2 \tag{5.49}$$

The third criterion pertains to attainment of a specified bubble diameter:

$$P = 0.033391\ U_{mf}(\alpha - 1)^{4/3}\ 2^{5/S-5} \tag{5.50}$$

The jetting condition at the gas-issuing orifices is influenced by the gas velocity and also by the orifice spacing. The fourth criterion, given below, is pertinent to this condition:

$$\frac{P}{d_o} \geq 6341\, d_o^{1.48}\left[\rho_g/(\rho_s-\rho_g)\left(U_{mf}^2\alpha^2/gd_o\right)\right]^{0.74} \tag{5.51}$$

Based on these four criteria, Sathiyamoorthy et al.[88] presented plots of the variation of P/d_o with operating velocity over the range $1U_{mf}$–$6U_{mf}$ to illustrate the trend. A typical plot corresponding to the first criterion is shown in Figure 5.9. It was found from the analysis that Ar = 1000 fits for all the criteria and can give a meaningful range of values from 1 to 6 for the practical operating velocity ratio U/U_{mf}.

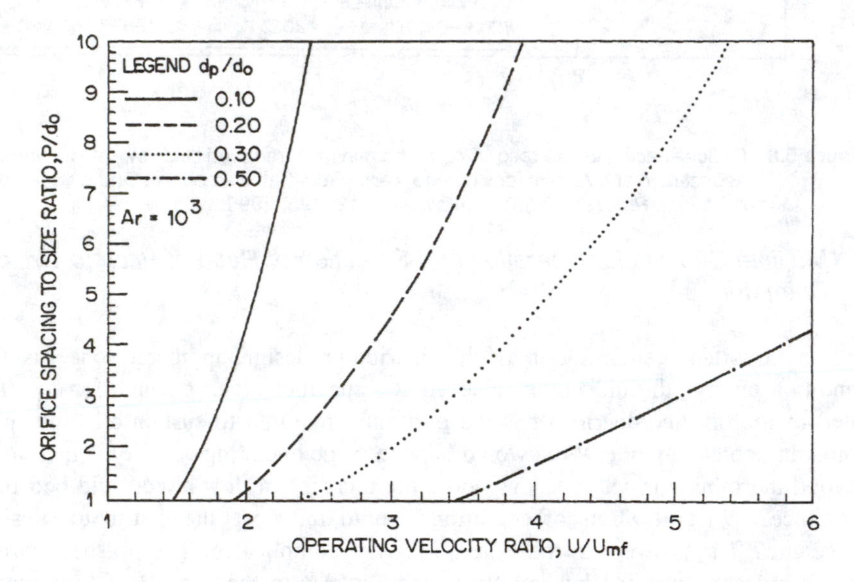

Figure 5.9 Criteria for selecting operating velocity and orifice spacing to avoid bubble coalescence near distributor in gas fluidized bed. (From Sathiyamoorthy, D., Geissen, S., and Vogelpohl, A., *Miner.Process. Extract. Metall. Rev.*, 9(1–4), 233, 1992. With permission.)

4. Distributor for Three-Phase Fluidization

The distributor design for a three-phase fluidized bed has not yet been given adequate consideration even though the proper distribution of both liquid and gas is more important and complex here than in the two-phase fluidized bed system. The literature on gas–liquid distributor design is scant. No standard procedure seems to have evolved for distributing the two fluids. Obviously, this kind of practical problem makes many research workers report their basic hydrodynamic data, which are often an anomaly compared with those reported by others. Among three-phase fluidized bed systems, the most commonly used or investigated system is that involving the cocurrent upflow of gas–liquid. In such a system, the liquid mainly serves to fluidize the solid, while the gas is just sparged to generate a well-dispersed system with three phases. In the absence of design information pertaining to the distributor in a three-phase fluidized bed reactor, the design criteria for a fluid distributor for a two-phase system can be extrapolated[88] with caution to a three-phase fluidized bed with a cocurrent upflow of gas and liquid. The basic parameters that must be evaluated in a three-phase fluidized bed reactor are U_{mf}, U_t, ϵ, and ΔP_b. Most of these parameters were described in the section on three-phase fluidization in Chapter 1. However, certain parameters not covered in Chapter 1 will be discussed here. The minimum fluidization velocity for a liquid–solid system is usually lowered when a third phase, namely, a gas phase, is introduced. The minimum fluidization velocity (U_{3mf}) for a three-phase system is usually expressed[89] as its ratio to the two-phase minimum fluidization velocity (U_{2mf}):

$$\frac{U_{3mf}}{U_{2mf}} = 1 - 376\ U_g^{0.327} \mu_l^{0.227} d_p^{0.213} (\rho_s - \rho_l)^{-0.423} \text{ (SI units)} \tag{5.52}$$

The correlation of Thonglimp[90] for Re_{mf} may be used to evaluate U_{2mf}.

$$\mathrm{Re}_{mf} = \frac{U_{2mf} \rho_l d_p}{\mu_l} = \left(31.6^2 + 0.0425\ \mathrm{Ar}\right)^2 - 31.6 \tag{5.53}$$

The bed pressure drop (ΔP_{b3}) for a three-phase fluidized bed is a function of the bed height and the individual phase holdup. The sum of the three holdup fractions — ϵ_s, ϵ_l, ϵ_g, and respectively, for solid, liquid, and gas, is equal to unity:

$$\epsilon_s + \epsilon_l + \epsilon_g = 1 \tag{5.54}$$

The voidage (ϵ_3) in a three-phase system is the sum of ϵ_l and ϵ_g. Although there are several correlations proposed in the literature for predicting ϵ_3, the one proposed by Begovich[89] is widely used. More information on this subject was presented in Chapter 1. Sathiyamoorthy and Vogelpohl[37] proposed the following method for calculating a parameter β_3 for a three-phase system:

$$\beta_3^1 = \left[\rho_s/\rho_l - \epsilon_l(\rho_s/\rho_l - 1) - \epsilon_g(\rho_s/\rho_l - \rho_g/\rho_l)\right]^{1/2} \Big/ \left\{(\mathrm{Fr}_D)^{1/2}(U_{3M}/U_{3mf})\left[(U_{3M}/U_{3mf}) - 1\right]\right\} \tag{5.55}$$

The correlation represented by Equation 5.55, when multiplied by the distributor constant (K_d), will give the parameter β_3 for a three-phase system, and it is similar to β_d for a two-phase system as given in Equation 5.45. Using β_3, the distributor plate design may proceed as illustrated for a two-phase system, but in this case β_3 does not depend on the gas–solid system alone and hence is not a constant as in the case of a two-phase system. Therefore, β_3 should be evaluated for various gas flow rates. As a consequence, the distributor design in terms of P/d_o and the operating velocity (depicted in Figures 5.8 and 5.9 for a two-phase system) will not result in a single plot but will give rise to a family of plots. All these parameters relating to distributor design for three-phase fluidization are presently only hypothetical and much remains to be done in this field. Nevertheless, the information provided here can be used as a rough guideline[37] by a designer in view of the paucity of literature on this topic. The three-phase fluidized bed reactor has recently emerged as a potential bioreactor. The solid particles in a biofluidized bed reactor are usually attached film bioparticles. More information on biofluidization can be found in a review paper by Schürgel.[91] For application in anaerobic waste treatment, Kim et al.[92] proposed a new type of distributor that has two plates, one above the other: the bottom one for liquid inlet and the top one for liquid redistribution. Sathiyamoorthy et al.[93] proposed a method for designing a distributor for a fluidized bed bioreactor, taking into account the properties of bioparticles.

B. Plenum Chamber

The section of a fluidized bed reactor below the distributor plate is called the plenum chamber or the wind box or calming section. This section plays a vital role in predistributing the fluid without maldistribution before it enters the distributor holes/nozzles. The need for gas predistribution was stressed long ago by Richardson.[72] It is usual practice to fill the plenum with packing material to have uniform predistribution of the fluid. Some distributor types such as pipe grid or double pipe do not require a plenum chamber. In such a distributor, it is essential that the fluid is discharged uniformly without any imbalance at the various fluid-distributing points, as demonstrated by Zenz[73] for a three-hole manifold pipe grid. The plenum chamber volume is usually chosen to be large enough to reduce the acceleration effects. Larger nozzles minimize entry–expansion losses. Litz[94] proposed criteria to locate gas entry points. He proposed the following empirical relations.

1. For horizontal stream entry:

$$H_{zh} = 0.2D + 0.5d_{noz} \text{ for } d_{noz} > D/100 \tag{5.56}$$

$$= 18d_{noz} \text{ for } d_{noz} < D/100 \tag{5.57}$$

2. For vertical stream entry:

$$\mathrm{H}_{zv} = 3(D - d_{noz}) \text{ for } d_{noz} > D/100 \tag{5.58}$$

$$= 100\, d_{noz} \text{ for } d_{noz} < D/100 \tag{5.59}$$

Commercial fluidized bed reactors are provided with plenum chambers in sections so as to effect elimination of the stagnant zone by increasing the pressure at the stagnant zone section at the expense of pressure in other sections. In general, sectioning of the plenum complicates the design. Baskakov et al.[83] do not support sectioning the plenum because it would cause surges and stalling in centrifugal blowers. More recent studies[95] showed that pressure fluctuations in the plenum chamber can be used to predict bubble formation and its eruption. This can be used in diagnose a fluidized bed reactor and control it. It was further shown that the frequency of bubble formation is controlled by the natural frequency of bed oscillations when the plenum chamber volume is kept large. In such a case, the fluidized bed is supposed to be in the resonant state and the bubbles are generated by bed oscillations. Because this finding could explain bubble formation and bubble dynamics, the plenum chamber is now likely to attract more interest; thus, a revival in research on this ignored bottom portion of the fluidization unit is expected.

C. Pumping Energy

Fluidized bed reactors have to be operated at higher pressure drops compared to packed bed or fixed bed reactors. It follows that the energy requirement for pumps or blowers is usually relatively high for the former. Sometimes the energy requirement may be so high that a fluidized bed operation may not be adopted in spite of its several advantages. Hence, it is necessary to estimate the pumping power required before adopting a fluidized bed for pilot- or commercial-scale operation. Let us consider the guidelines to predict the power requirement of a compressor or blower to achieve a gas fluidized bed operation. The gas at the blower inlet usually enters at atmospheric pressure (P_1) and has to be compressed to pressure (P_2) such that:

$$P_2 - P_1 = \Delta P_b + \Delta P_d + (\Delta P)_{\text{gas cleaning systems}} \tag{5.60}$$

The ideal shaft work for a compressor under adiabatic reversible operation, when kinetic and potential energy effects are negligible, can be evaluated from:

$$-W_{s,\text{ideal}} = \int_{P_1}^{P_2} V\, d_p \tag{5.61}$$

Assuming ideal gas behavior (i.e., $PV = nRT$) and eliminating the volume term in Equation 5.61, the ideal shaft work can be calculated by integration from the initial pressure (P_1) to the final required pressure (P_2):

$$-W_{\text{ideal}} = \frac{\gamma}{\gamma - 1} P_1 V_1 \left[\left(P_2 / P_1 \right)^{(\gamma-1)/\gamma} - 1 \right] = \frac{\gamma}{\gamma - 1} P_2 V_2 \left[1 - \left(P_1 / P_2 \right)^{(\gamma-1)/\gamma} \right] \tag{5.62}$$

The volume terms V_1 and V_2 correspond to the flow rate, and the constant γ = 1.4 for diatomic gases. For monoatomic and triatomic gases, γ is 1.67 and 1.33, respectively. The temperature rise that occurs during the adiabatic reversible process is given by:

$$T_2 - T_1 = (\Delta T)_{\text{ad}} = T_1 \left[\left(\frac{P_2}{P_1} \right)^{(\gamma-1)/\gamma} - 1 \right] \tag{5.63}$$

The actual magnitudes of the shaft work and the temperature rise are obtained from knowledge of the blower or the compressor efficiency (η). Thus,

$$-W_{s,\ \mathrm{actual}} = -W_{s,\ \mathrm{ideal}}/\eta \tag{5.64}$$

$$(\Delta T)_{\mathrm{actual}} = (\Delta T)_{\mathrm{ad}}/\eta \tag{5.65}$$

The efficiency factor varies from 0.55 to 0.75 for a turbo blower from 0.6 to 0.8 for a roots blower, and from 0.8 to 0.9 for an axial blower.

V. MODELING ASPECTS

A. Introduction

In this section, we will deal mainly with the various aspects of fluidized bed models. Over the past three decades, several models have been developed with regard to the gas fluidized bed. The purpose of each model is to analyze the gas–solid contact, delineate the exchange processes between the various phases as identified by each model, and study the effects of gas–solid properties and the operating velocity. Most of the models developed deal with a gas fluidized bed comprised of a simple cylindrical column. Models have not been developed for reactors that have tapered vessels or for a spout fluidized bed because of the complexity of the hydrodynamics involved in such systems. The hydrodynamics for a simple gas fluidized bed is so complex that even an isothermal nonreacting system has not reached a level of development that can assure confidence. Most of the models developed are confined to irreversible, solid catalyzed gas-phase reactions of the first order because of ease of mathematical formulation. Furthermore, any change in volume due to reactions to reactions is neglected. Models available in the literature for nonisothermal reactions are scant.

The objective of a fluidized bed model is to combine the chemistry of the reaction and several other hydrodynamic parameters mathematically to arrive at equations that are useful in estimating the degree of conversion and the size of the reactor. The diameter of a fluidized bed reactor depends on the stoichiometry of the reaction, the production rate, and the maximum allowable gas flow rate that would keep the solid inventory in the bed constant. The height of the reactor is determined from knowledge of the gas–solid contact and the chemical reaction rate. In addition to the above two factors, the available heat transfer area and the freeboard height for suppressing solid entrainment also play important roles. A fluidized bed reactor model should consider not only the single zone where a single pattern of gas–solid contact is prevalent but also all the possible zones which are responsible for the overall characteristics of a fluidized bed. A fluidized bed is made up of three zones: the grid zone where the gas enters through the distributor and emerges as bubbles or jets, the main bubbling bed including the bubble-erupting top surface, and the freeboard zone above the surface of the fluidized bed.

1. Grid Zone

The importance of the grid zone was covered in detail in the preceding section on distributors. The mechanism of gas entry into the fluidized bed from the gas-issuing point has been the subject of many research investigations. With regard to modeling a fluidized bed reactor, Yates et al.[44] found that around 70% of the gas that goes into the bubbles penetrates into the dense phase and has intimate contact with the solid particles during the period extending from bubble formation to bubble detachment. When the bubble is fully formed and detached, further gas–solid reaction occurs by interphase mass transfer. The gas from the dense phase subsequently enters the bubble phase through the wake below the bubble. Yang et al.[96] studied and developed a model for the prediction of butene conversion to butane at 300°C. This reaction is a fast oxidative type of dehydrogenation reaction. Their model prediction showed that butene conversion ranging from 40 to 70% can be obtained for a gas flow per orifice of up to 2000 cm³/s. The equilibrium time (t_e) required for the bubble to reach an equilibrium height H_{eq} was evaluated as:

$$t_e = H_{eq}(2/g)^{1/2}(\pi/8)^{1/6}\left(\frac{n_b}{x_o Q_{or}}\right)^{1/6} \tag{5.66}$$

where x_o is the fraction of orifice flow forming a visible bubble, n_b is the bubble frequency per orifice (s^{-1}), and Q_{or} is the volumetric flow rate per orifice (cm^3/s). A bubble equilibriation height of the order of 20 cm has been reported. There are other models[97,98] which deal with jets (and not the bubbles) at the distributor and thus bring out the influence of the distributor plate on fast reactions. However, much remains to be done in this field, as the bubble or gas jet characteristics at the distributor are essential in developing models.

2. Main Bubbling Bed

The bubbling bed is considered to begin above the distributor plate just at the point where the bubble or the gas jet reaches the equilibrium size. In other words, this zone is no longer influenced by or characterized in terms of the bubbles or jets that are present within the grid zone. The bubbling bed zone has been the subject of many model hypotheses. Many models now exist for a fluidized bed, and several are still undergoing further refinements. In principle, each model differs from the others in respect to the assumptions made regarding the gas–solid flow in the bed. By and large, almost all models uphold the conclusion that a bubbling bed consists of two phases: an emulsion phase rich in solids and a bubble phase with or without solid particles. Model development starts from the concept of the two-phase theory, which assumes that all gas in excess of that required for incipient fluidization appears in the bed as bubbles. This theory was originally proposed by Toomy and Johnstone.[99] A diagrammatic representation of this theory is depicted in Figure 5.10.

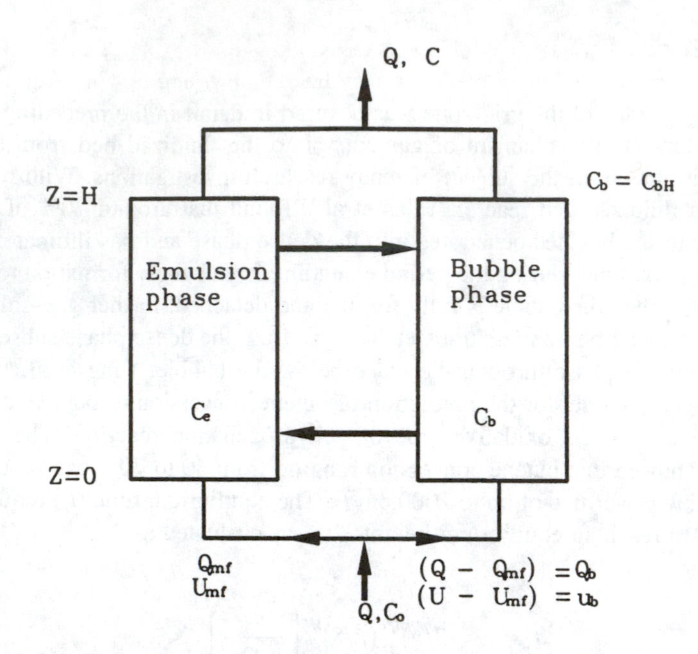

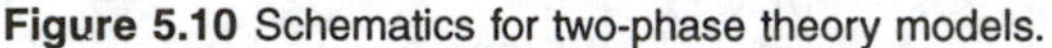

Figure 5.10 Schematics for two-phase theory models.

The reason for the development of a multiplicity of models to date lies in the various assumptions made with regard to visible bubble flow and actual bubble flow and the pattern and type of flow in the emulsion phase and the bubble phase. The types of flow normally assumed are plug, complete mixing, and axially dispersed. Furthermore, mass transfer between the two phases is assumed to take place by various modes such as convection, diffusion, and convective diffusion. In certain cases, the bubbles are considered to be devoid of solid particles, while in others they are assumed to contain a certain fraction of solids. All the above parameters, when combined in various manners, result in several models. Selection of a model for any specific application can be a difficult task for a beginner. Additional details pertaining to model types will be provided later in this chapter.

3. Freeboard Zone

Further reaction of the unreacted gas which escapes from the surface of a fluidized bed can continue in the freeboard zone, where entrained solid particles are present. Details on entrainment and solid particle elutriation from the bed surface were presented in Chapter 1. The idea of the occurrence of reactions in the freeboard zone was first conceived by Miyauchi and Furusaki.[100] Additional contributions to the literature on this subject were subsequently made by them[101] and their co-workers.[102] They measured the solid concentration in the freeboard and calculated the reactant conversion assuming piston flow (plug flow) for the gas. However, they did not attempt to analyze the mechanism by which solids appear in the freeboard. Later, Yates and Rowe[103] developed a hypothesis regarding particle ejection from the wake of the bubbles which erupt at the surface of the bed. This mechanistic model, along with a two-phase flow theory of gas–solid flow, can predict the rate of solid ejection from the surface of the fluidized bed. Using this solid ejection rate and assuming plug flow of gas, the fraction of unreacted gas at the exit can be expressed as:

$$C_h/C_s = \exp(-\alpha' h) \tag{5.67}$$

where

$$\alpha' = \frac{F}{3(U-U_t)}\left[\left(\frac{d_p}{6k_g(1-\epsilon_{mf})}+\frac{1}{K}\right)\right]^{-1} \tag{5.68}$$

Here, h is the height above the surface, C_h/C_s is the fraction of gas unconverted (relative to the surface concentration [C_s]), k_g is the gas mass transfer coefficient, K is the reaction rate constant, and F is the fraction of solid elutriated from the wake of the bubbles. In the development of the above equation, it was assumed that the particles are equidistant and each particle is surrounded by a cell of the reactant gas. In the case of gas flow in a perfectly mixed condition, the expression for the unconverted fraction of gas is

$$C_h/C_s = \frac{1}{(1+\alpha' h)} \tag{5.69}$$

The extent of freeboard reaction, as per Equation 5.67 was found to increase if the elutriation rate and the reaction rate increase. Later work by Chen and Wen,[104] using the axial dispersion model for chemical reaction and the elutriation model of Wen and Chen,[105] led to more realistic values compared to those obtained with other models. Freeboard conversion, especially for exothermic reactions such as the combustion of carbon in catalytic cracking, was found to have a profound effect as described by the model of de Lasa and Grace,[106] who assumed plug flow both for gas and solid and used the entrainment model of George and Grace.[107]

B. Some Basic Aspects of Modeling

Models pertaining to fluidized beds can generally be of two major types: models that consider only two phases and models that focus on a bubbling bed. In the former class of models, two parallel phases with cross-flow between them are considered. In a two-phase model, one phase is constituted of bubbles which are assumed, in general, to be devoid of solid particles and the second is a particle-rich dense phase or emulsion phase. In the simplest types (Level I) of two-phase models, the bubble-phase gas is attributed to the excess gas flow above that required to achieve minimum fluidization. In other words, the dense phase or the emulsion phase corresponds to a bed at the incipient state of fluidization. In a model of this type, parameters relating to bubble size variation or growth are not considered.

When the bubble size parameters are considered to be either constant or adjustable, the model is designated Level II. In models designated Level III, bubble size variation with bed height and bed diameter is also considered.

1. Two-Phase Model

A simple case of a two-phase model is shown in Figure 5.11a. Two-phase models are further refined if the grid region is also considered as a separate phase. Combinations of various types of models can be schematically represented, as in Figure 5.11b–d, especially if the grid zone is considered.

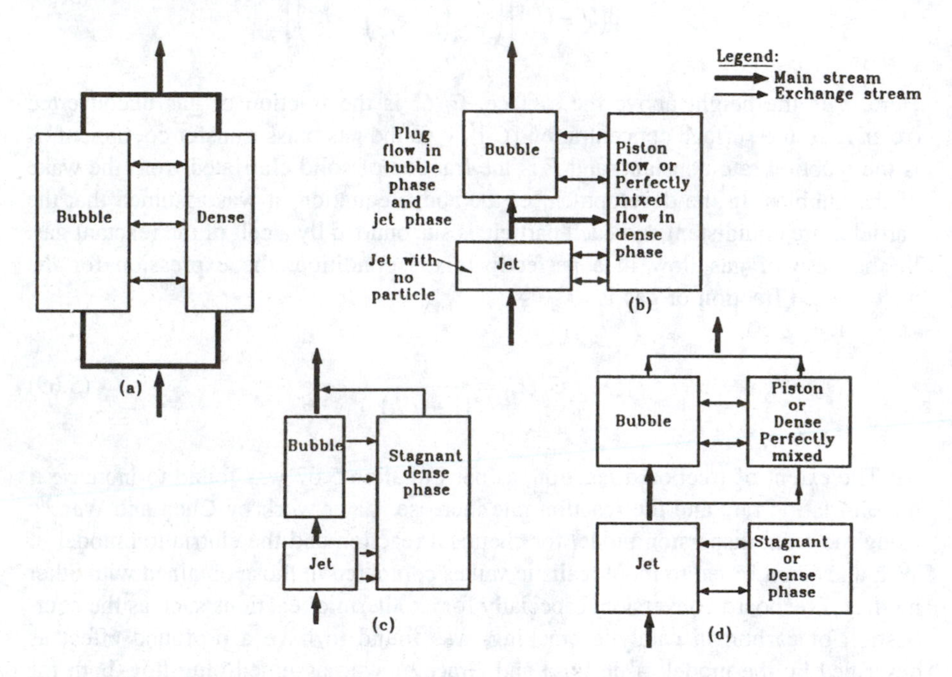

Figure 5.11 Schematic representation of two-phase model (a) and its extensions (b–d).

Grace and de Lasa[108] analyzed the simple two-phase model and extended it to include the grid zone, leading to various combinations as depicted in Figure 5.11; they considered either a perfectly mixed or a stagnant dense phase in the grid zone. Figure 5.11b corresponds to the Behie and Kehoe model,[109] which assumes the absence of particles in the jet, plug flow of gas in the jet and the bubbles, and perfectly mixed flow in the dense phase. For a constant volume of solid in a catalyzed first-order irreversible gas-phase reaction of type A → B, the mass balance equations under isothermal conditions can be formulated and the exit concentration can be predicted. The basic mass balance equations are as follows.

1. Case 1 (Figure 5.11b): No solid present in jet.
   - Jet phase ($0 \le x \le h$), plug flow:

$$U\frac{dc_j}{dz}+K_ja_j\left(c_j-c_d\right)=0 \tag{5.70}$$

- Bubble phase ($h < z \leq H$), plug flow:

$$\beta U\frac{dc_b}{dz}+K_ba_b\left(C_b-C_d\right)=0 \tag{5.71}$$

- Dense phase ($0 \leq z \leq H$), perfectly mixed flow:

$$(1-\beta)U\left(C_d-C_{jh}\right)+\int_0^{h'}K_ja_j\left(c_d-c_j\right)dz+\int_h^H K_ba_b\left(C_d-C_b\right)dz+K_rC_dH_{mf}=0 \tag{5.72}$$

2. Case 2 (Figure 5.11c)
   - Jet-phase ($0 < z \leq h$), solid fraction (σ), jet area fraction (ϵ_j):

$$U\frac{dc_j}{dz}+K_ja_j\left(C_j-C_d\right)+K_r\epsilon_j\sigma C_j/\left(1-\epsilon_{mf}\right) \tag{5.73}$$

   - Bubble phase: Conditions same as in Case 1. Equation 5.71 remains unchanged.
   - Dense phase is stagnant. Cross-flow occurs between jet phase and bubble phase:

$$K_ja_j(C_j-C_d)-K_r(1-\epsilon_j)C_d=0 \text{ for } (0\leq z\leq h) \tag{5.74}$$

$$K_ba_b(C_b-C_d)-K_r(1-\epsilon_b)c_d=0 \text{ for } (0\leq z\leq H) \tag{5.75}$$

3. Case 3 (Figure 5.11d)
   - Jet phase: Conditions same as in Case 2. Equation 5.73 remains unchanged.
   - Bubble phase: Conditions same as in Case 1. Equation 5.71 remains unchanged.
   - Dense phase: Composite of stagnant and completely mixed flow. Stagnant ($0 \leq z \leq h$): Equation 5.74 remains unchanged. Completely mixed ($h \leq z \leq H$): Equation 5.72, mass transfer from dense phase to jet neglected.

In all three cases, the exit concentration (C_{Ae}) can be computed from the mass balance:

$$C_{Ae}=\beta C_{bH}+(1-\beta)C_d \tag{5.76}$$

where β is the fraction flow in the bubble phase. The unconverted fraction (C_{Ae}/C_{Ao}) can be calculated in all three cases by using appropriate mass balances and evaluating C_j and C_b explicitly in terms of C_d. If X_A is the conversion for reactant gas A, then $X_A = 1 - (C_{Ae}/C_{Ao})$. The ratio C_{Ae}/C_{Ao} can be expressed in all three cases as follows:

Case 1:

$$\frac{C_{Ae}}{C_{Ao}} = \frac{1+\beta(K-1)\ \exp-\left(m_j+m_b\right)}{1+K-\beta\ \exp-\left(m_j+m_b\right)} \tag{5.77}$$

Case 2:

$$\frac{C_{Ae}}{C_{Ao}} = \exp\left[-\frac{m_j K_j}{m_j+K_j} - \frac{K_j \epsilon_j \sigma}{\left(1-\epsilon_j\right)\left(1-\epsilon_{mf}\right)} - \frac{m_b K_b}{m_b+K_b}\right] \tag{5.78}$$

Case 3:

$$\frac{C_{Ae}}{C_{Ao}} = \left[\frac{1+\beta\left(K_b-1\right)\exp-\left(m_b\right)}{1+K_b-\beta\exp-\left(m_b\right)}\right]\cdot\exp\left[-\frac{m_j K_j}{m_j+K_j} - \frac{K_j \epsilon_j \sigma}{\left(1-\epsilon_j\right)\left(1-\epsilon_{mf}\right)}\right] \tag{5.79}$$

where

$$K = K_r\, H_{mf}\big/U,\quad m_j = \frac{K_j a_j h}{U},\quad m_b = \frac{K_b a_b (H-h)}{\beta U}$$
$$K_r = \frac{K_r\left(1-\epsilon_j\right)h}{U},\quad K_b = \frac{K_r\left(1-\epsilon_b\right)(H-h)}{U} \tag{5.80}$$

and $\beta = (1 - U_{mf}/U)$, $0.001 << \sigma \leq 0.006$ (Merry[110]).

In all the above analyses, an isothermal condition is assumed even though the grid region has marked temperature gradients. A nonisothermal grid model based on certain hypotheses like those shown in Figure 5.11b was developed and tested[111] for industrial-type fluid bed reactors. The assumptions made are (1) the met phase is in piston flow, (2) the bubble phase is perfectly mixed, (3) the emulsion or dense phase can be approximated to continuously stirred tank reactor, (4) chemical reactions in the jet and bubble phases can be neglected because the particle concentration is negligible, and (5) the temperature gradient between phases is negligible as the heat capacity of the emulsion phase is very high. The model was used to simulate operation of the industrial unit at Destileria La Plata, Argentina. It is important to realize that experimental data for the jet-to-dense-phase transfer rate are scant in the literature. Furthermore, data for industrial-scale units are rare. In view of these facts, the grid region continues to be the subject of present and future research.

2. Bubbling Bed Models

The bubbling bed models constitute the second group of models and are based on bubble dynamics. It is interesting to note that bubbles induce mixing and are responsible for making it possible for the fluidized bed to remain in a well-agitated condition. At the same time, bubbles carry the reactant gas and thus allow it to bypass the feed gas. The bubbling bed models require data on the bubble rise velocity; the cloud of solids enveloping the bubble phase; the fraction of wake present at the rear of each rising bubble; the extent of gas flow that accounts for the bubble phase; the exchange coefficients between the bubble, cloud, and emulsion phases; and the size of the bubbles. A schematic of the bubbles present in the emulsion phase and the exchange processes around the bubbles is given in Figure 5.12.

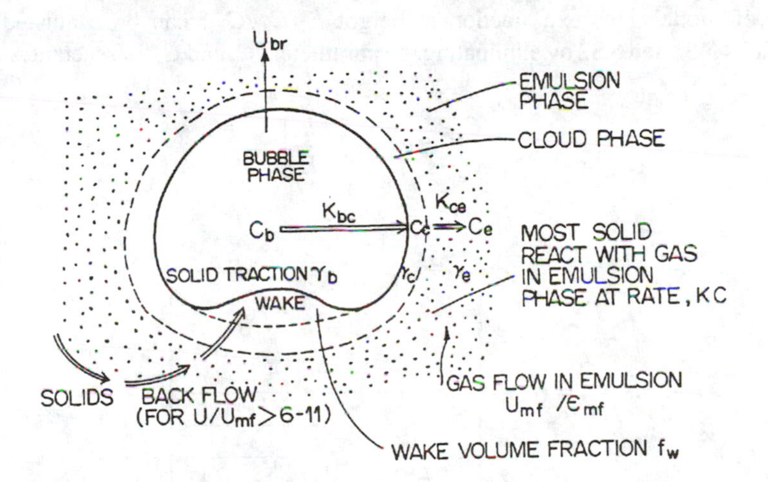

Figure 5.12 Bubbling bed model.

As depicted in Figure 5.12, a bubbling bed model essentially takes into consideration three distinct phases: the bubble phase, the cloud phase, and the emulsion phase. Each of these phases is assumed to have some solids, and the unreacted gas from each phase passes to the next phase in series; the gas is finally consumed by reaction with the solids present in the emulsion phase. The gas velocity in the emulsion phase is assumed to be U_{mf}/ϵ_{mf}. The solids are carried by the wake present at the bottom of each bubble. Because the solids are conserved in the bed so that there is no net flow, solids which are drawn up must be compensated for by the downward transport. These descending solids drag the gas and hence lead to backmixing of the gas, especially when $U/U_{mf} > 6$–11.

A typical bubbling bed model proposed by Kunii and Levenspiel[26] has the following mass balance for the three phases.

- Bubble phase:

$$U_B \frac{dc_b}{dx} = \gamma_b K_r C_b + K_{bc}(C_b - C_c) = 0 \tag{5.81}$$

- Cloud phase:

$$K_{bc}(C_b - C_c) = \gamma_c K_r C_c + K_{ce}(C_c - C_e) \tag{5.82}$$

- Emulsion phase:

$$K_c(C_c - C_e) = \gamma_e K_r C_e \tag{5.83}$$

- Exit fraction of gas:

For a type A → B reaction under the same conditions described for the first group of models, the exit fraction at height h (C_{Ae}/C_{Ao}) can be evaluated from Equations 5.81 and 5.82 by eliminating the parameters C_c and C_e for reactant A. Thus,

$$\frac{C_{Ae}}{C_{Ao}} = \exp\left[-\mathrm{K}\frac{h}{H}\right] \tag{5.84}$$

where

$$\mathrm{K} = H\frac{K_r}{U_{br}}\left[\gamma_b + \cfrac{1}{\cfrac{K_r}{K_{be}} + \cfrac{1}{\gamma_c + \cfrac{1}{\cfrac{K_r}{K_{ce}} + \cfrac{1}{\gamma_e}}}}\right] \tag{5.85}$$

Bubble interactions were not taken into consideration in the development of this model. Furthermore, the assumptions in reality will not hold to characterize reactor performance unless many parameters, such as K_r, K_{be}, and K_{ce}, are evaluated. Thus, the degree of accuracy of the predictions of the second category of models depends solely on the basic parameters, which should, in principle, correspond to realistic values.

C. Model Types

Models pertaining to fluidized beds are numerous, and the number is increasing as a result of the incorporation of refinements or advanced thinking on the basics of the subject. As seen in the preceding section, models can generally be grouped into two major types. Grace[112] grouped models as two-phase and bubbling bed. Horio and Wen[113] classified models into three categories:

1. Models that do not relate the parameters to bubble size and consider the parameters to be constant throughout the bed
2. Models with constant parameters related to bubble size
3. Models that use variable parameters through the bed related to the bubble size

The degree of sophistication of each model changes with the nature of the assumptions, and this results in a multiplicity of models. LaNauze[114] listed as many as 28 models for the fluidized bed combustion system.

Fluidized bed models have been analyzed and grouped in review papers.[42,115-119] The various hypotheses used by different researchers have resulted in several kinds of models. Wen[116] proposed a generalized material balance equation for a first-order reaction and suggested that the pertinent set of mass balance equations could account for almost all models. The set of generalized mass balance equations for the various phases are as follows.

1. Bubble phase (b):

$$U_b\delta_b \frac{dC_b}{dz} - K_{bc}\delta_b(C_b - C_c) - K_{bw}\delta_b(C_c - C_w) - K_r r_b C_b = 0 \tag{5.86}$$

2. Cloud phase (c):

$$E_x\delta_c\epsilon_e \frac{d^2C_c}{dz^2} - U_c\delta_c\epsilon_e \frac{dc_c}{dz} - K_{bc}\delta_b(C_b - C_e) - K_{ce}(C_c - C_e) - K_r\delta_c(1-\epsilon_e)C_c = 0 \tag{5.87}$$

3. Wake phase (w):

$$E_x\delta_w\epsilon_e \frac{d^2C_w}{dz^2} - U_w\delta_w\varepsilon_e \frac{dC_w}{dz} + K_{bw}\delta_b(C_b - C_w) - K_{we}\delta_w(C_w - C_e) - K_2\delta_w(1-\epsilon_e)C_w = 0 \tag{5.88}$$

4. Emulsion phase (e):

$$E_x\delta_e\epsilon_e \frac{d^2C_e}{dz^2} - U_e\delta_e\epsilon_e \frac{dc_e}{dz} + K_{ce}\delta_e(C_c - C_e) + K_{we}\delta_w(C_w - C_e) - K_r\delta_c(1-\epsilon_e)C_e = 0 \tag{5.89}$$

Various types of models and the related material balance equations along with the assumptions made in the development of these models are presented in Table 5.2. All the mass balance equations listed in Table 5.2 can be derived using the

Table 5.2 Fluid Bed Models and Mass Balance Equations

Model	Mass Balance Equations	Conditions	Remarks[a]
Shen and Johnstone[120]	$U_b dC_b/dz + K_{be}(C_b - C_e) = 0$ $U_e(1-\delta_b)\epsilon_e dC_e/dz - K_{be}\delta_b(C_b - C_e) + K_r(1-\delta_b)(1-\epsilon_e)C_e = 0$	$E_z = 0,\ a_b = 0$ $U_c = U_w = U_e$ $C_c = C_w = C_e$ $K_{be} = K_{bc} + K_{bw}$	B ↔ C + W + E Constant parameters
Lewis et al.[121]	$U_b\delta_b dC_b/dz + K_{be}\delta_b(C_b - C_e) + K_r a_b C_b = 0$ $K_{be}\delta_b(C_b - C_e) - K_r(1-\delta_b)(1-\epsilon_e)C_e = 0$	Same except $E_x \to \alpha$ $a_b \neq 0$	B ↔ C + W + E Constant parameters
Van Deemter[122]	$U_b dC_b/dz + K_{be}(C_b - C_e) = 0$ $E_x(1-\delta_b)\epsilon_e d^2C_e/dz^2 - U_e(1-\delta_b)\epsilon_e dC_e/dz + K_{be}\delta_b(C_b - C_e) - K_r(1-\delta_b)(1-\epsilon_e)C_e = 0$	Same except $E_x \neq 0$	B ↔ C + W + E Constant parameters
Davidson and Harrison[25]	$U_b dC_b/dz + K_{be}(C_b - C_e) = 0$ $U_e(1-\delta_b)\epsilon_e dC_e/dz - K_{be}\delta_b(C_b - C_e) + K_r(1-\delta_b)(1-\epsilon_e)C_e = 0$	Same	B ↔ C + W + E Plug flow in emulsion-phase parameters assumed for effective bubble diameter
Calderbark and Toor[123]	$U_b dC_b/dz + K_{be}(C_b - C_e) = 0$ $U_e(1-\delta_b)\epsilon_e dC_e/dz - K_{be}\delta_b(C_b - C_e) + K_r(1-\delta_b)(1-\epsilon_e)C_e = 0$	Same	B ↔ C + W + E Plug flow in emulsion phase, bubble diameter varies with bed height
Kato and Wen[124]	$U_b(\delta_b + \delta_c\epsilon_e)\,\Delta C_b/\Delta Z + K_{be}\delta_b(C_b - C_e) + K_r\delta_c(1-\epsilon_e) = 0$ $K_{be}\delta_b(C_b - C_e) = K_r(\delta_w + \delta_e)(1-\epsilon_e)C_e$	$E_z = 0,\ a_b = 0$ $U_w = U_e,\ C_b = C_e$ $U_b = U_c,\ C_w = C_e$	B + C ↔ W + E
Kunii and Levenspiel[26]	$U_b\delta_b dC_b/dz + K_{bc}'\delta_b(C_b - C_c) + K_r a_b C_b = 0$ $K'_{bc}\delta_b(C_b - C_c) - K'_{ce}(C_c - C_e) - K_r(\delta_c + \delta_w)(1-\epsilon_e)C_c = 0$ $K'_{ce}\delta_b(C_c - C_e) - K_r\delta_e(1-\epsilon_e)C_e = 0$	$E_z = 0$ $U_c = U_w = U_e = 0$ $C_c = C_w$ $K'_{bc} = K_{bc} + K_{bw}$ $K'_{ce}\delta_b = K_{ce}\delta_c + K_{wc}\delta_w$	B ↔ C ↔ E Constant parameters
Miyauchi and Morooka[125]	$U_b\delta_b dC_b/dz + K_{bc}\delta_b(C_b - C_c) - K_r a_b C_b = 0$ $K'_{bc}\delta_b(C_b - C_c) - K'_{ce}\delta_b(C_c - C_e) - K_r(\delta_c + \delta_w)(1-\epsilon_e)C_e = 0$ $K'_{ce}\delta_b(C_c - C_e) - K_r\delta_e(1-\epsilon_e)C_e = 0$	$E_z = 0,\ a_b = 0$ $U_c = U_w\ U_e < 0$ $C_c = C_w$ $K'_{bc} = K_{bc} + K_{bw}$ $K'_{ce} = K_{ce}\delta_c + K_{we}\delta_w$	B ↔ C + W + E Constant parameters
Gwyn et al.[126]	$U_b dC_b/dz + K_{be}(C_b - C_c) = 0$ $-U_c(\delta_c + \delta_w)\epsilon_e dC_c/dx + K_{bc}'\delta_b(C_b - C_c) - K_{ce}'\delta_b(C_c - C_e) - K_r(\delta_c + \delta_w)(1-\epsilon_e)C_c = 0$ $K'_{ce}\delta_b(C_c - C_e) - K_r\delta_e(1-\epsilon_e)C_c = 0$	$E_z = 0,\ a_b = 0$ $U_c = U_w\ U_e < 0$ $C_c = C_w$ $K'_{bc} = K_{bc} + K_{bw}$ $K'_{ce}\delta_b = K_{ce}\delta_c + K_{we}\delta_w$	B ↔ C + W + E Constant parameters
Fryer and Potter[127]	$U_b dC_b/dz + K'_{bc}(C_b - C_e) = 0$ $-U_c(\delta_c + \delta_w)\epsilon_e dC_c/dz + K'_{bc}\delta_b(C_b - C_c) - k'_{ce}\delta_b(C_c - C_e) - K_r(\delta_c + \delta_w)(1-\epsilon_e)C_c = 0$ $k'_{ce}\delta_b(C_c - C_e) - K_r\delta_e(1-\epsilon_e)C_e = 0$	$E_z = 0,\ a_b = 0$ $U_c = U_w\ U_e < 0$ $C_c = C_w$ $K'_{bc} = K_{bc} + K_{bw}$ $K'_{ce}\delta_b = K_{ce}\delta_c + K_{we}\delta_w$	B ← C + W → E
Werther[128]	$U_b\delta_b A_T dx dC_b/dz - D_G dA(\delta C_e/\delta y)_{y=0} = 0$ $D_G dA\delta^2 C_c/\delta y^2 - K_r A_T d\,x(\delta_c + \delta_w)(1-\epsilon_e)C_c = 0$ $-D_G dA(\delta C_e/\delta y)_{y=0} - K_r A_T dh\,\delta_e\epsilon_e C_c = 0$	$E_z = 0,\ a_b = 0$ $U_c = U_w = U_e = 0$ $C_c = C_w$ $-D_G dA(\delta C_e/\delta y)_{y=0} = (K_{bc} + K_{bw})\delta_b A_\tau dx(C_b - C_c)$	B ↔ C + W ↔ E´

B = bubble, C = cloud, W = wake, e = Emulsion.

generalized Equations 5.86–5.89. It should, however, be noted that in order to utilize the various models, several fundamentals parameters must be defined or evolved. Most of these parameters were generally touched upon in Chapter 1. To be more relevant to the models, certain other fundamental parameters that must, in principle, be understood and applied in fluidization research are introduced here.

1. Mass balance for total gas flow:

$$\begin{aligned}&[\text{Total gas flow, } U] = [\text{Net gas flow through bubbles, } U_b\delta_b]\\&+ [\text{Net gas flow through bubble wake, } U_b\delta_b\epsilon_{mf}(R + f_w - 1)]\\&+ [\text{Net gas flow emulsion, } \{1 - \delta_b(R + f_w)\}U_e\epsilon_{mf}]\end{aligned} \tag{5.90}$$

   where R and f_w are, respectively, the cloud and wake volumes expressed as ratios to the bubble volume.
2. Gas velocity in the emulsion phase (U_e):

$$U_e = (U_{mf}/\epsilon_{mf}) - U_s \tag{5.91}$$

   where U_s is the solids circulation velocity, which comes into the picture because a bubble drags a wake of solids up the bed. If solid circulation is neglected, then $U_s = 0$.
3. Excess gas velocity ($U - U_{mf}$): As per the two-phase theory of flow in fluidized solids, all gas in excess of U_{mf} passes as bubbles, that is,

$$(U - U_{mf}) = \delta_b U_b \tag{5.92}$$

Equation 5.92 appears to be valid for small particles or when U_{mf} is negligible compared to U_b. However, the deviation from this relationship is considerable when U_{mf} and δ_b are higher, which happens near the distributor plate with large particles. Horio and Wen[113] measured the visible bubble flow and found that $\delta_b U_b$ is less than ($U - U_{mf}$) and has the smallest value at the bottom of the bed; this value increases and approaches $U - U_{mf}$ in a deep bed that has slugging. The two-phase theory model is recommended for beds of small diameter.

Now let us consider the case of a bubbling bed model. It is obvious that the bubble flow pattern influences the solid movement, which in turn affects the percolating gas through the solid particles. When the bubble size is small, the bubble rises in the bed slowly, in which case the gas from the bubble can percolate through its permeable boundary. This can occur at a low gas flow rate when the gas velocity is increased. Solid particles which descend from the bed surface can drag the emulsion gas down and create gas backmixing. This can happen beyond a critical gas velocity (U_{cr}), which, according to Fryer and Potter,[127] is given by:

$$U_{cr}/U_{mf} = \left[1 + \frac{1}{\epsilon_{mf} f_w}\right]\left[1 + \delta_b\left(1 + f_w\right)\right] \tag{5.93}$$

When $U = U_{cr}$, backmixing of gas or flow reversal of gas in the emulsion phase occurs. In other words, $U_e < 0$. Fryer and Potter[127] measured the axial concentration profile of ozone in a three-dimensional fluidized bed and found a minimum at high gas velocities. They attributed this to the backmixing effect. For the typical range of $U_{cr}/U_{mf} > 6$–11, flow reversal of gas in the emulsion phase can occur as a result of the downward flow of solids. This kind of effect, which accounts for an axial variation of the concentration profile when $U > U_{cr}$, should be taken into consideration in a bubbling fluidized bed model.

D. Model Analysis

The analysis of models and the inferences that can be drawn from such analysis in terms of selection of a specific model for design applications is not a simple task. The main reasons are due to the many underlying principles that govern the development of models and the several assumptions that are not necessarily adopted in the same sense in all models. The models can be for a steady state or an unsteady state for either an isothermal or a nonisothermal reactor. The development of a model that covers all the complex hydrodynamic, thermal, and chemical characteristics of the reactor is a challenging task for a fluidization engineer. It is also not a simple job to analyze the numerous models and select a specific one for universal application. The basics of fluidization and many fundamental parameters are under constant review by researchers, and refinements are continuously being suggested. Modeling concepts and selection criteria still constitute a frontier in fluidization research. As seen in the preceding sections on the various aspects of modeling and the types of models, it is essential to study and further analyze the models. Although the preceding sections dealt with models qualitatively, the topic has not been treated from the standpoint of predicting the extent of conversion in a fluidized bed and comparing models to arrive at a selection criterion. In this section, specific attention is focused on four representative models which are well known in fluidization. The model development procedures to be followed in order to arrive at an equation to predict the exit concentration of the reactant gas have been discussed.

The four representative models considered here are those proposed by:

1. Davidson and Harrison[25]
2. Kunii and Levenspiel[26]
3. Kato and Wen[124]
4. Partridge and Rowe[131]

The development of each model is briefly outlined in the following sections in order to provide an overview of how models are built. The various assumptions and the basic parameters pertaining to each of these models are given in Table 5.3.

1. Davidson and Harrison Model[25]

This model is essentially similar to that of Orcutt et al.,[132] to which some details were added by Davidson and Harrison. Development of the model is based on the two-phase theory.

Table 5.3 Typical Models and Parameters[129,130]

Models Parameters	Davidson and Harrison[25]	Kunii and Levenspiel[26]	Kato and Wen[124]	Partridge and Rowe[131]
Phases considered	1. Solid-free bubble phase 2. Solid-rich emulsion phase	Three phases: (1) bubble, (2) cloud, and (3) emulsion	Fluidized bed is compartmented in series; two phases: (1) Bubble cloud and (2) Emulsion in each compartment	Two phases: (1) bubbles with cloud association and (2) emulsion phase
Gas-flow type	1. Plug flow in bubble phase 2. Plug flow and complete mixing in emulsion phase	Plug flow in bubble and backflow in emulsion phase	Perfect mixing in emulsion phase	Plug flow in emulsion phase
Gas flow-through phases	Gas in Excess of U_{mf} appears as bubbles	Gas flow in emulsion phase not considered	Gas flow in emulsion phase neglected	Visible gas flow for bubble phase
Assumption on bubble size	Constant	Constant	Varies along the bed height	Varies along the bed height
Transfer resistance	Resistance between cloud and emulsion interface neglected	Two resistances across the interfaces: (1) bubble to cloud wake and (2) cloud wake to emulsion	No transfer resistance across bubble/cloud interface	No transfer resistance across bubble/cloud interface
Basic parameters defined	Gas flow fraction for bubble: $\beta = 1 - U_{mf}/U$ Number of bubbles per unit bed volume: $N_b = (U - U_{mf})/U_bV_b$ Bubble volume: $V_b = \pi d_b^3/6$ Bubble surface area: πd_b^3 Bubble–emulsion interphase mass transfer coefficient: $K_{be} = 3/4\ U_{mf} + 0.975\ D_G^{1/2}(gd_b)^{1/4}$	Bubble volume fraction in the bed: $\delta = N_bV_b = U - U_{mf}/U_b$ Ratio of solid volume in cloud-wake phase to bubble-phase volume: $\gamma_e = (1 - \epsilon_{mf})\{[3U_{mf}/\epsilon_{mf}]/[U_{b\alpha} - (U_{mf}/\epsilon_{mf})] + f_w\}$ Ratio of solid volume in emulsion phase to bubble-phase volume: $\gamma_e = U_{b\alpha}[(1 - \epsilon_{mf})/(U - U_{mf})] - \gamma_e$, $f_w = V_w/V_b$, $U_{b\alpha} = 0.71(gd_b)^{1/2}$ Bubble cloud–wake mass transfer per unit bubble volume: $K_{be} = [4.5\ U_{mf} + 5.85D_G^{1/2}(gd_b^{-3/4})]d_b$ Cloud–wake emulsion, mass transfer coefficient per unit bubble volume: $K_{ce} = 6.78(\epsilon_{mf}D_GU_b)/d_b^{-3})^{1/2}$	Height of nth compartment: $\Delta h_n = 2d_{bo}(2 + \lambda)^n$, $\lambda = 1.4P_pd_pU/U_{mf}$ Bubble volume in nth compartment: $Vb_n = N_n(\pi/6)\Delta h_n^3$ N_n = number of bubbles in the nth compartment $(6s/\pi\Delta h^2)(H - H_{mf}/H)$ Cloud volume in nth compartment: $V_{e,n} = V_{bn}(3U_{mf}/\epsilon_{mf})/(U_{b\alpha} - U_{mf}/\epsilon_{mf})$ Gas-exchange coefficient per unit bubble volume: $F_o = 11/d_b$	Relative velocity: $R = U_b - (U_{mf}/\epsilon_{mf})$ Cloud diameter: $D_c = d_b[(V_b + V_c/V_b)]^{1/8}$ Cloud Reynolds number: $Re_c = D_c\rho_g(U_b - U_{mf}/\epsilon_{mf})/M_g$ Sherwood number: $Sh_c = K_{gc}D_c/D_G = 2 + 0.6Sc^{\frac{1}{3}}\ Rc_c^{1/2}$ where $Sc = \mu_g/P_gD_G$ Interphase exchange per unit volume of void unit: $Q_E = \pi\epsilon_{mf}D_cD_gSh_c/(V_b \cdot R)$

1. Mass balance at the exit: The mass balance for the reactant gas at the exit (see Figure 5.10) is

$$UC = U_{mf}C_e + (U - U_{mf})C_{bH} \tag{5.94}$$

Introducing the exit reactant gas concentration fraction (C/C_o), Equation 5.94 can be expressed as:

$$\frac{C}{C_o} = (1-\beta)\frac{C_e}{C_o} + \beta\frac{C_{bH}}{C_o} \tag{5.95}$$

where

$$\beta = 1 - U_{mf}/U$$

The dimensionless parameter $C_e' = C_e/C_o$ and $C_b' = C_{bH}/C_o$ should be evaluated to obtain $C' = C/C_o$. The parameter C_b can be evaluated from the mass balance across a bubble, and C_e can be evaluated from the mass balance across the emulsion phase.

2. Mass balance across a bubble:

$$\text{Rate of change of reactant concentration, } -\frac{dV_b}{dt}\cdot C_b = \text{Loss of reactant by exchange, } K_{be}\left(C_b - C_e\right) \tag{5.96}$$

Using $dt = dz/u_b$ and integrating and rearranging Equation 5.96, one obtains:

$$C_b' = C_e' + \left(1 - C_e'\right)\exp(-\eta) \tag{5.97}$$

where

$$\eta = K_{be}H/(U_bV_b) \tag{5.98}$$

3. Mass balance across the emulsion phase:

$$\begin{aligned}&\left(\text{Feed rate of reactant, } U_{mf}C_o\right) - \left(\text{exit flow rate of reactant, } U_{mf}C_e\right)\\ &- \left(\text{Loss of exchange in emulsion phase, } NK_{be}HC_e\right)\\ &+ \left(\text{Gain by exchange in bubble phase, } NK_{be}\int_o^H C_b dz\right)\\ &- \left[\text{Loss due to reaction, } KHC_e\left(1 - NV_b\right)\right] = 0\end{aligned} \tag{5.99}$$

Equation 5.99 can be solved for C_e', eliminating C_b by using Equation 5.97 and invoking the relation

$$\delta = NV_b = (H - H_{mf})/H = (U - U_{mf})/U_b \tag{5.100}$$

to give:

$$C'_e = \frac{\left(1 - \beta\, e^{-\eta}\right)}{1 - \beta\, e^{-\eta} + K'} \tag{5.101}$$

where

$$K' = KH_{mf}/U \tag{5.102}$$

Substitution of C_b' and C_e' in Equation 5.95 yields:

$$C'_e = \beta\, e^{-\eta} + \frac{\left(1 - \beta\, e^{-\eta}\right)^2}{1 - \beta\, e^{-\eta} + K'} \tag{5.103}$$

Equation 5.103 can be simplified in two cases. Case 1 is when the intergas exchange rate is very high (i.e., when $\eta \to \infty$):

$$C' = (1 + K')^{-1} \tag{5.104}$$

which is the case of a perfectly mixed system. Case 2 is when the reaction is fast (i.e., $K \to \infty$):

$$C' = \beta e^{-\eta} \tag{5.105}$$

From Equations 5.104 and 5.105, it can be assumed that there is a possibility that some reactant gas would always appear at the exit, whatever the conditions may be. This can be attributed to the bypassing of gas in the form of bubbles. The above treatment in the Davidson and Harrison (D-H) model corresponds to the condition where the emulsion phase is in a perfectly mixed state. If the condition corresponds to plug flow or piston flow, then

$$C' = \frac{1}{m_1 - m_2}\left[m_1 e^{-m_2 H}\left(1 - \frac{m_2 H U_{mf}}{U\eta}\right) - m_2 e^{-m_1 H}\left(1 - \frac{m_1 H U_{mf}}{U\eta}\right)\right] \tag{5.106}$$

where m_1 and m_2 are, respectively, the positive and negative values of m obtained from the equation

$$2H(1 - \delta)m = (\eta + K') \pm [(\eta + K')^2 - 4K'\eta(1 - \delta)]^{1/2} \tag{5.107}$$

The D-H model predicts C' for two types of gas flow, and the interphase exchange parameter (K_{be}) is assumed to be the sum of two terms: the convective term, q, and the diffusive terms, $K_{be} \cdot a$, where a is the interphase surface area.

2. Kunii and Levenspiel Model[26]

This model was analyzed and presented in an earlier section that dealt with bubbling bed models. Equations 5.81–5.85 provide the steps and Equation 5.84 predicts the exit gas concentration fraction. The various assumptions and parameters used in the development of this model are given in Table 5.3. The Kunii and Levenspiel (K-L) model has certain features in common with the D-H model in that it assumes a perfectly mixed emulsion phase and bubbles of constant size with a plug flow state throughout the bed. A major breakthrough associated with this model pertains to the development of a mechanism for the transfer or exchange of the reactant from the bubble phase to the emulsion phase. The transfer occurs in series from the bubble phase to its next phase, known as the cloud-wake phase, and then to the emulsion phase. There is no transfer of gas or reactant from the emulsion phase to the bubble phase and the reaction in the emulsion phase is considered. The overall resistance to mass transfer from the bubble to the emulsion phase can be expressed as:

$$\frac{1}{K_{be}} = \frac{1}{K_{bc}} + \frac{1}{K_{ce}} \tag{5.108}$$

The reaction rate constant (K_r) considered in the development of an expression for C' is for a first-order reaction. The relation governing the solid volume fraction in the three different phases is

$$(\gamma_b + \gamma_c + \gamma_e)f_w = (1 - \epsilon_{mf})\,(1 - f_w) \tag{5.109}$$

where $f_w = (U - U_{mf})/U_b$. The ratio of the wake volume to the bubble volume (V_w/V_b) ranges from 0.2 to 1 and is mostly taken as 0.5 from the empirical plots of Rowe and Partridge.[133] In the K-L model, the important basic parameter to be estimated is the bubble size, which is assumed to be constant irrespective of its level inside the bed. The treatment of the K-L model applies to a vigorously bubbling bed where the gas flow can be considered to occur mainly in the bubble form and the interstitial gas flow can be neglected.

A refinement of the K-L model, known as the countercurrent backmixing model, was presented by Fryer and Potter.[127] This model takes into account bubble size variation along the bed height and also the reaction in the emulsion phase. Bubble size variation was evaluated by using the relation

$$d_b = d_{bo} + mh \tag{5.110}$$

where the initial bubble size is given by:

$$d_{bo} = 1.08(U - U_{mf})^{0.3} \tag{5.111}$$

and the constant m

$$m = 0.0205U \quad \text{(in c.g.s. units)} \tag{5.112}$$

Because the exit concentration of the gas leaving the bed is mostly due to the bubbles, the K-L model uses the approximation $C' = C/C_o = C_{bH}/C_o$. Strictly speaking, the concentration is supposed to lie between the outlet bubble-phase and cloud-wake-phase concentrations. According to the K-L model,

$$C_c/C_o = C_c' = \frac{C_b' K_{bc}}{K_{bc} + \gamma_c K_r + K_{ce} - \dfrac{K_{ce}^2}{K_{ce} + \gamma_e K_r}} \tag{5.113}$$

where

$$C_b' = C_{bz}/C_o$$

and

$$C_e' = \frac{K_{ce} C_c'}{K_{ce} + \gamma_e K_r} \tag{5.114}$$

The gas transfer between a bubble and its cloud wake is assumed to occur by both molecular diffusion and bulk flow, as outlined by Orcutt et al.,[132] and the gas exchange from cloud wake to emulsion phase is assumed to occur by molecular diffusion, conforming to Higbie penetration theory. Reaction in the bubble phase due to catalyst particles raining through the bubble is neglected.

3. Kato and Wen Model[124]

The Kato and Wen (k-w) model is also known as the bubble assemblage model because here the fluidized bed is assumed to be made up of several compartments in series. The size of each compartment is considered to be equal to the bubble size at that height (z) which is obtained by the empirical correlation proposed by Kobayashi et al.[132]

$$d_b = 1.4\rho_b\, d_p(U/U_{mf})z \tag{5.115}$$

The initial diameter at the distributor was calculated by using the correlation of Cooke et al.:[135]

$$d_{bo} = \left[6\left(U - \frac{U_{mf}}{\pi n_o}\right)\right]^{0.4} g^{-0.2} \tag{5.116}$$

where n_o is the number of holes per unit surface area of the distributor. Each compartment is assumed to be made up of the bubble-phase (including voids) clouds and the emulsion phase. The reactant is considered to be in a perfectly mixed condition in each phase and backmixing of the bubble gas between the compartments is neglected. Gas percolation through the emulsion phase is also neglected. The expression for the interphase mass transfer is based on the work of Kobayashi et al.,[136] that is, $F_o = 11/d_p$ (CGS units). If Δh_n is the height of the nth compartment, then the mass balance can be used to predict $C_{b,n}$ and $C_{e,n}$. Figure 5.13 shows a schematic of the transfer of reactant mass in the nth compartment.

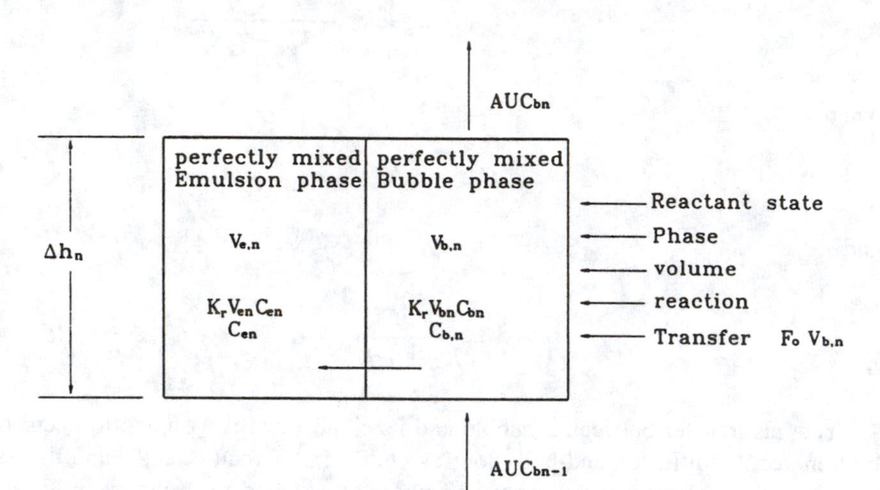

Figure 5.13 Representation of transport in the *n*th compartment.

1. Mass balance in the bubble phase:

$$\begin{aligned}&\text{Reactant input from the } (n-1)\text{th compartment, } AUC_{b,n-1}\\ &= [\text{Reactant output from bubble phase in the } n\text{th compartment, } AUC_{b,n}] + \\ &[\text{Reactant loss by reaction, } K_r V_{b,n} C_{b,n}]\\ &+ [\text{Reactant loss by transfer to emulsion phase, } F_{o,n} V_{b,n} (C_{b,n} - C_{e,n})]\end{aligned} \quad (5.117)$$

2. Mass balance in the emulsion phase:

$$\begin{aligned}&\text{Reactant loss by transfer from bubble phase, } F_{o,n} V_{b,n} (C_{b,n} - C_{p,n})\\ &= \text{Reactant consumption by reaction, } K_r V_{e,n} C_{e,n}\end{aligned} \quad (5.118)$$

From Equation 5.118, the concentration in the emulsion phase can be predicted as:

$$C_{e,n} = F_{o,n} V_{b,n} C_{b,n} / (K_r V_{b,n} + F_{o,n} V_{b,n}) \quad (5.119)$$

Eliminating $C_{e,n}$ by using Equation 5.119 in Equation 5.117, $C_{b,n}$ can be obtained as:

$$C_{b,n} = \frac{C_{b,n-1}(AU)}{\left[F_{o,n}V_{b,n} - \dfrac{\left(F_{o,n}V_{b,n}\right)^2}{K_r V_{e,n} + F_{o,n}V_{b,n}} + k_r V_{e,n} + AU\right]} \tag{5.120}$$

Because the exiting gas is mostly at the bubble-phase concentration (C_b) at height H, $C' = C/C_o = C_{bH}/C_o$ as in the K-L model.

4. Partridge and Rowe Model[131]

In this model, gas flow is assumed to be divided between two phases, one comprised of bubble-cloud void units (designated the cloud phase) and the other occupying the rest of the bed (termed the interstitial phase). The flow rate (Q_c) through the cloud phase is expressed in terms of the visible bubble flow (Q_b) obtained from the two-phase theory of Toomey and Johnstone[99] in the following manner:

$$Q_c = Q_b[1 + (R - 1)\epsilon_{mf}] \tag{5.121}$$

where

$$R = \frac{V_b + V_c}{V_b} = \frac{\alpha_R}{\alpha_R - 1} \quad \text{and} \quad \alpha_R = U_b\big/\left(U_{mf}/\epsilon_{mf}\right)$$

The cross-sectional area for cloud-phase flow is given by:

$$A_c = RA\epsilon_b = RQ_b/U_b \tag{5.122}$$

The flow through the interstitial phase is

$$Q_I = Q - Q_b \tag{5.123}$$

where $Q = AU$ and the cross-sectional area for interstitial-phase flow is

$$A_I = A - A_c \tag{5.124}$$

A diagrammatic representation of the Partridge and Rowe (P-R) model is shown in Figure 5.14. The exit reactant concentration can be estimated by a method similar to that adopted in the D-H model, that is, by using the relation

$$QC = Q_c C_{cH} + Q_I C_{IH} \tag{5.125}$$

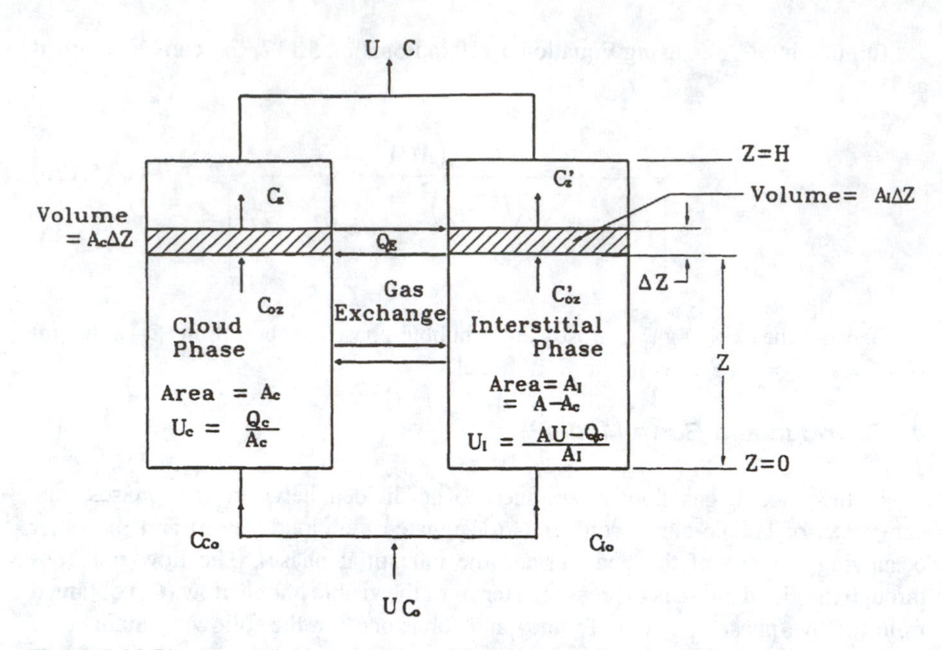

Figure 5.14 Diagrammatic representation of Partridge and Rowe model.

which requires the concentrations in the cloud phase and the interstitial phase to be evaluated by appropriate mass balance equations.

The mass balance equation across an element Δz in the cloud phase when $\Delta z \rightarrow 0$ is

$$U_c \frac{dC}{dz} + Q_E\left(C_{oz} - C_z\right) - K_c C_n = 0 \tag{5.126}$$

and for the interstitial phase is

$$U_I \frac{dC}{dz} + Q_E \frac{A_c}{A_I}\left(C'_{oz} - C'_z\right) + K_I C_n = 0 \tag{5.127}$$

where $K_c = K(R - 1)/R$ and $K_I = K$.

The pair of simultaneous Equations 5.126 and 5.127 can be solved numerically and the equations can be reduced to a linear second-order differential equation when $n = 1$ and can be solved to arrive at a solution for C:

$$C = A \exp(\alpha_c z) + B \exp(\beta_c z) \tag{5.128}$$

where $A = (L - \beta_c C_o)/(\alpha_c - \beta_c)$ and $B = (\alpha_c C_o - L)/(\alpha - \beta_c)$ are constants of integration which can be determined by using the relevant boundary conditions and α_c and β_c are the roots of the auxiliary equation pertinent to solution of the linear second-order differential equation. The mathematical details can be found in the original paper on the P-R model.[131] By making use of Equation 5.128, the exit concentration

(C) can be found after calculating C_{cH} and C_{IH} at height H and substituting for L for the evaluation of the constants A and B:

1. For the cloud phase:

$$L = L_c + \frac{1}{U_c}\left[Q_E C_{IO} - \left(Q_E + K_c\right)C_{co}\right] \tag{5.129}$$

2. For the interstitial bubble phase:

$$L = L_I = \frac{1}{U_I}\left[Q_E \frac{A_c}{A_I} C_{co} - \left(Q_E \frac{A_c}{A_I} + K_I C_{IO}\right)\right] \tag{5.130}$$

The applicability of the P-R model was tested in the case of the removal of sulfur dioxide from a gas stream at 400°C over a bed of cupric oxide supported[137] on alumina in a 10-cm-diameter fluidized bed. For a noncatalytic reaction of this type, P-R model led to good agreement between theoretical predictions and experimental results. However, the applicability of this model in the case of large-scale desulfurization has not yet been fully explored.

5. Comparison of Models

A comparative analysis of the four models described here was carried out by Chavarie and Grace[129] using their experimental data obtained in a two-dimensional reactor for the decomposition of ozone over a bed comprised of a mixture of glass beads and ferric chloride–coated alumina particles. Concentration profiles were obtained for the bubble phase by examining the absorption of an ultraviolet beam and for the emulsion phase by direct sampling. The investigations led to the following conclusions:

1. Both versions of the D-H model led to rather inaccurate estimations as a result of the high values of the interphase mass transfer rates.
2. The K-L model seemed to give the best predictions even though low interphase exchange values were used. This was attributed to the compensation provided by overestimation of the bubble volume fraction in the bed.
3. Although the K-W model led to good estimation of the bubble-phase concentration profile, it failed to predict the right values of the emulsion concentration and overall reactant conversion.
4. It appeared that the P-R model was inapplicable to two-dimensional setup as the bubble-cloud area calculated on the basis of this model exceeded the actual available area of the bed.

In view of the above observations, it was concluded that none of the four models is really satisfactory in its present form.

Later, a somewhat similar comparison was made by Stergiou et al.[130] in the context of a three-dimensional fluidized bed reactor 165 mm in diameter used to implement a complex reaction such as the ammoxidation of propylene to acrylonitrile. The results of this work can be summarized as follows:

1. Both versions of the D-H model overestimated the conversion due to overestimation of the interphase exchange coefficient.
2. The K-L model was found to be suitable for situations where $U/U_m > 7$. For $U/U_{mf} < 7$, its predictions were rather poor.
3. The K-W model led to the best predictions, provided the highest permissible mass of the catalyst was used.
4. The P-R model appeared to be acceptable for situations where $U/U_{mf} \leq 5$.

In view of these inferences, it can be concluded that none of the models is well suited for predictions over the entire operational range of a fluidized bed. A modified K-W model, which uses a reduced initial bubble size or considers the bubble wake as part of the cloud, was suggested to be acceptable. The performance of a bubble assemblage model that takes into account bubble growth and coalescence was investigated[138] with a view to simulating solid mixing and noncatalytic gas–solid reactions. The nature of the solids concentration profile for parallel as well as successive reaction was predicted by the bubble assemblage model. Estimations of concentration changes above the grid were found to be remarkable. Expansion of the experimental database in terms of the concentration changes caused by changes in the bed height and the bubble diameter was suggested for future experiments. Chiba and Kobayashi[139] proposed a bubble flow model which was conceptually akin to the K-W model but involved the calculation of bubble size in the ith compartment at height $h = (i - {}^1/_2)\,\Delta h$. This model was successfully tested for first-order irreversible catalytic reactions. However, the applicability of the model to fast and complex reactions could not be validated, mainly due to lack of experimental data.

A model analysis study by Corella and Bilbao,[140] which took into account axial variation of velocity inside the reactor due to gas expansion and also volume expansion due to the chemical reaction carried out in a fluidized bed reactor for the dehydration of 2-ethylhexanol, recommended the use of the K-L model in the context of a fluidized bed that is compartmented on the basis of bubble size variation along the bed.

Modeling studies on the chlorination[141] of rutile in a 15-cm-diameter fluidized bed reactor showed that the use of the K-W bubble assemblage model was appropriate in this case. In this specific study, no comparison with other models was made. It is generally observed from analyses of the models that it is not a simple task for any designer to choose a particular model for universal application. It is, however, clear that an improved model can be built by analyzing the pros and cons of existing ones. The K-L bubbling bed model was tested in several applications, whereas both versions of the D-H model were found to lead to unrealistic predictions due to the assumption of high mass transfer rates. The K-W bubble assemblage model takes into consideration the variation in bubble size. It should be possible to incorporate the strong points of the D-H model and the K-L model in conjunction with the concept of K-W bubble assemblage and also consider gas expansion inside the bed due to chemical reaction or axial variation of pressure inside the bed. In other words, a discretized bubble-phase model which takes into account all the positive but realistic features of models described earlier may result in better prediction. Furthermore, any model under consideration should take into consideration the grid zone for a refined approach. As pointed out earlier, the grid zone plays a vital role in determining the overall performance of a fluid bed reactor. Compartment sizing of the fluidized bed and the technique of determining the compartment size are also of great importance in modeling. Viswanathan et al.[142] proposed a compartment sizing method by defining a compartment size (Δh_i) equivalent to a hypothetical average bubble diameter (d_b) that would yield an exchange coefficient (K_{be}) equivalent to that obtained by considering all the bubbles present in that compartment.

Thus, for the *i*th compartment:

$$K_{bei}\delta_i \Delta h_i = \int_{h_{i-1}}^{h_i} K_{be}\delta \, dh \tag{5.131}$$

A graphical procedure was developed for evaluating the compartment size if the bubble sizes are actually known. Krishnamurthy and Sathiyamoorthy[143] applied the technique of compartment sizing and estimated the number of compartments and its influence on the gas-exchange coefficient, the height of each compartment, and the exit gas concentration in the specific case of ozone decomposition. The results estimated by using the four models in the case of a porous distributor for four typical flow rates ($U = 2U_{mf}$, $3U_{mf}$, $4U_{mf}$ and $5U_{mf}$) are presented in Figure 5.15.

The effect of distributor type on the extent of conversion in a fluidized bed was studied in detail by Walker.[144] His investigations showed that the initial bubble size was affected by the distributor, which in turn influenced the conversion. For a large-scale reactor where the gas flow rates would usually be high, the D-H model was recommended as the best. However, the merits of the other models were not brought out in this work. The data of Walker,[144] when applied to all four models, taking into consideration the expansion in the bed due to chemical reaction and axial pressure variation and also compartmenting the bed in serially connected sections as proposed Viswanathan et al.,[142] showed the extent of the conversion for the decomposition reaction of ozone on a catalyst bed; the results are depicted in Figure 5.16. It is evident from this figure that most models lead to unrealistic predictions in the case of a sieve plate fitted fluidized bed. Even for a porous distributor, which is supposed to generate smooth fluidization and hence maintain bed uniformity, the model predictions are not satisfactory. Thus, it is necessary to bring about further refinements in the existing models or to develop a hybrid model which would combine the merits of all the models and eliminate the discrepancies in model predictions.

E. Some Modern Models

At this point, it is appropriate to describe some models which have gained importance because of their merits pertaining to either their applicability or refinements in

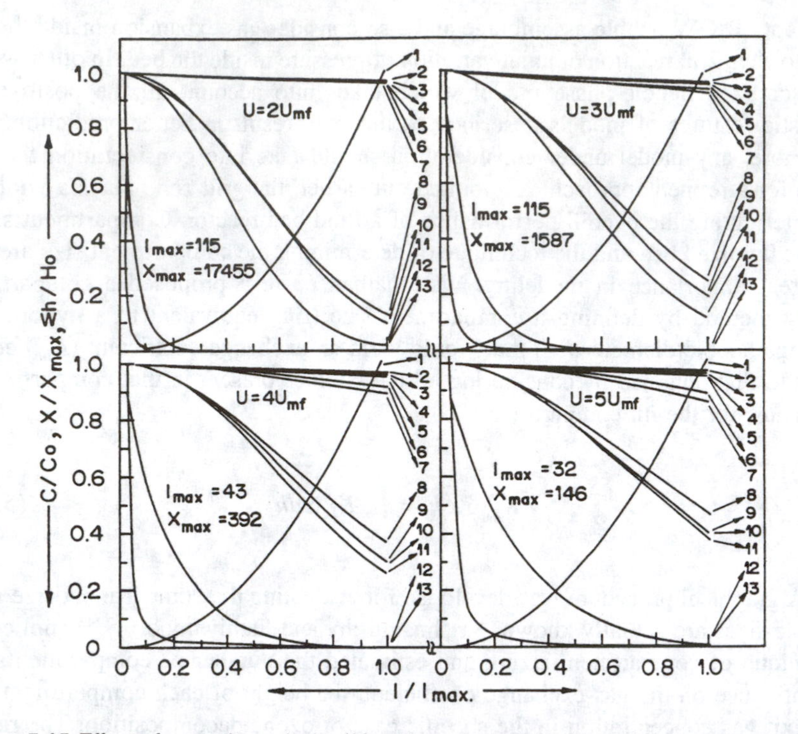

Figure 5.15 Effect of operating velocity (U) and compartment sizing (l/l_{max}) on gas-exchange rate (X/X_{max}) and exit gas conversion (C/C_o) with porous plate distributor. Key to curves: (1) $\Sigma h/H_o$; (2) bubble-phase DH-B; (3) bubble phase DH-A; (4) two-phase DH-E; (5) two-phase DH-B, E; (6) dense-phase DH-A; (7) dense-phase DH-B; (8) Kato and Wen model; (9) two-phase DH-A, C; (10) two-phase DH-B, C; (11) Kunii and Levenspiel model; (12) Partridge and Rowe model; (13) X/X_{max}. DH = Davidson Harrison model, A = perfectly mixed dense phase, B = plug flow in dense phase, E = across each compartment, and C = cumulative conversion. (From Krishnamurthy, N. and Sathiyamoorthy, D., in *Recent Trends in Chemical Reaction Engineering*, Wiley Eastern Ltd., New Delhi, 1987, 386. With permission.)

their underlying hypotheses. These models are not widely cited or discussed in the literature, although they have been proven to have considerable potential for application purposes. It should be recalled in this context that the original model prediction of May[145,146]/Van Deemter,[122] elaborated by VanSwaaij and Zuiderweg,[147] resulted in the equation

$$\frac{C}{C_o} = \exp\left[\left(-\frac{1}{N_r} + \frac{1}{N_a}\right)^{-1}\right] \tag{5.132}$$

where N_r is the number of reaction units, $= K(1-\epsilon_b)H/(U-U_{mf})$, and N_α is the number of mass transfer units = $K_m\, H/(U-U_{mf})$. For the largest possible value of N_α, namely, $N_\alpha = 4$ for industrial-scale reactors, the theoretical prediction of the extent of conversion, according to Equation 5.132, is 98% (i.e., $C/C_o = 0.02$). This shows that there is always some unreacted reactant at the exit. In reality, for many fluidized bed reactors the unreacted reactant at the exit can disappear. This contradiction is attributed to the

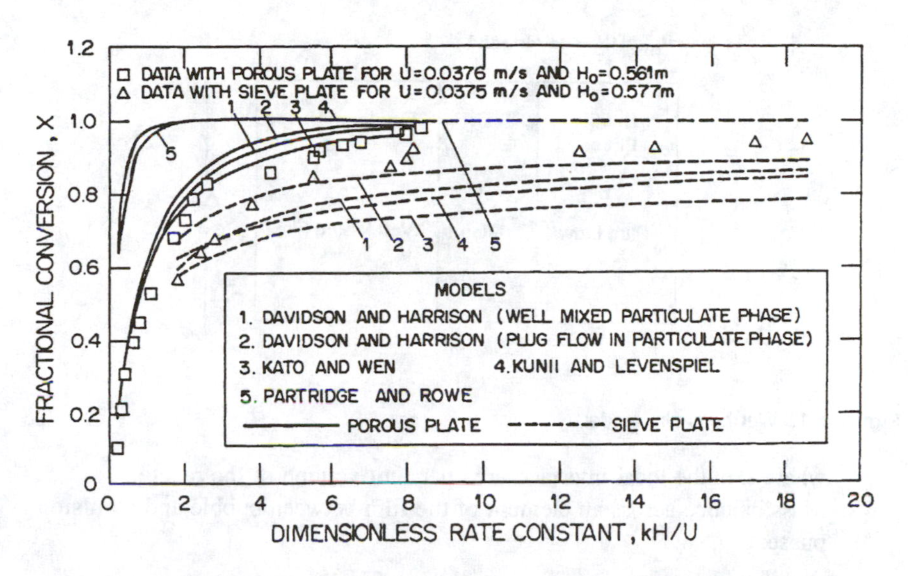

Figure 5.16 Effect of distributor on conversion of ozone in a fluidized bed. For catalyst type A, d_p = 53 µm, U_{mf} = 0.008 m/s, and ϵ_{mf} = 0.53. (From Sathiyamoorthy, D. and Vogelpohl, A., *Miner. Process. Extract. Metall. Rev.*, 12, 125, 1995. With permission.)

theory applicable only to slow reaction. The underestimation in the estimated extent of conversion was considered by Van Swaaij and Zuiderweg[147] to be also due to (1) an increased mass of the catalyst in the bubble or the cloud phase, (2) an acceleration in transport from the bubble to the dense phase by chemical reaction, and (3) the axial mass transfer profile caused by bubble growth from the distributor site onward.

1. Werther Model

Werther[148] developed a model based on the analogy of gas–liquid behavior with a gas fluidized bed. In this model, the reactant gas from the bubble phase is assumed to be transported in a manner similar to the diffusion of a gas through a thin film into a bulk of liquid in a gas–liquid reactor. This model is defined as the thin film model and the flow diagram corresponding to the model can be represented in the manner depicted in Figure 5.17.

With reference to the schematic diagram, the important features of the Werther model can be summarized as follows.

1. Mass balance in the bubble phase: Plug flow in the bubble phase is assumed, and flow through the emulsion phase, which consists of a film that offers resistance to the reactant gas from flowing into the emulsion phase, is neglected. Hence,

$$-\left(U-U_{mf}\right)\frac{\delta c_b}{\delta z}+d_G\epsilon_{mf}a\left[\frac{\delta \epsilon_e}{\delta y}\right]_{y=0}=0 \qquad (5.133)$$

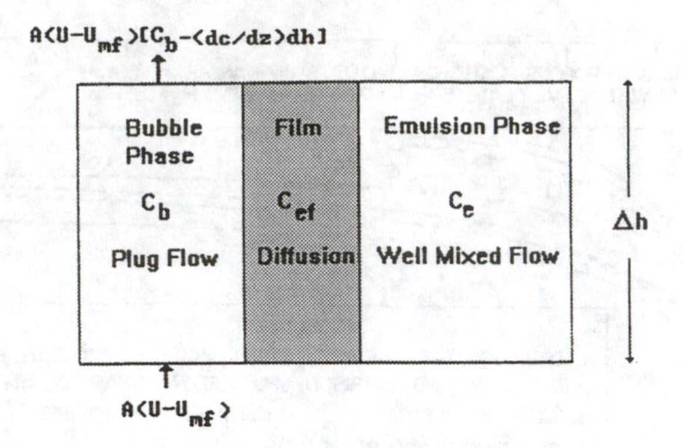

Figure 5.17 Werther's film model.

where a is the local interface area per unit volume of the reactor.

2. Mass balance across an element of the film between bubble and emulsion phase:

$$D_G \epsilon_{mf} \frac{\delta^2 C_e}{\delta y^2} - K\, C_e = 0 \tag{5.134}$$

3. Mass balance in the bulk of the emulsion phase:

$$-D_G \epsilon_{mf} \left[\frac{dC_e}{dy} \right]_{y=t_f} - K\left(1 - \epsilon_b - t_f a\right) a c_e = 0 \tag{5.135}$$

where t_f is the film thickness, given by:

$$t_f = \frac{\epsilon_{mf} D_G}{k_g} = \frac{D_{eff}}{k_g} \tag{5.136}$$

Integration of differential equations 5.133 and 5.134, where the boundary conditions are at $z = 0$, $C_b = C_o$ at $y = 0$, $C_e = C_b$; and at $y = t_f$, $C_e = C_{eb}$ (where C_{eb} is the concentration at the bulk of the emulsion phase), yields an equation for predicting the exit fraction of reactant gas:

$$C/C_o = \exp\left\{ -\left[\frac{\left(\phi^{-1}\right)\text{Ha} + \tan h\ (\text{Ha})}{\left(\phi^{-1} - 1\right)\text{Ha}\ \ \tan h\ (\text{Ha}) + 1} \right] H_\alpha N_\alpha \right\} \tag{5.137}$$

where Ha is the Hatta number, given by

$$\text{Ha} = \sqrt{\frac{k\, D_{\text{eff}}}{\text{k}_\text{g}}}$$

k is the reaction rate constant (s^{-1}), k_g is the mass transfer coefficient (m/s), D_{eff} is the effective diffusion coefficient (m^2/s), and $\phi = at_f/1 - \epsilon_b$).

In the case of a slow reaction (Ha << 1), Equation 5.137 reduces to Equation 5.132. Werther's thin film model was tested for several applications, such as low-temperature catalytic oxidation of ammonia, hydrogenation of ethylene, catalytic decomposition of nitrous oxide, and catalytic oxidation of benzene to maleic anhydride. The efficiency of the Werther model was demonstrated with regard to its applicability for fast reactions and for reactors of different diameters. This model was also analyzed by Werther and Schoessler,[42] who showed that it gives close agreement with experimental results. In view of these facts, it was suggested that heterogeneous catalytic gas-phase reactions in gas fluidized beds can be explained by this model, which was adapted from an analogy for gas–liquid behavior.

2. Fryer and Potter Model[149]

Fryer and Potter proposed a model that takes into account the backmixing in the emulsion phase; this model falls in the group of models of the convective type. The model of Fryer and Potter is referred to in the literature as the countercurrent backmixing model (CCBM). Particles similar in size to those of the cracking catalyst are often encountered in chemical industries, whereas particles with wide size distributions are quite common in metallurgical industries. This feature should be kept in mind when selecting a model. Fluidized bed modeling studies carried out by Mao and Potter[150] showed that both types of models (i.e., diffusive, known as dense-phase diffusion models, and CCBM) were applicable for fine particles (40–80 μm) and that agreement with experimental results deteriorated for coarser particles. The CCBM was suggested to be the best choice for particles of an intermediate sizes (100–200 μm) and for coarser particles. However, applicability of the model warrants further confirmation.

The CCBM considers the reversal of the emulsion gas. This model was proposed after experimental evidence of the existence of a minimum average gas concentration inside the reactor. The CCBM has not found wide application, mainly because of the difficulties associated with numerical solution of the governing equations, which correspond to a two-point boundary value problem. Development of this model will not be traced here, but a schematic representation of the CCBM is presented in Figure 5.18.

Buker[151] attempted to solve the governing equations pertinent to the Fryer and Potter model by using several numerical techniques such as shooting, superimposition, and finite difference methods. Accurate results could be obtained over an entire range of parameters by the method of finite difference. The compartment models which take into account variation in bubble size with bed height and the model modification which converts the boundary value problem to an initial value problem[152] are also relevant in the context of solving the CCBM equations. However, further studies are required to reach a conclusion as to which methods of solution of the CCBM equations are valid, especially for nonlinear reaction rate expressions and/or for nonisothermal fluid bed reactors.

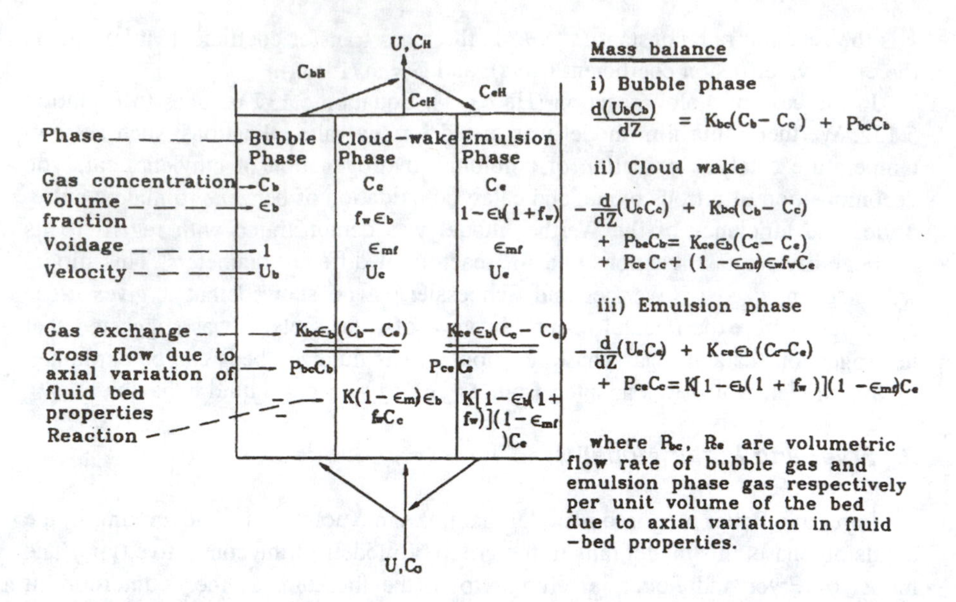

Figure 5.18 Countercurrent backmixing model: an explanation by a flow diagram.

3. *Peters et al.*[153] *Model*

Peters et al.[153] proposed a new three-phase model that takes into account the flow reversal of the emulsion gas, as does the Fryer and Potter CCBM model. These two models differ in their assumptions of boundary conditions and the division of gas flow. Peters et al.[153] model considers the variation in bubble size with height and the division of gas flow among the phases under question. A schematic of this model, with compartments in series with the three phases, namely, bubble, cloud, and emulsion phases, is shown in Figure 5.19.

4. *Slugging Bed Models*

Most of the models described so far cannot be applied to beds that operate in the slugging flow regime. The concepts and ideas used in the two-phase models or the bubbling bed models cannot be extended to slugging beds. Thus, much remains to be understood and sorted out in the area of modeling especially for fluidized bed reactors operating on slugging flow and fast fluidization conditions. Slugging occurs in a fluidized bed when the bubble size approaches the diameter of the bed. Although several criteria regarding this can be found in the literature,[154,155] the most common one[156] is expressed as

$$\mathrm{Fr}_{ms} = \frac{U_{ms}^2}{g \cdot d_p}\frac{\rho_g}{\Delta\rho} = 51.4\left(D_t/H_{mf}\right)^{1.79}\left(\rho_g/\rho_s\right)^{0.09} + 0.0042\ \mathrm{Ar}^{0.41} \tag{5.138}$$

where Fr_{ms} is the Froude number at the minimum slugging velocity.

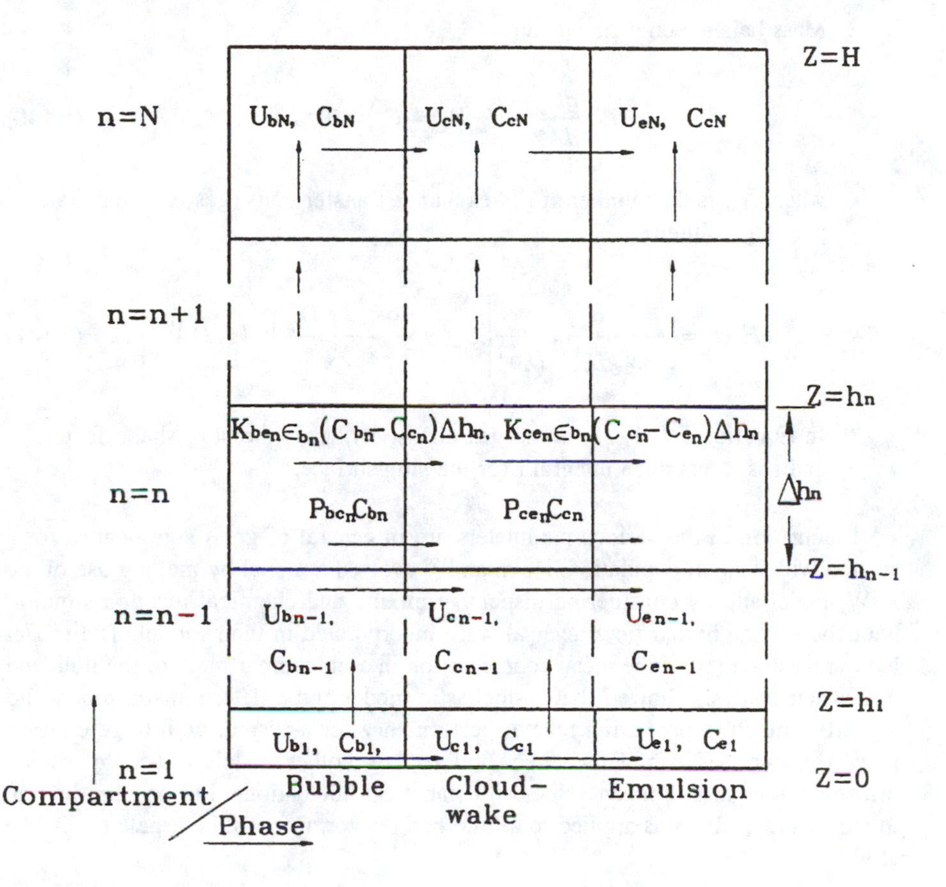

Figure 5.19 Peters et al. model.

Slugging is common in a small-diameter fluidized bed with a deep bed height. In many industrial reactors, when operated at high velocities such as the roaster used in metallurgical industries, slugging is inevitable.

A simple slugging bed model was presented by Hovmand and Davidson[157] for a catalyzed gas-phase reaction. This model was later extended by Raguraman and Potter[158] to describe the cyclic variation in a steady-state catalytic system. In the development of a model, the reaction in the slug phase is neglected and the gas flow through the emulsion phase is assumed to be at the incipient state of fluidization. The mass balances in a slugging bed can be formulated as detailed in the following.

1. Mass balance across the bed cross-section:

$$U_{mf}\frac{dC_e}{dZ^+}+\left(U-U_{mf}\right)\frac{dC_b}{dZ^+}+N_rC_e=0 \tag{5.139}$$

where $Z^+ = Z/H$ is the dimensionless bed height and N_r is the number of reaction units.

2. Mass balance on a rising slug:

$$\frac{dC_b}{dZ^+} + N_{ms}(C_b - C_e) = 0 \qquad (5.140)$$

where N_{ms} is the number of gas-exchange transfer units (gas-exchange rate per slug volume).

$$H/U_s = \frac{H_{mf}}{0.35(gD_t)^{1/2} D_t\phi_s}\left[U_{mf} + \frac{16\epsilon_{mf} I}{1+\epsilon_{mf}}\left(\frac{D_G}{\pi}\right)^{1/2} (g/D_t)^{1/4}\right] \qquad (5.141)$$

In Equation 5.141, D_G is the gas diffusivity, ϕ_σ is the slug shape factor, and I is the surface integral over the slug surface.

Fluctuation in the various parameters are in general of great significance for a system with a small population. Hovmand[159] evolved a model by making use of the K-W model and by considering dispersive mixing and chemical reaction simultaneously. Jet height and flow reversal were incorporated in their model. The model led to an estimation of the mean concentration in each compartment of the fluidized bed. Their analysis showed that a stochastic model and a deterministic model did not differ much in predicting reactor performance for a system with large entities. However, a stochastic model would be preferred over other models as it could provide information regarding mean values and statistical fluctuations. For an introduction to stochastic analysis as applied to a fluid bed reactor, refer to the article by Dabke et al.[160]

5. Stochastic Model

The fluidized bed reactor does not have constant values for many of its parameters. In reality, they fluctuate in nature between some boundary values. For example, the gas flow rate in the dense or emulsion phase and the bubble phase fluctuates and in a real sense it cannot be treated as a constant parameter. Many models treated thus far assume the basic parameters to be constant; a model based on such parameters is of the deterministic type. A model that takes into consideration the random nature of the various parameters is of the stochastic type. In bubbling bed models, the size and velocity of the bubbles in the fluidized bed play a central role. The size and spatial distribution of the bubbles determines the overall performance of the reactor. These parameters are at no time exactly repeatable. As a result, the exchange of gas between the bubble and the emulsion phase fluctuates within certain limits, and this has to be taken into account when assigning values to the relevant parameters. Buker[161] pointed out that no deterministic-type model can precisely describe actual reactor performance and thus the need to develop stochastic models. Krambeck et al.[162] proposed a stochastic mixing model for a gas fluidized bed. Despite the fact that this model could not accurately describe reactor behavior, it provided some evidence as to the influence of unsteady-state transport processes on reactor characteristics.

Ligon and Amundson[163] developed stochastic model by considering the fluctuating nature of a fluidized bed which was assumed to be composed of two well-mixed cells, one representing the bubble phase and the other the emulsion phase. A dimensionless mass transfer coefficient (K_{be} H/U), termed white stochastic noise, was evolved by computer simulation. The randomness in the mass transfer coefficient was found to affect reactor performance.

6. *Industrial-Scale Reactors*

Most of the fluidized bed reactor models were tested and simulated with reference to laboratory-scale or pilot-scale reactors and not industrial-scale reactors. Even when a model was tested on an industrial-scale reactor, the testing was confined to a set of fixed parameters which corresponded to the operating conditions of the reactor. Johnson et al.[164] evaluated the K-W model, the K-L model, and the Grace model[115] by testing them with respect to a 2.13-m-diameter industrial fluidized bed with an expanded bed height of 7.9 m. This reactor was used for the manufacture of phthalic anhydride by the catalytic oxidation of naphthalene. The models were tested with reference to a bed operated at a superficial gas velocity of 0.43 m/s using a catalyst with a mean particle diameter of 53 μm at 636 and 266 kPa. Published data from the literature were used to evaluate most of the relevant hydrodynamic parameters. Reactor performance predicted by all three models showed good agreement. The result appeared to depend crucially on the kinetic expression and the bubble diameter used. It was also shown that there was no need to pay special attention to the grid and freeboard regions as these regions exerted little influence on reactor performance. This kind of study on model testing with regard to industrial-scale reactors appeared to indicate that the bubbling bed model can be used to analyze reactor performance for complex reactions carried out on an industrial scale. However, further tests of a similar nature on large-scale reactors need be conducted in order to assess the reliability of these models.

7. *Fluidized Bed Reactor Efficiency*

The efficiency of a fluidized bed reactor can be estimated by using the gas tracer technique. A large-scale reactor performs in a different way compared to a laboratory-scale reactor. Hence, scaling up and determination of reactor efficiency are among the usual problems encountered when dealing with the design of industrial-scale fluidized bed reactors. Figure 5.20 shows qualitatively that reactor efficiency falls with increasing reactor diameter. Gas tracer experiments are used to generate data pertinent to the evaluation of reactor efficiency in large-scale reactors. The tracers used in general are nonabsorbing. In a heterogenous system such as a fluid bed reactor, the residence time data usually lead to overestimation of the actual residence time and also the conversion. Krambeck et al.[165] used adsorbing-type tracers such as SF_6 to test fluid bed reactor efficiency. SF_6 adsorptivity was varied over a wide range by changing the moisture content of the catalyst. This tracer

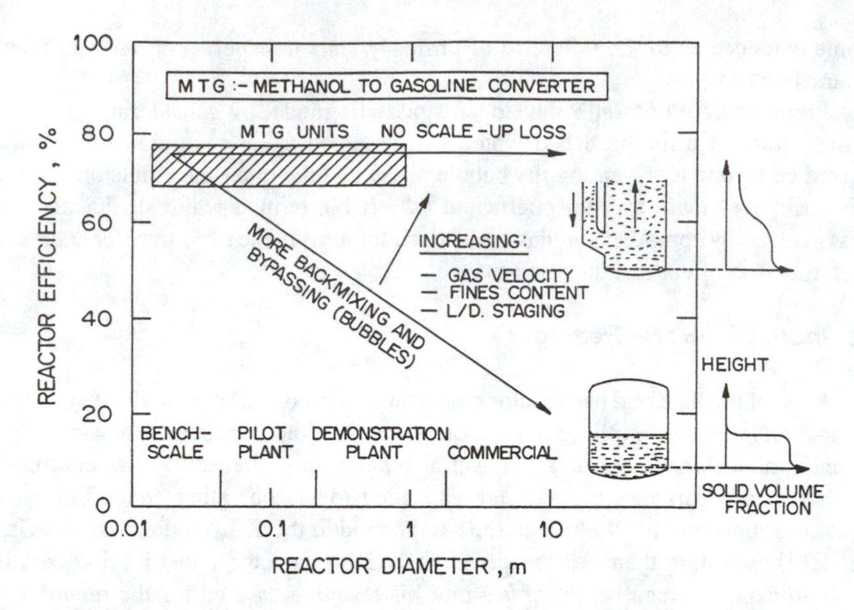

Figure 5.20 Potential effect of scaleup on fluid bed reactor efficiency. (From Krambeck, F.J., Avidan, A.A., Lee, C.K., and Lo, M.N., *AIChE J.*, 33(10), 1727, 1987. With permission.)

technique was used in scaleup studies pertaining to a methanol-to-gasoline conversion process. When the fluidized bed reactor approached homogeneity under turbulent conditions, the apparent reactor efficiency, as determined by this tracer technique, was close to the value predicted by two-phase models. The constraint associated with the use of SF_6 or a similar adsorbing catalyst is that the reactant should have the same adsorption behavior as the tracer.

Kunii and Levenspiel[166] pointed out that the gases that adsorb on the solid surface enhance the bubble–emulsion gas interchange up to ten times, and in view of this the necessary corrections should be applied to the bubbling bed model tested using an adsorbing-type tracer. A theory developed by these authors showed that the interchange coefficient is strongly affected by the gas–solid adsorption equilibrium constant. In a bubbling bed, even a small fraction of solid dispersed in a bubble acts as a carrier vehicle for the adsorbing gas and enhances the interchange coefficient significantly.

The pros and cons of fluidized bed reactor models with regard to their adoption for industrial-scale reactors were discussed by Bolthrunis.[167] He stressed the need for a fluid bed model which would be amendable to easy scaleup and be less dependent on past experience. It was suggested that a bubbling bed model can predict reactor behavior properly because many fluidized bed reactors are of the bubbling type. At the same time, it was pointed out that bubbling bed models are not appropriate to describe most of the commercial fluid bed reactors that usually operate in the turbulent regime. Plug flow models, when corrected for the kinetic constant, can perform as satisfactorily as bubbling bed models. It appears that there is no correlation available in the literature for predicting the Peclect number for a fluidized bed on the basis of simple hydrodynamic considerations. It is possible that a novel approach for formulating a fluid bed reactor model applicable to industrial reactors will evolve as a result of continued research in this area.

NOMENCLATURE

A	area of the base of reactor (m^2)
a	local interfacial area available for the transfer operation (m^2)
a'	surface area per unit volume of the bubble phase (Equation 5.25) (m^{-1})
A_b	surface area of the bubble (m)
a_b	interfacial area of bubble phase per unit bed volume (m^2/m^3)
A_c	area of the part of bed cross-section within gas clouds at any instant (m^2)
A_I	area of the part of bed cross-section outside gas clouds at any instant available for interstitial phase flow (m^2)
a_j	interfacial area of jet phase per unit bed volume (1/m)
a_o	area of an orifice (m^2)
Ar	Archimedes number ($d_p^3\ \rho_f(\rho_s - \rho_f)/g/\mu_f^2$ (–)
a_R	local interface area per unit volume of the reactor (m^{-1})
A_t	cross-section area of the column or tube (m)
C	concentration of reactant (mol/l)
c	a constant (–)
C'	concentration of the reactant in the interstitial gas phase (mol/m^3)
C_{Ae}	exit concentration of reactant A (mol/l)
C_{Ao}	initial concentration of reactant A (mol/m^3)
C_b	concentration of reactant in bubble phase (mol/m^3)
C_{bH}	concentration of the reactant in the bubble at height H (mol/l)
C_{bh}	concentration of reactant in bubble phase at height h (mol/m^3)
C_{bz}	concentration of the reactant at level z in the bed (mol/m^3)
C_c	concentration of reactant in cloud phase (mol/m^3)
C_{co}	concentration of reactant in cloud phase at $Z = 0$ (mol/m^3)
C_d	concentration of reactant in dense phase (mol/l)
C_{do}	orifice-discharged coefficient (–)
C_e	concentration of reactant in emulsion phase (mol/l)
$C_{e,n}$	concentration of reactant in emulsion phase in the nth compartment (mol/l)
c_h	concentration at height h from the surface of a fluidized bed (mol/l)
C_{IO}	concentration of interstitial phase at $Z = 0$ (mol/m^3)
C_j	concentration of the reactant in the jet phase (mol/l)
C_o	initial concentration of reactant (mol/l)
C_p	factor dependent on the arrangement of orifice spacing in Equation 5.40
c_p	specific heat at constant pressure (J/kg · K)

C_s	concentration at the surface of a fluidized bed (mol/l)
c_v	specific heat at constant volume (J/kg · K)
C_w	concentration of reactant in wake phase (mol/m^3)
D	bed diameter (m)
D_G	gas diffusivity (m^2/s)
d_B	bubble diameter (m)
d_{BM}	maximum bubble diameter (m)
d_{Bo}	initial bubble diameter (m)
d_{BS}	maximum stable bubble size (m)
D_{eff}	effective diffusion coefficient (m^2/s)
d_{noz}	nozzle diameter (m)
d_o	orifice diameter (m)
d_p	particle diameter (m)
D_s	height of fluidized zone near distributor (m)
D_t	diameter of the tube or column (m)
d_t	column or tube diameter (m)
d_v	volume diameter of bubble (m)
e	efficiency factor for a plug flow reactor in Equation 5.34 (–)
E_x	axial dispersion coefficient (m^2/s)
F	fraction of solid elutriated from the wake of the bubbles (–)
F_o	interphase mass transfer coefficient ($F_o = \pi/d_p$) (CGS units)
Fr_{ms}	Froude number at the minimum slugging velocity (–)
f_w	volume fraction of wake region, $(U - U_{mf})/U_b$ (–)
g	acceleration due to gravity (m/s^2)
H	bed height up to surface of the fluidized bed (m)
h	bed height ($<H$) (m)
H_α	height of mass transfer unit (m)
Ha	Hatta number, $\sqrt{(kD_{\text{eff}})}/(k_g)$ (–)
h/D_t	aspect ratio (–)
H_e	expanded bed height (m)
H_{eq}	equilibrium height (m)
H_{mf}	bed height at minimum fluidization velocity (U_{mf}) (m)
H_s	static bed height (m)
H_{sp}	spout height above the operating orifice (m)
H_{zh}	height of location of nozzles below the distributor for horizontal stream entry (m)
H_{zv}	height of location of nozzles below the distributor for horizontal stream entry (m)
I	surface integral over the slug surface (Equation 5.41) (–)
K	first-order reaction rate constant (s^{-1})
k_{bc}	bubble-to-cloud-phase mass transfer coefficient (s^{-1})
K_{be}	bubble-to-emulsion exchange coefficient (m/s)
K_{bw}	bubble-to-wake mass transfer coefficient (m/s)
k_c	rate constant for cloud phase in Equation 5.217 (l/s)
k_{ce}	mass transfer coefficient between cloud and emulsion phase (s^{-1})
K_D	distributor flow factor (–)

K_d distributor constant in Equation 5.46
k_g mass transfer coefficient (m/s)
K_I rate constant for interstitial gas in Equation 5.127 (s^{-1})
K_j rate constant in jet region (s^{-1})
K_r first-order rate constant per unit volume dense phase (s^{-1})
K_{we} mass transfer coefficient between wake and emulsion phase (m/s)
l depth of the orifice up to which particles spilling occurs (m)
m a constant used in Equation 5.107 (–)
m_j, m_b dimensionless mass transfer parameter as defined in Equation 5.80 (–)
N number of tuyeres or orifices (–)
n number of bubbles (–)
$\bar{n}$ exponent in Richardson–Zaki equation (–)
N_α number of mass transfer units (–)
n_b bubble frequency per orifice (s^{-1})
N_{ms} number of gas-exchange transfer units (–)
n_o number of holes per unit surface area of the distributor (–)
N_r number of reaction units (–)
P orifice spacing (pitch) (m)
P_s pressure drop at stagnation zone (kPa)
Q volumetric gas flow rate (m^3/s)
Q_b visible bubble flow rate (m^3/s)
Q_c gas flow rate through the cloud phase (m^3/s)
Q_{cH} concentration of the reactant in the cloud phase at height H (mol/m^3)
Q_E flow rate of clould gas per unit volume of cloud phase (s^{-1})
Q_I gas flow rate through interstitial area (m^3/s)
Q_{IH} concentration of the reactant in the interstitial gas phase at height H (mol/m^3)
Q_M volumetric flow rate of gas when all the gas-diffusing sites are fully operational (m^3)
Q_{or} volumetric flow rate per ofifice (m^3/s)
R ratio of volume of the displaced emulsion phase to the bubble volume (Equation 5.11) (–)
r rate of reaction (mol/l · s)
r_c volume of solids dispersed in cloud wake region, emulsion phase divided by volume of bubbles (–)
Re_M Reynolds number corresponding to U_M (–)
Re_{mf} Reynolds number at minimum fluidization velocity (U_{mf}) (–)
Re_{opt} Reynolds number corresponding to optimum gas velocity (U_{opt}) (–)
Re_p particle Reynolds number (–)
s number of coalescence of bubbles in lateral direction (–)
t distributor plate thickness (m)
t_e equilinrium time required for the bubble to reach an equilibrium height H_e (s)
t_f film thickness (m)
U operating velocity (m/s)

U_{2mf} minimum fluidization velocity for a two-phase system (m/s)
U_{3M} velocity at which all fluid distribution nozzles are fully operational in a three-phase fluidized bed (m/s)
U_{3mf} minimum fluidization velocity for a three-phase system (m/s)
U_b bubble velocity (m/s)
U_{br} rising velocity of the bubble (m/s)
U_c critical velocity for complete fluidization in Equation 5.6 (m/s)
U_{cr} critical gas velocity (m/s)
U_e emulsion phase velocity (m/s)
U_e gas velocity in the emulsion phase (m/s)
U_g operating velocity of gas (m/s)
U_I gas operating velocity through interstitital area, Q_I/A_I (m/s)
U_M operating velocity at which all the gas-issuing sites are fully operational (m/s)
U_m velocity of gas to sustain the operation of all tuyeres or orifices (m/s)
U_{mf} minimum fluidization velocity (m/s)
U_{ms} minimum slugging velocity (m/s)
U_o velocity through the orifice (m/s)
$U_{o,m}$ orifice velocity when $U = U_M$ in the bed (m/s)
U_{opt} optimum gas velocity (m/s)
U_s solid circulation velocity (m/s)
U_t terminal velocity (m/s)
V volume of the reactor (m^3)
v volumetric flow rate of reactant (m^3/s)
V_b bubble volume (m^3)
V_o hole velocity (m/s)
W solids weeping rate (kg/m^2/s)
W_{bed} weight of the bed (kg)
$W_{s,\text{idl}}$ ideal shaft work for a compressor under adiabatic reversible operation, when kinetic and potential energy effects are negligible (J)
x conversion (–)
X_A conversion of reactant gas A (–)
x_o fraction of orifice flow forming a visible bubble (–)
Z height above the distributor (m)
z depth up to which solid particles penetrate an orifice of diameter d_o (m)
Z^+ dimensionless bed height (–)

Greek Symbols

α velocity ratio, U/U_{mf} (–)
α_d distributor parameter (Equation 5.15) (–)
α_R bubble interstitial gas velocity ration, $U_b/(U_{mf}/\epsilon_{mf})$ (–)
α' parameter in Equation 5.68 as defined by Equation 5.69
β fraction of gas flow in bubble or jet phase (–)

β_d distributor parameter in Equation 5.45 for two-phase fluidization (–)
β_3' distributor parameter in Equation 5.55 for three-phase fluidization (–)
γ ratio of specific heats at constant pressure and constant volume (–)
γ_b volume of solids dispersed in bubbles divided by volume of bubbles (–)
γ_c volume of solids dispersed in cloud wake region, emulsion phase (m^3)
γ_e volume fraction of solid in emulsion phase (m^3)
ΔP_b pressure drop across the bed (kPa)
$\Delta P_{b,mf}$ pressure drop across the bed at minimum fluidization velocity (kPa)
ΔP_d pressure drop across the distributor (kPa)
ΔP_{ds} pressure drop caused by stagnant zones (kPa)
$\Delta P_{d,z}$ excess pressure drop required by nonoperating orifice (kPa)
δ_b bubble volume fraction (–)
δ_c cloud volume fraction (–)
ϵ voidage at the operating velocity (U) (–)
ϵ_b bubble holdup (–)
ϵ_b volume area fraction of bubble phase (–)
ϵ_e voidage in emulsion phase (–)
ϵ_g gas holdup (–)
ϵ_j area fraction of the jet phase (–)
ϵ_l liquid holdup (–)
ϵ_M voidage in a gas fluidized bed when all the gas-diffusing sites are fully operational (–)
ϵ_{mf} voidage at the minimum fluidization velocity (U_{mf}) (–)
ϵ_s solid holdup (–)
η dimensionless parameter (Equation 5.98) (–)
λ constant in Equation 5.25
μ_l viscosity of liquid (kg/m · s)
ρ_b density of particle (kg/m^3)
ρ_f density of fluid (kg/m^3)
ρ_g density of gas (kg/m^3)
ρ_l liquid density (kg/m^3)
ρ_s density of solid (kg/m^3)
σ solid fraction (–)
τ residence time (s)
ϕ fraction of free open area (–)
ϕ a $t_f/(1 - \epsilon_b)$ (Equation 5.137)
ϕ_s slug shape factor in Equation 5.41
Ψ nonequality factor in Equation 5.26

REFERENCES

1. Siegel, R., Effect of distributor plate to bed resistance ratio on onset of fluidized bed channelling, *AIChE J.*, 22, 590, 1976.

2. Richardson, J.F. and Zaki, W.N., Sedimentation and fluidization, *Trans. Inst. Chem. Eng.*, 32, 35, 1954.
3. Shi, Y.R. and Fan, L.T., Effect of distributor to bed resistance on uniformity of fluidization, *AIChE J.*, 30, 860, 1984.
4. Fakhimi, S. and Harrison, D., Multi-orifice distributors in fluidized beds — a guide to design, in *Chemeca-70*, Butterworths, Australia; Inst. Chemical Engineers Symp. Ser. 33, London, 1970, 29.
5. Sathiyamoorthy, D. and Rao, Ch.S., Gas distributors in fluidized bed, *Powder Technol.*, 20(2), 47, 1978.
6. Sathiyamoorthy, D. and Rao, Ch.S., Multi-orifice distributors in gas fluidized beds — a model for design of distributors, *Powder Technol.*, 24, 215, 1979.
7. Whitehead, A.B. and Dent, D.C., Behaviour of multiple tuyere assemblies, in *Proc. Int. Symp. on Fluidization*, Drinkenburg, A.A.H., Ed., Netherlands University Press, Amsterdam, 1967, 802.
8. Sathiyamoorthy, D. and Rao, Ch.S., The choice of distributor to bed pressure drop ratio in gas fluidized beds, *Powder Technol.*, 30, 139, 1981.
9. Mori, S. and Moriyama, A., Criteria for uniform fluidization of nonaggregative particle, *Int. Chem. Eng.*, 18(2), 245, 1978.
10. Chyang, C.S. and Huang, C.C., Pressure drop across a perforated plate distributor in a gas-fluidized bed, *J. Chem. Eng. Jpn.*, 24(2), 249, 1991.
11. Baskakov, A.P., Tuponogov, V.G., and Philippovsky, N.F., Uniformity of fluidization on a multi-orifice gas distributor, *Can. J. Chem. Eng.*, 63, 886, 1985.
12. Stephens, G.K., Sinclair, R.J., and Potter, O.E., Gas exchange between bubbles and dense phase in a fluidized bed, *Powder Technol.*, 1, 157, 1967.
13. Sathiyamoorthy, D., Rao, Ch.S., and Raja Rao, M., Prediction of operating velocity in a gas–solid fluidized bed system, *Indian Chem. Eng.*, 27(4), 25, 1985.
14. Kassim, W.M.S., Flowback of Solids through Distribution Plate of Fluidized Bed, Ph.D. thesis, University of Aston, Birmingham, 1972; cited in Botterill, J.S.M., Ed., *Fluid Bed, Heat Transfer*, Academic Press, London, 1975, 81.
15. Brien, C.L., Bergougnov, M.A., and Baker, C.G.J., Leakage of solids (weeping and dumping) at the grid of 0.6m diameter gas fluidized beds, in *2nd Int. Conf. on Fluidization*, Cambridge University Press, Cambridge, 1978.
16. Whitehead, A.B., Dent, D.C., and Gartside, G., Flow and pressure maldistribution at the distributor level of gas–solid fluidized bed, *Chem. Eng. J.*, 1, 175, 1970.
17. Brien, C.L., Bergougnov, M.A., and Baker, C.G.J., Grid leakage (weeping and dumping particle back flow) in gas fluidized beds, in *Fluidization*, Grace, J.R. and Metsen, J.M., Eds., Plenum Press, New York, 1980, 201.
18. Serviant, G.A., Grid leakage in fluidized beds, *Can. J. Chem. Eng.*, 48, 496, 1970.
19. Otero, A.R. and Munoz, R.C., Fluidized bed gas distributors of bubble cap type, *Powder Technol.*, 9, 279, 1974.
20. Biddulph, M.W., Oscillating behaviour on distillation trays. II, *AIChE J.*, 21, 41, 1975.
21. Levoy, P., Cohen, De., and Lara, G., *Rev. Metall.*, 55, 75, 1958.
22. Sathiyamoorthy, D., Studies on Multi-Orifice Distributors in Gas–Solid Fluidized Bed, Ph.D. thesis, IIT, Bombay, 1983.
23. Sathiyamoorthy, D., Rao, Ch.S., and Raja Rao, M., Flow back of solids through distributors in gas–solid fluidized beds, in *Proc. Int. Symp. on Recent Advances in Particulate Science and Technology,* IIT, Madras, 1982, 675.
24. Harrison, D., Davidson, J.F., and de Kock, J.W., On the nature of aggregative and particulate fluidization, *Trans. Inst. Chem. Eng.*, 39, 202, 1961.

25. Davidson, J.F. and Harrison, D., *Fluidized Particles*, Cambridge, University Press, Cambridge, 1963.
26. Kunii, D. and Levenspiel, O., *Fluidization Engineering*, John Wiley, New York, 1969.
27. Mori, S. and Wen, C.Y., Estimation of bubble diameter in gaseous fluidized beds, *AIChE J.*, 21, 109, 1975.
28. Sathiyamoorthy, D., Rao, Ch.S., and Raja Rao, M., Effect of distributors on heat transfer in gas fluidized beds, *Chem. Eng. J.*, 37, 149, 1988.
29. Mazumdar, P. and Ganguly, U.P., Effect of aspect ratio on bed expansion in particulate fluidization, *Can. J. Chem. Eng.*, 63, 850, 1985.
30. Porter, K.E., Quasion, H.A., Hassan, A.O., and Aryan, F.A., Gas distribution in shallow packed beds, *Ind. Eng. Chem. Res.*, 32(10), 2408, 1993.
31. Qureshi, A.E. and Creasy, D.E., Fluidized bed gas distributors, *Powder Technol.*, 22, 113, 1979.
32. Geldart, D. and Baeyens, J., The design of distributors for gas fluidized beds, *Powder Technol.*, 42, 67, 1985.
33. Geldart, D. and Kelsey, J.R., in Proc. of Tripartite Chem. Eng. Conf., Montreal Inst. Chem. Eng., London, 1968, 90.
34. Agarwal, J.C., Davis, W.L., and King, D.T., Fluidized bed coal dryer, *Chem. Eng. Prog.*, 58(11), 85, 1962.
35. Whitehead, A.B., in *Fluidization*, Davidson, J.F. and Harrison, D., Eds., Academic Press, London, 1971, chap. 19.
36. Saxena, S.C., Sathiyamoorthy, D., and Sundaram, C.V., Design principles and characteristics of distributors in gas fluidized beds, in *Transport Processes in Fluidized Bed Reactors*, Doraiswamy, L.K. and Kulkarni, B.P., Eds., Wiley Eastern Ltd., New Delhi, 1987, chap. 6.
37. Sathiyamoorthy, D. and Vogelpohl, A., On the distributors and design criteria for gas–solid and gas–liquid–solid fluidized beds, *Miner. Process. Extract. Metall. Rev.*, 12, 125, 1995.
38. Botterill, J.S.M., *Fluid-Bed Heat Transfer*, Academic Press, London, 1975.
39. Basu, P., Design of distributors for fluid bed boilers, in *Fluidized Bed Boilers: Design and Application*, Basu, P., Ed., Pergamon Press, New York, 1984, 45.
40. Porter, K.E., Ali, Q.H., Hassan, O.A., and Aryan, A.F., Gas distribution in shallow packed beds, *Ind. Eng. Chem. Res.*, 32(10), 2408, 1993.
41. de Lasa, H., Role of end zones in the design and operation of fluidized catalytic reactors, in *Transport Processes in Fluidized Bed Reactors*, Doraiswamy, L.K. and Kulkarni, B.D., Eds., Wiley Eastern Ltd., New Delhi, 1987, chap. 7.
42. Werther, J. and Schoessler, M., The analogy between bubbling fluidized beds and gas/liquid systems as a basis for modeling technical fluid bed reactors, in *Transport Processes in Fluidized Bed Reactors*, Doraiswamy, L.K. and Kulkarni, B.D., Eds., Wiley Eastern Ltd., New Delhi, 1987, chap. 3.
43. Werther, J., Effect of gas distributor on hydrodynamics of gas fluidized beds, *Ger. Chem. Eng.*, 1, 168, 1978.
44. Yates, J.G., Rowe, P.N., and Cheesman, D.J., Gas entry effects in fluidized bed reactors, *AIChE J.*, 30(6), 890, 1984.
45. Rowe, P.N., Prediction of bubble size in gas fluidized bed, *Chem. Eng. Sci.*, 31, 285, 1976.
46. Baur, W., Dr. Ing. dissertation, University of Erlangen-Nurnburg, Germany, 1980.
47. Asif, M., Kalogerakis, N., and Behie, L.A., Hydrodynamics of liquid fluidized beds including distributor region, *Chem. Eng. Sci.*, 47(15/16), 4155, 1992.

48. Alvarez-Cuenca, M., Oxygen Mass Transfer in Bubble Columns and Three Phase Fluidized Beds, Ph.D. thesis, University of Western Ontario, Canada, 1979.
49. Alvarez-Cuenca, M., Nerenberg, M.A., and Asfour, A.F.A., Mass transfer effects near distributor of three phase fluidized beds, *Ind. Eng. Chem. Fundam.*, 23, 381, 1984.
50. Asfour, A.F.A. and Nhaesi, A.H., An improved model for mass transfer in three phase fluidized beds, *Chem. Eng. Sci.*, 45, 2895, 1990.
51. Lee, S.L.P., Soria, A., and de Lasa, H.I., Evolution of bubble length distribution in three phase fluidized bed, *AIChE J.*, 36, 1763, 1990.
52. Tang, W.T. and Fan, L.S., Gas–liquid mass transfer in a three phase fluidized bed containing low density particles, *Ind. Eng. Chem. Res.*, 29, 128, 1990.
53. Jamialahmadi, M. and Muller-Steinhagen, H., Effect of solid particles on gas holdup in bubble columns, *Can. J. Chem. Eng.*, 69, 390, 1991.
54. Kimura, T. and Kojima, T., Evaluation of contribution of grid zone in silicon production via fluidized bed CVD-monosilane and argon system, *Kogaku Ronbunshu*, 17, 81, 1991.
55. Saxena, S.C. Upadhyay, S.N., and Ravetto, F.T., Performance characteristics of multi jet tuyeres distributor plate, *Powder Technol.*, 30, 155, 1981.
56. Fan, L.S., *Gas–Liquid–Solid Fluidization Engineering*, Butterworths, Boston, 1989.
57. Hirata, A., Hosaka, Y., and Umezawa, H., Characteristics of simultaneous utilization of oxygen and substrate in a three phase fluidized bed bioreactor, *J. Chem. Eng. Jpn.*, 23(39), 303, 1990.
58. Farag, I.H., Ettouney, H.M., and Raj, C.B.C., Modelling of ethanol bioproduction in three phase fluidization bed reactors, *Chem. Eng. Commun.*, 79, 47, 1989.
59. Costa, E., Lucas, A.D., and Garela, P., Fluid dynamics of gas–solid–liquid fluidized beds, *Chem. Prog. Des. Dev.*, 25, 849, 1986.
60. Dhanuka, V.R. and Stepanek, J.B., Simultaneous measurement of interfacial area and mass transfer coefficient in three phase fluidized beds, *AIChE J.*, 26, 1029, 1980.
61. Kim, J.O. and Kim S.P., Gas–liquid mass transfer in a three phase fluidized bed with floating bubble breakers, *Can. Chem. Eng.*, 68, 368, 1990.
62. Fan, L.S., Bavarian, F., Gorowara, R.L., and Kriescher, B.E., Hydrodynamics of gas–liquid–solid fluidization under high hold-up conditions, *Powder Technol.*, 53, 285, 1987.
63. Fan, L.S., Muroyama, K., and Chern, S.H., Hydrodynamics characteristics of inverse fluidization in liquid–solid and gas–liquid–solid system, *Chem. Eng. J.*, 24, 143, 1982.
64. Kono, H., A new concept for three phase fluidized beds, *Hydrocarbon Process.*, 59(1), 123, 1980.
65. Saxena, S.C., Vadivel, R., and Saxena, A.C., Gas holdup and heat transfer from immersed surfaces in two and three phase systems in bubble columns, *Chem. Eng. Commun.*, 85, 6, 1989.
66. Hiby, J.W., Critical minimum pressure drop of gas distribution plate in fluidized bed units, *Chem. Ing. Tech.*, 36, 328, 1964.
67. Hiby, J.W., Mindestdruckverlust des Anstrombodens bei Fluidatbetten, *Chem. Ing. Tech.*, 39(10), 1125, 1967.
68. Gregory, S.A., The distributors plate problem, in *Proc. Int. Symp. on Fluidization*, Drinkenburg, A.A.H., Ed., Netherlands University Press, Amsterdam, 1967, 751.
69. Rowson, H.M., Fluid bed adsorption of carbon-disulphide, *Br. Chem. Eng.*, 8(3), 180, 1963.
70. Baskakov, A.P. and Kozin, V.E., On the investigation of gas distribution in granular material fluid-bed set up, *Khim. Prom.*, 6, 448, 1965.

71. Gvozdev, V.D., Salnikov, A.A., Fomichev, A.G., Tickhonov, V.A., and Vasilnov, A.S., Design and construction of equipment with a fluidized bed granular material. 1. Gas distributor grids, *Int. Chem. Eng.*, 3(4), 562, 1963.
72. Richardson, D.R., How to design fluid flow distributors, *Chem. Eng.*, 68, 83, 1961.
73. Zenz, F.A., Minimize manifold pressure drop, *Pet. Refiner,* 41(12), 125, 1962.
74. Avery, D.A. and Tracey, D.H., The application of fluidized beds of activated carbon to solvent recovery from air or gas streams, Tripartite Chem. Eng. Conf. on Fluidization, Montreal, Inst. Chem. Eng., London, 1968, 28.
75. Medlin, J., Wong, H.W., and Jackson, R., Fluid mechanical description of fluidized bed, convective instability in bounded beds, *Ind. Eng. Chem. Fundam.*, 13, 247, 1974.
76. Zuiderweg, F.J., Design report on fluidization, in *Proc. Int. Conf. on Fluidization,* Drinkenburg, A.A.H., Ed., Netherlands University Press, Amsterdam, 1967, 751.
77. Otero, A.R., Rof, J.S., and Lago, E.C., Fluidized bed calcination of uranyl nitrate solutions, in *Proc. Int. Conf. on Fluidization,* Drinkenburg, A.A.H., Ed., Netherlands University Press, Amsterdam, 1967, 769.
78. Whitehead, A.B. and Dent, D.C., Behaviour of multiple tuyere assemblies, in *Proc. Int. Symp. on Fluidization*, Drikenburg, A.A.H., Ed., Netherland University Press, Amsterdam, 1967, 802.
79. Zabrodsky, S.S., *Hydrodynamics Heat Transfer in Fluidized Bed*, MIT Press, Cambridge, MA, 1966.
80. Zabrodsky, S.S., On operation of gas distributor in fluidized bed equipment, in Proc. Int. Symp. on Fluidization and its Application, Toulouse, France, 1973, 69.
81. Zenz, F.A., Bubble formnation and grid design, in Tripartite Chem. Eng. Conf. Session 32, Montreal, 1968, 136.
82. Kelsey, J.R., Discussion of papers presented at the first session of symposium on fluidization, *Inst. Chem. Eng. Symp. Ser.*, 30, 86, 1968.
83. Baskakov, A.P., Tuponogov, V.G., and Phillippovsky, N.F., Uniformity of fluidization on multi-orifice distributor, *Can. J. Chem. Eng.*, 63, 886, 1985.
84. Whitehead, A.B., Some problems in large scale fluidization, in *Fluidization*, Davidson, J.F. and Harrison, D., Eds., Academic Press, London, 1971, 781.
85. Fakhimi, S., Sorabhi, M., and Harrison, D., Entrance effects at a multiorifice distributor in gas fluidized beds, *Can. J. Chem. Eng.*, 61, 364, 1983.
86. Yue, P.L. and Kolaczkowski, J.A., Multi-orifice distributor design for fluidized beds, *Trans. Inst. Chem. Eng.*, 60, 164, 1982.
87. Geldart, D., The effect of particle size distribution on the behaviour of gas fluidized beds, *Powder Technol.*, 6, 201, 1972.
88. Sathiyamoorthy, D., Geissen, S., and Vogelpohl, A., Optimum design approach for gas fluidized bed, *Miner.Process. Extract. Metall. Rev.*, 9(1–4), 233, 1992.
89. Begovich, J.M., Hydrodynamics of Three Phase Fluidized Beds, M.Sc. thesis, Oak Ridge National Laboratory/TM 6448, 1978.
90. Thonglimp, V., Docteur-Ingenieur thesis, Institut National Polytechnique, Toulouse, France, 1981.
91. Schürgel, K., Biofluidization: application of the fluidization technique in biotechnology, *Can. J. Chem. Eng.*, 58, 299, 1980.
92. Kim, P., Luerweg, M., Striegel, B., Aivasidis, A., and Wandrey, C., Typical application and scale up of fluidized bed reactors with porous sintered glass spheres in anaerobic waste water treatment, *Chem. Ing. Tech.*, 66, 336, 1990.
93. Sathiyamoorthy, D., Geissen, S., and Vogelpohl, A., A design approach for liquid distributor in fluidized bed bio-reactor, in *4th World Congress of Chem. Engineering*. Behrens, D., Ed., DECHEMA Kartsruche, Germany, 1991, Session 7–2, Paper 14.

94. Litz, W.J., Design of gas distributors, *Chem. Eng.*, 79(25), 162, 1972.
95. Kage, H., Yamaguchi, N., and Matsuno, Y., Frequency analysis of pressure fluctuations in fluidized bed plenum, *J. Chem. Eng. Jpn.*, 24(1), 76, 1991.
96. Yang, X.T., Horne, D.G., Yates, J.G., and Rowe, P.N., Distributor-zone reaction in a gas fluidized bed, *AIChE Symp. Ser.*, 80(241), 41, 1984.
97. Behie, L.A. and Kehoe, P., The grid region in a fluidized bed reactor, *AIChE J.*, 19, 1070, 1973.
98. Grace, J.R. and de Lasa, H.I., Reaction near the grid in fluidized beds, *AIChE J.*, 24, 364, 1978.
99. Toomey, R.D. and Johnstone, H.F., Gas fluidization of solid particles, *Chem. Eng. Prog.*, 48, 220, 1952.
100. Miyauchi, T. and Furusaki, S., Relative contribution of variables affecting the reaction in fluid bed contractors, *AIChE J.*, 20, 1087, 1974.
101. Miyauchi, T. and Furusaki, S., Concept of successive contact mechanism for catalytic reaction in fluid beds, *J. Chem. Eng. Jpn.*, 7, 201, 1974.
102. Furusaki, S., Kikuchi, T., and Miyauchi, T., Axial distribution of reactivity inside a fluid bed contactor, *AIChE J.*, 22, 354, 1976.
103. Yates, J.G. and Rowe, P.N., A model for chemical reaction in the freeboard region above a fluidized bed, *Trans. Inst. Chem. Eng.*, 55, 137, 1977.
104. Chen, L.H. and Wen, L.Y., Model of solid gas reaction phenomena in the fluidized bed free board, *AIChE J.*, 28, 1019, 1982.
105. Wen, C.Y. and Chen, L.H., Fluidized bed free board phenomena. Entrainment and elutriation, *AIChE J.*, 28, 117, 1982.
106. de Lasa, H.I. and Grace, J.R., The influence of the freeboard region in a fluidized bed catalytic cracking regenerator, *AIChE J.*, 25, 984, 1979.
107. George, S.E. and Grace, J.R., Entrainment of particles from aggregative fluidized beds, *AIChE J. Symp. Ser.*, 74(176), 67, 1978.
108. Grace, J.R. and de Lasa, H.I., Reaction near the grid in fluidized beds, *AIChE J.*, 24(2), 364, 1978.
109. Behie, L.A. and Kehoe, P., The grid region in a fluidized bed reactor, *AIChE J.*, 19, 1070, 1973.
110. Merry, J.M.D., Fluid and particle entrainment into vertical jets in fluidized beds, *AIChE J.*, 22, 315, 1976.
111. Errazu, A.F., de Lasa, H.I., and Sarti, F., A fluidized bed catalytic cracking regenerator model: Grid effects, *Can. J. Chem. Eng.*, 57, 191, 1979.
112. Grace, J.R., An evaluation of models for fluidized bed reactors, *AIChE Symp. Ser.*, 67, 159, 1971.
113. Horio, M. and Wen, C.Y., An assessment of fluidized bed modelling *AIChE Symp. Ser.*, 73, 9, 1977.
114. LaNauze, R.D., Fundamentals of coal combustion in fluidized beds, *Chem. Eng. Res. Des.*, 63, 3, 1985.
115. Grace, R., Generalized models for isothermal fluidized bed reactors, in *Recent Advances in the Engineering Analyses of Chemically Reacting Systems*, Doraiswamy, L.K., Ed., Wiley Eastern Ltd., New Delhi, 1984, chap. 13.
116. Wen, C.Y., Flow regimes and flow models for fluidized bed reactors, in *Recent Advances in the Engineering Analyses of Chemically Reacting Systems*, Doraiswamy, L.K., Ed., Wiley Eastern Ltd., New Delhi, 1984, chap. 14.
117. Yates, J.G., Modelling fluidized bed reactors, in *Transport Processes in Fluidized Bed Reactors*, Doraiswamy, L.K. and Kulkarni, B.D., Eds., Wiley Eastern Ltd., New Delhi, 1987, chap. 2.

118. Buker, B.D., Modelling of nonisothermal fluid bed reactors, in *Transport Processes in Fluidized Bed Reactors*, Doraiswamy, L.K. and Kulkarni, B.D., Eds., Wiley Eastern Ltd., New Delhi, 1987, chap. 5.
119. Chen, L.H. and Too, J.R., Fluidized bed models, in *Encyclopedia of Fluid Mechanics*, Vol. 4, Cheremisioff, N.P., Ed., Gulf Publishing, Houston, 1985, chap. 38.
120. Shen, C.Y. and Johnstone, H.F., Gas–solid contact in fluidized beds, *AIChE J.*, 1, 349, 1955.
121. Lewis, W.K., Gilliland, E.R., and Glass, W., Solid catalysed reaction in a fluidized bed, *AIChE J.*, 5, 419, 1959.
122. Van Deemter, J.J., Mixing and contacting in gas–solid fluidized beds, *Chem. Eng. Sci.*, 13, 143, 1961.
123. Calderbank, P.H. and Toor, F.D., Reaction kinetics in gas fluidized catalyst beds–experimental, in *Proc. Int. Symp. on Fluidization*, Drikenburg, A.A.H., Ed., Netherlands University Press, Amsterdam, 1967, 652.
124. Kato, K. and Wen, C.Y., Bubble assemblage model for fluidized catalytic cracker, *Chem. Eng. Sci.*, 24, 135, 1969.
125. Miyauchi, T. and Morooka, S., Mass transfer rate between bubble and emulsion phase in fluid bed, *J. Chem. Eng. Jpn.*, 33, 369, 1969.
126. Gwyn, J.E., Moser, J.H., and Parker, W.A., A three phase model for gas–solid fluidized beds, *Chem. Eng. Prog. Symp. Ser.*, 66, 19, 1970.
127. Fryer, C. and Potter, O.E., Experimental investigation of models for fluidized bed catalytic reaction, *AIChE J.*, 22, 38, 1976.
128. Wether, J., Flow fundamental of fluidized-bed engineering, Werther, Joachim, *Chem. Ing. Tech.*, 49(3), 193, 1977.
129. Chavasrie, C. and Grace, J.R., Performance analysis of a fluidized bed reactor. III. Modification and extension of conventional two phase models, *Ind. Eng. Chem. Fundam.*, 14(2), 86, 1975.
130. Stergiou, L., Laguerie, C., and Gilot, B., A discrimination between some fluidized bed reactor models for ammoxidation of propylene to acrylonitrile, *Chem. Eng. Sci.*, 39(2), 713, 1984.
131. Partridge, B.A. and Rowe, P.N., Chemical reaction in bubbling gas-fluidized bed, *Trans. Inst. Chem. Eng.*, 44, T335, 1966.
132. Orcutt, J.C., Davidson, J.F., and Pigford, R.E., Reaction time distributions in fluidized catalytic reactors, *Chem. Eng. Prog. Symp. Ser.*, 58(38), 1, 1962.
133. Rowe, P.N. and Partridge, B.A., A x-ray study of bubbles in fluidized beds, *Trans. Inst. Chem. Eng.*, 43, 157, 1965.
134. Kobayashi, H., Arai, F., and Chiba, T., Behavior of bubbles in gas fluidized bed, *Chem. Eng. Tokyo*, 29, 858, 1965.
135. Cooke, M.J., Harris, W., Highley, J., and William, D.F., Kinetics of oxygen consumption in fluidized bed carbonizers in Proc. Tripartite Chem. Eng. Conf., Montreal, 1968, 21.
136. Kobayashi, H., Arai, F., and Chiba, T., Determination of gas cross-flow coefficient between the bubble and emulsion phases by measuring residence-time distribution of fluid in a fluidized bed, *Chem. Eng. Tokyo*, 31, 239, 1967.
137. Best, J.R., and Yates, J.G., Removal of sulfur dioxide from a gas stream in a fluidized bed reactor, *Ind. Eng. Chem. Process Des. Dev.*, 16(3), 347, 1977.
138. Yoshida, K. and Wen, C.Y., Behaviour of fluidized beds based on the bubble assemblage model, *AIChE Symp. Ser.*, 67(116), 151, 1971.
139. Chiba, T. and Kobayahsi, H., Modeling of catalytic fluidized bed reactors, in *Proc. Int. Conf. Fluidization and Its Application*, Cepaudes, Toulouse, France, 1973, 468.

140. Corella, J. and Bilbao, R., The effect of variation of the gas volume and of the bubble size on the conversion in a fluidized bed, *Int. Chem. Eng.*, 24(2), 302, 1984.
141. Youn, I.J. and Park, K.Y., Modeling of fluidized bed chlorination of rutile, *Metall. Trans. B*, 20B, 959, 1989.
142. Viswanathan, K., Ramakrishnan, T.S., and Subba, Rao, D., Compartment sizing for fluidized bed reactors, *Indian Chem. Eng.*, 24(4), 28, 1982.
143. Krishnamurthy, N. and Sathiyamoorthy, D., Effect of compartment sizing and distributors on the models of gas fluidised beds, in *Recent Trends in Chemical Reaction Engineering*, Kulkarni, B.D., Mashelkar, R.A., and Sharma, M.M., Eds., Wiley Eastern Ltd., New Delhi, 1987, 386.
144. Walker, B.V., The effect of distributor type on conversion in a bubbling fluidized bed, *Chem. Eng. J.*, 9, 49, 1975.
145. May, W.G., Fluidized bed reactor studies, *Chem. Eng. Prog.*, 55, 49, 1959.
146. May, W.G., Dechema Monagraph 32, Verlag Chemie, Wienheim, 1959, 471.
147. Van Swaaij, W.P.M. and Zuiderweg, F.J., Generating hydrogen-rich gas, in Proc. 5th Eur. Symp. on Reaction Engineering, Amsterdam, 1972, 89.
148. Werther, J., Modelling and scale up of industrial fluidized bed reactor, *Chem. Eng. Sci.*, 35, 372, 1980.
149. Fryer, C. and Potter, O.E., Counter current back mixing model for fluidized bed catalytic reactors. Applicability of simplified solutions, *Ind. Eng. Chem. Fundam.*, 11, 338, 1972.
150. Mao, Q.M. and Potter, O.E., Fluidized bed reactor modelling, *AIChE Symp. Ser.*, 80(241), 65, 1984.
151. Buker, D.B., The counter-current backmixing model for fluid bed reactors — computational aspects and model modifications, *Sadhana*, 10(1–2), 13, 1987.
152. Jayaraman, V.K., Kulkarni, B.D., and Doraiswamy, L.K., An initial value approach to counter current back mixing model of the fluid bed, *ACS Symp. Ser.*, 168, 19, 1981.
153. Peters, M.H., Fan, L.S., and Sweeney, T.L., Reactant dynamics in catalytic fluidized bed reactors with flow reversal of gas in the emulsion phase, *Chem. Eng. Sci.*, 37, 553, 1972.
154. Stewart, P.S.B. and Davidson, J.F., Slug flow in fluidized beds, *Powder Technol.*, 1, 157, 1967.
155. Darton, R.C. LaNauze, R.D., Davidson, J.R., and Harrison, D., Bubble growth due to coalescence in fluidized beds, *Trans. Inst. Chem. Eng.*, 55, 274, 1977.
156. Broadhurst, T.E. and Becker, H.A., Onset of fluidization and slusgging in beds of uniform particles, *AIChE J.*, 21(2), 238, 1975.
157. Hovmand, S. and Davidson, J.F., Pilot plant and laboratory scale fluidized reactors at high gas velocities; the relevance of slug flow, in *Fluidization*, Davidson, J.F. and Harrison, D., Eds., Academic Press, London, 1971, 193.
158. Raguraman, J. and Potter, O.E., Modelling the slugging fluidized bed reactor, *ACS Symp. Ser.*, 65, 400, 1978.
159. Hovmand, S., Chemical Reaction in Slugging Fluidized Beds, Ph.D thesis, University of Cambridge, Cambridge, 1968.
160. Dabke, N.S., Kulkarni, B.D., and Doraiswamy, L.K., Stochastic analysis of fluid bed reactors, in *Transport Processes in Fluidized Bed Reactors*, Doraiswamy, L.K. and Kulkarni, B.D., Eds., Wiley Eastern Ltd., New Delhi, 1987, chap. 4.
161. Bukur, D.B., Caram, H.S., and Amundson, N.R., in *Chemical Reactor Theory; A Review*, Lapidus, S.L. and Amundson, N.R., Eds., Prentice-Hall, Englewood Cliffs, NJ, 1977, 686.

162. Krambeck, F.J., Katz, S., and Shinar, R., A stochastic model for fluidized beds, *Chem. Eng. Sci.*, 24, 1497, 1969.
163. Ligon, J.R. and Amundson, N.R., Modelling of fluidized bed reactors, an isothermal bed with stochastic bubbles, *Chem. Eng. Sci.*, 36, 653, 1981.
164. Johnson, J.E., Grace, J.R., and Graham, J.J., Fluidized-bed reactor model verification on a reactor of industrial scale, *AIChE J.*, 33(4), 619, 1987.
165. Krambeck, F.J., Avidan, A.A., Lee, C.K., and Lo, M.N., Predicting fluid bed reactor efficiency using adsorbing gas tracers, *AIChE J.*, 33(10), 1727, 1987.
166. Kunii, D. and Levenspiel, D., Effect of particles in bubbles on fluidized bed mass and heat transfer kinetics, *J. Chem. Eng. Jpn.*, 24(2), 183, 1991.
167. Bolthrunis, C.O., An industrial perspective of fluid bed reactor models, *Chem. Eng. Prog.*, 85(5), 51, 1989.

CHAPTER 6

Some Advanced Application Areas of Fluidization

I. NEW TECHNIQUE

A. Magnetically Stabilized Fluidized Beds

1. *Introduction*

One important innovation in fluidization technology with promising applications for magnetizable particulate solids is the magnetically stabilized fluidized bed (MSB). In gas fluidized beds, discussed in the earlier chapters, the gas in excess of that corresponding to the minimum fluidization velocity passes as bubbles, and this allows the reactant gas to bypass as bubbles through the bed without reaction. Furthermore, the wide variation in the gas–solid residence time and the backmixing in a conventional fluidized bed ultimately result in poor performance of the reactor. These drawbacks can be overcome by fluidizing a bed of ferromagnetic or magnetizable solids under the influence of an AC or DC magnetic field. The first studies in this area were carried out by Filippov[1,2] and Kirko and Filippov.[3] An MSB is nonbubbling and possesses flowability with constant pressure drops which are independent of the fluidizing fluid and the size of the solid particles. Rosensweig[4,5] reported several features of and phenomena relating to an MSB. An MSB is said to combine the advantage of a fixed bed and a fluidized bed.[6] Gas bypassing and backmixing and attrition of solids, which are common in fluidized beds but are not exhibited by a fixed bed, are eliminated in an MSB, which also has such advantages of fluidized beds as constant pressure drop and continuous solids throughput. Another advantage of an MSB which is not necessarily associated with a fixed or conventional fluidized bed is the trapping of fine particulate solids. This advantage has brought the MSB into the limelight as a proven particulate filter.[7]

2. Basics

The minimum fluidization velocity is theoretically not affected by magnetization of the bed of solids. However, there seems to be a transition velocity for each applied field of magnetization above which the magnetized bed bubbles, thereby producing a state of instability. A phase diagram for the MSB generated for a bed of steel spheres in the size range 177–250 μm indicates that the transition velocity increases as the applied magnetic field strength is increased. Figure 6.1 shows the schematic of a bubbling unstabilized fluidized bed (Figure 6.1a) and an MSB (Figure 6.1b) and the phase diagram (Figure 6.1c) for a fluosolid magnetic fluidized bed. The phase diagram in Figure 6.1c consists of three zones: an unfluidized bed below the minimum fluidization velocity (U_{mf}), a stabilized bed between the minimum fluidization velocity and the transition velocity, and an unstable bed beyond the transition velocity. Most studies on MSBs are concerned with the application of a uniform magnetic field to a bed of partly or fully magnetizable solids, where the magnetic

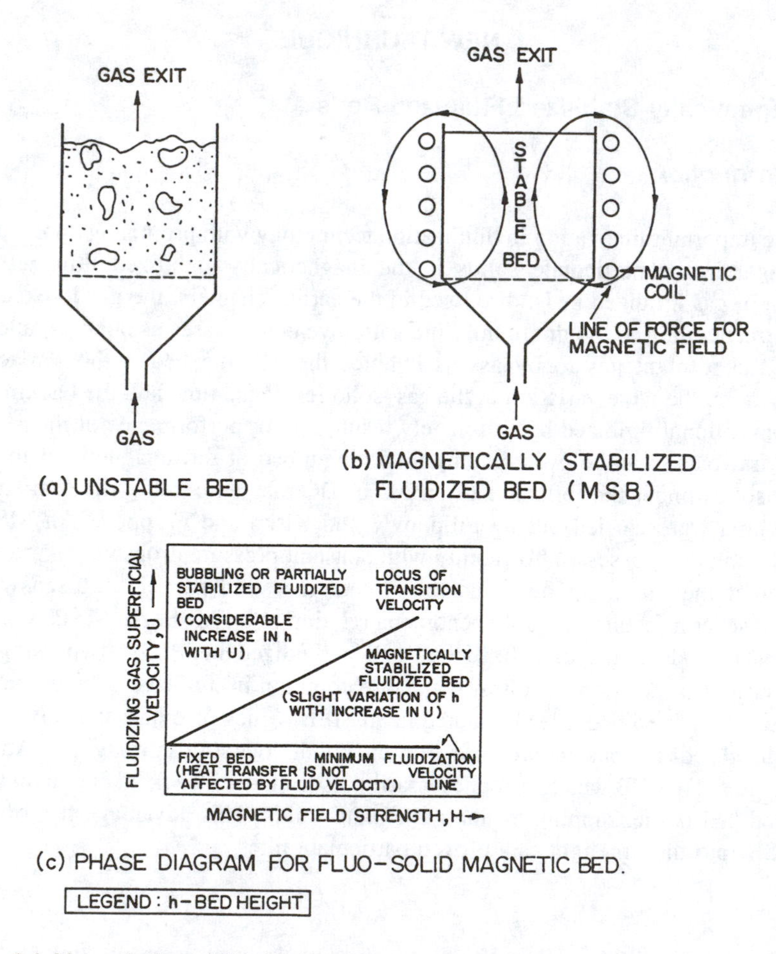

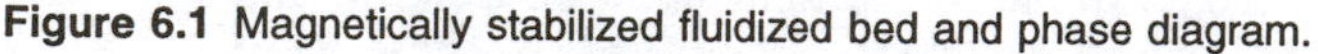

Figure 6.1 Magnetically stabilized fluidized bed and phase diagram.

field is collinear with the fluidizing gas. The applied uniform magnetic field does not exert any net force but enhances the interparticle forces. The solids flow behavior in an MSB is similar to plug flow, and thus high bed turnover is possible. Studies on the MSB by Rosensweig et al.[6] showed that radial dispersion decreases, the Nusselt number varies as the square of the particle Reynolds number, and a high pressure drop and bypassing of the gas are eliminated. Studies on the rheology of the MSB by Wei[8] showed an extremely slow settling rate of the bed, indicating a high level of bed viscosity. The MSB exhibits viscoelastic behavior, and thus the magnetically stabilized solids possess a yield stress. Magnetic field strength, bed voidage, and bed depth are the prime factors that directly influence rheology, whereas the size and density of the particles and gas velocity exert an indirect influence.

The cross-flow MSB was described by Siegell and Coulaloglou.[9] In a cross-flow MSB, continuous solids movement transverse to the ascending flow of the fluidized bed is effected, and cross-flow of solids approaching near plug flow condition was achieved even adjacent to the wall without solids backmixing both in the vertical and horizontal directions. Fluid bypassing in the form of bubbles was also eliminated. In the cross-flow configuration, the continuous solids throughput is not affected by a transition from the stabilized to the bubbling regime. The solids velocity profile in a cross-flow MSB corresponds to a plug flow pattern which is promoted at high magnetic fields and low superficial velocities. The cross-flow of solids in an MSB can be increased by tilting the grid plate as well as the bed surface. A tilt of 5.36° to the distributor plate was found to increase the solids cross-flow rate substantially. The cross-flow MSB was found[9] to be an efficient filter for particulate-laden gases. A cross-flow MSB with an unstabilized portion can be used to filter and elutriate the trapped fine particles continuously. As the specific gravity of the MSB can be controlled by adjusting the fluid velocity, the magnetic field, and the particle size, such a system can be used to separate a mixture of solids of different specific gravities. The light solids float and the heavy solids sink, and they are separated continuously in a cross-flow MSB. Siegell and Coulaloglou[9] demonstrated the separation of a limestone and coal mixture in an MSB. Chromatographic separation of gases was also demonstrated. Separation of helium and carbon dioxide and hydrocarbon gaseous mixtures is possible. Thus, the MSB is an alternative to conventional mechanized chromatographic separating systems, which are costly.

3. Characteristics

Because the MSB has the advantages of a fixed bed as well as a fluidized bed, it can be used as an efficient gas–solid contacting device for continuous countercurrent operation under a wide range of operating conditions. In such an operation for countercurrent gas–solid contact, studies relating to the mixing behavior are scant. When heterogeneous normal fluidization is turned into stable fluidization under the influence of a magnetic field, the mixing behavior changes dramatically. Hence, the residence time of the fluidizing gas and its radial as well as axial mixing also are correspondingly influenced by the magnetic field strength. Geuzens and Thoenes[10] studied axial and radial mixing in the MSB and reported that both axial and radial mixing are suppressed significantly compared to the free bubbling state of fluidization. A stretched mixing cell model was proposed to explain the mixing behavior. The radial mixing in the MSB corresponds to a packed bed, while the axial mixing is between packed and normal fluidized beds.

The basic aspects of the MSB are still evolving. Studies by Sonoliker et al.[11] on the influence of a magnetic field on the fluidization of iron particles showed an increase in minimum fluidization velocity (U_{mf}) with magnetic field strength. However, Rosensweig et al.[6] reported no change in U_{mf} values in an MSB. Arnaldos et al.[12] investigated heat and mass transfer in the MSB and found that the heat transfer coefficient from an immersed-type heater decreases with an increase in magnetic field strength as well as an increase in the concentration of the magnetizable fraction of the bed material. The heat transfer coefficient (h) was observed to increase with fluid velocity in a partially stabilized MSB (i.e., bubbling bed) but showed only a slight increase in the MSB. Arnaldos et al.[12] proposed correlations to predict the heat transfer coefficient in an MSB. Mass transfer results obtained by drying air in an MSB with steel (350–420 μm) and alumina (630–890 μm) particles showed that the operating efficiency decreases with increasing air velocity at a constant magnetic field strength and the drying efficiency increases with an increase in magnetic field strength. The bypassing of gas as bubbles during fluidization at high fluid velocities is attributed to low operational efficiency.

B. Semifluidized Bed

1. Description

In the preceding section, we saw that a state of fluidization with the advantages of both fluidized and fixed beds is achieved in an MSB. This becomes possible with a magnetic field coupled with a bed of ferromagnetic materials, which are the essential part of the system. A novel technique which does not require either a magnetic coil or a ferromagnetic bed material, but can add the advantages of both fluidized and packed beds while minimizing their defects, is semifluidization. A schematic of a semifluidized bed is shown in Figure 6.2a. A semifluidized bed is simply a fluidized bed with restricted expansion brought about by a porous or sieve plate positioned firmly above the expanding head of the fluidized bed. The sieve plate allows the gas, but not the solid particles, to pass through it. In other words, it acts as a filter with a packed layer of solids below it. A zone of fluidized bed exists between the packed layer of solids and the gas distributor plate. A semifluidized bed can be visualized as a combination of a mixed and a tubular reactor in series. The fluidized zone above the distributor corresponds to a mixed reactor and the packed layer above it acts as a tubular reactor. This combination is very conducive for an optimal exothermic reaction. Cholette and Blanchet[13] and Cholette and Cloutier[14] showed that a mixed and tubular combination is more efficient than the respective individual types. For example, for an exothermic reaction under adiabatic conditions, the reaction rate initially increases due to adiabatic temperature rise; the reaction rate subsequently decreases due to the

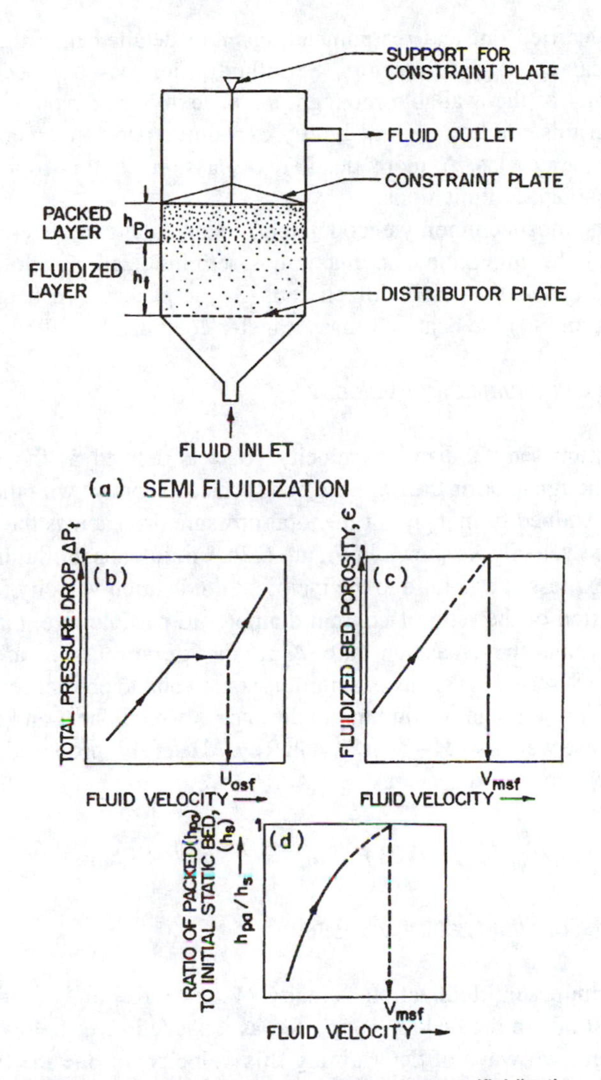

Figure 6.2 Semifluidized bed and determination of minimum semifluidization operating velocity (U_{osf}) and maximum semifluidization velocity (V_{msf}). (a) Semifluidization, (b) total pressure drop versus fluid velocity, (c) fluidized bed porosity versus fluid velocity, and (d) ratio of packed bed to initial bed versus fluid velocity.

depletion of the reactants. In such a situation, a mixed reactor (i.e., a fluidized zone in a semifluidized bed) can be used to carry out the increasing rate of reaction, followed by a tubular reactor (i.e., a packed layer in a semifluidized bed) to accomplish the remaining reaction which occurs at relatively low rates. Thus, a semifluidized bed reactor is ideally suited for implementing an exothermic type of reaction.

The subject of semifluidization was first introduced by Fan et al.[15] Since then, ample literature has emerged on this technique. Murthy and Roy[16] published a comprehensive review on this subject, highlighting the pros and cons of the various investigations carried out and recommending more detailed investigations on the design and scaleup of such reactors. Semifluidization has not been extensively discussed in any of the available monographs or textbooks devoted exclusively to fluidization. In this context, we will briefly explain certain fundamental aspects in the next few paragraphs. A more detailed discussion on this topic will not be presented due to space limitations.

Some of the most commonly encountered fundamental parameters in semifluidization are (1) the minimum and maximum semifluidization velocities, (2) the pressure drops across the semifluidized bed, (3) the packed bed height below the restraint plate, and (4) the heat and mass transfer coefficients.

a. Minimum Semifluidization Velocity

The minimum semifluidization velocity (U_{osf}) is defined as the fluid velocity which is just enough to bring the fluidized bed surface in contact with the top restraint plate. It is determined from a plot of the total pressure drop across the semifluidizer against the fluid velocity, as shown in Figure 6.2b. Minimum semifluidization velocity is usually expressed as a ratio to the incipient fluidization velocity (U_{mf}), and this ratio is a function of the ratio of column diameter to particle size, particle density to fluid density, and the expansion ratio. A number of correlations are available in the literature to describe the status of semifluidization and to predict semifluidization velocities at minimum and maximum conditions, as well as heat and mass transfer coefficients; these were listed by Murthy and Roy.[16] He et al.[17] proposed the following correlation for a gas–solid system:

$$U_{osf}/U_{mf} = 0.08\, R^{1.09} d_p^{-0.75} \rho_s^{-0.06} h_s^{-0.12} \text{(SI units)} \tag{6.1}$$

b. Maximum Semifluidization Velocity

The maximum semifluidization velocity (V_{msf}) corresponds to the condition where all the solids in the bed are supported as a packed layer below the restraint plate. There are two ways of determining this velocity. In one method, a plot is made of fluidized bed porosity versus fluid velocity, and the fluid velocity obtained by extrapolation (as shown in Figure 6.2c) to the maximum bed porosity (i.e., unity) is taken to be the maximum semifluidization velocity. In the other method, a plot is made of the ratio of the packed bed to the static bed height (h_s) versus fluid velocity, and the fluid velocity obtained by extrapolation (as shown in Figure 6.2d) of this ratio to its maximum value (i.e., unity) is considered to be the maximum semifluidization velocity (V_{msf}).

Maximum semifluidization velocity is in general a function of the Archimedes (or Galileo) number. Some correlations establish V_{msf} as a function of the ratio of column diameter to particle size. Murthy and Roy[16] surveyed the literature and listed correlations for estimating V_{msf}. They also presented useful charts to predict both U_{osf} and V_{msf}.

c. Pressure Drop and Voidage

Total pressure drop (ΔP_t) in a semifluidized bed is the sum of the pressure drops across the fluidized bed (i.e., ΔP_f), the packed bed (i.e., ΔP_{pa}), and the constraint plate (ΔP_c), that is,

$$\Delta P_t = \Delta P_f + \Delta P_{pa} + \Delta P_c \tag{6.2}$$

The pressure drop across the packed bed can be predicted for a known value of the packed bed height using the Ergun equation (see Equation 1.60 at minimum fluidization condition), and the pressure drop across the constraint plate can be determined by fluid mechanics using the flow area fraction and the superficial semifluidization velocity (U). The pressure drop across the fluidized bed section with voidage ϵ_f is

$$\Delta P_{bf} = (1 - \epsilon_f)\,(\rho_s - \rho_g)\,(h - h_{pa}) \tag{6.3}$$

where h is the semifluidized bed height. The voidage (ϵ_f) can be estimated from the mass balance of the solids. If h_s is the initial static bed height with a voidage ϵ_s and ϵ_{osf} is the voidage at the onset of semifluidization, then:

$$h(1 - \epsilon_{osf}) = (h - h_{pa})\,(1 - \epsilon_f) + h_{pa}\,(1 - \epsilon_{pa}) \tag{6.4}$$

Upon rearranging,

$$\epsilon_f = \epsilon_{osf} + \frac{\epsilon_{osf} - \epsilon_{pa}}{h/h_{pa} - 1} \tag{6.5}$$

where

$$\epsilon_{osf} = 1 - (h_s/h)\,(1 - \epsilon_{mf}) \tag{6.6}$$

It is necessary to be able to predict the packed bed height in a semifluidized bed; otherwise, it has to be determined by experiment. There are several correlations available in the literature,[16] and they were listed by Murthy and Roy.[16] For a gas–solid fluidized bed, Ho et al.[17] proposed a simple correlation for the packed bed height (h_{pa}):

$$(h - h_s)/(h - h_{pa}) = [(U - U_{mf})/(U_t - U_{mf})]^{0.38} \tag{6.7}$$

d. Limiting Factors and Applications

Predictions of heat and mass transfer coefficients were discussed and correlations tabulated in the literature.[16] For a liquid semifluidized bed, Varma et al.[18] found that the wall-to-fluid heat transfer coefficient increases with the solids concentration and decreases with increasing particle diameter and particle thermal conductivity. Because of its complex hydrodynamic features, knowledge pertaining to the design of the semifluidized bed reactor is still far from adequate, and applications of this novel reactor have not yet been fully exploited despite the fact that it offers several advantages. Semifluidized bed reactor applications were discussed[16] in the context of ion exchange, filtration, and bioreactors. In an ion-exchange operation, a semifluidized bed offers good resin–liquid bed contact, smaller pressure drops, high resin utilization efficiency, and prevention of resin loss. Gaseous streams emerging from hot flue gases of power plants and insoluble suspensions in liquids can be filtered efficiently in a semifluidized bed. An attached film bioreactor using the principle of semifluidization can be operated economically without loss of microorganisms from the attached surface due to elutriation.

C. Spout Fluid Bed

1. General Description

In the preceding two sections, we saw that the MSB and the semifluidized bed have emerged as a combination of packed and fluidized beds, enabling the benefits of both of these techniques to be realized in a single unit. In another combination, a fluidized bed is hybridized with a spouted bed, thereby giving rise to a new generation of fluid–solid contractors, termed spout fluid reactors. Normal fluidization fails when the particles are large and coarse. In a spouted bed, local mixing is not as good as in a fluidized bed due to the fact that solid particles in the annular region between the spout and the wall of the equipment move in a packed layer. However, the solids turnover is very high due to a high circulation rate of solids. As most of the fluid passes through the central core of the spouted bed, the short residence time of the spout fluid necessitates a high recycle ratio. In a spout fluid bed, particulate solids with a wide range in size from fine (micron) to coarse (millimeter) can be successfully brought into intimate contact with the fluidizing fluid. In other words, fluidizable and aeratable particles (i.e., Group A and B Geldart particles) as well as spoutable particles (Group D Geldart particles) can be brought into intimate contact with the fluid in a spout fluid bed reactor. A spout fluid bed essentially consists of a fluidized bed in which a spout is generated by an independently monitored flow of a spouting gas or liquid through the spouting nozzle/orifice fitted at the center of the distributor plate of the fluidized bed. In such a configuration (shown in Figure 6.3a), the annular region of the spout (i.e., the space between the spout and the wall) is fluidized. Chatterjee[19] referred to this type of bed as a spout fluidized bed. If the annular region around the central spout is not fluidized but just aerated (as the configuration shown in Figure 6.3b), the condition is quite different from spout fluidization. This type of operation is termed spouting with aeration. Littman et al.[20] and Vukovic et al.[21] also referred to such a bed as a spout fluidized bed, whereas Sutanto[22] clarified the two terms and distinguished between the two types of beds.

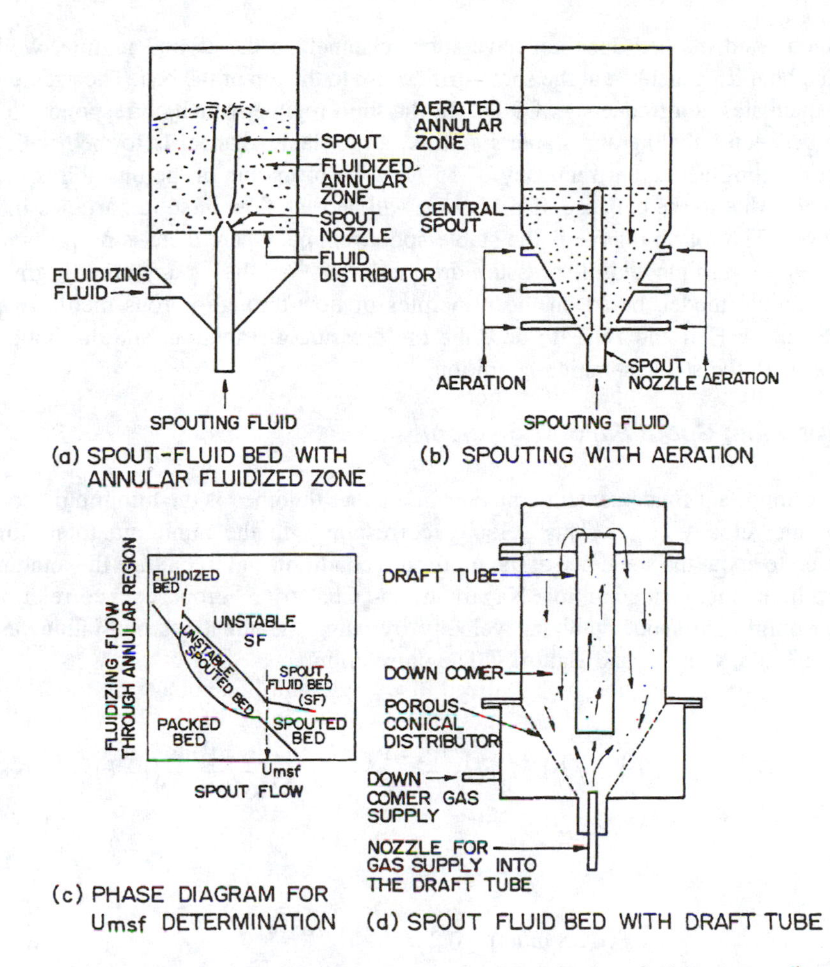

Figure 6.3 Spout fluid beds and phase diagram for determination of minimum spouting fluidization velocity (U_{msf}).[19-21,23,31]

2. Hydrodynamics

In a spout fluidized bed, mixed particles of different sizes or densities are not stratified; this is particularly advantageous in operations such as roasting, calcination, combustion, and gasification, where the particles sink due to shrinking or disintegration. Spout fluidization is a new and improved version of the fluidization technique and remains unexplored in process and extraction metallurgy. The basics of the spout fluid bed, especially pressure and flow characteristics, and its mechanics were discussed by Nagarkatti and Chatterjee[23] and Hadzisdmajlovic et al.[24] The pressure distribution in a spout fluid reactor was studied by Heil and Tels.[25] The spout fluid bed operation is very flexible in that it is possible to independently control the flow rates into the fluidized as well as the spouted regions and achieve different flow regimes. Heil and Tels[25] observed four different flow regimes. In the first regime, where the pressure drop is highest, the flow through the bed is by percolation from the spout, and the bed does not have spout channel. In the second regime, well-defined bubbles generated at the spout orifice rise to the top of the bed. The pressure drop fluctuates at a frequency of 5 Hz. In the third regime, which corresponds to a stage between bubbling and stable spouting, an unstable channel is formed with a pressure drop fluctuation frequency of 15 Hz. The formation of the unstable spout channel is due to the collapse of the spout wall and the flow of solid particles into the spout. The fourth regime is the stable spout regime, where there is no pressure fluctuation with time and the pressure drop is lower than that in any other regime. A theoretical model, based on the principles of flow-through porous media, was developed by Heil and Tels[25] to describe the pressure distribution, and this model conforms to the stable spouting condition.

3. *Minimum Spout Fluidizing Velocity*

An important fundamental parameter in a spout fluid bed is the minimum spout fluidizing velocity (U_{msf}). This velocity corresponds to the minimum total flow required to spout the static bed at the minimum condition and to fluidize the annular region in the incipient condition. Nagarkatti and Chatterjee[23] proposed a correlation for the minimum spout fluidizing velocity by using the equations of Mathur and Gishler[26] and Mamuro and Hattori.[27] The correlation is

$$U_{msf} = \left(d_p/d_c\right)\left(d_o/d_c\right)^{1/3} 2\, gH_o \frac{\left(\rho_s - \rho_f\right)^{0.5}}{\rho_f} + \frac{d_p^2\left(\rho_s - \rho_f\right)g}{1650\,\mu}(1-\theta) \tag{6.8}$$

where

$$\theta(\text{CGS units}) = 0.2\, d_p^{-0.320} d_o^{0.235} H_o^{0.160} \tag{6.9}$$

and d_o, d_c, and H_o correspond, respectively, to the spout orifice diameter, the column diameter, and the initial bed height. U_{msf} remains unaffected with regard to fluid flow through the spout. It may be noted here that U_{msf} cannot be determined experimentally using ΔP versus flow rate data by a method similar to that used to determine U_{mf}. It can be determined from the phase diagram constructed on the basis of information regarding stable and unstable spout flow obtained from the flow variations in the spout and the annular packed bed of solids. The transition point from unstable to stable spout flow is taken as U_{msf} and is often difficult to determine. In the case of spout flow at the minimum condition, flow variations through the annular layer can bring about a transition from an unstable to the stable condition, and the flow rate at the transition is taken as the minimum spout fluidizing velocity; the phase diagram is shown in Figure 6.3c. An unstable spout fluid bed does not have a well-defined spout and particle movement is erratic. Details of the experiments and the methodology pertinent to the evaluation of U_{msf} were presented by Nagarkatti and Chatterjee.[23] The distributor type affects U_{msf}. U_{msf} values are higher for flat-type distributors than for conical-type distributors.

Studies[26] on the wall-to-bed heat transfer coefficient (h) in spout field beds revealed that h increases with increasing mass velocity and particle diameter and decreases with increasing bed height. Under identical flow conditions, the maximum value of h was found to be 30% higher in spout flow than in a fluidized bed.

4. Spout Generation

Spout fluid beds can be of various configurations depending on the type of distributor plate used for the fluidizing fluid and the nature of the spout fluid inlet. Chatterjee et al.[28] described four types of spout fluid beds used in applications such as coal combustion, heating, and quenching and recirculation of particulate solids. The distributor plate can be constituted of a packed layer of large particles, a sieve plate, or a perforated conical bottom. The spout fluid can be injected through a single nozzle or orifice located centrally in the distributor. The spouts can be of multiple types, each spout emerging out of the spout nozzle and located axially one above the other. The spout can be generated by injecting the spout fluid through a spout tube centrally immersed by suspension into the bed. A spout tube of this type is useful for flexible operation; it can be immersed into the bed to the desired depth, and the solid particles can be fed along with the spout fluid into the reactor.

5. Draft Tube

A common fluidized bed for solid recirculation inside the bed makes use of a draft tube through which a spout fluid is injected, thereby enabling high solid circulation as well as good mixing. Figure 6.3d shows a schematic of a spout fluid bed with a draft tube. Such a configuration is believed to have been first used by Taskaev and Kozhima[29] as early as 1956. A review of the application of the draft tube within the fluidized bed was published by Yang and Keairns.[30] The desired solid circulation in a fluidized bed can be achieved by passing the spout fluid into the draft tube starting from the flow rate required to just fluidize the solids up to the flow rate required to pneumatically carry the solid particles out of the draft tube. Such a concept can be applied for various liquid–solid as well as liquid–gas–solid contacts. In a spout fluid bed with a draft tube, the minimum spout fluidizing velocity is lowered because the spout fluid is confined to the tube; in a normal spout fluid bed, the spout fluid leaks through the spout wall into the fluidizing zone. There is no limit to the spoutable bed height. Some of the important design parameters of a spout fluid bed with a draft tube are the gas bypassing characteristics of the distributor plate, the ratio of the diameter of the draft tube to the diameter of the gas-supplying orifice or nozzle, the ratio of the area between the downcomer (i.e., annular zone) and the draft tube, the ratio of the diameter between the draft tube and the draft tube gas supply, the distance between the distributor plate and the inlet point of the draft tube, and the ratio of the area of the draft tube gas supply and the concentric solid feeder. Yang and Keairns[30] proposed some correlations based on experimental data. The same authors[31] also investigated the solid circulation rate and the gas bypass characteristics for different distributor plates. They found that the solid circulation rate is strongly affected by the design configuration at the bottom of the draft tube and the downcomer, and this was attributed to the gas bypass characteristics. A spout fluid bed with a draft tube is flexible in operation in that the gas supply into the annular region (or downcomer) and through the draft tube can be controlled independently. Hence, a spout fluid bed with a draft tube can be used for several types of liquid–solid or liquid–gas–solid contacting.

D. Centrifugal Fluidized Bed

1. *Concept*

The centrifugal fluidized bed (CFB), wherein fluidization of particulate solid materials is achieved by passing the fluidizing fluid radially through a rotating cylindrical perforated distributor, is a new concept. The particulate solids form a layer on the inner portion of the rotating distributor under centrifugal action. The fluidizing fluid enters a plenum and is then distributed to fluidize the solids. A schematic of a CFB is shown in Figure 6.4a. Unlike a conventional fluidized bed, radial acceleration can be varied by varying the rotation (i.e., angular velocity, ω), and a centrifugal force several times higher than the gravitational force can be brought into play. By this method, the minimum fluidization velocity required to fluidize a fine particulate solid can be increased severalfold without allowing the fines to be elutriated out of the system. Relatively high operating velocities along with high solids concentration and good fluid–solid contact make this type of reactor attractive; in addition it has the advantage of compact size and flexibility in operation.

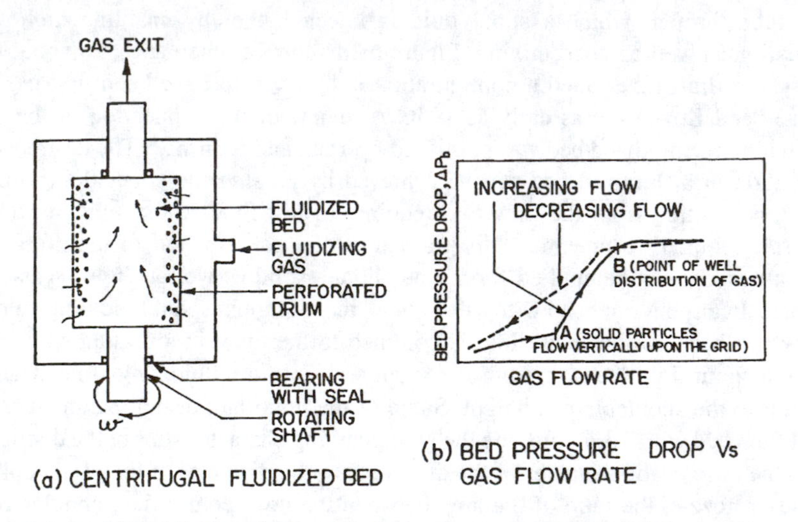

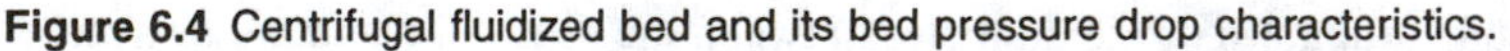

Figure 6.4 Centrifugal fluidized bed and its bed pressure drop characteristics.

2. *Basics*

CFBs were reported[32,33] to have promising applications in coal combustion when large capacities are required and in the handling of fine coal particles. These reactors are also used in food drying.[34] CFBs are also recommended[35] for the combustion of high-sulfur coal using a bed of dolomite or limestone. A wide range of power output from a CFB can be achieved by manipulating the angular velocity (rotation), bed temperature, and operating velocity. The literature on CFBs is scant, and the technique has not yet been exploited in large-scale operations. The reactor can be a potential candidate for the halogenation of fine solid particles, which are prone to elutriation in conventional fluidization. The basics of this class of reactors were investigated by Levy et al.,[35] who described methods for determining the minimum fluidization velocity experimentally and for predicting the same theoretically. Fluidization in a CFB is not achieved at start-up. A critical angular velocity is required to bring the initial bed of solids into a layer on the distributor; fluidization of this layer is then achieved by the radial flow of the fluidizing fluid against the centrifugal force. A typical variation in bed pressure drop with gas flow rate for a CFB is shown in Figure 6.4b. During the flow of the fluidizing gas, when it is gradually increased, the bed flows up vertically along the grid (point A) and the gas is then well distributed at a gas flow rate corresponding to point B. The bed pressure drop follows the same trend as in a conventional bed for decreasing flow, as depicted by the dotted line in Figure 6.4b. Features like start-up, minimum fluidization, and particle elutriation are influenced in a very complex manner by parameters such as bed geometry, distributor type, particle size, particle density, distributor pressure drop, and angular velocity. Levy et al.[35] proposed the following relationship to predict bed pressure drop, assuming the tangential velocity is independent of radius:

$$\Delta P_b = (\rho_s - \rho_f)\,(1 - \epsilon)\,\omega^2 r_o \ln (r_o / r_i) \tag{6.10}$$

where r_o and r_i are, respectively, the outer and inner radii of the bed. It was also proposed that the radial velocity for a thin bed at the minimum fluidization state can be computed by the equation

$$\mathrm{Ar} = \left[150\left(1-\epsilon_{mf}\right)/\left(\epsilon_{mf}^3 \phi_s^2\right)\right]\mathrm{Re}_{mf} + \left[\,1.75/\left(\epsilon_{mf}^3 \phi_s\right)\right]\mathrm{Re}_{mf}^2 \tag{6.11}$$

Equation 6.11 is the same as Equation 1.60. A conical (tapered) distributor with a half cone angle varying from 2 to 22°C was suggestged[35] to achieve the formation of a uniform solid layer over the grid. Because centrifugal fluidization is achieved at a high flow rate of the gas even at the incipient state, particle loss by elutriation is an important factor in design considerations. Levy et al.[36] attempted to predict the elutriation loss and proposed a model to estimate the air flow rate at which elutriation occurs. The elutriation rate was found to be a function of the gas flow rate, angular velocity, and bed thickness. Terminal velocity is a strong function of angular velocity and bed thickness. The flow is predicted to be inviscid and irrotational except near the axis of rotation and within the end wall boundary layers.

Rietema and Mutsers[37] reported experimental results obtained by varying the bed porosity and apparent gravity. A human centrifuge, generally used for testing and training pilots, was employed to house a fluidized bed apparatus and vary the apparent gravity. The results showed that the elasticity of the dense phase is of mechanical origin and is of considerable importance in that it brings the stability of homogeneous fluidization. A CFB is very different from the one investigated by Rietema and Mutsers[37] because a fluidization experiment carried out in a human centrifuge is not the same as in a CFB. A horizontally rotating nozzle gas distributor was used in the experiments conducted by Novosad and Kostelkova[38] to fluidize cohesive powders. However, a system of this type, in principle, is different from a CFB.

3. Applications

The CFB (also referred to as a rotating fluidized bed) was shown by Pfeffer et al.[39] to have high efficiency in filtering dust and moisture from a gas stream. The dust-laden fluidizing gas is filtered by the collector particles which are in a fluidized state. The conventional fluidized bed has been used for dust filtration or capture for a long time. However, the operating velocity range with such a bed is limited (1.2–1.5 U_{mf}) as gas bubbles start forming above the minimum fluidization velocity. The compact size, the high gas throughput, the economically acceptable pressure drop (20–30 cm H_2O), and the ability to filter particles in the size range 0.3–10 μm make the CFB filter attractive for industrial applications. In many pyrometallurgical operations, especially where dust produces fumes at high temperatures, such filters can be employed continuously. The dust particles in centrifugal filters are collected by inertial impact. The speed of rotation, gas velocity, gas humidity, nature and size of the collector particles and dust particles, and bed thickness are the parameters that affect the filtration efficiency. The CFB has not yet emerged as a special topic of research, although its application in coal combustion is established and it is now finding use in many other areas. Due to its application as a high-efficiency filter, it can emerge as important pollution control equipment for cleaning corrosive high-temperature gases.

E. Compartmented Fluidized Bed

1. Interconnected Operation

Fluidized beds in certain cases are operated by interconnecting them. One such use which has proved to be a great success is fluid catalytic cracking. In this process, one reactor is used to crack the hydrocarbons using a catalyst, while the other is used to regenerate the spent catalyst and transfer it back to the first reactor. It has been demonstrated that such a combination of two separate fluidized beds, one for carrying out the reaction and the other for regeneration, as depicted in Figure 6.5a, is commercially viable. Continuous and independent control of each of these externally interconnected fluid beds is possible. An improved version of such a twin-bed operation is the compact circulating fluidized bed, which uses a single vessel but contains two fluidized beds separated by a partition plate. There are provisions for circulating the solids between the fluidized bed reactors. Kuramoto et al.[40] developed a system for circulating fluidized particles within a single vessel, and Fox et al.[41] investigated the control mechanism for the circulation of solids between adjacent vessels.

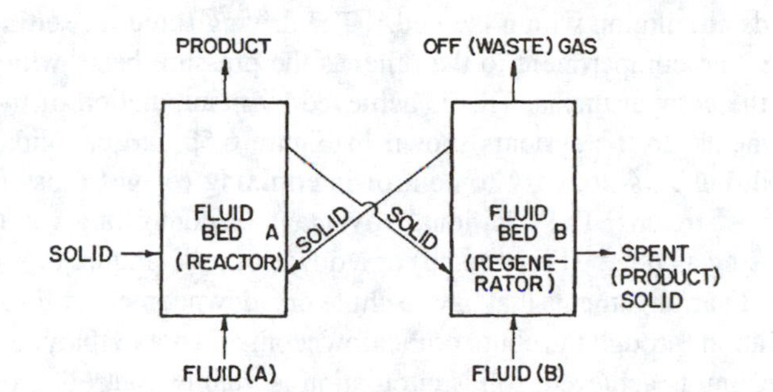

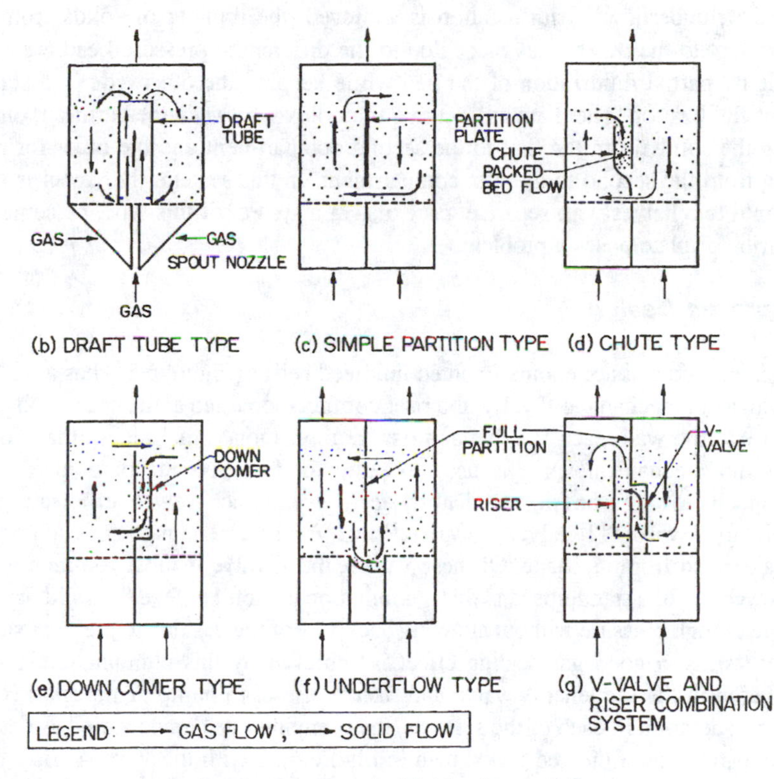

Figure 6.5 Interconnected and compartmented fluidized beds.[40-48]

2. *Compact Twin Beds*

The design of a compact twin fluidized bed was discussed by Masson.[42] There are several means of circulating fluidized solids within a single vessel. A simple and widely used type[30,31,43-45] makes use of a draft tube. A draft tube mounted coaxially (as shown in Figure 6.5b) just above the distributor in a cylindrical fluidized bed can induce solids circulation. The solids ascend through the draft tube along with the gas jet introduced by a spout nozzle and then are recirculated by downflow through the annular zone. A rectangular vessel with a central partition plate (as illustrated in Figure 6.5c) can induce solids circulation within the bed.[42] The driving force for setting the solids in motion from one compartment to the other is the pressure head, which has to be high in one of the compartments. This is achieved by manipulation of the fluid flow in the compartments. In the systems shown in Figure 6.5b and c, solids circulation can be achieved, but it is not easy to control intermixing (or gas cross-flow) of the two fluidizing gas streams. The additional advantage of controlling gas leakage can be realized by using a chute[46] (Figure 6.5d) or a downcomer[42] (Figure 6.5e) along with a partition plate. In arrangements that have a chute or a downcomer, additional aeration to induce circulation through the chute or the downcomer is not employed. As a result, although gas sealing is achieved, solids circulation is mainly controlled by the solids flow characteristics through the chute or downcomer. In the combination shown in Figure 6.5f, underflow[47] with aeration is achieved; the transfer of solids from one compartment to the other takes place due to the differential pressure head created at one side by partial fluidization of the bed while keeping the other side in a state of vigorous fluidization. There are at least two underflows between the partition, one for drawing the solids from the first to the second compartment and the other for recirculating from the second to the first compartment. In this system, the vessel is partitioned into two halves with separate gas exits. In a system of this type, placement of the distributor plate poses a problem.

3. *Improved Design*

An improved class of compartmented fluidized beds[48] (Figure 6.5g) has a partition plate with a nonmechanical V valve and riser combination fitted at the bottom on either side close to the wall; each V valve and riser combination is the mirror image of the other. A nonmechanical valve is used to pump or transport the particulate solids pneumatically without any mechanical device. Several such valves are discussed in the literature: J valve,[49] L valve,[50] O valve,[42] and V valve.[51] The name of each type of valve is derived from its shape. Of these valves, the V valve is most commonly used at the lower end of a standpipe, and this combination is useful to feed the solid particles to a zone of high pressure without allowing backflow of the gas due to pressure surges. In other words, a good gas sealing effect is achieved by this combination. Such a combination in the sequence V valve and riser is useful to pump or transport solids. By varying aeration in each of the systems, the pumping rate can be varied as desired. The compartmented fluidized bed shown in Figure 6.5g with the V valve–riser combination was studied by Sathiyamoorthy and Rudolph[48] with the objective of evaluating the pressure drop characteristics of the V valve and the riser for the circulation of various solids such as ilmenite, zircon, and sand. Flow through a V valve and riser combination seals gas leakage. The hydrodynamics of such a compartmented fluidized bed was studied in the dense-phase circulating condition by He,[52] and models for gas cross-flow and solids circulation were proposed.

4. *Application*

The dense-phase compartment fluidized bed has several advantages in that it can be used for two independent gas–solid reactions, where the solid is the same for both. One such example is the combustion and gasification of coal. Coal burned in one compartment using air, when circulated to the other compartment, allows gasification, with steam receiving the necessary heat of reaction from the combustion zone. Here, the coal is circulated without mixing air and steam, and thus it is possible to obtain pure and high-caloric-value gases at the exit of the gasifier. In other words, the substantial amounts of nitrogen and carbon dioxide formed in the combustor are not mixed to reduce the caloric value of the gasifier product gases, which are mainly hydrogen and carbon monoxide. Such a compartmented fluidized bed can be successfully used to carry out many pyrometallurgical operations. For example, it is often necessary to dry the feed material and subsequently implement the gas–solid reaction. These two processes can be carried out in a single reactor, utilizing the heat of reaction for drying the feed material. In this manner, UF_4 can be prepared, starting from U_3O_8, in the same reactor; U_3O_8 is reduced by hydrogen gas in one compartment and the resultant UO_2 is hydrofluorinated in the other compartment to obtain UF_4. In both cases, the product is solid, and this helps achieve the two reactions in a compact reactor without any elutriation, which otherwise is a predominant factor in a conventional fluidized bed. Rudolph[53] explained how this class of reactors can be used for the preparation of molybdenum metal powder starting from ammonium paramolybdate or molybdenum disulfide. One compartment can be effectively used for calcination of ammonium paramolybdate (or roasting MoS_2) using air and the other compartment can be used to reduce the oxide. The reduction of MoS_2 to Mo usually is carried out in two steps, as pointed out in Chapter 2. This two-step reduction can be carried out independently but in a continuous manner in a single-compartment vessel. Thus it is clear that there is ample room to explore use of this new technique in extraction and process metallurgy.

II. FLUIDIZED ELECTRODE CELLS

A. Basics of Fluidized Electrodes

1. Need for Fluidized Electrodes

The fluidized electrode was first discovered[54,55] in the late 1960s, and its application since then has been well accepted in electrochemical processes. In such a process, the chemical reaction occurs on the electrode surface, and thus the capacity of the equipment depends on the electrode surface area. In a conventional electrolytic cell, the electrodes used are usually of the slab type and have only limited surface area. If the electrode is porous or is made up of a bed of electrically conducting particulate solids, the surface area per unit volume available for the electrochemical reaction is large. In other words, the space–time yield of an electrolytic cell provided with fluidized electrodes is significantly increased. In a cell meant for the electrowinning of a metal from an aqueous solution, the space–time yield is typically of the order of 0.07 tons of metal per cubic meter of the cell per day, a figure which obviously is very low. Hence, in an effort to increase the space–time yield of electrolytic cells, fluidized electrodes have been used, with some radical changes in cell design. Flett[56] suggested that the ratio of the costs of different capacities for a chemical process is proportional to the ratio of capacities raised to a power of 0.6–0.7. As the capacities, in turn, are affected by the surface available for the chemical reaction, the fluidized electrode, with its large surface area, plays an important role in determining the capacity as well as the economics of the cell. Excellent mass transport characteristics and mobility of bed due to fluidization facilitate continuous deposition and extraction of metals. The use of the fluidized electrode in extractive metallurgy was summarized by Flett[57] two decades ago.

2. Description

A simple electrolytic cell provided with a fluidized cathode in a side-by-side electrode design is depicted in Figure 6.6a. Electrically conducting particles, which have continuous contact with a current feeder due to fluidization by the upflowing catholyte, act as the cathode. Short-circuiting of current flow by the fluidized particles by their contact with the anode is arrested by a semipermeable or porous membrane which allows the electrolyte, but not the fluidized particles, to flow. The anode can be a simple slab or may be of any other type, and the anolyte can be stationary or continuously flowing. It is advantageous to operate the cell at maximum current density to maximize the output of the cell and to minimize capital costs. A cell working at a low current density, on the other hand, is very close to chemical equilibrium and necessitates the use of lean/dilute solutions for efficient operation. However, a cell operating at maximum current density preferably requires purified and concentrated solutions.

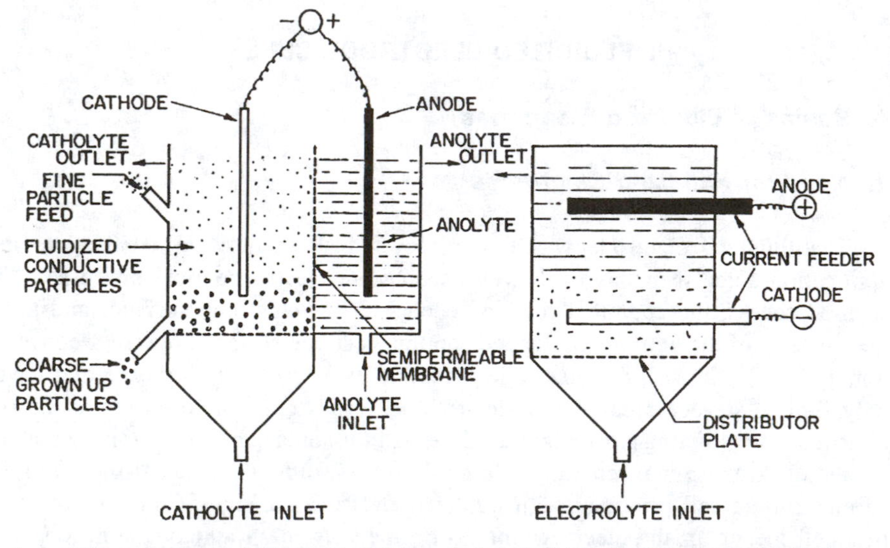

Figure 6.6 Fluidized electrode cells.

3. Electrical Conduction

Electrical conduction in the fluidized electrode was shown to influence bed behavior.[58] The chain of contacts established by the particles and the current feeder is assumed to be responsible for electronic conduction. Current flow from the diaphragm to the current feeder is affected by expansion of the bed due to the flow of the electrolyte. A fluidized electrode is considered to be a superimposed continuum of two entities: the particulate bed and the electrolyte within it. When electrolyte resistance is lower than bed resistance (which happens at high bed expansions), current flow through the electrolyte and electrodeposition of the metal take place on the current feeder. At the other extreme, when bed resistance is lower than electrolyte resistance (which happens at low bed expansions), electrodeposition occurs in the vicinity of the diaphragm. To avoid these two extremes, electrolyte flow should be kept at some intermediate flow rate for uniform electrodeposition on the particulate solid. Maintaining differential pressure near the diaphragm, by passing the electrolyte from the anode to the cathode, was found to eliminate electrodeposition near the diaphragm.[59]

4. Electrowinning Cells

Electrowinning using the fluidized electrode is reported[60] to have, in general, three problems: (1) excessive cell voltage or DC energy consumption, (2) defluidization due to adherence of particles on the current feeder, and (3) short-circuiting caused by the metal dendrites that adhere to the diaphragm and traverse the anode chamber. Particle sizes below 400 μm have not been used successfully. The power consumption to pump the electrolyte increases with particle size. Although this energy consumption is relatively small for small particles, the energy consumed by the cell increases as the particle size grows due to electrodeposition. The cell depicted in Figure 6.6a is a side-by-side electrolyte design; here, the current flows perpendicular to the flow direction of the electrolyte. If the anode is positioned in a parallel orientation above the cathode as shown in Figure 6.6b (known as parallel electrodes design), the resulting arrangement is suitable for laboratory tests. In a comprehensive review on fluidized electrowinning, Dubrovsky et al.[60] pointed out the pros and cons of this technique and also discussed the cell configuration that would work at a reduced energy consumption. The subject of fluidized electrodes has grown to such an extent that it cannot be discussed at length in this section. Several papers have been published on this topic based on work carried out at the University of California, Berkeley. The reader is directed to the work of J.W. Evans and co-workers,[58-60] which is cited in the following discussion of the salient features of fluidized electrodes as applied to extraction metallurgy, in particular the importance of cell design.

5. Selection

In addition to application in electrodeposition, fluidized electrodes are also used in the synthesis[61] of organic chemicals and in fuel cells.[62] Because it is useful to remove metal ions from dilute waste solutions, fluidized electrodes have gained importance in the purification of waste streams. The fluidized electrode has also been explored as an alternative to conventional electrowinning cells for the extraction of copper, nickel, cobalt, and a host of other metals, including precious metals such as gold and silver. Significant amounts of copper are treated and recovered by hydrometallurgy. In dump leaching, *in situ* leaching, and also copper-bearing acid waste streams, metal recovery using the conventional electrolytic cell poses problems due to the early onset of concentration polarization. Hence, other methods such as solvent extraction and ion exchange have to be used. In a fluidized electrode cell, the direct electrolysis of metal-bearing lean solutions is feasible using a comparable cell current at a reduced current density. This prevents rapid depletion of the Nernst layer, a factor responsible for the onset of polarization. Flett[56] proposed a flowsheet for the recovery of copper from dilute leach solutions by using a fluidized cathode for electrowinning and a fluidized anode for electrorefining of particulate copper obtained from an electrowinning cell. The acid produced in the electrowinning cell can be recycled for leaching, and a fraction of the pure particulate copper particles obtained from the refining cell can be used as seed particles in the electrowinning cell. Flett[56] pointed out that in cathodic operation, the current efficiency increases with an increase in cell current. However, it is possible to achieve a current efficiency in excess of 100% independent of cell current. This anomaly was not explained.

B. Cell Types and Design

1. *Criteria for Cell Geometry*

A detailed review on electrolytic cell design by Marshall and Walsh[63] gives an account of the various types of cells and their inherent merits and drawbacks. Features pertaining to hydrodynamics, potential distribution, mass transport, and cell construction are covered in this review. The cells are classified into three categories based on their geometry: stationary, rotating, and three-dimensional electrodes. Such electrochemical cells are mainly used for the removal of metals from aqueous solutions by cathodic deposition. They are also used in inorganic and organic synthesis as well as chloralkali industries. By and large, the dominant factor in all designs and especially in metal decomposition reactions is the rate of product removal. The parameters often considered in evaluating cell performance are current efficiency, cell voltage (i.e., energy consumption), anolyte as well as catholyte flow, particle size, concentration, temperature, anode geometry, cell orientation (i.e., inclined or vertical), and diaphragm material. Most researchers evaluate the data using combinations of such operating parameters of the cell over a limited period. However, during prolonged operation of industrial cells, metals are deposited on the current feeder, which necessitates frequent shutdown of the cell to clean up the current feeder. Metal deposition on the current feeder can be avoided by tilting the cell. Goodridge and Vance[64] investigated the electrowinning of zinc using a circulating bed electrode by titling the cell 21°. Circulation of the fluidized bed with the dense phase close to the wall and the lean phase close to the diaphragm of the cell can increase the deposition of zinc from an acid solution with an improved current efficiency at a high cathode density. Higher yields than in a conventional cell at comparable energy consumption can also be achieved.

2. *Anode Chamber and Electrolyte*

In an electrolytic cell used by Sabacky and Evans,[59] the anode side of the cell consumed 65% of the electrical energy as calculated by Ziegler et al.[65] This was attributed to high current densities. Oxygen bubbles tend to fill the anode chamber even when the anolyte is kept rapidly moving up, thereby causing a high I-R drop. Hence, modification of the anode chamber is thought to be essential to reduce energy consumption. This was achieved by Dubrovsky et al.[66] by placing the anode directly on the diaphragm. The anode thus serves as a support to the diaphragm, and the oxygen bubbles escape through the mesh. Dubrovsky and Evans[67] employed a highly acidic anolyte so as to have low anolyte resistivity and less energy consumption and a neutral or low acidic catholyte so as to achieve a high current efficiency. It is also possible to have two-zone fluidization in the fluidized cathode by skillfully injecting the catholyte either into an electrolytic cell which is inclinded[68] (as shown in Figure 6.7a) or by the spouting bed[69,70] technique (as shown in Figure 6.7b). In such cases,

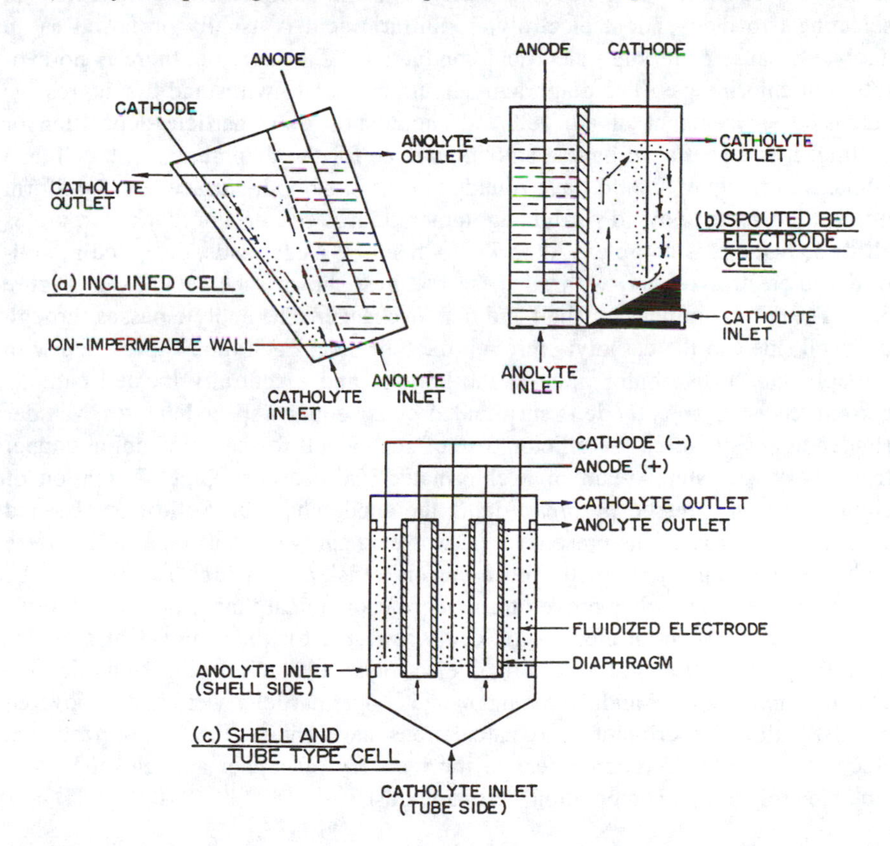

Figure 6.7 Schematics of some modified electrode cells.[68-70,72]

a dense phase of fluidized solids descends along the current feeder, thus establishing good electrical contact. A lean solid phase moves upward along the diaphragm, thereby eliminating any solid agglomeration on the surface of the diaphragm. Another way of making a dynamic electrode is by inert gas sparging,[71] which can reduce bed expansion without significantly affecting the solids concentration and increase the mass transfer rate due to the turbulence created by gas sparging.

3. Construction

Electrolytic cell construction and process control should be simple, and any scaleup should be economically favorable. The most commonly encountered problem in scaleup of an electrolytic cell is selection of the diaphragm. A diaphragm should have enough mechanical strength to withstand the fluidization pressure as well as the pressure on the anolyte chamber. In many chloride hydrometallurgical operations, electrolysis is accompanied by chlorine evolution. This can be suppressed by pressurizing the anode chamber, thereby arresting movement of the chlorine ions toward the anode. The diaphragm should also have low permeability so that the catholyte does not mix easily with the anolyte. This would help in selecting two independent electrolytes. Sulfuric acid is usually preferred as an anolyte because of its high electrical conductivity. Furthermore, there is no evolution of chlorine gas. The diaphragm should be able to withstand the aggressive corrosion environment of the cell and should not allow particle deposition or scaling. It is essential to have a smooth surface for the diaphragm. The cell as a whole, in principle, should have rounded corners, smooth surfaces, and uniform fluidization so as to avoid particle agglomeration. A cell design that incorporates all these features is complex. Akzo Zout Chemie,[72] Nederlands, designed a fluidized bed electrolytic cell with all these features, shown schematically in Figure 6.7c. The cell is similar to a shell and tube exchanger. The anolyte passes through the shell side and the catholyte through the tube side. Each tube is provided with a diaphragm, a distributor plate at the bottom, and a centrally located cathode current feeder. Each cathode is surrounded by six equally spaced anodes. Van der Heiden et al.[72] reported the efficient use of such a cell for electrowinning copper from the wastewater stream of a chlorinated hydrocarbon plant. Formation of chlorine was suppressed by pressurizing the anode chamber. Solid particles and chlorinated hydrocarbons present in the waste stream were not found to interfere with the operation of the cell, and a current efficiency as high as 70% can be achieved. The process has proven to be more economical than a precipitation/filtration process. Such an electrolytic cell was tested for the removal of mercury from the process stream of a chloralkali electrolysis plant (Hengelo, Netherlands). The mercury deposited and the amalgamated copper particles were later recovered by distillation. Experiments and calculations have shown that a fluidized bed electrode cell is more economical in the recovery of copper and nickel from a zinc electrolyte than cementation with zinc dust.

C. Modeling of Fluidized Electrodes

1. Cell Parameters

The space–time yield of an electrolytic cell on the basis of Faraday's law is given by:

$$Y_t = A_s \frac{i\beta M}{ZF} \tag{6.12}$$

It is the product of two terms: (1) the specific electrode area (A_s), which is dependent on the cell type and its construction, and (2) $i\beta M/ZF$, which represents the reaction that takes place in the electrolytic cell. If the reaction is slow (i.e., kinetics controlled), the space–time yield has to be increased by increasing the surface area available for the reaction. Flow-through porous electrodes, packed bed electrodes, or fluidized electrodes can increase A_s. It has been established that such electrodes are useful for the electrolysis of low-concentration solutions, waste streams, and downstream solutions of conventional electrolytic cells that process concentrated solutions. Because it is three-dimensional, the fluidized electrode is complex in that the current and the potential distributions in the electrode are nonuniform. The electrochemical rate equation for a plane electrode is a function of overvoltage, bulk concentration (c) in the electrolyte, flow velocity (u), and temperature. In a fluidized electrode cell, additional parameters such as particle size (d_p), particulate bed conductivity (K_p), electrolyte conductivity (K_l), bed voidage (ϵ), and bed depth (L) in the direction of current flow are important. Kreysa[73] suggested that the rate equation for a plane electrode is based on microkinetics, whereas it is based on macrokinetics for a three-dimensional fluidized electrode. A complex but important parameter pertinent to predictions regarding the fluidized electrode is the cathode current density (i_b), but few experimental data are available in the literature to test the validity of any predictive model.

2. Potential and Current Distribution

Goodridge[74] pointed out that a fluidized electrode can be monopolar or bipolar. In a monopolar fluidized bed, the electronic current (I_p) enters or leaves the bed through the current feeder, while the same is done by the ionic current (I_e) at the opposing boundary. Goodridge et al.[75] proposed equations for I_p and I_e:

$$I_p = -K_p A \frac{d\phi_p}{dX} \tag{6.13}$$

$$I_e = -K_e A \frac{d\phi_e}{dX} \tag{6.14}$$

where K_p and K_e are the specific electrical conductivities of the particulate bed and the electrolyte, respectively. The quantity K_p is influenced by hydrodynamic conditions. If $K_p >> K_e$, reaction occurs far away from the current feeder. An idealized electrode potential and current distribution in such a case are shown in Figure 6.8a. If $K_p \approx K_e$, there is an inactive zone near the center, as can be seen from the idealized ϕ (and I_x/I_L) versus X/L plot shown in Figure 6.8b. In a bipolar fluidized electrode, the potential on each particle is constant (i.e., resting potential), and a steep potential gradient imposed on the electrolyte renders the particle cathodic on one side and anodic on the other. Because the particles are in an agitated state, polarity frequently is reversed, and hence they are subjected to a self-cleaning action.

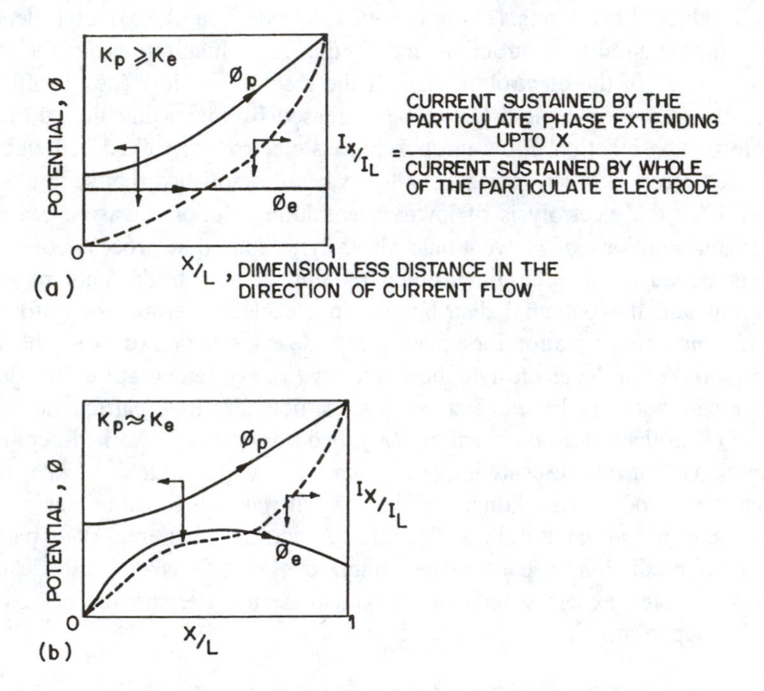

Figure 6.8 Potential and current distribution in idealized particulate electrode. Current feeder at $x/L = 0$ and diaphragm at $X/L = 1$. Active zone exists away from current feeder when $K_p \geq K_e$ and an inactive zone exists near the center when $K_p \approx K_e$.

3. *Charge Transport*

A macroscopic investigation of fluid flow through a porous electrode by Newman and Tobias[76] formed the basis of subsequent theoretical investigations on and modeling of fluidized electrodes. Trainham and Newman[77] developed a model for flow through a porous electrode as applied to iron removal from dilute solutions and predicted the current and potential distributions as well as the concentration in the presence of side reactions above and below a limiting current. It is necessary to carry out an electrochemical reaction below a certain limiting current density in order to avoid undesirable chemical reactions. Sabacky and Evans[58] presented models following the theory of Newman and Tobias and proposed expressions for the particulate current, electrolyte current, and reaction rate. The transport of ions within the electrolyte is assumed to be a combination of migration under electric field, diffusion/dispersion, and bulk flow. The mechanism that governs charge transport is electronic conduction through the chain of conducting particles which lie in contact with each other. Sabacky and Evans[58] suggested that the effective electrical conductivity of the particulate bed is a key parameter affecting bed performance. Poor electrical conductivity leads to a low current and poor power efficiency. Operational difficulties are encountered with low as well as high electrical conductivity beds. Bed conductivity is decreased by bed expansion but remains unaffected by any change in particle size and electrolyte conductivity. Sabacky and Evans[59] also developed a mathematical model to describe the behavior of fluidized cathodes used for the electrodeposition of metals. The model covers the effects of electrode and electrolyte resistivities on cell voltage, power consumption, and bed effectiveness and is applicable to multicomponent electrolytes and to electrolytes that contain oxidizing agents. This model can predict current efficiency. Sabacky and Evans[78] later tested their model for electrodeposition of copper. The mathematical model was found to be more or less consistent with the results of polarization and electrodeposition experiments. Power consumption at the anode chamber was found to be excessive.

4. *Current Density*

a. *Model Development*

When a fluidized electrode cell is used for electrowinning, the important parameter to be evaluated is the cathode current density, which is also an important parameter in determining the space–time yield. The cathode current density (i_b) can be evaluated[73] by the equation

$$i_b = A_s \int_0^L i\,\eta(x)dx \tag{6.15}$$

where

$$\eta(x) = \phi_p(x) - \phi_e(x) \tag{6.16}$$

and the specific electrode surface area (A_s) is given by:

$$A_s = 6(1 - \epsilon)/d_p \tag{6.17}$$

In order to use Equation 6.15, it is necessary to have knowledge of the potentials ϕ_p and ϕ_e. If, on the other hand, the total current (I) and the surface area of the particles are known, then, according to LeRoy,[79]

$$i_b = I/S \tag{6.18}$$

where

$$S = N\pi \int_0^\infty d^2 P(d) \cdot d(d) \qquad (6.19)$$

The function $P(d)$ represents the fraction of the particles with sizes in the range of d and $d + d(d)$. The number N in an electrode is calculated from knowledge of the weight (W) of the particles, their density (ρ_p), and their size distribution. Thus,

$$N = \frac{6W}{\pi \cdot \rho} \int_0^\infty d_p^3 P(d) d(d) \qquad (6.20)$$

Kreysa[73] developed an expression to predict the optimum bed depth at which the space–time yield is maximal. Details of the development of the model and the relevant scaleup procedure are beyond the scope of this book.

b. Particle Size Effect

Fluidized electrodes, unlike porous or packed electrodes, have gained prominence in the electrowinning of metals because the electrodeposited metal does not clog or hinder the operation. The electrolyte flow can be kept in a turbulent state through interstitial flow. Furthermore, the electrowinning cell can be operated on either a semicontinuous or a continuous basis. The cathode current density in a batch fluidized electrode has a tendency to increase as the specific surface area of the particles keep decreasing due to particle growth caused by electrodeposition. The growing particles tend to settle, thereby defluidizing the electrode. Hence, a fluidized bed electrode, when employed for long runs in an electrowinning operation, should be attended to periodically to remove particles that have grown quite large and to replace them with fresh seed particles. Thus, it is imperative that a fluidized bed electrowinning operation be carried out at a constant bed current density; to achieve this, it is necessary to have a constant particle size, a constant particle number, and a constant bed weight of the fluidized electrode. LeRoy[79,80] analyzed this aspect by periodically removing particles (accomplished either by elutriation of the fines or by gravity flow of coarse particles) from the bed and adding a certain original size fraction. A detailed analysis of the particle size distribution and the methods to achieve constant current density in fluidized electrowinning was provided by LeRoy.[80]

5. Current Feeder

In a fluidized electrode, the particles fluidized are usually electrically conducting, and a current feeder is normally used to create electrical contact with the fluidized bed in order to render the bed electrochemically active. However, there are situations where the current feeder such as a plate or grid is immersed in the electrolyte which fluidizes an inert bed of solids like glass ballotini. The purpose here is to enhance mass transport and therefore deposition by using an electrochemically inert fluidized bed. This was the basis of the Chemelec cell developed at Electricity Research Station, Capenhurst, England, the details of which can be found in a publication by Lopez-Cacicedo.[81] Mass transfer data in the literature for inert fluidized bed electrodes of this type are scant. Walker and Wragg[82] gave a brief account of the work carried out in this area and also proposed correlations for predicting the mass transfer rate. They proposed correlations to estimate mass transport in cathodic deposition of copper. In the case of mass transfer at a plane wall in an electrochemically inert fluidized electrode, the correlation gives

$$j_D \epsilon = 0.138\,[\mathrm{Re}_p/1-\epsilon]^{-0.39}\,(d_p/d_e)^{-0.39} \tag{6.21}$$

and is valid for

$$0.936 < [\mathrm{Re}_p/1-\epsilon] < 67$$

For mass transfer between the electrolyte and the particle within an active fluidized bed of conducting copper particles, the correlation is

$$j_p = \frac{k}{u}\mathrm{Sc}^{2/3} = 1.55\left[\frac{\mathrm{Re}_p}{1-\epsilon}\right]^{-0.49} \tag{6.22}$$

and is valid for

$$2.6 < \frac{\mathrm{Re}}{1-\epsilon} < 30$$

D. Three-Phase Fluidized Electrodes

1. *Role of the Third Phase*

So far, we have discussed fluidizing electrodes where a particulate bed of solids is fluidized by a liquid catholyte. This class of electrodes is associated with two-phase fluidization. If a third phase, that is, a gas, is also introduced, then a different type of electrode, known as a three-phase fluidized electrode, is formed. The gas introduced may be inert or reactive. Kusakabe et al.[83] introduced nitrogen gas in a two-phase fluidized electrode (copper particles fluidized by an aqueous solution containing copper). They used a rectangular-shaped three-phase fluidized electrode chamber, and the dimensions of the cathode chamber were 65 cm (height) × 15 cm (length) × 22 cm (width). The areas of the current feeders used were 90, 150, and 300 cm^2, and the fluidized bed heights studied were 6, 10, and 20 cm. The studies revealed that the potential difference between the particles and the electrolyte solution phase near the current feeder in the cathode chamber increased as the bed expansion increased. This indicated the presence of an electrochemically active zone near the current feeder which resulted in an increased rate of copper reduction. Increased bed expansion can also cause an increase in bed resistance. The nitrogen gas introduced into the bed increased mass transport from liquid to solid and enhanced the rate of electrodeposition of copper.

2. Cells

a. Third-Phase-Injected Type

The three-phase fluidized electrode was employed by Oloman and Watkinson[84] for the electroreduction of oxygen. Details pertaining to the electroreduction of oxygen in alkaline solutions can be found in the literature.[85] The reduction of oxygen by carbon is favored in alkaline solutions. The oxygen can be supplied into the cell either by presaturating the electrolyte or by directly sparging air into the cell. The latter case uses a three-phase fluidized electrode (i.e., air–electrolyte–carbon-particle fluidized electrode). The three-phase fluidized electrode cell studied by Oloman and Wstkinson[84] consisted of a cathode chamber 42 cm (height) × 5 cm (length) × 1 cm (width) with a current feeder plate with a 20-cm^2 surface area. The anode and the cathode chambers were separated by a diaphragm made of 100-mesh nylon backed by 0.31-cm-thick Plexiglass. The catholyte was distributed through a block of 0.63-cm-thick Plexiglass with 30 0.04-cm holes, and oxygen was distributed through 12 pinholes (~0.01 cm) drilled in an 80-cm-long, 0.16-cm-outer-diameter Teflon tube. In general, bed expansion in a fluidized bed has an adverse effect, as the potential distribution is unfavorably disturbed. An increase in the flow rate of oxygen in the three-phase electrode at a given current density increases hydrogen peroxide production. At high electrolyte flow rates, the introduction of oxygen brings about a large contraction in the bed, thereby creating a high concentration of oxygen; this is responsible for yielding hydrogen peroxide at a relatively higher rate compared to a two-phase fluidized bed under the same operating conditions. Particle size plays a key role in a three-phase fluidized bed. As particle size is increased, gas–liquid mass transfer increases, but liquid-to-solid mass transfer is lowered due to reduction of the surface area of the solid. For any given oxygen gas flow rate, increasing catholyte flow increases production.

b. Third-Phase-Generated Type

In this type of three-phase fluidized electrode, the gas phase, instead of being introduced, evolves in the reactor due to the reaction. Photoelectrolysis or photocatalysis by radiant energy[86] is a pertinent example. Another typical example is the photosynthesis of ethane and hydrogen using platinum as the cathode and a particulate bed of single-crystal titania (TiO_2) as the anode. The electrolyte solution is ethanoic acid (CH_3COOH). The gases evolved are hydrogen and carbon dioxide at the cathode and the anode, respectively. Three-phase fluidized electrodes are still largely of academic interest.

E. Applications

Because an electrically conducting mobile particulate bed in an electrolyte increases the specific surface area of the electrode, enhances ionic transport as well as mass transport from the liquid to the particles, reduces the optimum size of the cell, and improves and the space–time yield, it has stimulated research interest in electrochemical engineering with a view to adopting this technique in electrowinning[57-59,64,66,67,72,78-80,87,88] and electrorefining[57] of metals, fuel cells,[62,89] organic chemical reactions,[54,88,90] inorganic chemical reactions,[91,92] and in wastewater cleanup.[72] This technique is advantageous in electrochemical extraction of metals because the metal particles deposited by electrolysis are highly pure, spherical in shape, and can be discharged continuously. Short-circuiting of the electrodes due to the bridging effect of the deposited metal, which poses problems with conventional planar electrodes, is eliminated in a fluidized bed electrode (FBE). The use of FBEs for the production of powders of copper and other nonferrous metals by direct reduction was proposed long ago by Scuffham and Rowden,[92] and Coeuret[93] discussed the use of such electrodes in sulfide mineral oxidation/electrodissolution.

The use of fluidized electrodes for electrowinning and electrorefining has been recognized in the context of commercially important metals such as copper, zinc, cobalt, and nickel. However, studies on the extraction of these metals by this technique have mostly been confined to the laboratory. Industrial-scale plants that make use of this route for the recovery of such key metals appear to be very few in number. The details of commercial units, if any, are not published in the open literature, and many pertinent technical details are patented or remain proprietary. The following discussion highlights a few studies that have demonstrated the applications of fluidized electrodes in the extractive metallurgy of copper, cobalt, nickel, zinc, and silver and in waste stream processing.

The use of the fluidized electrode in electrowinning/electrorefining is perhaps the oldest of its applications. Dubrovsky et al.[60] included in their review an account of the work carried out up to the early 1980s on the electrowinning of metals. Since then, there have been radical changes and improvements in this area as a result of continued research. Some salient features of a few select examples of the application of FBEs are presented next.

1. Fluidized Bed Electrodes in Copper Extraction

Goodridge and Vance[94] described copper deposition using a pilot-scale fluidized bed cell. The cell was made of Perspex and was rectangular in shape, with 500- to 700-μm copper particles fluidized by copper sulfate solution in a catholyte chamber. The current was fed to the catholyte chamber through a copper sheet that was kept flush with the wall of the cell on the catholyte side. The anolyte chamber was provided with a platinized titanium sheet, and this chamber was supported by a 0.2-m^2 cationic membrane on a polyethylene-coated metal grid. Power to the cell was supplied from a 1200-A, 150-V rectifier. The current density to the diaphragm ranged from 1000 to 5000 A/m^2. The electrolyte to the cell was distributed through the holes of a distributor plate made of Perspex. Experimental results showed that a current efficiency close to 100% could be obtained when the diaphragm current density was above 2000 A/m^2. The current efficiency remained the same irrespective of the electrolyte concentration when the latter exceeded 500 ppm of copper. Current efficiency was influenced by the electrolyte flow rate (or percent bed expansion); the lower the bed expansion, the higher the current efficiency. A mathematical model developed to determine the current distribution and electrode potential in a three-dimensional FBE was found to fit well with experimental results and useful with regard to further scaleup of the cell.

Van der Heiden et al.[72] explained the tremendous cost saving that can be achieved by the use of FBEs to remove copper content in sulfuric acid solution from 300 to 5 mg/l. A fluidized electrode cell unit to remove copper at a rate of 1 kg/h from aqueous solution containing 100 g/l of sulfuric acid was set up at Wuppertal in West Germany. The volumetric flow rate of the sulfuric acid solution was maintained at 3 m^3/h. The sulfuric acid could be regenerated and recycled by this technique. The conventional ion-exchange technique used ammonia for neutralization, thus adding to the cost through the loss of both sulfuric acid and ammonia. The estimated annual saving at 1978 prices was $200,000 per FBE. It should be mentioned here that investment costs for ion exchange and for FBE are in the same range. Sabacky and Evans[78] studied copper deposition in an FBE, using superficial current densities in the range 200–600 A ft^{-2}, from an electrolyte solution containing 100 g/l sulfuric acid and 0–20 g/l of copper. The studies were carried out on a laboratory-scale experimental unit with the goal of developing relevant mathematical models. Mass transfer studies on the electrodeposition of copper from aqueous solutions containing sulfuric acid were reported by Walker and Wragg,[82] who proposed correlations to predict the mass transport factor (j_D) for copper. Masterson and Evans[95] investigated fluidized bed electrowinning of copper using cylindrical and rectangular cells operating at current densities of 150 and 1000 A. A small unit was used to measure current efficiency and energy consumption and to investigate the effects of several variables such as diaphragm material, anolyte geometry, anolyte flow, fluidized particle size, and temperature. The feasibility of electrowinning in a larger cell was demonstrated, and a modification to the anode was suggested to reduce energy consumption in the cell.

2. *Fluidized Bed Electrodes in Nickel Extraction*

Sherwood et al.[96] described the work carried out at Amax Extractive Research and Development, Inc. on fluid bed electrolysis of nickel. One of the objectives of this work was to recover nickel from the process liquors generated from sulfuric acid pressure leaching of nickeliferrous laterite containing 8 g/l of nickel. The investigation was actually carried out with a synthetic feed liquor containing 8 g/l nickel and 60 g/l sodium sulfate. The cell used was cylindrical in shape with electrodes of the side-by-side design. The cell had two concentric compartments divided by a porous cylindrical polypropylene diaphragm. The outer cell was a 6.6-cm-OD and 2.5-cm-ID catholyte chamber, and the central compartment was the anolyte chamber. The fluidized cathode materials tested were titanium, steel shots, graphite, sand, nickel shots, and nickel oxide. Studies were carried out to examine the effects of several parameters that affect the electrodeposition of nickel. The parameters studied included current density, pH, temperature, nickel concentration, cathode current feeder design, flow distribution system, electrolyte conductivity, and impurities in the electrolyte. The series of experiments carried out on the electrodeposition of nickel on fluidized bed cathodes revealed that nickel can be recovered effectively from dilute electrolyte solutions but at a relatively low current efficiency of 50%. The current efficiency increased at higher current densities and at higher pH values. The rate of nickel deposition increased at a higher temperature (e.g., 70°C). Current efficiency increased at higher current densities even at parts-per-million levels of nickel concentration in the electrolyte. Cathode materials such as hydrogen-reduced nickel oxide pellets and nickel electroless plated steel spheres were suggested from the standpoint of their effectiveness and cost. The use of nylon membranes that are consistent with cell operation at low voltage and high current efficiencies without pore clogging was suggested for successful cell design. Impurities such as magnesium sulfate and chromium can interfere with the electrodeposition of nickel. Magnesium sulfate can increase the conductivity of the electrolyte and trivalent chromium can deposit on the cathode current feeder, inhibiting current flow to the particulate bed. Hence, any scaleup of FBE cells for the electroextraction of nickel should give due consideration to the selection of several parameters for successful operation and economical product collection.

3. Electrowinning of Cobalt, Silver, and Zinc

a. Cobalt

The electrodeposition of cobalt from an aqueous solution is accompanied by the simultaneous evolution of hydrogen. The electrochemical reaction and the electrode potentials at 25°C are

$$Co^{++} + 2e^- = Co \qquad E_o = -0.27\ V \tag{6.23}$$

$$2H^+ + 2e^- = H_2 \qquad E_o = 0\ V \tag{6.24}$$

$$2\ H_2O + 2e^- = H_2 + 2\ OH^- \qquad E_o = -0.414\ V\ (\text{at pH 7}) \tag{6.25}$$

Hydrogen evolution results in the formation of metal hydroxide at the cathode,[97] and a low current density is required to prevent this adverse cathode reaction. Under high current densities at low cobalt concentration and high pH values, the electrode is polarized with respect to Reactions 6.23 and 6.24, and hence Reaction 6.25 will occur, leading to the formation of cobalt hydroxide on the electrode. This undesirable situation must be avoided in order to maintain a cathode of good quality. FBEs, by virtue of their ability to establish low current densities, eliminate hydrogen evolution and allow metal deposition to occur continuously. This favorable feature, together with the added advantage of automated operation, has made the use of fluidized electrodes in the electrowinning of cobalt attractive. Dubrovsky and Evans[98] reported their investigations on the electrowinning of cobalt accomplished by FBE in 50- and 100-A cells. The cell was fitted with an impermeable diaphragm to permit the use of an anolyte and a catholyte whose compositions could be varied independently. The cobalt content of the catholyte varied from 4.88 to 100 g/l and the pH from neutral (6.0) to highly acidic (1.0) conditions. Superficial current densities up to 1.09 A/cm^2 were used to study cobalt deposition. Current efficiency, energy consumption of the cell, and deposit quality were examined. Fluidized electrodes were found suitable for the electrowinning of cobalt down to 17 g/l from a catholyte in the pH range 1.8–3.0 and at current densities up to 0.5 A/m^2. Even though the electrode potentials for hydrogen evolution and cobalt deposition are quite close, the feasibility of carrying out electrowinning of cobalt without the apparent deterioration of cathode quality or significant loss of current efficiency was demonstrated. However, energy consumption in fluidized bed electrowinning cells is higher than in conventional cells, but this is not due to the choice of the FBE. Resistance at the anode chamber and the diaphragm appears to be responsible for this high energy consumption. Dubrovsky and Evans[98] recommend the development of an improved fluidized bed electrowinning cell where this resistance loss is minimized.

b. Silver

Silver is electrowon from its aqueous cyanide solution in the Zadra[99] cell, which used an extended surface area cathode with steel wool packing. This kind of cell has to be operated batchwise, and on prolonged operation the cathode undergoes degradation due to cementation with silver solution. Also, the electrowon silver is contaminated with steel. The fluidized electrode is an alternative to the Zadra cell. Kreysa and Heitz[100] investigated the electrowinning of silver using a low current density. Fluidized electrowinning of silver was also explored by Huh et al.[101] The electrolytic cell designed by Huh et al.[101] consisted of a cathode chamber of rectangular cross-section (63 mm × 19 mm) in the direction of the current flow; the bed height used was 150 mm with 25% expansion. The cathode chamber was provided with a 4-mm-thick graphite current feeder and was separated from the anode chamber by a low-permeability diaphragm. The catholyte used was a silver- (200 mg/l) containing solution mixed with 15 g/l of silver cyanide solution. The cathode-fluidized particles were silver particles 1 mm in diameter and also silver-coated copper particles. The anolyte used was an alkaline solution (sodium hydroxide dissolved in distilled water and pH adjusted). The use of an acidic electrolyte was avoided for safety reasons. Six different anode materials were used: Diamond Shamrock dimensionally stable anode (DSA), pure lead, 304 stainless-steel mesh, platinum-coated titanium mesh, 316 stainless-steel plate with drilled holes 3.97 mm in diameter, and grooved graphite plate. The mesh, the drilled holes, and the grooves were made to facilitate easy escape of oxygen from the anode, thereby reducing the cell voltage and decreasing the energy consumption of the cell. The 304 stainless-steel and lead anodes and the DSA were not found to be suitable due to their rapid deterioration. The electrowinning of silver was carried out batchwise, and the experiments were stopped when the hydrogen evolution in the cathode was intense. The experiments showed that silver can be electrowon successfully in a manner similar to any other commercial unit that employs fluidized electrodes. Current efficiency was found to be acceptable except when the cell was operated at high currents and low silver concentrations. The catholyte can be stripped of silver down to parts-per-million levels. Impurities such as copper, zinc, and iron do not seem to interfere with the quality of the electrodeposited silver. Cell voltage was found to be influenced by the pH of the anolyte. There seems to be considerable potential for the use of fluidized electrodes to electrowin silver from lean solutions economically without compromising product quality.

c. *Zinc*

Although fluidized electrodes have been tested and accepted for the electrowinning of copper, cobalt, nickel, and silver, application of this technique for the electrowinning of zinc has not been explored in detail. The pertinent literature is scant. Only a few investigations[64,102] have been conducted on a laboratory scale to examine the electrodeposition of zinc on the FBE from acid sulfate solutions. Conventional cells are mainly used in industrial practice. There are some practical problems associated with the process of electrodeposition of zinc. One relates to the proximity of the electrode potentials of zinc and hydrogen. Thermodynamically, hydrogen deposition is more favorable than zinc deposition, but many commercial cells are successfully electrowinning zinc. This is due to the kinetics, which favors the deposition of zinc rather than hydrogen on a zinc surface. In other words, there is a large overpotential for hydrogen deposition on zinc, and this in turn implies that any mechanism that would lower this overpotential would result in the evolution of hydrogen, simultaneously bringing down the current efficiency of the cell. A fluidized electrode cell can be sealed at the top and the hydrogen can be collected as a by-product. In a conventional cell, acid mist is a major environmental problem, but it is avoided in a fluidized electrode. An important problem often encountered during the electrodeposition of zinc is the presence of metallic impurities which are more electropositive than zinc. Some impurities, such as antimony, arsenic, germanium, selenium, and tellurium, act in a complex way. The mechanism of impurity interaction during the electrowinning of zinc has not been fully explored. However, certain impurities which form hydrides can initiate localized corrosion or trigger the dissolution of the electrodeposited zinc even when the cathode is polarized. The advantages of the fluidized electrode in electrowinning are mainly due to high space–time yields, automation, continuous operation, and reduced electricity consumption. These advantages have encouraged some basic research activities in this field. Goodridge and Vance[64] used a circulating bed fluidized electrode with an inclination of 21° and studied the electrowinning of zinc. The inclination was provided to induce circulation within the cell and to create a lean phase near the diaphragm of the cell so as to avoid any deposition there. The diaphragm used was a 10^{-2}-m^2 semipermeable anionic membrane. The rectangular-shaped cell was made of Perspex and was 0.025 m wide in the direction of the current flow. Zinc deposition was achieved on a fluidized cathode of zinc particles 500–700 μm in size. The electrolytes used in the cell were kept at 35°C and were free from impurities that would interfere with the electrodeposition of zinc.

Experimental results have shown the possibility of electrowinning zinc using a circulating FBE at a high current efficiency and space–time yield. The investigations of Jiricny and Evans[102] on fluidized bed electrowinning of zinc involved varying the zinc concentration from 50 to 75 g/l and the sulfuric acid concentration from 0 to 150 g/l. These ranges of variation correspond to normal plant operation. The studies were conducted to optimize these ranges of concentrations and also the impurity levels (i.e., cobalt, nickel, antimony, and glue) and to examine their effects on current efficiency, cell voltage, and power consumption. The cell current used was 50 A and the cell was similar to the one already described for the electrowinning of silver. The fluidized particles were zinc-coated copper particles 400–600 μm in diameter, and the catholyte was hydrated zinc sulfate ($ZnSO_4 \cdot 7H_2O$) dissolved in sulfuric acid (AR grade). The range of concentrations of zinc and the acid used corresponded to normal plant operating conditions, as mentioned previously. The effects of metal impurities such as cobalt, nickel, and antimony and of flue gas were studied and the cell was operated over a wide range of current densities (100–10,000 A/m^2). Experimental findings showed that zinc could be electrowon with a high space–time yield by using a fluidized cathode, but energy consumption was high and was affected by the acid concentration and the impurities. The high energy consumption was associated with fluidized bed electrowinning of zinc but not copper. Thus, it appeared that an FBE, if operated to derive the benefit of a high space–time yield (i.e., low capital cost), has the disadvantage of high energy consumption. Because these two factors vary with time and geographical location, Jiricny and Evans[102] did not comment on economic aspects pertinent to the selection of fluidized bed electrowinning of zinc. Considerable developmental work is required before this technique can be used at a current consumption equivalent to that of a conventional cell. Low acidic content and glue additions to the electrolyte are recommended to lower energy consumption, but implementation of these suggestions in actual practice warrants an altogether different electrolytic plant. It should be noted that Van der Heiden et al.[72] gave a cost comparison and found the use of the FBE for the removal of copper and cadmium from zinc electrolytes to be more economical compared to cementation with zinc dust.

4. *Fluidized Bed Electrodes in Aqueous Waste Treatment*

Many industrial waste streams contain toxic metals in very low concentrations. The automobile industry uses huge amounts of rinsing water for metal plating operations. The chlorinated hydrocarbon industry releases chloride and metal wastes, and the chloralkali industry waste stream contains mercury. It is necessary to detoxify these wastes and recover the metals before disposal of the waste. These days, there is considerable concern regarding environmental pollution, and it is now imperative that each industry remove these toxic materials from the waste stream and dispose of them safely. A single-step clean process for recovering metals from lean waste solutions is more desirable than any downstream processing. It is worth mentioning here that electrochemistry is often used to analyze the metal values present in 0.01-ppb levels. Hence, the electrochemical route for the removal of trace metal values from waste solutions is obviously accepted. Conventional methods of subjecting the waste to precipitation, cementation, solvent extraction, and ion exchange require subsequent additional operations to recover the marketable metal. For example, precipitation and cementation end up with sludges or cakes, while ion exchange and solvent extraction end up with concentrated solutions, thus necessitating further operations for the recovery of valuable metals. Akzo Zout Chemie[72] (Netherlands) was probably the first to develop an FBE plant for the recovery of copper from the waste stream of a plant that processed chlorinated hydrocarbons. The copper content was brought down from 100 to 1 mg/l in a single pass. A cell current efficiency of up to 70% without chlorine evolution at the anode chamber (by pressurization) was achieved. This technique proved to be more economical than precipitation/filtration in terms of capital and operational costs. The FBE was successfully tested at Hengelo (Netherlands) to remove mercury from the brine stream of the mercury cell of a chloralkali electrolysis plant; the mercury level in the stream was brought down from 5 to 0.05 mg/l. Mercury was electrodeposited over fluidized copper particles, and the amalgamated mercury was recovered by distillation from the copper particles. Van der Heiden et al.[72] listed a few industrially important applications of the fluidized electrode, especially for metal extraction. Fluidized bed electrolysis proved to be viable for the separation of copper from a nickel solution using nickel electrolyte (100 g/l Cl^{-1}, pH 1). The copper content was brought down from 2 g/l in 50-g/l nickel solution to 0.1 mg/1 and 99% electrodeposited copper was obtained. The FBE was also shown to be efficient in removing arsenic from copper, which is useful in the hydrometallurgy of copper as the presence of arsenic leads to the evolution of arsenic hydride during copper electrolysis.

5. Miscellaneous Applications

In fluidized beds of electronically conducting particles, the polarity changes frequently, and hence the bipolar FBE concept was used[91] for the electrolysis of seawater, production of hypobromite, and synthesis of dimethyl sebacate. The electrolysis of water using a bed of solids such as graphite or baked carbon particles results in the generation of hypochlorite, which inhibits the accumulation of microorganisms used in cooling water tanks of power stations. In a conventional cell, the major problem is passivation of the electrode material with the precipitation of magnesium hydroxide. In a bipolar fluidized electrode, this problem can be eliminated, and prolonged electrolysis of seawater without deposition of magnesium hydroxide can be accomplished. It is appropriate to recall here that electrochemical oxidation of sulfides[93] with *in situ* leaching is accomplished using the fluidized electrode. In this respect, fluidized bipolar cells have potential application in the recovery of metal values from sulfide concentrates (e.g., molybdenum from molybdenum disulfide), simultaneously eliminating the pollution due to sulfur, which in a roast leach operation appears as sulfur dioxide in the off gas, necessitating the use of post-reactor pollution control equipment. An important area of application of fluidized electrodes relates to fuel cells.[62] A fuel cell is an electrochemical power source where reactants such as oxygen, hydrogen, ammonia, methanol, and hydrazine (i.e., gas or liquid fuels) are continuously fed and current-producing reactions are accomplished on the surface of a current-collecting electrode. The electrode is not directly involved in the reaction and remains inert. However, a suitable coating of catalytic materials improves cell performance tremendously. The reaction sites at the intersection of the three phases (gas, liquid, and solid) are improved by the use of highly porous electrically conducting materials such as nickel, carbon, and platinum. A fluidized electrode of such material is an ideal choice. The use of gaseous fuel poses problems such as stabilizing the three-phase fluidization, loss of electrical conduction due to bubbles that coalesce rapidly, and transport of the electrode materials from the bed.

Electrochemical reactions in organic chemicals require current densities of the order of 1–30 mA/cm^2 compared to 100 mA/cm^2 for inorganic electrochemical reactions. This implies the use of a very large planar-type electrode, which is not economical. A fluidized electrode is an alternative to this and is the best choice for organic electrochemical reactions. Fluidized bed electrolytes are also employed to study mass transfer and then to simulate heat transfer characteristics, based on analogy. One such application involves the use of the limiting diffusion current technique for accurate and rapid cold modeling of heat transfer in situations such as downstream flow of nozzles and within nuclear reactor core blockages. Tucker and Wragg[103] investigated the mass transfer rates from a liquid fluidized bed of ballotini particles to immersed tubular nickel electrodes to simulate heat transfer characteristics by analogy. Recently, fluidized electrodes were also employed in the conditioning of the scrub solutions[104] obtained from the flue gas scrubbers of waste or toxic incinerators. The scrub solution is often loaded with heavy metals and is also contaminated with hydrochloric acid. The conventional method, comprised of neutralization, precipitation, and filtration, results in large quantities of sludge that contain heavy metals, and disposal of the sludge poses a formidable environmental problem. In a fluidized electrolytic cell, the hydrochloric acid is split into hydrogen and chlorine, simultaneously depositing the heavy metal at the cathode.

III. FLUIDIZED BED BIOPROCESSING

A. Introduction

Our mineral resources are rapidly being depleted and it is unlikely that new deposits will be found to meet the increasing demand. There have been continued efforts to process the available resources efficiently and to utilize them optimally. In recent times, in order to fill the gap between demand and supply of minerals, it has become necessary to devise methods to gainfully process low-grade or complex ores and ore fines. In pursuit of new technologies consistent with economic and environmental imperatives, bioprocessing of minerals has emerged as a promising alternative to hitherto practiced chemical methods of mineral processing and metal extraction. Biotechnology is becoming increasingly important in food processing, pharmaceuticals, agriculture, and wastewater treatment, but it has not yet been fully exploited in the mineral industries. The literature on bioprocessing of minerals is scant, and research in this area has not been intensive. In our presentation on this subject, we will deal with some important aspects of bioprocessing of minerals and then touch upon the use of fluidized bioreactors for the same.

B. Bioassisted Processes

1. Microorganisms

The details of conventional bioleaching and its success were well reviewed by Torma[105,106] and Groudeva and Groudeva.[107] In the area of mineral processing, bioleaching is employed for sulfide and nonsulfide minerals. It is also employed in mineral flocculation and flotation. A recent trend is to treat wastewater from mineral industries by biological methods. The use of microorganisms such as *Thiobacillus ferroxidans* is well known for oxidation of sulfur, sulfides, and Fe^{2+}. It was recently shown[108] that these microorganisms can also oxidize Cu^+, UO_2, and Mo^{5+}. Bioleaching studies were proposed in the context of processing polymetallic ores[107] and also platinum group metals.[109]

With the advent of genetic engineering, genetically improved strains of autotrophic bacteria have been developed to meet the host of environmental conditions that prevail during the bioprocessing of minerals. The microorganisms used in mineral processing and metal extraction should be able to sustain their activities in media that contain cyanides, chlorides, alkalies, and viruses that can kill them. Good attachment to the minerals and a sustained ability to leach out precious components of the minerals in hot or cold environments and to tolerate increased metal ion and surfactant concentrations are some of the most desirable attributes of the microorganisms used in bioprocessing of minerals. Such microorganisms are usually developed by chemical mutagenesis, conjugation and transduction, and modern genetic engineering. The details of the steps involved in genetic manipulations of *Thiobacillus ferroxidans* and its application in biohydrometallurgy were described by Holmes and Yates.[110] Heterotrophs useful for bioleaching are shown in Figure 6.9 with typical examples.

2. Mineral and Metal Extraction

The applications of biologically assisted processes in the treatment of various minerals and of the waste streams from mineral industries are listed in Table 6.1. The information in Table 6.1 is not exhaustive but rather is a list of representative examples to show the important contribution of microorganisms not only to mineral and metal extraction but also to the preservation of clean ecosystems. The biological methods for treating the waste organic chemical streams from flotation and hydrometallurgical plants are increasingly drawing the attention of many researchers. The conventional methods of treating these complex organic wastes are difficult to adopt and also are not economically viable. In this regard, biogradation is an alternative and promising way of disposing of the organic effluents from mineral plants and the waste from the milling and mining equipment that generates large volumes of petroleum waste. Although many investigations on the biogradation of a variety of

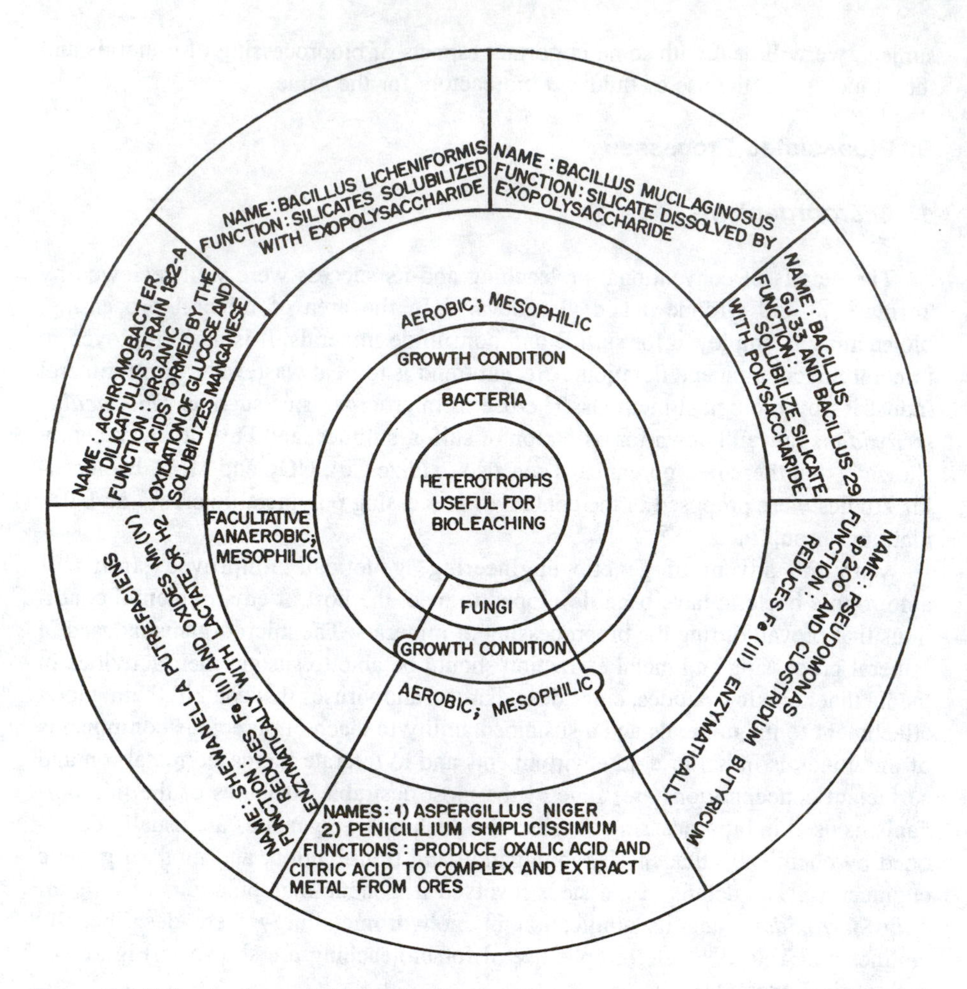

Figure 6.9 Heterotrophs useful for bioleaching.

organic chemicals have been carried out, not much work has been done on the disposal of flotation collectors. Studies[124] on biogradation of collectors showed that flotation collectors can be effectively destroyed and disposed of by biogradation methods. The commercial exploitation of such an economically viable process is still unrealized and today's atmosphere of environmental awareness serves as strong incentive to look into the possible application of this biological method of waste disposal. It should be noted that frothers and modifiers used in flotation cells are also potential threats to the environment because they are toxic and always appear in waste streams in nontrivial amounts. It appears that no work has been carried out on the biogradation of such frothers and modifiers. Although bacteria and algae are known to concentrate metal ions and clean waste aqueous solutions that contain heavy metal ions, not much is known about the influence of such absorbed or adsorbed metal ions on microorganisms, specifically with respect to their electrokinetic and hydrophobic flotation and flocculation behaviors. Microorganisms can be

Table 6.1 Bioprocessing in Modern Mineral and Metal Extraction

Area of Application	Processes	Remarks	Ref.
Biooxidation of sulfide ores (explored as an alternate route for roasting and preoxidation)	Cyanide extraction of arsenopyrite gold ore; biotreatment followed by cyanidation	Gold recovery enhanced to 84% compared to 54% obtained without biotreatment but with 24-hr cyanidation	111
	Combined chemical and bioleaching (using a mixed culture) of polymetallic sulfide ore	Metals extracted: Zn, 77%; Cu, 73%; Pb, 71%; Au, 63%; Ag, 57%; recovery of these metals was higher than those obtained without biotreatment	107
	Biologically assisted leaching of Greek laterites containing 0.73–4.27% Ni; microorganisms such as *Aspergillus* and *Penicilliium* were used	Useful for economical recovery of Ni from low-grade lateritic deposits (<1% Ni); *Penicillium* is effective in solubilizing Ni	112
	As recovery by cyanidation after biotreatment of Arizona manganese–silver ore for 14 days with 2-l 15% pulp density concentrate	Solubilized Mn and Ag, respectively, 99.8 and 92.5% by *Bacillus* strain MBX 1	113
Bioleaching of Mn ore	Leaching rate of Mn by citric acid is confirmed and hence Mn-bearing material was subjected to bioleaching with factory-grade as well as food-grade molasses	Column leach with periodic medium replacement can yield 99% Mn recovery in 51 weeks; heap leach proved to be less efficient	114
Bioprocessing of phosphate ore	Solubilization of phosphate from ores tested with 860 microbial isolates; solubilization of PO_4 occurred	Some organisms were inhibited in their action with soluble PO_4; solubilization of PO_4 in rock phosphate is possible	115
Biological treatment of mineral aqueous wastes from mineral industries	Biosorbtion for accumulation of heavy metals like U (i.e., UO_4^{2+}) and Th and desorption of accumulated metal by treatment with Na_2CO_3 solution	Excellent accumulation of metal; useful for cleanup of mine and effluent wastewater	116, 117
	Metabolic process	Accumulation of Pb 11	117
	Biofix beads with microorganism fixed in the porous polysulfone matrix used in column contactor; commercially adopted for metal cleanup and recovery of microorganisms	The beads removed heavy metal from low levels; the metal ions eluted and beds reused; presence of Ca(II) and Mg(II) did not affect the process	118, 119
	Organic mats that contain three layers composed of blue-green algae at the top, grass in the middle, and bacteria at the bottom when used on pond surface could remove metals such as Se	Se present as selenate; bacteria use algae as food and reduce selenate to Se, which can be harvested from the mat which has structural integrity	120,121

Table 6.1 Bioprocessing in Modern Mineral and Metal Extraction (continued)

Area of Application	Processes	Remarks	Ref.
	Bacterial deoxidation of Hg waste; inorganic Hg that transformed to alkylated species by biocatalyzed process is volatilized by genetically amplified microorganisms (mer A)	Methylated form of Hg is highly toxic and bioaccumulates in aquatic organisms, causing serious problems in aquatic ecosystems	122
	Reduction of toxic Cr(VI) to less toxic Cr(III) by microorganisms like *Pseudomonas aeruginosa* PA01	Hexavalent Cr(VI) is toxic and a carcinogen; the reducing capability of microorganism is usually inhibited as Cr(VI) concentration is increased	123
Mineral flocculation by microorganism	Depending on the microorganism's hydrophobicity and electrokinetic behavior and also the character of the mineral, microorganism adheres to mineral surface and flocculates as finely divided mineral suspensions	Microorganism *Mycobacterium phlei* can selectively flocculate coal from pyrite; it flocculates hematite by adhering readily, but not with quartz	124
Microorganism in froth flotation	Microorganisms work as collector and float the mineral value; some microorganisms work as depressants in flotation of multicomponents (e.g., pyrite and coal, arsenic and bismuth ore)	The size and type of microorganism are important as they are usually larger than conventional flocculants and collectors; *M. phlei* can float hematite from quartz	125–127

readily floated and harvested, and anionic as well as cationic collectors are capable of collecting them. Surfactants have not been studied to date as a collecting agent. Microorganisms, during their culture, need nutrients such as starch, sugar, and protein. Smith[127] reviewed the floatability of microorganisms and stressed the need for future research in this area. So far, we have highlighted the use of biochemical techniques in mineral industries and presented some essential information that can be used to pursue further study in this field. To become familiar with recent advances in this area, refer to the current literature, in particular the articles by Smith et al.[124] and Smith and Misra.[128]

C. Bioreactors

1. Requirements

A bioreactor is a reaction vessel in which feed materials are biologically converted into specific products through the use of microorganisms or plant, animal, or human cells or enzymes. Unlike chemical reactors, bioreactors require close control of temperature, pH, and shear stress. In order to implement a monoseptic process, sterilization is important. A bioreactor maintains a condition conducive to growth of the desired microorganisms and the relevant metabolic activities. A bioreactor, therefore, requires suitable nutrients such as salts, vitamins, oxygen (for aerobic reactors), and sugars. There are two types of microbial processes: aerobic (the organisms require oxygen) and anaerobic (no oxygen required). A bioreactor that maintains optimum values of temperature, pH, dissolved oxygen, and concentrations of substrate can yield optimum bioconversion. The type of biochemical reactor selected also depends on the characteristic features of the bioorganism. Bioreactors are normally classified into two categories (for both aerobic and anaerobic processes): submerged and surface. In a submerged reactor, microorganisms grow and function in a liquid suspension, whereas in a surface or attached film reactor, a carrier helps in providing a film for the microorganism to attach onto and grow. In order to have good biological activity in a bioreactor, good mass transport from the liquid to the bioparticle (in an attached film reactor), a good supply of nutrients, and uniformity of temperature as well as control of pH and nutrient concentrations are essential. In an aerobic bioreactor, the transfer of oxygen from air to the liquid medium should be achieved by dispersing the gas efficiently. Fine dispersion of the gas in the reactor is required to create the maximum interfacial surface area and hence to obtain the maximum volumetric mass transfer coefficient ($k_l a$), where k_l is the liquid-side mass transfer coefficient and a is the specific surface area. The extent of energy input to the reactor is a measure of the surface area created inside the reactor for an efficient mass transfer operation.

2. Types

The classification of bioreactors is shown in Figure 6.10a, and the energy input methods to bioreactors are depicted in Figure 6.10b. Bioreactors are broadly classified as aerobic and anaerobic with further subclassification as submerged and surface (or attached) in each of the two main classifications. Schugerl[129] described various types of bioreactors and their characteristic features. Depending on the mode of energy input, bioreactors are classified into two main groups: mechanically agitated and nonmechanically agitated. There are several types of mechanically agitated bioreactors, such as simple vertical models that have impellers of various designs and horizontal types with aerating tubes, agitators, and rotating discs. The various types of mechanical agitators are shown in Figure 6.11a. A mechanically agitated reactor is associated with a high power requirement, construction problems, high capital cost, and high energy consumption. As a result, an alternative type of reactor, where energy is imported by nonmechanical agitation, has emerged. These are classified into two major groups based on the mechanism of energy input: (1) liquid jet and (2) compressed gas expansion. The various types of nonmechanically agitated reactors are shown in Figure 6.11b. Energy input into the reactor by means of a liquid jet is accomplished in tower reactors, which can be single jet, loop flow, or coaxial draft tube. These reactors are also known as loop reactors, with liquid jet as the main loop driver. The details of multiphase loop reactors were presented by Vogelpohl.[130] In another group of reactors, known as downflow reactors, energy is supplied by means of a liquid jet directed downward from the top of the reactor. Typical examples are draft tube plunging jet, tower, and immersed reactors. A list

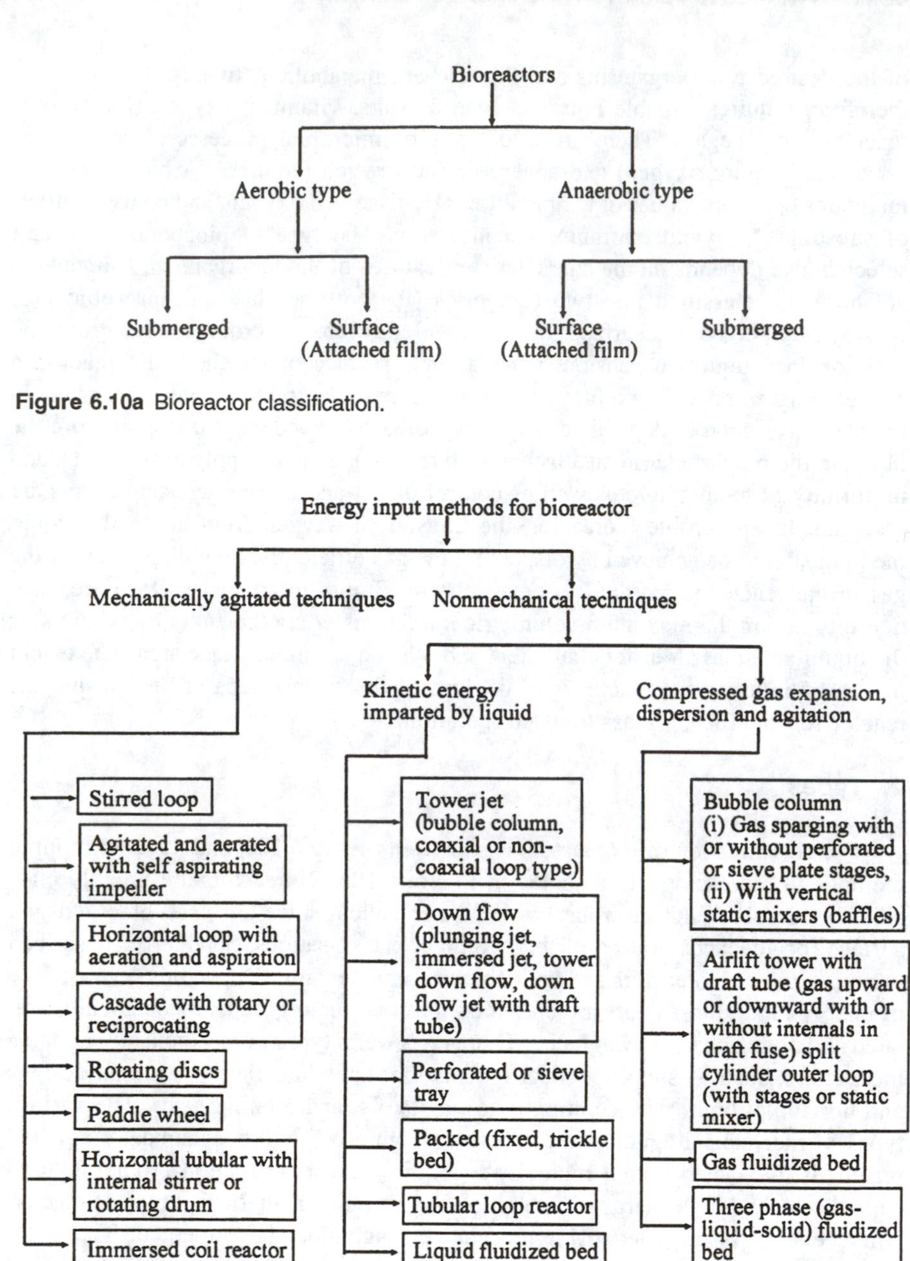

Figure 6.10a Bioreactor classification.

Figure 6.10b Energy input methods for bioreactors.

of reactors that belong to the group in which energy input is achieved by expansion of compressed gas is given in Figure 6.10b; typical examples (shown in Figure 6.11b) of reactors in this class are packed (fixed or trickle) bed, two-phase or three-phase fluidized bed, and tubular loop reactors. Optimum gas distribution in a two-phase as well as a three-phase fluidized bed warrants specialized skill and the

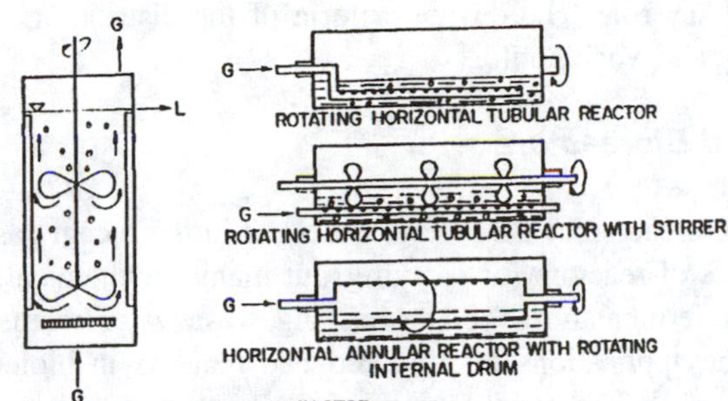

(a) MECHANICALLY AGITATED REACTORS

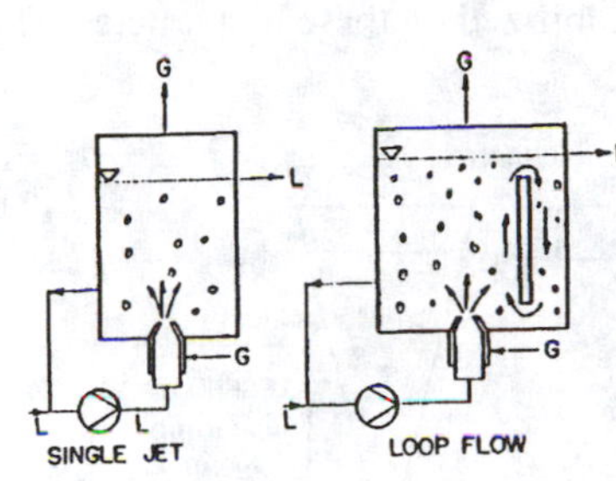

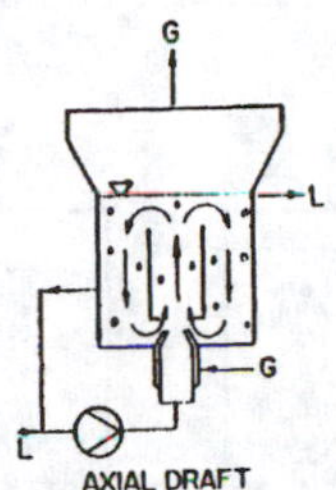

TOWER REACTORS

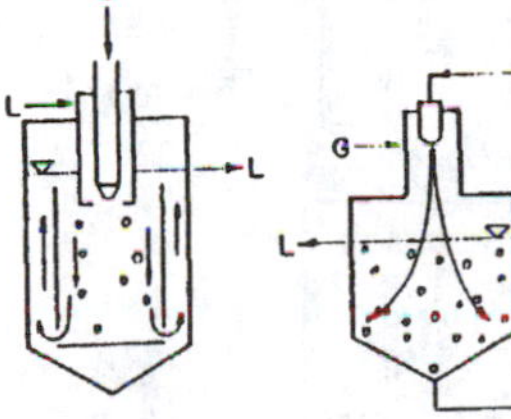

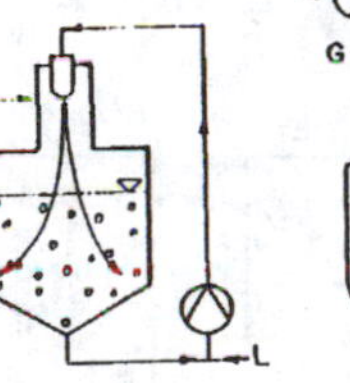

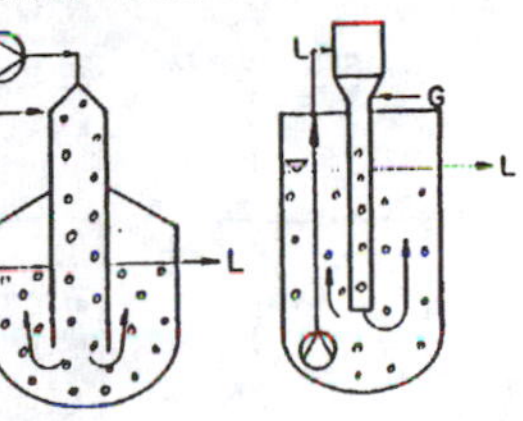

DRAFT TUBE PLUNGING JET TOWER TYPE IMMERSED JET

DOWN FLOW REACTORS

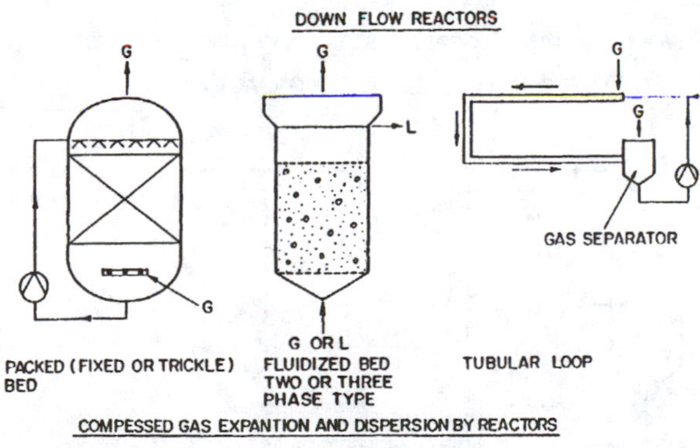

COMPESSED GAS EXPANTION AND DISPERSION BY REACTORS

(b) NONMECHANICALLY AGITATED REACTORS

Figure 6.11 Types of bioreactors. G = gas, L = liquid.

distributors play a key role. The design criteria of the distributors were discussed by Sathiyamoorthy and Vogelpohl.[131]

3. *Fluidized Bed Bioreactors*

Fluidized bed bioreactors (FBBs) have emerged in the recent past as one of the most efficient forms of reactors for carrying out many biochemical reactions such as those pertinent to fermentation, food processing, wastewater treatment, and phenol biodegradation. The applications of fluidized bed reactors in biotechnology were reviewed by Schügerl.[132] The various types of FBBs are shown in Figure 6.12. The fluidized solids, also called bioparticles in FBBs, serve as carrier particles for the microorganisms which can also be immobilized on these bioparticles. The metabolic

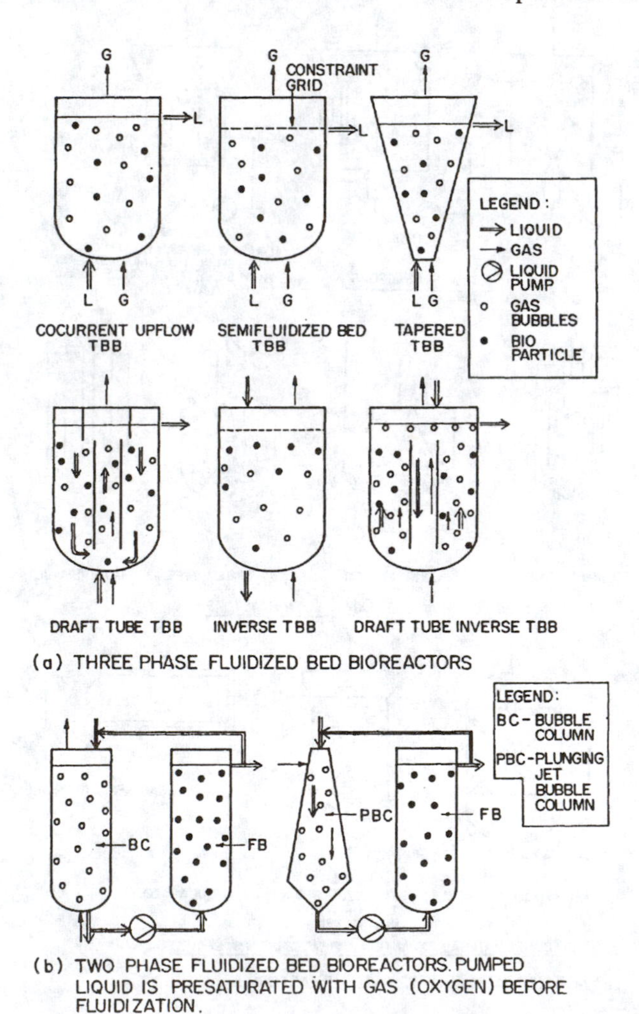

Figure 6.12 Various types of three-phase fluidized bed bioreactors. BC = bubble column, TBB = three-phase fluid bed, PBC = plunging jet bubble column.[130,135]

activities of an immobilized cell system can decrease, remain the same, or increase depending on the immobilization technique, the microbial species, and the nature of the biochemical reaction. FBBs are being increasingly used in aerobic processes, in particular in wastewater treatment. The main reasons for the selection of FBBs or conventional reactors such as activated sludge aeration tanks, trickling filters, rotating biological contractors, and trickle bed reactors are: (1) a large solid–liquid contact area, (2) low washout of microbes from the system, (3) biomass recovery at high substrate loading, (4) minimum sludge recycle, and (5) clog-free operation during biomass accumulation. Application of the FBB in water treatment was reviewed by Miller[133] and Oppelt and Smith.[134] The operating conditions of a wide variety of FBBs for aerobic wastewater treatment were summarized by Fan.[135] The selection and design of aerobic bioreactors, including FBBs, were discussed by Mersmann et al.[136] In an FBB, the particles selected should have properties such as minimum variation in size and shape, low cost, appropriate specific gravity for good fluidization, good abrasive resistance, and a surface conducive to cell immobilization. Generally, activated carbon, anthracite, and sand particles are used in the FBB. The purpose of this concluding section is to highlight the importance of biochemical processes that have emerged which use the FBB successfully in commercial-scale operations worldwide. Among the various FBBs, the simplest is the conventional countercurrent three-phase fluidized bed reactor. A semifluidized bed reactor is useful in minimizing particle elutriation, and some details of a semifluidized bed were presented in the beginning of this chapter. The use of a tapered fluidized bed as a bioreactor for continuous operation was demonstrated by Scott and Hancher.[137] In order to have a high degree of internal circulation, a draft tube FBB[138] is recommended. Fan[135] also mentioned the draft tube FBB and pointed out its commercial success. When the bioparticle density is less than the liquid density, the direction of liquid flow is downward, countercurrent to the gas flow. Here, fluidization is reversed, and the particles are dragged by the fluid along with the gravitational force. The inverse FBB was described by Karamanev et al.[139] In many cases, the liquid in an aerobic process can be presaturated with oxygen and the resultant liquid can be used to fluidize bioparticles, thereby achieving a simple two-phase FBB. In Figure 6.12b, two such two-phase FBBs with different gas presaturation techniques are shown.

It is obvious from the foregoing account that FBBs have emerged as a promising alternative to many of the conventional bioreactor systems. Their application in wastewater treatment has already demonstrated their potential with regard to commercial-scale operation. However, application of the FBB in mineral and metal extraction has not yet made an impact either on a laboratory scale or in industry. It is worthwhile to mention here one of the important developments in flotation cells brought about by Prof. Jameson.[140] The improved flotation cell, named the Jameson cell, is represented schematically in Figure 6.13 and has now widely replaced the bubble column type of cells. This cell has been successfully tested in a number of flotation plants around the world. A mineral slurry, when pumped through a downcomer, traps air, obviating the need for a compressor or a blower. The mineral slurry and air, in the presence of a frother, generate enormous amounts of froth for collecting the ore concentrate. This cell is operated continuously and is simple to adopt.

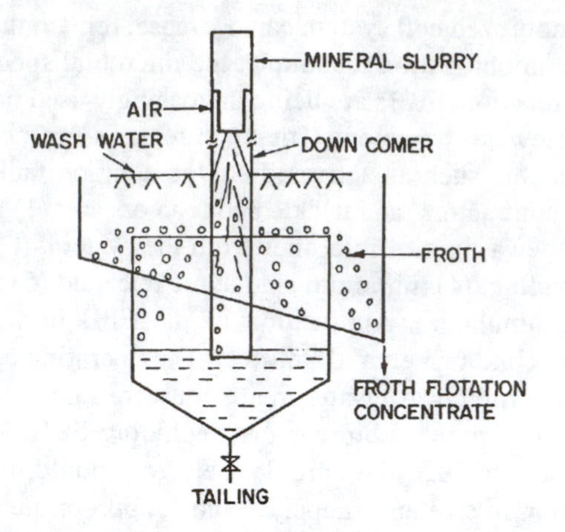

Figure 6.13 Jameson cell for froth flotation.

In principle, this cell is similar to a downflow reactor and works similar to a plunging jet reactor that has an extended downcomer. This type of reactor has the potential to be adopted in mineral processing for application in areas such as leaching and froth flotation. Future research in this line is worth pursuing in order to exploit the microbial activities in mineral processing and metal extraction.

NOMENCLATURE

A	cross-sectional area in direction normal to current flow (m^2)
Ar	Archimedes number, $d_p^3(\rho_s - \rho_f)\rho_f g/\mu_f^2$
A_s	specific electrode surface area (m^2/m^3)
C_b	cathode current density (A/m^2)
d_c	column diameter (m)
d_e	channel equivalent diameter (m)
d_o	diameter of spout orifice (m)
d_p	particle size (m)
E_o	electrode potential at 25°C (V)
F	Faraday's constant, 96,522 Coulomb-g equivalent
h	semifluidized bed height (m)
H_o	initial bed height (m)
h_{pa}	packed bed height in a semifluidized bed (m)
h_s	initial static bed height (m)
I	current density (A/m^2)
I_e	ionic current in particulate electrode (A)
I_L	current sustained by the whole of the particulate electrode of length L (A/m^2)
I_p	electronic current in particulate electrode (A)
I_x	current sustained by the particulate phase extending up to length X (A/m^2)

j_D	j factor for mass transfer at a plane wall for the case of electrochemically inert fluidized electrode (Equation 6.21) (–)
j_p	j factor for mass transfer between electrolyte and particle with an active fluidized bed of conducting particles (Equation 6.22) (–)
K_e	effective specific electrolyte conductivity of electrolyte (Ω^{-1}/m), mho/m
k_l	liquid-side gas–liquid mass transfer coefficient (m/s)
$k_l a$	volumetric liquid-side gas–liquid mass transfer coefficient (l/s)
K_p	effective specific electrical conductivity of dispersed particulate phase (mho/m)
M	molar concentration of electrolytic solution (mol/cc)
N	number of particles (–)
R	expansion ratio, h/h_s (–)
Re_{mf}	Reynolds number at minimum fluidization velocity, $U_{mf} d_p \rho_f / \mu$
r_i	inner radius of the bed (m)
r_o	outer radius of the bed (m)
S	surface area of the particle (m^2)
U	superficial fluid velocity (m/s)
U_{mf}	minimum fluidization velocity (m/s)
U_{msf}	maximum semifluidization velocity (m/s)
U_{osf}	minimum semifluidization velocity (m/s)
U_t	free-fall terminal velocity of particle (m/s)
X	distance in the direction of current in particulate electrode (m)
Y_t	space–time yield (mol/$m^3 \cdot$ s)
Z	number of electrons taking part in the reaction (–)

Greek Symbols

β	inverse molar density (m^3/mol)
ΔP_c	pressure drop across the constraint plate in the semifluidized bed (kPa)
ΔP_f	pressure drop across the fluidized section of the semifluidized bed (Equation 6.2) (kPa)
ΔP_{pa}	pressure drop across the packed section of the semifluidized bed (kPa)
ΔP_t	total pressure drop across the semifluidized bed (kPa)
ϵ	bed voidage (–)
ϵ_f	bed voidage in the fluidized section of the semifluidized bed (–)
ϵ_{mf}	voidage at the minimum fluidization velocity (U_{mf}) (–)
ϵ_{osf}	voidage at the onset of semifluidization (–)
ϵ_{pa}	bed voidage in the packed bed section of the semifluidized bed (–)
η	overvoltage (V)
θ	as defined by Equation 6.9
μ	viscosity of fluid (Pa $\cdot$ s)
ρ_f	density of fluid (kg/m^3)

ρ_g density of gas (kg/m^3)
ρ_s density of solid (kg/m^3)
ϕ sphericity or shape factor of the particle (–)
ϕ_e potential in the ionically conducting liquid (electrolyte) phase (V)
ϕ_p potential of electronically conducting particle (V)
ω angular velocity (rad/s)

REFERENCES

1. Filippov, M.V., *Prik. Magnitogridrodin, Tr. Inst. Fiz. Akad. Nauk Latv. SSR (USSR)*, 12, 215, 1960.
2. Filippov, M.V., *Izv. Akad. Nauk Latv. SSR (USSR)*, 12, 47, 1961.
3. Kirko, I.M. and Filippov, M.V., *Zh. Tekh. Fiz. (USSR)*, 30(9), 1081, 1960.
4. Rosensweig, R.E., Fluidization: hydrodynamic stabilisation with a magnetic field, *Science*, 204, 57, 1979.
5. Rosensweig, R.E., Magnetic stabilization of the state of uniform fluidization, *Ind. Eng. Chem. Fundam.*, 18(3), 260, 1979.
6. Rosensweig, R.E., Siegell, J.H., Lee, W.K., and Mikus, T., Magnetically stabilized fluidized solids, *AIChE Symp. Ser.*, 77(205), 8, 1981.
7. Luccesi, P.J., Hatch, W.H., Mayer, F.X., and Rosensweig, R.E., Magnetically stabilised beds — new gas–solids contacting technology, in *Proc. 10th World Petroleum Congress*, Vo. 14, Hyden & Sons, Philadelphia, 1–7.
8. Wei, K.L., The rheology of magnetically stabilised fluidized solids, *AIChE Symp. Ser.*, 79(222), 87, 1983.
9. Siegell, J.H. and Coulaloglou, C.A., Cross flow magnetically stabilised fluidized beds, *AIChE Symp. Ser.*, 80(241), 129, 1984.
10. Geuzens, P.L. and Thoenes, D., Axial and radiala gas mixing in a magnetically stabilised fluidized bed, in *Heat and Mass Transfer in Fixed and Fluidized Beds,* Van Swaaij, W.P.M. and Afgan, N.H., Eds., Hemisphere Publishing, Washington, D.C., 1986, 697.
11. Sonoliker, R.L., Ingle, S.G., Giradkar, J.R., and Mene, P.S., *Indian J. Technol.*, 10, 377, 1972.
12. Arnaldos, J., Puigjaner, L., and Casal, J., Heat and mass transfer in fixed and fluidized beds, in *Fluidization V,* Ostergaard, V.K. and Sorensen, A., Eds., Engineering Foundation, New York, 1986, 425.
13. Cholette, A. and Blanchet, J., Optimum performance of combined flow reactors under adiabatic conditions, *Can. J. Chem. Eng.*, 39, 192, 1961.
14. Cholette, A. and Cloutier, L., Mixing efficiency determination for continuous flow systems, *Can. J. Chem. Eng.*, 37, 105, 1959.
15. Fan, L.T., Yang, Y.C., and Wen, C.Y., Semifluidization: mass transfer in semifluidized beds, *AIChE J.*, 5, 407, 1959.
16. Murthy, J.S.N. and Roy, G.K., Semifluidization: a review, *Indian Chem. Eng.*, 29(2), 9, 1986.
17. Ho, T.C., Yau, S.J., and Hopper, J.R., Hydrodynamics of semifluidization in gas–solid systems, *Powder Technol.*, 50, 25, 1987.
18. Varma, R.L., Pandey, G.N., and Tripathi, G., Heat transfer in semifluidized beds, *Indian J. Technol.*, 19, 11, 1972.

19. Chatterjee, A., Spout fluid bed technique, *Ind. Eng. Chem. Process Des. Dev.*, 9, 340, 1970.
20. Littman, H., Vukovic, D.V., Zdanski, F.K., and Grabavcic, Z.B., Pressure drop and flow rate characteristics of a liquid phase spout fluid bed at the minimum spout fluid flow rate, *Can. J. Chem. Eng.*, 52, 174, 1974.
21. Vukovic, D.V., Hadzismajlovic, Dz.E., Grbavcic, R.B., Garic, R.V., and Littman, H., Regime maps for two phase fluid solids mobile beds in a vertical column with nozzle and annular flow, in Proc. 2nd Int. Symp. on Spouted Beds, Canadian Society for Chemical Engineering, Vancouver, October 1982, 93.
22. Sutanto, W., Hydrodynamics of Spout Fluid Beds, M.A.Sc. dissertation, University of British Columbia, Vancouver, 1983.
23. Nagarkatti, A. and Chatterjee, A., Pressure and flow characteristics of gas phase spout fluid bed and the minimum spout fluid condition, *Can. J. Chem. Eng.*, 52, 185, 1974.
24. Hadzisdmajlovic, Dz.E., Grabavcic, Z.B., Vukovic, D.V., and Littman, H., The mechanics of spout fluid beds at the minimum spout fluid flow rate, *Can. J. Chem. Eng.*, 61, 343, 1983.
25. Heil, C. and Tels, M., Pressure distribution in spout fluid bed reactors, *Can. J. Chem. Eng.*, 61, 331, 1983.
26. Mathur, K.B. and Gishler, P.E., A technique for contacting gases with coarse solid particles, *AIChE J.*, 1, 157, 1955.
27. Mamuro, T. and Hattori, H., Flow pattern of fluid in spouted bed, *J. Chem. Eng. Jpn.*, 1, 1, 1968.
28. Chatterjee, A., Adusumilli, R.S.S., and Deshmukh, A.V., Wall to bed heat transfer characteristics of spout fluid beds, *Can. J. Chem. Eng.*, 61, 390, 1983.
29. Taskaev, N.D. and Kozhima, M.I., Semicoking of Kok-Yangak coal in a circulating bed, *Tr. Akad. Nauk Kirgiz. SSR*, 7, 109, 1965.
30. Yang, W.S. and Keairns, B.L., Design of recirculatory fluidized beds for commercial applications, *AIChE Symp. Ser.*, 74(176), 218, 1978.
31. Yang, W.E. and Keairn, B.L., Studies on the solid circulation rate and gas by passing in spouted fluid bed with a draft tube, *Can. J. Chem. Eng.*, 61, 349, 1983.
32. Metcalfe, C.I. and Howard, J.R., Towards higher intensity combustion — rotating fluidized bed, in *Fluidization*, Davidson, J.F. and Keairns, D.L., Eds., Cambridge University Press, Cambridge, 1978, 278.
33. Demircan, N., Gibbs, B.M., Switthenbank, J., and Taylor, D.S., Rotating fluidized bed combustor, in *Fluidization*, Davidson, J.F. and Keairns, D.L., Eds., Cambridge University Press, Cambridge, 1978, 270.
34. Lazer, M.E. and Farkas, D.F., The centrifugal fluidized bed. 2. Drying studies on piece form foods, *J. Food Sci.*, 36, 315, 1971.
35. Levy, E., Martin, N., and Chen, J., Minimum fluidization and startup of a centrifugal fluidized bed, in *Fluidization*, Davidson, J.F. and Keairns, D.L., Eds., Cambridge University Press, Cambridge, 1978, 71.
36. Levy, E.K., Shakespeare, W.J., Tabatabaie-Raissi, A., and Chen, J.C., Particle elutriation from centrifugal fluidized beds, *AIChE Symp. Ser.*, 77(205), 87, 1981.
37. Rietema, K. and Mutsers, S.M.P., The effect of gravity upon the stability of a homogeneously fluidized bed investigated in a centrifugal fluidized bed, in *Fluidization*, Davidson, J.F. and Keairns, D.L., Eds., Cambridge University Press, Cambridge, 1978, 81.
38. Novosad, J. and Kostelkova, E., Incipient fluidization of cohesive powders by horizontal rotating nozzle, in *Fluidization*, Davidson, J.F. and Keairns, D.L., Eds., Cambridge University Press, Cambridge, 1978, 87.

39. Pfeffer, R., Tardos, G.I., and Gal, E., The use of a rotating fluidized bed as a high efficiency filter, in *Fluidization V*, Ostergaard, V.K. and Sorensen, A., Eds., Engineering Foundation, New York, 1986, 667.
40. Kuramoto, M., Furusama, T., and Kunii, D., Development of a new system for circulating fluidized particles within a single vessel, *Powder Technol.*, 44, 77, 1985.
41. Fox, D., Molodtsof, Y., and Large, J.F., Control mechanisms of fluidized solids circulation between adjacent vessels, *AIChE J.*, 35(12), 11933, 1989.
42. Masson, I.H.A., Design of a compact twinned fluidized bed systgem, in *Fluidization VI*, Grace, J.R., Schemilt, L.W., and Bergougnou, M.A., Eds., Engineering Foundation, New York, 1989, 383.
43. La Nauze, R.D., A circulating fluidized bed, *Powder Technol.*, 15, 117, 1976
44. Rudolph, V. and Judd, M.R., Circulation and slugging in a fluid bed gasifier fitted with a draft tube, in *Circulating Fluidized Bed Technology*, Basu, P., Ed., Pergamon Press, New York, 1985, 437.
45. Yang, Y.L., Jin, Y., Yu, Z.Q., and Wang, Z.W., Investigation on slip velocity distributions in the riser of dilute circulating fluidized bed, *Powder Technol.*, 73, 67, 1992.
46. Chong, Y.O., O'Dea, D.P., Leung, L.S., Nicklin, D.J., and Lottes, J., Design of standpipe and nonmechanical V valve for a circulating fluidized bed, in *Circulating Fluidized Bed Technology II*, Basu, P. and Large, P.F., Ed., Pergamon Press, Oxford, 1988, 493.
47. Rudolph, V., Rei, M.H., and Lin, S.Y., Horizontally circulating fluid bed for catalytic reaction and regeneration: transport phenomena in a cold test model, in Proc. 4th R.O.C. Symp. on Catalysis and Reaction Eng., Kaosiung, Taiwan, 1985, 100.
48. Sathiyamoorthy, D. and Rudolph, V., Hydrodynamics of dense phase circulating gas fluidized bed, in *Circulating Fluidized Bed Technology III*, Basu, P., Horio, M., and Hasatani, M., Eds., Pergamon Press, New York, 1990, 505.
49. Zenz, F.A. and Othmer, O.F., *Fluidization and Fluid Particle Systems*, Reinhold Publishing, New York, 1960.
50. Knowlton, T.M., Hirson, I., and Leung, L.S., The effect of aeration tap location on the performance of a J-valve, in *Fluidization*, Davidson, J.F. and Keairns, D.L., Eds., Cambridge University Press, Cambridge, 1978, 128.
51. Liu, D., Li, X.G., and Kwauk, M., Pneumatically controlled multistage fluidized beds, in *Fluidization*, Grace, J.R. and Matsen, J.M., Eds., Plenum Press, New York, 1980, 485.
52. He, Y., Hydrodynamics of a Compartmented Dense Phase Circulating Fluidized Bed, Ph.D. dissertation, University of Queensland, Australia, 1993.
53. Rudolph, V., Selection and application of gas–fluid bed reactors in the nonferrous metals industry, *Miner. Process. Extract. Metall. Rev.*, 10, 87, 1992.
54. Backhurst, J.R., Coulson, J.M., Goodridge, F., Plimles, R.E., and Fleischmann, M., A preliminary investigation of fluidized electrode, *J. Electrochem. Soc.*, 116, 1600, 1969.
55. Backhurst, J.R., Goodridge, F., Plimles, R.E., and Fleischmann, M., Some aspects of a fluidized zinc/oxygen electrode system, *Nature*, 221, 55, 1969.
56. Flett, D.S., The fluidized bed electrode in extractive metallurgy, *Chem. Ind.*, 16, 983, 1972.
57. Flett, D.S., The electrowinning of copper from dilute copper sulphate solutions with a fluidized bed cathode, *Chem. Ind.*, 23, 300, 1971.
58. Sabacky, B.J. and Evans, J.W., The electrical conductivity of fluidized electrodes — its significance and some experimental measurement, *Metall. Trans. B*, 8B, 5, 1977.

59. Sabacky, B.J. and Evans, J.W., Electrodeposition of metals in fluidized bed electrodes. I. Mathematical model, *J. Electrochem. Soc.*, 126, 1176, 1979.
60. Dubrovsky, M., Huh, T., Evans, J.W., and Carey, C.D., Fluidized electrowinning of metals — a review, in *Proc. 3rd Int. Symp. on Hydrometallurgy*, Assare, K.O. and Miller, J.D., Eds., TMS-AIME, Atlanta, GA, 1983, 759.
61. Janssen, L.J., Oxygen reduction at a fluidized bed electrode, *Electrochim. Acta*, 16, 151, 1971.
62. Berent, L.J., Mason, R., and Fells, I., Fluidized fuel cell electrodes, *J. Appl. Chem. Biotechnol.*, 21, 71, 1971.
63. Marshall, R.J. and Walsh, F.E., A review of some recent electrolytic cell designs, *Surf. Technol.*, 24, 45, 1985.
64. Goodridge, F. and Vance, C.J., The electrowinning of zinc using a circulating bed electrode, *Electrochim. Acta*, 22, 1073, 1977.
65. Ziegler, D.P., Dubrovsky, M., and Evans, J.W., A preliminary investigation of some anode for use in fluidized bed electrodeposition of metals, *J. Appl. Electrochem.*, 11, 625, 1981.
66. Dubrovsky, M., Ziegler, D., Materson, I.F., and Evans, J.W., Electrowinning of copper and cobalt using fluidized bed cathodes, in *Extraction Metallurgy 81*, Inst. Min. and Metall., London, 1981, 91.
67. Dubrovsky, M. and Evans, J.W., An investigation of fluidized bed electrowinning of cobalt using 50 and 100 amp cells, *Metall. Trans. B*, 13B, 293, 1982.
68. James, G.S., Denar, B.I., Moergeli, W.R., and Parel, S.A., U.S. Patent 3,974,049, 1976.
69. Steppke, H.D. and Kammel, R., Electrolysis with fluidized bed electrode, *Erzmetall*, 26, 533, 1973.
70. Goodridge, F. and Vance, C.J., *Electrochim. Acta*, 22, 1073, 1977.
71. Avedesian, M.M. and Holko, A.P., Norando Mines Ltd., U.S. Patent 4,141, 804, 1979.
72. Van der Heiden, G., Raats, C.M.S., and Boon, H.F., Fluidized bed electrolysis for removal or recovery of metals from dilute solutions, *Chem. Ind.,* 465, 1, 1978.
73. Kreysa, G., Kinetic behaviour of packed and fluidized bed electrodes, *Electrochim. Acta*, 23, 1351, 1978.
74. Goodridge, F., Some recent developments in monopolar and bipolar fluidized bed electrodes, *Electrochim. Act*, 22, 929, 1977.
75. Goodridge, F., Holden, D.I., Murray, H.D., and Plimley, R.F., Fluidized electrodes: mathematical model of the fluidized bed electrode, *Trans. Inst. Chem. Eng.*, 49(128), 136, 1971.
76. Newman, J. and Tobias, C., Theoretical analysis of current distribution in porous electrodes, *J. Electrochem. Soc.*, 109, 1183, 1962.
77. Trainham, J.A. and Newman, J., A flow through porous electrode model: application to metal-ion removal from dilute streams, *J. Electrochem. Soc.*, 124(10), 1528, 1977.
78. Sabacky, B.J. and Evans, J.W., Electrodeposition of metals in fluidized bed electrodes. II. An experimental investigation of copper electrodeposition at high current density, *J. Electrochem. Soc.*, 126(7), 118, 1979.
79. LeRoy, R.L., Fluidized electrowinning. I. General modes of opration, *Electrochim. Acta*, 23, 815, 1978.
80. LeRoy, R.L., Fluidized bed electrowinning. II. Operation at constant current density, *Electrochim. Acta*, 23, 827, 1978.
81. Lopez-Cacicedo, C.L., The recovery of metals from rinse waters in "Chemetec" electrolytic cell, *Trans. Inst. Met. Finish.*, 53, 74, 1975.
82. Walker, A.T.S. and Wragg, A.A., Mass transfer in fluidized bed electrochemical reactors, *Electrochim. Acta*, 25, 323, 1980.

83. Kusakabe, K., Morroka, S., and Kato, Y., Mass transfer coefficient at the wall of a rectangular fluidized bed for liquid–solid and gas–liquid–solid systems, *J. Chem. Eng. Jpn.*, 20, 433, 1980.
84. Oloman, C. and Watkinson, A.P., The electroreduction of oxygen to hydrogen peroxide on fluidized cathodes, *Can. J. Chem. Eng.*, 53, 268, 1975.
85. Latimer, W.M., *The Oxidation States of the Elements and Their Potentials in Aqueous Solutions*, 2nd ed., Prentice-Hall, New York, 1952.
86. Augugliaro, V., D'Alba, F., Rizzuti, L., Schiavello, M., and Sclafani, A., Conversion of solar energy to chemical energy by photo assisted process. II. Influence of the iron content on the activity of doped titanium dioxide catalysts for ammonia photo production, *Int. J. Hydrogen Energy*, 7(11), 851, 1982.
87. Wilkinson, J.A.E., The electrolytic recovery of metal values using the fluidized bed electrode, *Trans. Inst. Met. Finish.*, 4l9, 16, 1971.
88. Wilkinson, J.A.E. and Haines, K.P., Feasibility study on the electrowinning of Cu with fluidized bed electrodes, *Trans. Inst. Min. Metall.*, 81, C157, 1972.
89. Bockris, J., O'M. and Srinivasan, S., Eds., *Fuel Cells: Their Electrochemistry*, McGraw-Hill, New York, 1969, chap. 7.
90. Fry, A.J., *Synthetic Organic Electrochemistry*, Harper and Row, New York, 1972.
91. Goodridge, F., King, C.J.H., and Wright, A.R., Performance studies on a bipolar fluidized bed electrode, *Electrochim. Acta*, 22, 1087, 1977.
92. Scuffham, F.B. and Rowden, G.A., Production of metal powders by direct reduction, *Chem. Eng.*, 268, 444, 1972.
93. Coeuret, F., The fluidized bed electrode for the continuous recovery of metals, *J. Appl. Electrochem.*, 10(6), 687, 1980.
94. Goodridge, F. and Vance, C.J., Copper deposition in a pilot plant scale fluidized bed cell, *Electrochim. Acta*, 24, 1237, 1979.
95. Masterson, I.F. and Evans, J.W., Fluidized electrowinning of copper experiments using 150 A and 1000 A cells and some mathematical modelling, *Metall. Trans. B*, 13B, 3, 1982.
96. Sherwood, W.G., Queneau, P.B., Nikolic, C., and Hodges, D.R., Fluid bed electrolysis of nickel, *Metall. Trans. B*, 10B, 659, 1979.
97. Nakahara, S. and Mahanjan, S., The influence of solution of pH on microstructure of electrodeposited cobalt, *J. Electrochem. Soc.*, 127, 283, 1980.
98. Dubrovsky, M. and Evans, J.W., An investigation of fluidized bed electrowinning of cobalt using 50 and 1000 amp cells, *Metall. Trans. B*, 13B, 293, 1982.
99. Hall, K.B., *World Min.*, 27, 44, 1974.
100. Kreysa, G. and Heitz, E., Kinetic investigation into silver deposition in a fluidized and packed particle electrode, *Chem. Ind.*, April 19, p. 332, 1975.
101. Huh, T., Evans, J.W., and Carey, C.D., The fluidized bed electrowinning of silver, *Metall. Trans. B*, 14B, 353, 1983.
102. Jiricny, V. and Evans, J.W., Fluidized bed electrodeposition of zinc, *Metall. Trans. B*, 15B, 623, 1984.
103. Tucker, R.F. and Wragg, A.A., Simulation of heat transfer to horizontal tubes in a fluidized bed using the electrochemical mass transfer modelling technique, in *Heat and Mass Transfer in Fixed and Fluidized Beds*, Vanswaaij, W.P.M. and Afgan, N.H., Eds., Hemisphere Publishing, New York, 1986, 589.
104. Haertel, G. and Lindner, I., Conditioning of waste hydrochloric acid, in Removal of Pollutants and Extraction of Valuable Materials, Transaction 1, 1st Status Seminar, Wilderer, P.A., Potzel, U., and Doeller, J., Eds., Technical University of Munich, 1992, 133.

105. Torma, A.E., Leaching of metals, in *Bioleaching, Comprehensive Treatise*, Vol. 6a, Microbial Processes, Rehn, H.J., Ed., VCH Publishers, Berlin, 1988, 367.
106. Torma, A.E., Impact of biotechnology on metal extractions, *Miner. Process. Extract. Metall. Rev.*, 2, 289, 1987.
107. Groudeva, V.I. and Groudev, S.N., Combined bacterial and chemical leaching of a polymetallic sulphide ore, in *Mineral Bioprocessing,* Smith, R.W. and Misra, M., Eds., The Minerals, Metals and Materials Society (TMS), Warrendale, PA, 1991, 153.
108. Sugii, T., Hirayama, K., Inagaki, K., Tanaka, H., and Tano, T., Molybdenum oxidation by *Thiobacillus ferroxidans, Appl. Environ. Microbiol.*, 58, 1768, 1992.
109. Yopps, D.L. and Baglin, E.G., Extracting platinum group metals from still water complex flotation concentrate by a two stage bacterial oxidation/chemical treatment process, in *Mineral Bioprocessing*, Smith, R.W. and Misra, M., Eds., TMS, Warrendale, PA, 1991, 247.
110. Holmes, D.S. and Yates, J.R., Basic principles of genetic manipulation of *Thiobacillus ferroxidans* for bio-hydrometallurgical applications, in *Microbial Mineral Recovery,* Ehrlich, H.L. and Brierley, C.L., Eds., McGraw-Hill, New York, 1990, 29.
111. Paponetti, B.A., Ubaldini, S., Abbruzzese, C., and Toro, L., Biometallurgy for the recovery of gold from arsenopyrite ores, in *Mineral Bioprocessing*, Smith, R.W. and Misra, M., Eds., TMS, Warrendale, PA, 1991, 163.
112. Alibhai, K., Leak, D.J., Dudeney, A.W.L., Agatzini, S., and Tzeferis, P., Microbial leaching of nickel from low grade Greek laterite ores, in *Mineral Bioprocessing*, Smith, R.W. and Misra, M., Eds., TMS, Warrendale, PA, 1991, 191.
113. Rusin, P.A. and Sharp, J.E., Enhanced recovery of manganese and silver from refractory ore through biotreatment, in *Mineral Bioprocessing*, Smith, R.W. and Misra, M., Eds., TMS, Warrendale, PA, 1991, 207.
114. Noble, E.G., Baglin, E.G., Lamshire, D.L., and Eisele, J.A., Bioleaching of manganese from ores using heterotrophic microorganisms, in *Mineral Bioprocessing*, Smith, R.W. and Misra, M., Eds., TMS, Warrendale, PA, 1991, 233.
115. Rogers, R.D. and Wolfram, J.H., Biological separation of phosphate from ores using heterotrophic microorganisms, in *Mineral Bioprocessing*, Smith, R.W. and Misra, M., Eds., TMS, Warrendale, PA, 1991, 219.
116. Sakaguchi, T. and Nakajima, A., Accumulation of heavy metals such as uranium and thorium by microorganism, in *Mineral Bioprocessing*, Smith, R.W. and Misra, M., Eds., TMS, Warrendale, PA, 1991, 309.
117. Golab, Z. and Smith, R.W., Accumulation of lead in two fresh water algae, *Miner. Eng.*, 5, 1003, 1992.
118. Jeffers, T.H., Ferguson, C.R., and Bennett, P.G., Biosorption of metal contaminants from acidic mine waters, in *Mineral Bioprocessing*, Smith, R.W. and Misra, M., Eds., TMS, Warrendale, PA, 1991, 289.
119. Bennett, P.G. and Jeffers, T.H., Removal of metal contaminants from a waste stream using bio-fix beads containing sphagnum moss, in *Mining and Minerals in Processing Wastes*, Doyle, F.M., Ed., SME/AIME, Salt Lake City, 1990, 279.
120. Archibold, E.R., Use of mixed microbial ecosystems and the removal of heavy metals from contaminated soil and water, presented at the 201st Americal Chemical Society Meeting, Atlanta, April 19, 1991.
121. Bender, J. and Gould, J., Uptake, transformation and deposit of Se(VI) by a mixed selenium tolerant ecosystem, presented at the 201st American Chemical Society Meeting, Atlanta, April 19, 1991.

122. Ogunseitan, O.A. and Olson, B.H., Potential for genetic enhancement of bacterial detoxification of mercury waste, in *Mineral Bioprocessing*, Smith, R.W. and Misra, M., Eds., TMS, Warrendale, PA, 1991, 325.
123. Apel, W.A. and Turick, C.E., Bioredediation of hexavalent chromium by bacterial reduction, in *Mineral Bioprocessing*, Smith, R.W. and Misra, M., Eds., TMS, Warrendale, PA, 1991, 376.
124. Smith, R.W., Misra, M., and Dubel, J., Mineral bioprocessing and the future, *Miner. Eng.*, 4(7–11), 1127, 1991.
125. Smith, R.W., Misra, M., and Chen, S., Adsorption of a hydrophobic bacterium on to hematite: implications in the froth flotation of mineral, *J. Ind. Microbiol.*, 11(2), 63, 1993.
126. Solozhenkin, P.M. and Lyubavina, L., The bacterial leaching of antimony and bismsuth bearing ores and the utilisation of sewage wastes, in *Biogeochemistry of Ancient and Modern Environments*, Proc. 4th Int. Symp. on Environmental Geodensity, Trudinger, P.A., Walter, M.R., and Ralph, B.J., Eds., Springer, New York, 1980, 615.
127. Smith, R.W., Flotation of algae, bacteria, and other microorganisms, *Miner. Process. Extract. Metall. Rev.*, 4, 279, 1989.
128. Smith, R.W. and Misra, M., Recent developments in the bioprocessing of minral, *Miner. Process. Extract. Metall. Rev.*, 12, 37, 1993.
129. Schugerl, K., *Bioreaction Engineering*, Vol. 2 (translated from German by Valerie Cottrell), John Wiley & Sons, Chichester, England, 1991, chap. 1.
130. Vogelpohl, A., Multiphase flow in loop reactors, in *Proc. 4th World Congress of Chemical Eng.*, Behrens, D., Ed., DECHEMA, Frankfurt Main, 1992, 1109.
131. Sathiyamoorthy, D. and Vogelpohl, A., On the distributors and design criteria for gas–solid and gas–liquid–solid fluidized beds, *Miner. Process. Extract. Metall. Rev.*, 12, 125, 1995.
132. Schugerl, K., Biofluidization: application of fluidizaton technique in biotechnology, *Can. J. Chem. Eng.*, 67, 178, 1989.
133. Miller, D.G., in *Biological Fluidized Bed Treatment of Water and Waste Water*, Cooper, P.R. and Atkinson, B., Eds., Ellis Horwood, Chichester, England, 1981, 35.
134. Oppelt, E.T. and Smith, J.M., in *Biological Fluidized Bed Treatment of Water and Waste Water*, Cooper, P.R. and Atkinson, B., Ellis Horwood, Chichester, England, 1981, 165.
135. Fan, L.S., *Gas–Liquid Solid Fluidization Engineering*, Butterworth, Boston, 1989, chap. 8.
136. Mersmann, A., Schneider, G., Voit, H., and Wenzig, E., Selection and design of aerobic bioreactors, *Chem. Eng. Technol.*, 13, 357, 1990.
137. Scott, C.D. and Hancher, C.W., Use of a tapered fluidized bed as a continuous bioreactor, *Biotech. Bioeng.*, 18, 1393, 1976.
138. Mahajan, S.P., Khilar, K.C., and Narjari, N.K., Biological denitrification in a fluidized bed, *Biotech. Bioeng.*, 26, 1445, 1984.
139. Karamanev, D., Nikolov, L., and Chavarie, C., A new bioreactor — the inverse fluidized bed biofilm reactor, in *Proc. IV European Congress on Biotechnol.*, Vol. 1, Neijssel, O.M., Vander Meer, R.R., and Luyben, K.Ch.A.m., Eds., Elsevier Science, Amsterdam, 1987, 328.
140. Clayton, R., Jameson, G.J., and Manlapig, E.V., The development and application of the Jameson cell, *Miner. Eng.*, 4(7–11), 925, 1991.

Index

A

B

C

D

E

F

G

H

I

J

K

L

M

N

O

P

R